Ore Deposit Models

Edited by RG Roberts
and PA Sheahan

Reprint Series 3

Canadian Cataloguing in Publication Data
Main entry under title:
Ore deposit models

(Geoscience Canada reprint series; 3)
Includes bibliographical references.
ISBN 0-919216-34-X

1. Ore deposits. 2. Geological modeling.
I. Roberts, R. G. (Robert Gwilym) II. Sheahan, Patricia III. Geological Association of Canada.
IV. Series.

TN263.074 1988 553.4 C89-004120-2

Additional copies may be obtained by writing to:

Geological Association of Canada
Publications
Department of Earth Science
Memorial University of Newfoundland
A1B 3X5

ISBN 0-919216-34-X

Front cover design: Peter Russell

Typesetting:
Geological Association of Canada
and Typeline, Mississauga, Ontario

Printing: Love Printing Service Ltd.
Ottawa, Ontario

First Printing May 1988
Second Printing Jan 1990

CONTENTS

PREFACE

There are two components to an ore deposit model: the empirical model, which consists of an assemblage of data, including observational data, which characterizes the deposit; and a conceptual model that attempts to interpret the data through a unifying theory of genesis. It is probable that many geologists associate the term "ore deposit model" with the conceptual model, thinking in terms of the magmatic model for nickel deposits, or the syngenetic model for volcanic-associated massive sulphide deposits.

The empirical model is developed from the comparison of data from a large number of examples of the ore-type in order to establish the common geological factors. The selection of data calls for judgement from the geologist, but he may also be influenced by his own expertise which may result in emphasis being placed on geochemical data at the expense of field data. This emphasis or bias may become even more pronounced with the development of the conceptual model, and may result in the development of a "structural model", or a "geochemical model" for a deposit.

The empirical model is the data base for the conceptual model, which attempts to provide a coherent interpretation for all the events involved in the formation of the deposit. It is, in fact, a causal model, a description of the processes that resulted in the observational data. At best the conceptual model provides only a partial explanation of the data since such models are continually updated and refined by new information and by the reinterpretation of old information in the light of progress in the science.

The sophistication and the level of development of ore type models, particularly the conceptual aspects of the models, is very variable, and is certainly a reflection of the cumulative research effort. For many ore deposit geologists, and particularly for exploration geologists, the most important aspect of the model is the description of the temporal relationships of the ore type, and its relationship to the host rocks. These general aspects of the geology define the framework within which the refinements to the model must operate. For example, it is now well established from the interpretation of geochemical data and observational data at all scales that chromite deposits in mafic and ultramafic rocks formed by magmatic processes; and there is a general consensus that volcanic-associated massive sulphide deposits, principally from the evidence of their morphology, the structural relationships to their host rocks, and from comparison with modern analogues, were deposited at and near the sea floor. The refinements to the conceptual model for these deposits are developed within these generally accepted constraints. But for some deposits, notably Archean lode gold deposits, model development is still at the stage where such fundamental geological constraints are the subject of debate. Consequently, descriptions of individual deposits of this type, and summaries of the deposits, tend to be directed towards answering these fundamental questions.

The form that a model takes may also be influenced by how it is used. The principal value of a conceptual model is that it directs the geologist to questions that would otherwise not be asked, which may result in data that would otherwise remain ignored or unavailable. The danger in a conceptual model is that it provides a convenient explanation which may promote a reluctance to revise or discard the model regardless of the significance of new data. In effect, the model is used to explain away the data regardless of inconsistencies. This may reflect a bias related to the geologist's expertise which may be outside that required to evaluate critical data. The recognition of this danger may lead an exploration geologist to mistrust theories of ore genesis. Thus Ridge (1983), while acknowledging that such an idea flies in the face of convention , argues strongly against the use of genetic concepts or theories of ore genesis as an aid in exploration. He suggests, for example, that for an exploration geologist, "The fact that many massive sulphide deposits are found in volcanic rocks is important, but what genetic relationships, if any, may exist between these rocks is not". On the other hand Valliant (1985), in a discussion of Lac Mineral's exploration successes, while crediting an empirical model developed from experience of the Bousquet and Doyon deposits in Quebec as a basis for the assessment of the potential of the Hemlo district, he acknowledges the influence of the conceptual model for syngenetic stratabound gold deposits developed by Ridler (1970) and Hutchinson (1976) for the interpretation of the Bousquet deposit and the deposits of the Malartic district.

J.M. Allen stated in his introduction to this series of papers in Geoscience Canada, that their purpose was "to present descriptions of models currently in use". This has been achieved by the authors with varying degrees of emphasis on the different aspects of the models. With the exception of the papers on uranium deposits and on granophile mineral deposits by Tilsley and Strong, respectively, the authors have summarized the ore-types within the framework of empirical and conceptual models. In the two articles referred to, the emphasis is on theoretical aspects of the desposits. The content of the articles is aimed at students, and at professional geologists. It is hoped that the reader will use the articles as a point of view, or a way of perceiving, the deposits here described; and from there he would be advised to create his own models.

The series on ore deposit models was initiated by J.M. Allen in his capacity as the Chairman of the Publications Committee of the Mineral Deposits Division of the Geological Association of Canada. The publication of this reprint collection was funded by the Geological Association of Canada and the Mineral Deposits Division. We would like to thank Lee Barker for the valuable contributions he made in sustaining the series, and Monica Easton for her support and for her work in the preparation of manuscripts for the publisher. The index was compiled by Patricia Sheahan.

References

Hutchinson, R.W., 1976, Lode gold deposits - the case for volcanogenic derivation: Pacific Northwest Metals and Mining Conference, State of Oregon Department of Geology and Mining Industries, Fifth Gold and Money Session and Gold Technical Session, p. 64-105.

Ridler, R.H., 1970, Relationship of mineralization to volcanic stratigraphy in the Kirkland-Larder Lakes area, Ontario: Geological Association of Canada, Proceedings, v. 21, p. 33-42.

Valliant, R., 1985, The Lac discoveries: Canadian Mining Journal, May, p. 117-128.

R. Gwilym Roberts
Patricia A. Sheahan

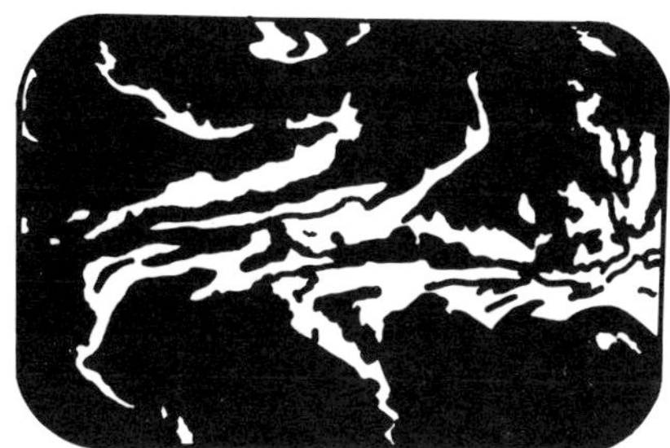

Archean Lode Gold Deposits

R. Gwilym Roberts
Department of Earth Sciences
University of Waterloo
Waterloo, Ontario N2L 3G1

Introduction

Approximately 60% of the world's cumulative gold production has come from rocks that are more than 2500 million years old. Eighteen percent of the cumulative production is from Archean lode deposits, and 40% from the paleoplacer deposits of the Witwatersrand of South Africa which were probably formed from the erosion of lode deposits.

The greenstone belts of all Archean shield areas are characterized by numerous lode gold deposits. The Superior Province is the largest and most productive of the Archean cratons. It has yielded 170 million ounces of gold from hundreds of deposits (Hodgson and MacGeehan, 1982), and the production from other Archean greenstone terranes is approximately proportional to their size. Examples of major producers occur in all the Archean shield areas of the world: the Superior and Slave Provinces of Canada (the mines of the Porcupine, Red Lake, Kirkland Lake, and Yellowknife districts); the early Precambrian Shield of Montana (the Homestake Mine); the Brazilian Shield (the Morro Velho and Raposo Mines); the Kaapvaal Craton and the Zambian Craton (the deposits of the Barberton Mountain Land and the deposits of the Gwanda and Midland belts); the shield areas of western Australia (the Golden Mile, Kalgoorlie); and the Indian Shield (the Kolar gold fields).

It is generally accepted that, compared to the Archean, lode deposits are scarce in the Proterozoic. However, the comparatively small number of deposits in the Aphebian greenstone belts of the Canadian Shield, may be due to a lower level of exploration activity. Deposits with many of the characteristics of Archean lode deposits occur in the Mesozoic volcanic-sedimentary terranes of British Columbia (deposits of the Coquihalla, Bralorne-Pioneer, Cariboo, and Cassiar districts) and California (the Mother Lode), and in the Cambro-Ordovician greywackes and shales of Nova Scotia (deposits in the Meguma group) and Victoria State, Australia (Ballarat-Bendigo district).

Lithology of the Deposits

The deposits include a wide range of lithological and structural types. However, in general, the ores consist of veins (open space filling) and altered wall rock (replacement or metasomatism). The veins generally consist of coarse or "cherty" quartz with lesser amounts of albite and carbonate (typically ferroan dolomite), tourmaline, sericite and chlorite. In some systems, tourmaline or carbonate may be the principal constituent of the veins. Opaque minerals rarely constitute more than 5% of a vein. Pyrite is invariably present and is the most abundant sulphide; pyrrhotite and arsenopyrite are common, and other opaque minerals may include galena, sphalerite, chalcopyrite, molybdenite, stibnite, tellurides and scheelite. In greenschist facies rocks, the altered wall rock immediately adjacent to the veins, is characterized by minerals that also occur in the veins: carbonates, quartz, sericite, albite and pyrite.

Typically, ore grade gold occurs in the veins, generally in small fractures in quartz, and in the wall rock where it is usually associated with iron sulphides. It is not uncommon for most or all the gold of an ore zone to be contained in wall rocks immediately adjacent to veins.

In replacement orebodies, such as occur in the Campbell and A.A. White Mines in the Red Lake district, (Andrews *et al.*, 1986), and in the Golden Mile deposits, Kalgoorlie (Phillips, 1986), quartz veins are a minor component. The ore zones consist of silicified mafic rock containing disseminated pyrite and arsenopyrite replacing and forming stringers in the mafic host. Disseminated gold also occurs in association with iron sulphide in magnetite-rich iron formations (the Lupin Mine and B-zone Cullaton Lake in the Northwest Territories), and less frequently, in association with carbonates in chemical sediments (the Homestake deposit, South Dakota). In this type, gold-bearing quartz veins generally constitute a minor part of the ore.

Regional Setting

Lithological and Stratigraphic Relationships. The relationship of gold deposits to host lithology and stratigraphic position has been addressed by Hutchinson (1976), Hutchinson and Burlington (1984), Hodgson and MacGeehan (1982) and Hodgson (1983). The deposits are confined to the volcanic-intrusive-sedimentary rocks of greenstone belts, and do not normally occur in the enclosing paragneiss. (The Renabie Mine, near Wawa, Ontario, may be an exception.) Within the greenstone belt all lithologies are capable of hosting individual orebodies, but the assemblage that most characterizes a gold mining district is mafic volcanic rocks with significant amounts of ultramafic komatiitic flows, and sedimentary rocks of the greywacke-shale association. Felsic volcanic rocks are not normally significant, although there are exceptions such as the Red Lake district, Ontario, and the Malartic district in Quebec. However, felsic intrusive rocks, although not volumetrically significant at the scale of a mining district, have long been appreciated by prospectors for their spatial association with gold deposits. Hodgson and MacGeehan (1982) estimate that more than 90% of the larger deposits of the Superior Province (deposits with production greater than one million ounces) are hosted by, or are immediately adjacent to, felsic porphyries. Cherry (1983) estimates that 70% of the gold of the Kirkland Lake-Larder Lake district, Ontario, and 60% of the gold of the Porcupine district, Ontario, is from deposits with a strong association with felsic intrusions. A similar association has been noted from the Red Lake district, Ontario (Pirie, 1981), Malartic - Val d'Or, Quebec (Latulippe, 1982) and the Yellowknife district, Northwest Territories (Boyle, 1961; Helmstaedt and Padgham, 1986). In the Kirkland Lake district, Ontario, a suite of syenitic intrusions that host the deposits, intrude trachyte flows and trachytic tuffs, which, on the basis of chemical composition are described by

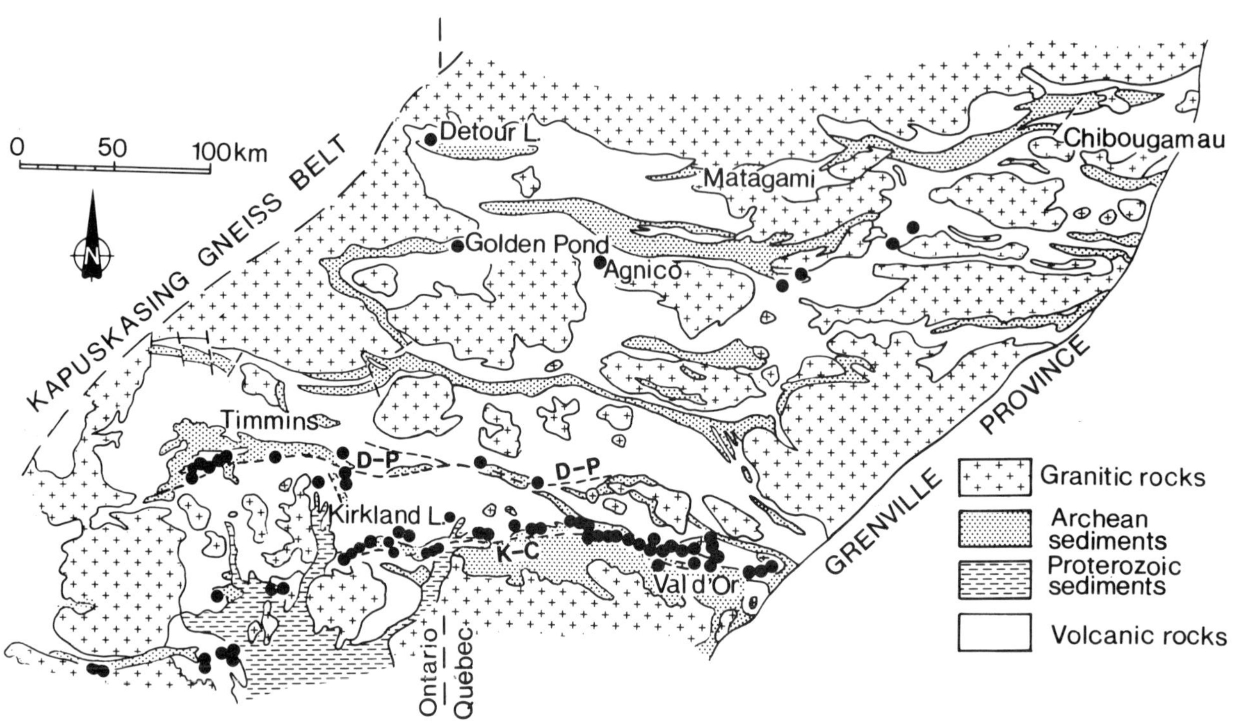

Figure 1 *The distribution of gold deposits in the Abitibi greenstone belt.* ***D-P:*** *the Destor-Porcupine Fault;* ***K-C:*** *the Kirkland Lake-Cadillac break. (After Goodwin and Ridler, 1970; MERQ-OGS, 1983; Latulippe, 1982; and other sources).*

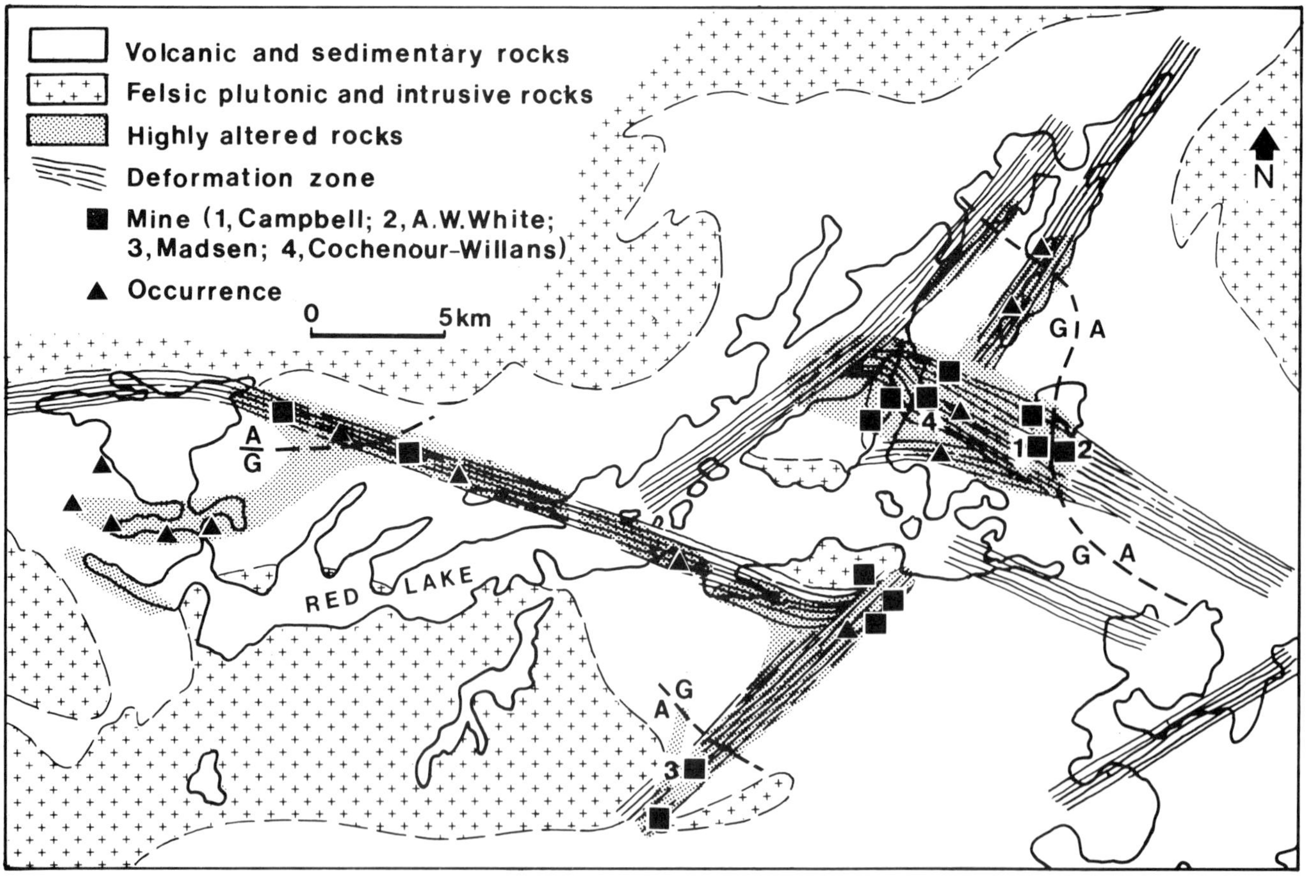

Volcanic and sedimentary rocks
Felsic plutonic and intrusive rocks
Highly altered rocks
Deformation zone
Mine (1, Campbell; 2, A.W.White; 3, Madsen; 4, Cochenour-Willans)
Occurrence
0 5km
N
A/G
G/A
G A
A G
RED LAKE
1
2
3
4

Figure 2 *The relationships of the gold deposits and regional alteration to deformation zones and major felsic plutons in the Red Lake district, Ontario, Canada.* ***G/A*** *indicates the greenschist-amphibolite facies isograd. (After Andrews* et al., *1986).*

Cooke and Moorhouse (1969) and Kerrich and Watson (1984) as being co-magmatic. A comparable co-magmatic relationship, on the basis of petrographic and geochemical similarities, has been proposed for the pyroclastic Krist Formation and the quartz porphyries associated with many of the deposits of the Timmins district, Ontario (Karvinen, 1981; Pyke, 1982; Gibson *et al.*, 1982). However, recent age dating, quoted in Wood *et al.* (1986), indicates that approximately 15 m.y. elapsed between the end of volcanic activity and the emplacement of the porphyries.

There is no evidence of stratigraphic control to gold deposits within the komatiitic-tholeiitic successions. In the Timmins district, the deposits are distributed throughout the komatiitic to iron-rich tholeiitic flow rocks of the Tisdale Group (over 3500 m thick), and into the sedimentary rocks of the overlying Porcupine Group. A similar lack of stratigraphic control was noted by Hutchinson and Burlington (1984) for deposits adjacent to the Cadillac break in the Noranda district of Quebec.

Gold deposits in iron formations may be stratabound in as much as each individual deposit may be more or less restricted to a sedimentary unit or facies, but there is no evidence that the deposits of a district are restricted to a stratigraphic interval in the way that volcanogenic massive sulphide deposits are. For example, gold-bearing iron formations of the Point Lake basin of the Slave Province occur throughout the Contwoyto Formation (Gibbins, 1981).

Regional Structural Relationships

The association of gold deposits with regional faults or "breaks" was appreciated by early explorationists. It is illustrated by the concentrations of deposits about the Destor-Porcupine and Kirkland Lake-Cadillac Fault systems in the Abitibi Belt of Ontario and Quebec (Figure 1), and about the regional faults in the Wabigoon Province in western Ontario (Poulsen, 1983). It is now appreciated that the gold deposits are related to steeply dipping planar shear zones of brittle to ductile deformation, and that the regional faults are a manifestation of brittle deformation within these zones of anomalously high strain. The significance of this is demonstrated by the distribution of deposits in the Red Lake district, Ontario, (Figure 2) where, until recently, it had not been possible to show a relationship between regional structures, specifically faults, and gold deposits (Pirie, 1981; MacGeehan and Hodgson, 1982). However, Andrews and Wallace (1983) and Andrews *et al.* (1986) have shown that the deposits are related to planar, deformation zones, or shear zones, of brittle-ductile strain.

The shear zones are regional structures, generally sub-parallel to the volcanic stratigraphy, up to several kilometres wide and may be well over 100 km long. They consist of zones of faulting and intense shearing that may be sub-parallel and relatively continuous or anastomosing, enclosing islands of relatively unstrained rocks. The kinematic and strain indicators, such as foliations, stretch lineations and asymmetrical strain shadows around minerals indicate that deformation has been principally by simple shear strain.

The volcanic strata in the terranes between the shear zones, are typically sub-vertical and folded, but the foliations and lineations, indicative of tectonic flattening and stretching are, as a rule, only locally developed. Deformation culminated in the Kenoran Orogeny with folding and tectonic shortening across the volcanic basin. In western and northwestern Ontario, and in the Yellowknife district, the deformation has been ascribed to the diapiric rise of granitic plutons (Schwerdtner *et al.*, 1979; Boyle, 1961; Helmstaedt and Padgham, 1986; Andrews *et al.*, 1986). In these districts, the gold-bearing shear zones are related to the same period of crustal shortening, and developed as conjugate systems in which the principal direction of shortening is into the obtuse angle of the intersecting shears.

The regional structures of the Abitibi Belt have a somewhat different history of deformation. Dimroth *et al.* (1982) and Jensen (1985) describe the Destor-Porcupine and Kirkland Lake-Cadillac Fault systems as growth faults, or normal faults, that developed during the early stages of volcanic activity, presumably under an extensional regime. They suggest that the associated belts of turbidite-greywacke sediments are related to the history of movements on the faults. As growth faults, they may well define the limits of volcanic terranes that developed with varying degrees of independence such that volcanic and sedimentary units periodically overlapped them. Archambault (1985) has proposed that during the Kenoran Orogeny, shortening across the Abitibi Belt, in a north-south stress field, was achieved by folding of the volcanic terrane about east-trending axes, followed by faulting and shearing with reverse sense of movement on the Destor-Porcupine and Kirkland Lake-Cadillac Fault zones as the isoclinal folds became locked in, and finally by the development of east-northeast and west-southwest conjugate shear and fault systems within the volcanic terrane. Thus, the principal gold-bearing shear zones of the Abitibi Belt are similar to those of the greenstone belts of northwestern Ontario and the Yellowknife district in that they are late tectonic structures, but differ in that their location is apparently controlled by the earlier (probably volcanic) structures.

It is probable that other sedimentary belts at the margins of the Abitibi greenstone belt and within the greenstone belt, indicate growth faults, and have the same significance for the location of gold deposits as the Destor-Porcupine Fault. The locations of the Golden Pond and Detour Lake deposits appears to support such a supposition (Figure 1).

The porphyries in the Abitibi Belt are concentrated in the deformation zones. This is particularly evident in the Porcupine district where several porphyry bodies are precisely located in structural discontinuities.

Structural Geology of the Vein Systems

The structural geology of the lode deposits is, in many respects, a scaled-down version of the structural geology of the greenstone belt. The vein systems occur in the central parts of discrete shear zones within the larger regional shear zones where rotational or simple shear strain predominates, but individual veins may extend laterally beyond the sub-vertical shear zone, for a limited distance, into the enclosing, less deformed rock where deformation is probably by pure shear strain. Veins may also occur in dilation zones associated with folding.

Vein systems are tabular, sub-vertical structures. Typically, the thickness of a vein system is measured in metres, and its strike and dip dimensions measured in tens or hundreds of metres (Figure 3). The economically viable part of the vein system may be considerably smaller. The vein system, in turn, may be part of a larger geological structure which consists of a system of discrete shears each hosting a vein system. For example, the seven mines of the Kirkland Lake district, Ontario, are located on a continuous vein-bearing shear system that has a strike length of 5 km, a width of 450 m, and extends for a vertical depth of at least 2 km.

Vein Systems in Shear Zones. The significance of shear zones to the location and geometry of vein systems was recognized by many of the early investigators, but most of the early studies did not explain the geometry of the veins, and the minor structures of the shear zones in terms of the assumed deformation process. Fortunately, the recent renewal of interest in Archean lode gold deposits has coincided with a resurgence of shear zone studies. The application of the results of these studies has confirmed the findings of earlier geologists, and has shown that the internal structures and geometry of many vein systems may be explained in terms of the development of a simple shear strain system.

Ramsay (1980) has classified shear zones into brittle, brittle-ductile and ductile. Brittle shears are characterized by the abrupt offset of markers, and are associated with faults and breccias. Ductile shears are characterized by continuous offset, and the development of mylonites. Brittle-ductile shear zones are associated with the development of strong foliation. In as much as mylonite is a cataclastic rock, it may be argued that the distinction between brittle and ductile is a question of the scale of observation. Most shear zones show evidence of both brittle deformation and ductile deformation, developed at different times in the history of the shear zone. The transition from ductile to brittle may occur as a consequence of an increase in the competency of the rock through which the shear zone passes, an increase in the rate of strain, or an increase in the pore pressure in the shear zone. The development of fabrics (and their kinematic significance) in a brittle-ductile shear zone, culminating in a mylonitic zone is illustrated in Figure 4.

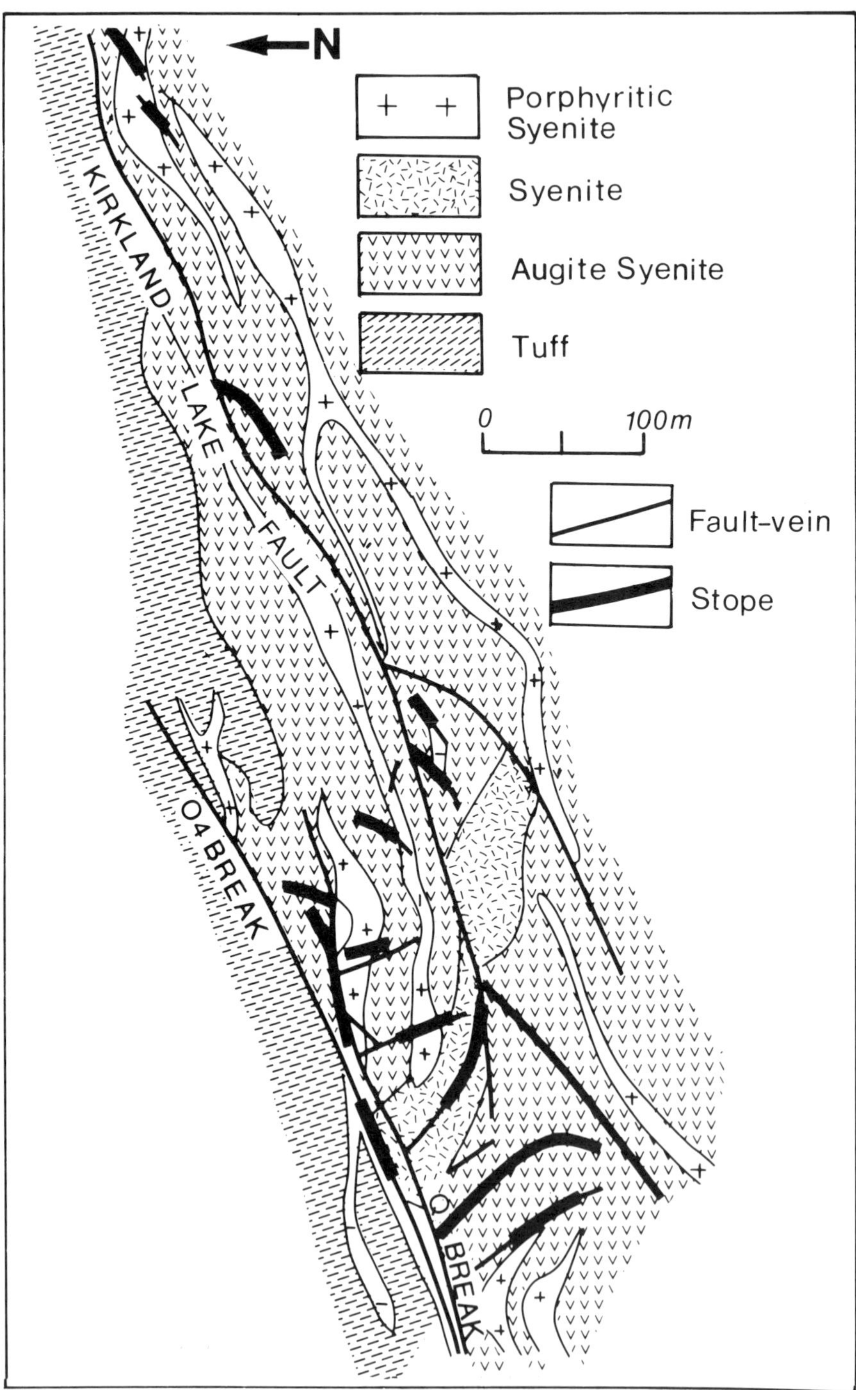

Figure 3 *A section through the Macassa deposit, Kirkland Lake, Ontario (looking eastward). The top of the section is approximately 850 m below surface. (After Charlewood, 1964).*

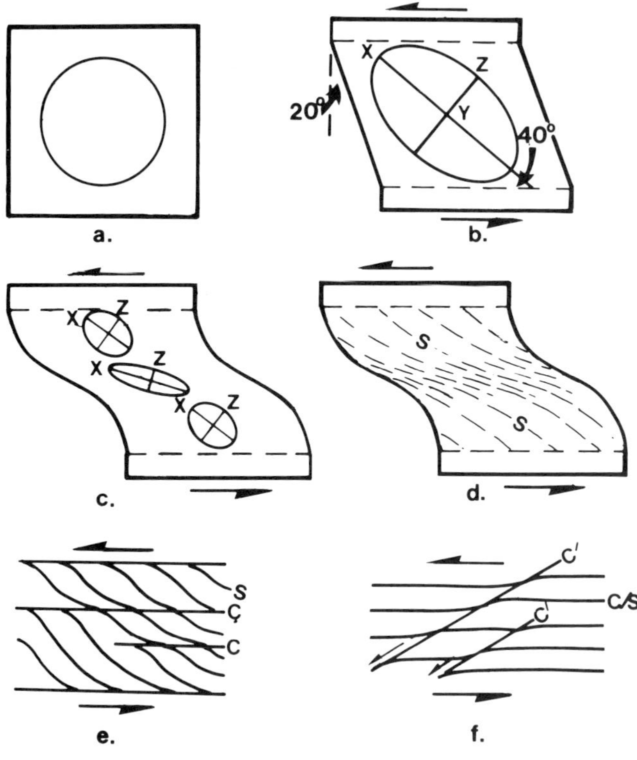

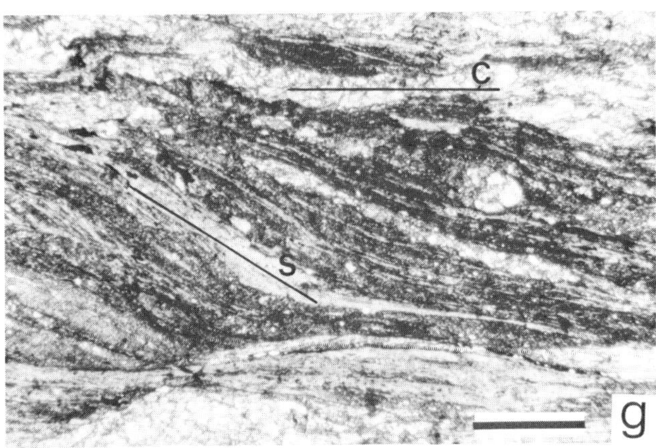

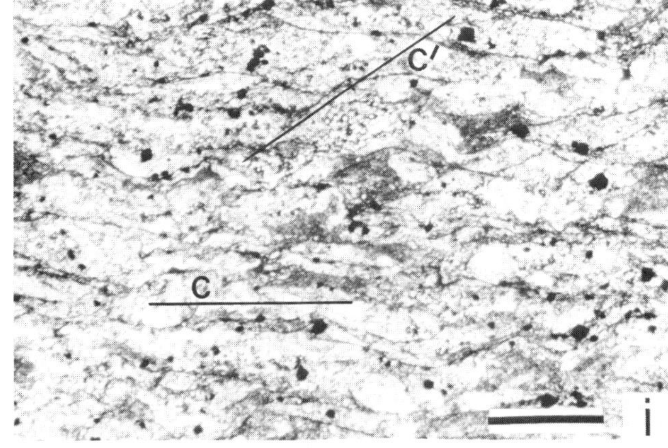

Figure 4 *The development of foliations by simple shear strain in a ductile shear zone.*

(a) *The undeformed state.*

(b) *Deformed by homogeneous simple shear strain, showing the orientation of the finite strain ellipsoid. A penetrative foliation forms in the XY plane of the strain ellipsoid, the plane of flattening.*

(c) and (d) *Heterogeneous simple shear under conditions of low bulk strain.*

(c) *The variation of the orientation and shape of the finite strain ellipsoids through the shear zone.*

(d) *The variation of attitude and intensity of the penetrative foliation S, developed in the planes of flattening of the finite strain ellipsoids across the shear zone.*

(e) and (f) *Heterogeneous simple shear strain under conditions of high bulk strain. (After Simpson, 1983).*

(e) *The C foliation, a spaced foliation, developed as planes of shear parallel to the walls of the shear zone. S is rotated into C.*

(f) *The development of C ' (shear banding); the spaced foliation transects C foliation in highly strained rocks.*

(g) to (i) *Photomicrographs of thin sections illustrating the foliations developed around the H-zone, Upper Canada Mine, Ontario; scale bars are 5 mm. Negatives of the photomicrographs are reversed to illustrate a sinistral shear movement.*

(g) *C and S foliations developed in sheared trachyte enclosing the ore body. Plane light; dark areas are mainly chlorite, light areas are feldspar, quartz and minor carbonate.*

(h) *The H-zone ore, protomylonitic trachyte, is at the most highly strained part of the shear zone; the C foliation is defined by the fine-grained cataclastic quartz and feldspar. Crossed nicols.*

(i) *A prominent fabric interpreted as C ' foliation developed in the protomylonitic trachyte of the H-zone ore. Plane light.*

The sequential formation of fractures in a shear zone, and their orientation, were first described in 1929 by Riedel in a series of experiments, which were later confirmed by Tchalenko (1968). Although the experiments modelled brittle shear, the structures can be recognized in brittle-ductile regimes. The orientation of the shear fractures, and the sense of movement across them, are illustrated in Figure 5. The low-angle Riedel shears (R) and the high-angle Riedel shears (R′) are the first fractures to form, followed by reverse shears or pressure shears (P), and finally the D shears which occupy the central part of the shear zone. Extensional fractures (T in Figure 5) form during shearing in a plane perpendicular to the X-axis of the strain ellipsoid at any instant of strain. They will thus tend to transect the shear zone and be oriented at a high angle to the foliation. Veins in the shear fractures are generally larger and, economically, more significant than extension veins.

In several recent studies of vein systems, the veins have been identified as being in the position of the D, P, R and R′ shears. Examples include: the Duport Mine, Ontario (Smith, 1986), the Sigma Mine, Quebec (Robert *et al.*, 1983), various deposits in the Red Lake district (Hugon and Schwerdtner, 1985; Andrews *et al.*, 1986), Cochenour-Willans Mine, Red Lake (Sanborn and Schwerdtner, 1986) and in the Sturgeon Lake District (Poulsen and Franklin, 1981).

The interpretation of the fabrics and the vein geometry in terms of the shear zone model in the above examples is significant because it identifies the fundamental nature of the deformation process, and may be of value in predicting the attitude and even the location of the ore shoots. The long dimensions of some of the orebodies described by Andrews *et al.* (1986) from the A.A. White, Cochenour-Willans and Campbell Mines are parallel to the stretch lineations (the X-axis of the strain ellipsoid, Figure 6). By contrast, the ore shoots in shear zones in the Star Lake pluton, Saskatchewan, closely parallel the intermediate axis of strain (Poulsen *et al.*, 1986). This direction is parallel to the lineation formed by the intersection of the S and C foliations in Figure 6, and would also parallel the intersection of any subsidiary shear with the main shear zone (Poulsen *et al.*, 1986). In the Cameron Lake deposit, Ontario, Melling *et al.* (1986) have shown that the location and attitude of the gold zones are controlled by the intersection of bedding-controlled splay shears with the Cameron Lake Shear Zone.

Generally, the pattern of the vein system is complicated by the formation of different types of veins (and consequently at different attitudes), more or less contemporaneously, and by the sequential emplacement of veins as deformation progresses. The formation of veins in extension fractures, contemporaneously with veins in shear fractures is illustrated in Figure 7a. The extension veins may terminate against the larger veins in shear fractures (generally in P or D shears), or merge continuously with them to form a discrete vein or a layer within a vein. Robert *et al.* (1983) describe sub-horizontal extension veins in the Sigma Mine that extend beyond the limits of the sub-vertical shear zones for as much as 75 m. The authors suggest that the extension veins form in the YZ plane of the strain ellipse in the region of plane strain, outside the shear zones (Figures 7b and 3).

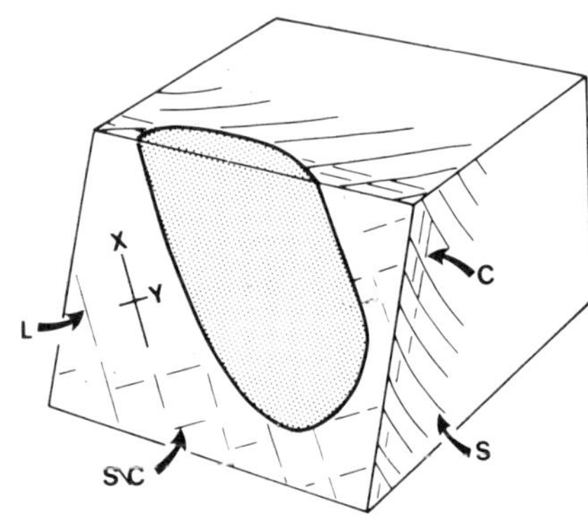

Figure 6 *The relationship between the fabrics (the axes of the finite strain ellipsoid) and the ore shoot. The ore shoot is parallel to the X-axis of the finite strain ellipsoid) and the ore shoot. The ore shoot is parallel to the X-axis of the finite strain ellipsoid, i.e., the stretch lineation. See text for examples of ore shoots parallel to the lineation formed by the intersection of S and C foliations.*

Figure 7 ***(opposite page)* (a) *(upper right)*** *Vein in extensional fracture, T, merging with veins in the sub-vertical shear zone (Sigma Mine, Quebec, Robert* et al.*, 1983). Bar is one metre.*
(b) ***(middle right)*** *The relationship between the flat veins and the bulk strain at the Sigma Mine. Veins in the extension fractures pass from the shear zone into the area of irrotational strain. The veins are perpendicular to the X-axis of the strain ellipse, and contain the Z-axis, which bisects the obtuse angle of the conjugate shears.*
(c) ***(lower right)*** *A vein system at the Hollinger Mine, Ontario (drawing from a photograph in Michie, 1967). Extensional veins pass into veins that are oblique to the attitude of the main vein system. The oblique veins are interpreted as P veins. Bar is one metre.*

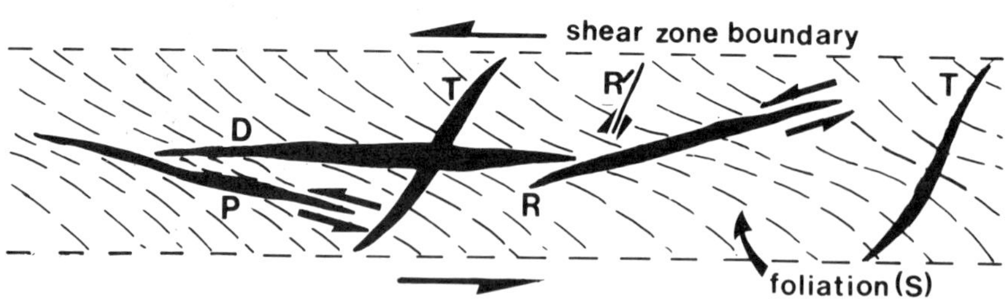

Figure 5 *The orientation of shear fractures and extension fractures in a brittle-ductile shear zone. The fractures are shown occupied by veins.* **R**, *low-angle Riedel shear fractures, 15 degrees to shear zone boundary;* **R′**, *high-angle Riedel shear fractures, 75 degrees to SZB;* **P**, *shear fractures or reverse fractures, 15 degrees to SZB;* **D**, *principal shear fractures, parallel to SZB;* **T**, *extension fractures, form in YZ plane of the strain ellipse, perpendicular to the S foliation.*

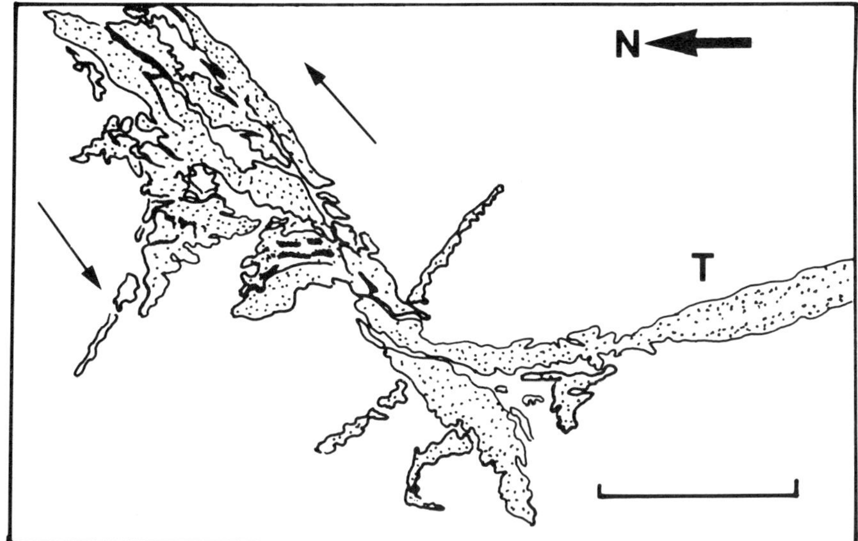

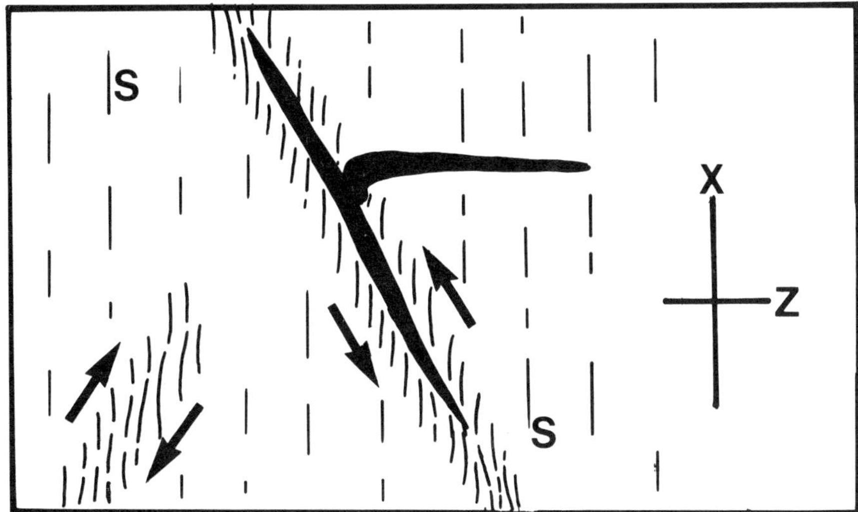

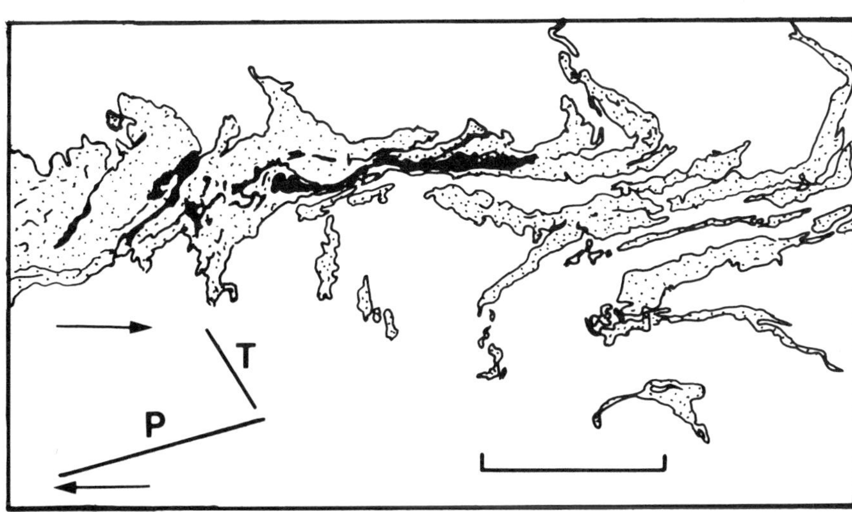

Early formed veins may be deformed as shearing progresses, and may be cut by later generations of veins. Veins oblique to the direction of shear are rotated, boudinaged and folded by the mechanisms shown in Figure 8, and they may also form similar folds as the result of differential shear. The larger veins in many shear systems are in P or D shear fractures, and, because they are in the plane of shearing, or approximately in the plane of shearing, they are rarely folded, although they may be pulled apart and boudinaged. The net result is that the shear zone may contain folded veins, boudinaged veins, folded and boudinaged veins, together with veins that are apparently not strained.

Veins in shear fractures are typically layered, and the layering is usually ascribed to repeated opening and filling of the structure. Veins in extension fractures are generally not layered suggesting that they formed as a single event. Extension fractures may be expected to dilate since they originate in a plane perpendicular to the direction of maximum extension of the strain ellipsoid. The P and R shear fractures are oblique to the principal direction of shearing, and as such, have the potential for dilation and vein formation, as shearing progresses. Guha *et al.* (1983) have shown that voids and open spaces may be created in ductile shear zones during the overriding and irregularities of different scales as shearing proceeds.

In some ore zones, veins are not developed. The gold occurs as pervasive disseminations and its location is controlled by the degree of strain in the rocks. For example, the H-zone of the Upper Canada Mine, Kirkland Lake, is in a ductile shear zone in which the highest grade ore is confined to a comparatively narrow mylonitic rock (Figure 4h) enclosed in less strained rocks in which S and C foliations are developed (Figure 4g). Hemlo and the replacement bodies in the Red Lake district may be "other" examples of gold orebodies in ductile deformation zones. Orebodies of this type suggest that grain size reduction and the fracturing of grains associated with mylonitization, may promote permeability in the rocks.

From the above description, it is apparent that the formation of veins is a part of the deformation process. Shearing does not appear to be a ground preparation process during which open structures are formed, and into which veins are later emplaced; rather, shearing and veining are parts of a continuous process. The continuity of vein formation throughout the history of the shear zones implies fluid circulation, or the availability of the fluid, over a long period of time. It is also apparent that the shear zones that host the gold deposits do not involve large displacements; even though in some situations, they may be adjacent to regional structures. This is illustrated in the Dome Mine, Ontario, where vein-bearing shear zones transect an unconformity at a high angle, with negligible displacement. It is probably more reasonable to consider the shear zones as structures that accommodated the shortening of the greenstone belt, rather than structures on which large displacements occurred. In this sense, they are analogous to folds.

A structural environment of simple shearing should not always be assumed. These processes have not been recognized in the vein deposits associated with the Paleozoic shales and greywackes of the Meguma Group in Nova Scotia, and the Ballarat-Bendigo-district of Victoria, Australia. In these deposits, the veins have been shown to be related to the fold history of the region under compressive stresses. Bedding-parallel veins in the Meguma Group formed at the onset of horizontal compression at the early stages of deformation, and vein formation continued to the later stages of folding (Graves and Zentilli, 1982). The saddle reefs of the Ballarat gold field formed in zones of dilation produced by delamination of beds at the hinges of folds (Baragwanath, 1953). The Camlaren deposit in greywackes of the Yellowknife Supergroup, is an example of gold-bearing veins in saddle reef structures in Archean rocks (Boyle, 1979; Padgham, 1981).

Gold in Chemical Sedimentary Rocks. The lithologies that have been interpreted as evidence for syngenetic or exhalative gold, precipitated in association with chemical sediments, generally fall into two categories: gold-bearing, laminated units in volcanic rocks; and gold-bearing, iron sulphide-rich units in banded iron formation. In Canada, examples of gold-bearing laminated units that have been interpreted as sediments include: the ankerite units at the Dome and Aunor Mines (Fryer and Hutchinson, 1976; Fryer *et al.*, 1979; Karvinen, 1981; Roberts, 1981); the ESC ore zone, Dickenson Mine (Crocket *et al.*, 1981; Kerrich *et al.*, 1981); the Agnico-Eagle deposit (Barnett *et al.*, 1982); and the Bousquet deposit (Valliant *et al.*, 1982). Similar gold-bearing units have been described from the Barberton Mountain Land (Viljoen, 1984; Ward, 1984) and Western Australia (Fehlberg and Giles, 1984). The sedimentary origin for the laminated structure of these units has been brought into question by the re-interpretation of the ESC zone at Dickenson as a shear zone (Rigg and Helmstaedt, 1981; MacGeehan and Hodgson, 1982; Mathieson and Hodgson, 1984), and on the basis of sulphur isotopic studies by Lavigne and Crocket (1983). The syngenetic interpretation of the ankerite units at the Dome and Aunor Mines has been questioned by Hodgson (1983) and Macdonald (1984).

The ankerite units at the Dome Mine consist of layers, typically 1-15 cm thick, of ferroan dolomite, chert, tourmaline-quartz, pyrite, and schistose, altered mafic material. Pyrite generally occurs as thin laminations in schistose, altered mafic rock. The layers are parallel to the foliation of the enclosing schistose, altered mafic rock, and have been interpreted as primary sedimentary structures. Coarse-grained quartz veins, with minor amounts of tourmaline, carbonate and pyrite occur in the units, both conformable and disconformable to the laminations of the units. In terms of the syngenetic model, the coarse quartz veins are assumed to have been emplaced during the post-depositional history of the units, principally during metamorphism (Hutchinson and Burlington, 1984; Roberts, 1981). However, the structures of the laminated ankerite unit shown in Figure 9 are more reasonably interpreted as a vein system. The ankerite unit is a narrow, ductile shear zone. The foliation in the wall rock (S foliation) is deflected into the schistosity of the ankerite unit which is the C foliation of the shear zone. The attitudes of the foliations indicate a left-handed horizontal component. The folds in the C foliation of the ankerite unit are comparable to sheath folds developed in a ductile shear (Platt, 1983). The folds also indicate a left-handed sense of movement. The layers of ankerite, chert, *etc.*, which characterize the unit, were emplaced parallel to the C fabric during the ductile shear deformation, and are incorporated into the sheath folds. The coarse quartz veins were emplaced at various stages, during a later brittle-ductile phase of shearing.

The interpretation of the ankerite units at the Dome Mine, and other gold-bearing laminated units referred to above as vein systems in shear zones, places in doubt the sedimentary status of other, similar structures. No doubt these will be re-examined with the shear zone model in mind.

Gold in banded iron formation presents a different problem in that there is generally no question of the sedimentary character of the host rock. Gold occurs in association with iron sulphides in the

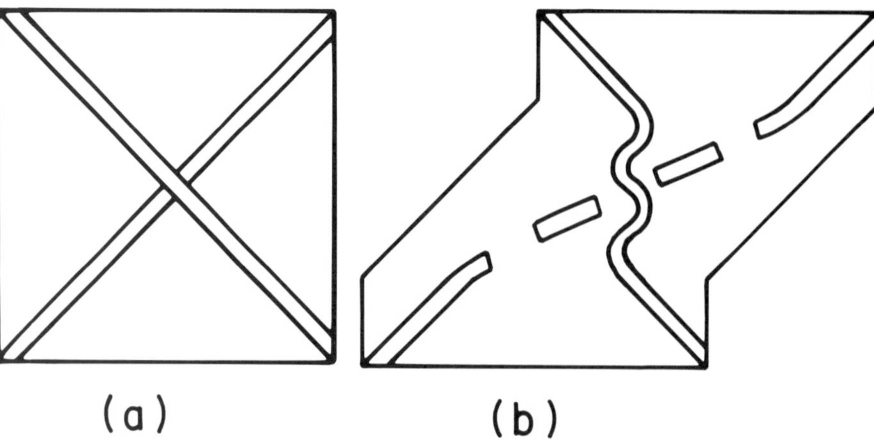

Figure 8 (a) *and* **(b)** *show how veins may be boudinaged or folded by buckling within in the shear zone. With continued deformation, the folded vein will become boudinaged. (After Ramsay, 1980).*

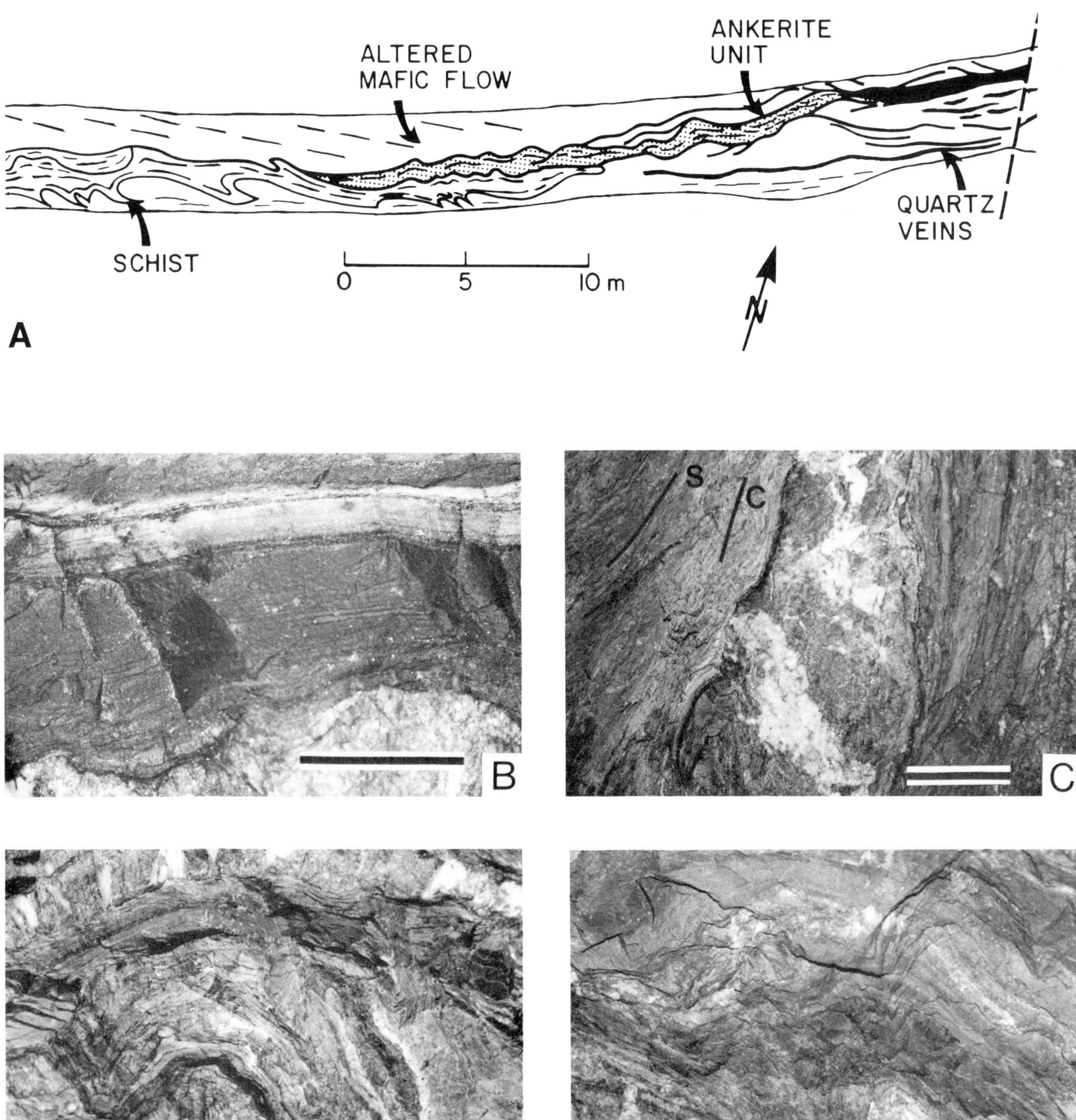

Figure 9 (A) *Map of an ankerite unit in pillowed mafic flows, in a drift on the 12th level, Dome Mine, Timmins. Foliation is indicated by dashed lines. The foliation in the ankerite unit and the host schist is a C foliation. The foliation in altered mafic flow is a S foliation. The attitude of the two foliations indicates a shear with a left-handed component of movement in the horizontal plane. The geometry of the folds in the ankerite unit and host schist are compatible with the same sense of shear.*

(B) *The laminated structure of an ankerite unit; bar is 25 cm. The light coloured layer (upper) consists of interlaminated tourmaline and chert; dark layer (centre) consists of schistose, altered mafic material and laminations of pyrite; light coloured layer (lower) consists of ankerite (ferroan dolomite) and quartz.*

(C) *Horizontal section of an ankerite unit; bar 10 cm. The attitudes of the S and C foliations indicate a left-handed shear. The white quartz veins in a layer of ferroan dolomite, occupy extension fractures which also indicate a left handed sense of shear.*

(D) and (E) *Folds in the ankerite unit; bar 10 cm. The sense of overturning of the folds is the same as that illustrated in* **(A)**; *The fold in* **(D)** *is from the east end of the drift, and* **(E)** *is from the west end of the drift.*

iron formation and more rarely with iron oxides or carbonates. Examples include: the Lupin deposit, Contwoyto Lake (Kerswell, 1984); the Vubichikewe deposit, Zimbabwe (Fripp, 1976); the B-zone, Cullaton Lake (Page and Roberts, 1984). The precise association of gold with iron sulphides, and the occurrence of the sulphides in sedimentary laminations, has been interpreted as evidence for co-precipitation of gold and iron sulphides in the sedimentary environment. Phillips *et al.* (1984), however, have documented the replacement of beds of magnetite in oxide iron formation, by gold-bearing iron sulphides, adjacent to quartz veins. The iron sulphides are clearly not sedimentary. Similar structures have been described from the MacLeod-Cockshutt deposit in Geraldton, Ontario (Macdonald, 1984), and the Timmins district (Fyon *et al.*, 1983b). Macdonald (1984) points out that the gold at MacLeod-Cockshutt is in quartz-ankerite veins, and in iron formation within 50 cm of the veins. The veins occupy zones of brittle failure associated with folds in the iron formation. Macdonald (1984) also describes a spatial relationship between alteration of the iron formation and the veins. These relationships have not been recorded from the Lupin deposit and the B-zone, Cullaton Lake, but nevertheless, they seriously question the interpretation of syngenesis of gold associated with banded iron formations.

Wall Rock Alteration

The alteration of igneous rocks, associated with deposits in low-grade greenschist facies rocks, is characterized by hydrolysis and carbonatization of ferromagnesian minerals and oxides. In mafic igneous rocks, the greenschist assemblage of actinolite-epidote-albite-quartz gives way to an outer chlorite alteration zone characterized by the assemblage chlorite-calcite, and an inner carbonate zone of ferroan dolomite-sericite-pyrite-quartz. The sequence of mineral changes for ultramafic and felsic igneous rocks is very similar (Figure 10). The chlorite alteration zone is of regional extent, with diffuse outer boundaries. It encompasses the linear deformation zones that host the deposits (Figure 2), and extends into the surrounding, less deformed rocks. The carbonate alteration forms comparatively narrow haloes, between less than one metre and tens of metres thick, around individual veins or vein systems. Extensive carbonate alteration is generally due to overlapping of such haloes.

The major components added to the altered rocks are: CO_2, K, S and H_2O. The trace elements include: Au, B, As, Rb, W, Mo, Ba and Sb.

The chemical equations given in Table 1 describe the reactions for the transformation of actinolite and epidote to chlorite and calcite in the chlorite zone, and the formation of sericite, ferroan dolomite and iron sulphides in the carbonate zone. The assemblages indicate increasing activity of CO_2 in the fluid in equilibrium with the wall rock. In general, magnesium, calcium, iron and aluminum of the ferromagnesium minerals enter the alteration mineral assemblage with no contribution of these elements from the fluids. Consequently, the original ferro-magnesian mineral content of the rock is a factor in controlling the maximum CO_2 added to the rock, and hence, the degree of carbonatization of the rock.

Sodium is essentially immobile throughout the zones of alteration in the dolerite at the Golden Mile, Kalgoorlie (Phillips, 1986), and is strongly depleted in the carbonate-rich altered mafic rocks at the Dome Mine, Timmins (Roberts and Reading, 1981). However, in the McIntyre-Hollinger ore zone, albite (with sericite and ankerite)

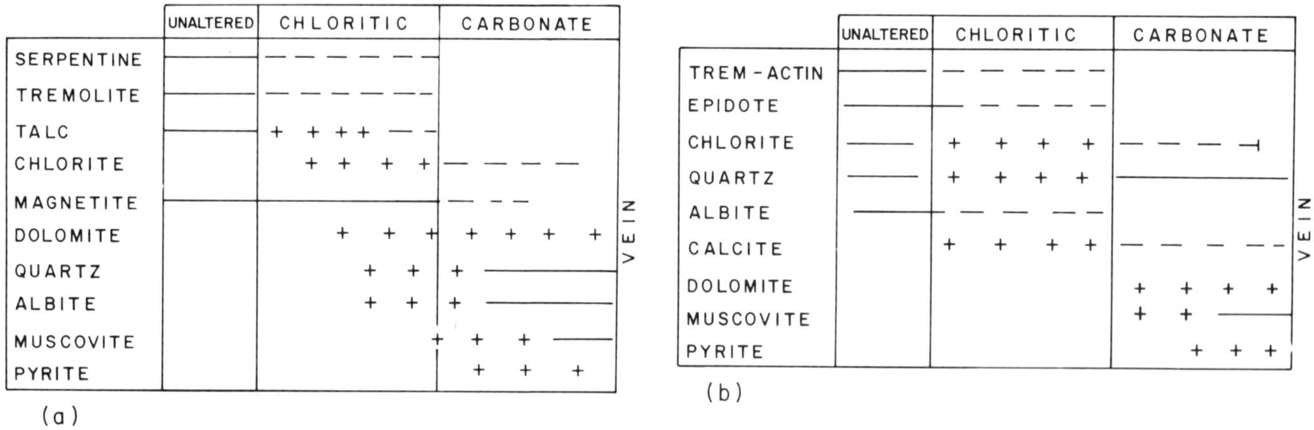

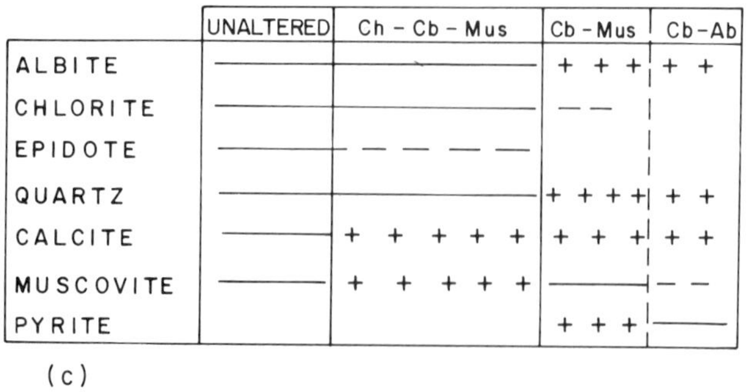

Figure 10 *Stability ranges of minerals in alteration zones in greenschist facies rocks.*
(a) ***(upper left)*** *ultramafic host rocks;*
(b) ***(above)*** *mafic host rocks. (After Fyon and Crocket, 1982);*
(c) ***(lower left)*** *felsic host rocks. (After Robert and Brown, 1986).*
Ch-Cb-Mus: *chlorite-carbonate-white mica zone;*
Cb-Mus: *carbonate-white mica zone;*
Cb-Ab: *carbonate albite zone.*

characterizes the most strongly altered mafic rocks (Smith and Kesler, 1985), and in the Sigma Mine, hydrothermal albite replaces white mica (with a concomittant increase in sodium) in the alteration zone immediately adjacent to the veins. Furthermore, albite is a common accessory in veins, and therefore, it is probable that the hydrothermal fluids contain significant sodium. The replacement of sericite by albite (Equation 3, Table 1) is favoured by relatively high a_{Na^+}/a_{K^+} in the fluids, and by higher temperatures (Hemley and Jones, 1964). These conditions are most likely to be found in the immediate proximity of the vein. Elsewhere in the alteration zone, the comparatively lower activity of sodium may result in the dissolution of albite.

The reactions given in Table 1 imply that the major components (with the exceptions of CO_2, K, S, H_2O, and Na^+) are little changed in the alteration zones. This holds true for the Golden Mile, Kalgoorlie (Phillips, 1986), the mafic rocks in the Dome Mine (Roberts and Reading, 1981), and probably other alteration zones in low-grade greenschist facies rocks. In these deposits, major element abundances or ratios may be used with reliability to determine the original rock composition. However, hydrolysis of the ferromagnesian minerals and reaction with carbonate and other ionic species in the fluids to form new minerals, requires that the major elements pass into solution, thus providing the potential for their redistribution along concentration gradients within the hydrothermal system. In a careful study of mass balance involved in alteration at the Sigma Mine, Quebec, Robert and Brown (1986) show that significant amounts of Fe, Mg, Al and Ti are lost from the alteration zones immediately adjacent to the veins. Since these elements are not added to the outer alteration zones, they conclude that they were flushed into the veins where they combined with Si, B and S of the fluids to precipitate minerals such as tourmaline, iron sulphides, chlorite and sericite. Depletions of major cations, in this case Ca, Mg and Na, are also reported from the more highly altered zones (the ferroan dolomite domains) in greenschist facies rocks in the Red Lake district (Andrews *et al.*, 1986; MacGeehan and Hodgson, 1982; Pirie, 1981). Andrews *et al.* (1986) suggest that the extensive silicification in the altered rocks of the Red Lake district may be due to depletion of Ca, Mg and Na.

The few studies available suggest that alteration in amphibolite-grade metamorphic rocks is represented either by mineral assemblages of lower P-T conditions than the metamorphic grade (*i.e.* a retrograde assemblage), or by assemblages that reflect the metamorphic grade of the host rocks. An example of the former occurs in the Duport deposit, Ontario, where chlorite,

Table 1 Chemical equations summarizing some of the reactions that take place in alteration zones in deposits in greenschist facies rocks. (From Kerrich, 1983; Phillips, 1986).

(1) $3Ca_2(Mg,Fe)_5Si_8O_{22}(OH)_2 + 2Ca_2Al_3Si_3O_{12}(OH) + 10CO_2 + 8H_2O \rightarrow 3(Mg,Fe)_5Al_2Si_3O_{10}(OH)_8 + 10CaCO_3 + 21SiO_2$

actinolite; epidote; fluid; chlorite; calcite

(2) $3(Mg,Fe)_5Al_2Si_3O_{10}(OH)_8 + 15CaCO_3 + 2K^+ + 15CO_2 \rightarrow 2KAl_3Si_3O_{10}(OH)_2 + 15Ca(Mg,Fe)(CO_3)_2 + 3SiO_2 + 9H_2O + 2H^+$

chlorite; calcite; fluid; sericite; ferroan dolomite; fluid

(3) $2Fe_3O_4 + 6CO_2 \rightarrow 6FeCO_3 + O_2$

magnetite; siderite

(4) $FeCO_3 + 2H_2S \rightarrow FeS_2 + CO_2 + 2H_2O$

siderite; fluid; pyrite; fluid

(5) $Fe_3O_4 + 6H_2S + O_2 \rightarrow 3FeS_2 + 6H_2O$

magnetite; fluid; pyrite

(6) $3NaAlSi_3O_8 + K^+ + 2H^+ \rightarrow KAl_3Si_3O_{10}(OH)_2 + 6SiO_2 + 3Na^+$

albite; fluid; sericite; fluid

quartz and carbonate associated with the emplacement of gold, overprints and replaces amphiboles of the earlier high-grade metamorphism (Smith, 1986). Examples of the latter occur in the Dickenson Mine, Red Lake district, Ontario (Mathieson and Hodgson, 1984), the Hemlo deposits (Kuhns *et al.*, 1986) and the Kolar gold fields, India (Narayanswami *et al.*, 1960). The alteration in these deposits contains minerals such as biotite, garnet, anthophyllite-cummingtonite, cordierite, gedrite and alumino silicates such as andalusite and staurolite. Carbonatization is rare, and calcite, the predominant carbonate, is principally in the veins. The low carbonate content and the predominance of calcite over dolomite reflects the higher metamorphic grade. The few available data indicate that the alteration zones have been severely depleted in Na, Ca and Mg with enrichment (or residual enrichment) in Si and Al (Mathieson and Hodgson, 1984; Kuhns *et al.*, 1986; Cameron and Hattori, 1985; Andrews *et al.*, 1986).

The fact that the geochemistry of alteration in amphibolite facies rocks differs from that of low-grade greenschist metamorphic terranes suggests that the alteration assemblages were formed by rock-water reactions at the higher P-T conditions of metamorphism, rather than by simple upgrading of previously altered rocks. Andrews *et al.* (1986) have documented this relationship between alteration and metamorphism, on the regional scale, for the Red Lake district. The alteration assemblages throughout the district reflect increasing thermal gradients toward the batholith contacts, and the authors conclude that "contact thermal metamorphism and hydrothermal alteration occurred as one and the same process" (Andrews *et al.*, 1986, p. 20).

Fluid Inclusion and Isotope Data

The composition of the ore-forming fluid, and a possible source for the fluid, may be deduced from the evidence of fluid inclusions in vein material, and the isotopic compositions of the hydrothermal minerals.

Fluid Inclusions. Detailed studies of the fluid inclusions in individual deposits are rare. Reconnaissance studies on Canadian deposits (Kerrich, 1983; Walsh *et al.*, 1984) and Australian deposits (Phillips and Groves, 1983; Donnelly *et al.*, 1977), show that inclusions are consistently CO_2-H_2O bearing. There is a wide range of CO_2:H_2O ratios in co-existing inclusions from CO_2-poor to CO_2-dominated (Wood *et al.*, 1986; Guha *et al.*, 1982). Wood *et al.* (1986), in their study of the Hollinger-McIntyre vein system, concluded that the wide variation of compositions was due to phase separation during the main stage of mineralization (unmixing of the original fluid to form two immiscible phases) as a result of pressure decrease within the vein system. Thus, they were able to equate homogenization temperatures with trapping temperatures. They give an average temperature of 277 ± 48°C for the main period of mineralization. Homogenization temperatures (minimum temperatures) in other deposits vary from 200 to 490°C.

The estimated salinities of the inclusion fluid are low: < 2 wt.% NaCl equivalent (Kerrich, 1983); < 2 to < 4 wt.% NaCl equivalent, (Phillips and Groves, 1983); approximately 1 wt.% NaCl equivalent (Wood *et al.*, 1986).

Light Stable Isotope Compositions.

Oxygen isotopes. With few exceptions, the $\delta^{18}O$ of vein quartz falls consistently in the narrow range of + 10 to + 16‰ (Fyon *et al.*, 1983a; Kerrich, 1983; Golding and Wilson, 1983; Kerrich and Watson, 1984), and is typically identical to the quartz of the altered host rock (Kerrich, 1983). The uniformity of the isotopic composition of the quartz of the vein and altered rocks indicates a uniform source for the solutions, external to the environment of the deposit. The $\delta^{18}O$ of carbonates reported by Fyon *et al.* (1983a) from the Timmins region ranges from + 9 to + 14.5‰, and by Golding and Wilson (1983) from Kalgoorlie is 12.4 ± 0.5‰.

The isotopic composition of the fluid may be estimated from the fractionation curves for dolomite-water and quartz-water and used in combination with the temperature of formation of the mineral. Estimates of $\delta^{18}O$ of fluids for a number of Archean deposits range from + 4.5 to + 12‰ (Colvine *et al.*, 1984; Wood *et al.*, 1986). These values fall within, or are close to, the generally accepted values for magmatic fluids (+ 5.5 to + 10‰; Taylor, 1979), but they also fall within the more extensive range of $\delta^{18}O$ values for metamorphic fluids. The usefulness of estimates of fluid compositions is also tempered by the fact that the $\delta^{18}O$ of minerals may be modified by later water-rock reactions, and by the difficulty in obtaining reliable temperature estimates.

Carbon isotopes. The carbon isotopic composition of hydrothermal carbonates in the ore are of particular significance since the ore-forming fluids are characterized by CO_2, and because it is generally accepted that the carbon isotopic compositions of minerals are less vulnerable to modification by later rock-water reactions than oxygen isotopes.

The averaged $\delta^{13}C$ values for hydrothermal dolomite from 13 deposits in the Timmins area, not associated with carbonaceous sediments, range from -2.4 to -4.4‰ (Fyon *et al.*, 1983a; Wood *et al.*, 1986). The average value for carbonate from the No. 4 vein at Kalgoorlie, is -3.6‰ (Golding and Wilson, 1983).

The calculated $\delta^{13}C$ for hydrothermal solutions in the temperature range of 200-500°C, under oxidizing conditions such that graphite is absent, and neutral pH, is approximately that of the ferroan dolomite in equilibrium with it (Ohmoto and Rye, 1979). Thus the $\delta^{13}C$ values of the carbonates quoted above are approximately the same as the $\delta^{13}C$ of the solutions from which they were precipitated.

Hydrogen isotopes. Hydrogen isotopic data on vein quartz, minerals and fluid inclusions are reported from the Timmins district (Fyon *et al.*, 1983a), the Homestake Mine (Rye and Rye, 1974), and the Macassa Mine, Kirkland Lake (Kerrich and Watson, 1984). The δD values are as follows: Timmins: + 6 to -50‰; Homestake: -55.8 to -12‰; Macassa Mine: -60 to -105‰. The data at the Macassa Mine were obtained in conjunction with oxygen isotope data, and Kerrich and Watson (1984) conclude that the ores were precipitated from solutions with $\delta^{18}O$ of + 7 to + 9.6‰, and δD values of -35 to -85‰, over the temperature range of 380-490°C.

Sulphur isotopes. Sulphur isotope data give some indication of the chemistry of the ore fluids and the conditions of deposition of gold. They provide very limited information on the source of the fluids. Values of $\delta^{34}S$ of iron sulphides from Canadian and Zimbabwean deposits, and the Homestake Mine (Crocket and Lavigne, 1984; Wanless *et al.*, 1960; Rye and Rye, 1974; Lambert *et* al., 1984; Wood *et al.*, 1986; Pattison *et al.*, 1986) fall within the narrow range of -0.7 to +7‰. The isotopic values are not unique in as much as they fall within the broad range of many metal sulphide deposit types, but the restricted range of values is unique. For example, Thode and Goodwin (1983) found that the $\delta^{34}S$ values in carbonaceous chert and siderite facies iron formation from the Helen Mine, Michipicoten, Ontario, vary by as much as 30‰ S, and approximately half of the values are isotopically light sulphur. The narrow range around 0‰ suggests that the sulphur of the fluids was in a reduced form.

In contrast to the results for the gold deposits given above, Lambert *et al.* (1984) record values of -2 to -8‰ at the Golden Mile, Kalgoorlie; and in the Hemlo deposit, Ontario, Cameron and Hattori (1985) report values that range from approximately 0 to -17.5‰. In the Hemlo deposit, there is a remarkably strong correlation between the abundance of gold and the depletion of ^{34}S in pyrite and barite (increasing negative values). This is convincing evidence that pyrite and gold were deposited contemporaneously (Cameron and Hattori, 1985). (However, there is no correlation between the abundance of gold and the abundance of pyrite.) The evidence from Hemlo and Kalgoorlie suggests that gold was precipitated from a partially oxidated solution, in which ^{34}S is fractionated into sulphate, and iron sulphide is accordingly depleted in the heavier isotope.

Source of the Fluids. The proposed sources for the hydrothermal fluids are: metamorphic fluids (Kerrich, 1983; Phillips and Groves, 1983, 1984); juvenile fluids formed by granulitization of the lower crust and/or degassing of the upper mantle (Colvine *et al.*, 1984); magmatic hydrothermal fluids (Wood *et al.*, 1984; Burrows and Spooner, 1985; Hodgson and MacGeehan, 1982; Hodgson, 1982); recirculated seawater (Hutchinson and Burlington, 1984).

The low chloride content and high carbonate content of the solutions make it very unlikely that the fluids are recirculated seawater or connate water. Seawater, trapped in volcanic rocks or recharging into a thermal system, would quickly lose carbon dioxide by reactions with calcium-bearing clay minerals (Thompson, 1971), and the salinity of the water (3.6% in modern seawater) would tend to increase as a result of water-rock reactions.

The compositions of fluids in metamorphic rocks are comparable in several respects with ore-forming fluids. Salinities of fluids from inclusions in metamorphic rocks are generally low, and the CO_2:H_2O ratio increases from the greenschist facies, where the fluids are H_2O dominated, to the granulite facies, where they are almost pure CO_2 (Crawford, 1981; Touret, 1981). The CO_2 of metamorphic fluids is either of internal origin (produced by decarbonation reactions, or by the oxidation of carbonaceous material), or it is derived from an external deep-seated source: the degassing of the mantle (Crawford, 1981). Carbon dioxide generated by prograde decarbonation reactions is enriched in ^{13}C by 3-5‰ with respect to the starting material (Sheppard and Schwarcz, 1970). Thus CO_2 from calcite precipitated at low temperatures in volcanic rocks on the sea floor (0 ± 3‰), will be enriched to between +3 and +5‰ (Colvine *et al.*, 1984). In order to attain the estimated carbon isotopic composition of gold ore fluids (-2.4 to -4.4‰), the fluid formed by decarbonation would be required to mix with fluids from other sources or reservoirs (Kerrich, 1983). The uniformity of the isotopic composition of hydrothermal carbonates argues against such a fluid mixing model (Colvine *et al.*, 1984).

The fluids associated with granulization are a potential source of hydrothermal fluids. CO_2-rich inclusions are ubiquitous in granulites over a wide range of rock compositions, prompting Touret (1981) to suggest an external source (juvenile fluids from the mantle) as a control to the fluid phase in these high-grade metamorphic rocks. The compositions of the granulite-associated fluids are comparable with the more CO_2-rich fluids recognized in gold deposits. However, if these are the result of phase separation from a more H_2O-rich parent fluid (Wood *et al.*, 1986), then the transformation of the metamorphic fluids to ore-forming fluids will require significant dilution.

The CO_2 of juvenile fluids, based on the isotopic studies of carbonatites and diamonds, is generally assumed to have $\delta^{13}C$ values of between -4 and -8‰ (Taylor *et al.*, 1967; Deines, 1970; Deines and Gold, 1973). Pineau *et al.* (1976) interpreted carbon ($\delta^{13}C$ = -7.6‰), in fluid inclusions from the Mid-Atlantic ridge, to be of deep-seated origin. The isotopic composition of granulite-associated fluids and the identification of the isotopic signature of the proposed juvenile fluid is not unequivocal. Kruelen (1980), in a study of the Naxos metamorphic terrane in Greece, identified a population of fluids, which occur in low-grade to high-grade schists, and which have $\delta^{13}C$ of -1 to -5‰, as probably of external mantle origin. These values for $\delta^{13}C$ of juvenile fluids are comparable to the deduced value of -2.4 to -4.4‰ for ore fluids.

Magmatic hydrothermal fluids are not generally considered to have the low salinities, and the high CO_2 contents associated with the fluids of gold deposits. However, Wood *et al.* (1984) point out that the Boss Mountain deposit, British Columbia (molybdenum), the Logtung deposit, Yukon (tungsten, molybdenum), the Mink Lake deposit, Ontario (molybdenum), and Tanco, Manitoba (tantalum, tin), all of which are generally believed to be of magmatic-hydrothermal origin, contain CO_2-bearing fluid inclusions. Higgins (1980) also notes an association of CO_2 fluids with lode tungsten deposits. As a generalization, CO_2-rich magmatic hydrothermal fluids appear to be associated with molybdenum and tungsten, and these are common trace elements in gold deposits.

Burrows and Spooner (1985) describe alteration zones in the Mink Lake deposit, that are closely associated with igneous intrusions, which consist of carbonate alteration and involve the enrichment of sodium and potassium. The $\delta^{13}C$ values of carbonates in the altered rocks (-2.7 to -3.9‰) are within the range of values from carbonates in gold deposits. Indeed, the deduced carbon isotopic composition of the ore-forming fluids is more consistent with a magmatic reservoir (-3 to -7‰; Hoefs, 1980) than any other probable source.

The available data is not sufficient to define a source for the mineralizing fluids of Archean lode gold deposits with much conviction. However, the compositions of the fluid inclusions and the compositions of the fluids as deduced from isotopic data point to a magmatic-hydrothermal source.

Solubility of Gold

Gold chloride complexes are stable in strong saline brines, and their stability increases with temperature (Henley, 1973). However, the chloride-poor fluid inclusions and the lack of base metals in the deposits argue against chloride brines as the ore-forming solutions. Seward (1973, 1984) has shown experimentally that gold thio-complexes are stable to at least 300°C and that they dominate the transport of gold in geothermal fields in New Zealand. Data on the stability of base metal thio-complexes are few, but the available data indicate that in alkali solutions, with low sulphur content, base metals will not be as soluble as gold, thus providing a mechanism for the separation of base metals and gold in hydrothermal solutions passing through mafic rocks (Hodgson and MacGeehan, 1982).

The calculated solubilities for gold in Figure 11 indicate that neutral solutions with low concentrations of total sulphur are capable of transporting gold in comparatively high concentrations as the thio-complex. The conditions that bring about saturation and precipitation are probably more critical. The solubility of gold in the thio-complex state increases with the fugacity of oxygen up to the boundary between the fields of reduced and oxidized sulphur (Figure 11). At higher oxygen fugacities, sulphate is the dominant species, and with the decrease in the activity of reduced sulphur species, the solubility of gold in the thio-complex decreases markedly. Oxidation of the fluids is therefore an efficient mechanism for the precipitation of gold. Gold may also be precipitated by changes of pH (an increase in the pH of the solution may occur as the result of loss of carbon dioxide by pressure decrease); a decrease of temperature; or by the reduction in the activity of reduced sulphur by reaction with iron in the host rock to precipitate iron sulphide, and consequent destabilization of the thio-complex.

Source of Gold

With the exception of the magmatic hydrothermal model, none of the various models for the genesis of the ore-forming solutions includes a specific source of gold. The most obvious source is the rocks of the greenstone belt. Pyke (1976), noting the proximity of gold deposits to ultramafic rocks, suggested these rocks as the "source bed" for gold. This is not supported by the data on unaltered rocks; the abundance of gold in ultramafic komatiites is of the same order as other primary igneous rocks: 0.5-2 ppb (Tilling *et al.*, 1973; Anhauesser *et al.*, 1975; Kwong and Crocket, 1978). However, the distribution of gold in Precambrian volcanic rocks is not uniform. Saager and Meyer (1984) and Saager *et al.* (1982) identified two statistical populations: a background population of approximately 1 ppb, in which gold is probably associated with silicates and oxides; and an excess value population of higher, more variable values, in which gold is associated with sulphides. In volcanic rocks, the threshold value is 6.8 ppb, and 19% of the data set are in the excess value population.

An explanation for the bimodal distribution of gold may be found in the work of Keays and Scott (1976) who showed that in modern oceanic basalts, loosely attached, intracrystalline gold and sulphur from magmatic sulphides are mobilized, by the reaction of seawater with the still hot lava, from the crystalline interiors of pillows to the glassy rims. Keays (1984) concluded that similar processes operated in the Precambrian, and that the gold leached from the volcanics was probably fixed in sulphide-bearing, interflow sediments. Bavinton and Keays (1978) report as much as 150 ppb Au in sediments at Kambalda, and similarly elevated values are recorded from sediments in southern Africa (Saager *et al.*, 1982) and the Red Lake district (MacGeehan and Hodgson, 1982; Lavigne and Crocket, 1983). Keays (1984), however, does not suggest that remobilized gold on the sea floor would normally concentrate as ore deposits, but that it would provide elevated concentrations of gold in association with sulphides, and as such it would be readily accessible for later concentration.

Keays (1984) has suggested that Archean komatiitic liquids had a greater capacity than younger, lower temperature magmatic systems, to dissolve sulphur (and with it gold) and thus transfer these elements from the mantle into the crust.

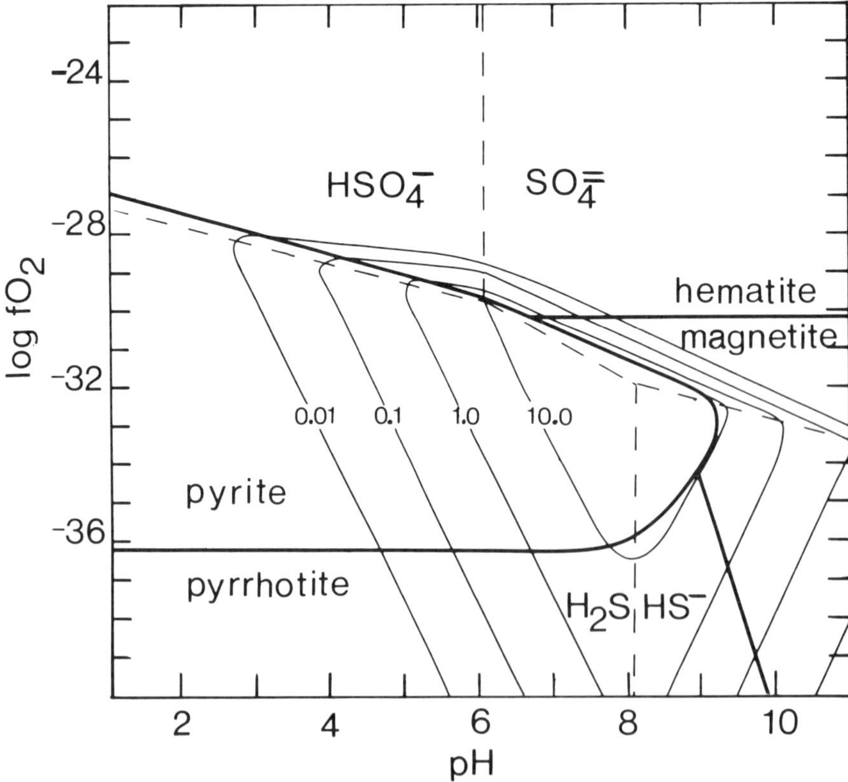

Figure 11 *Calculated gold solubilities as a function of* pH *and* log fO_2. *Gold in the thio-complex* $Au(HS)_2^-$*: Contours are in ppm. Temperature* = 300°C; *total sulphur 0.05 m. (Data from Seward 1973, 1974).*

Discussion

As a result of an increase in exploration activity and research, a large amount of information on gold deposits has been published recently. Consequently considerable progress has been made in defining the basic geological setting of the deposits, but, understandably, hypotheses of genesis are in a state of flux. In the following, aspects of the principal genetic models are briefly reviewed.

The volcanic syngenetic-remobilization model is summarized by Hutchinson and Burlington (1984). According to the model, gold is concentrated in the crust by volcanic-related processes, principally by chemical precipitation in exhalite sediments and Algoman-type banded iron formation. The wall rock alteration is assumed to be related to this primary process. Gold, quartz and associated minerals are remobilized from the sediments to vein structures by post-depositional processes which may include: compaction and diagenesis of the auriferous sediments; deformation and metamorphism; and the deformation-metamorphic processes associated with the emplacement of igneous intrusions.

The model has been brought into question by the following: (a) the reinterpretation of gold-bearing layered units, which were previously considered to be sediments, as sheared rocks with introduced vein material; (b) the recognition that in many deposits in Algoman-type iron formation, primary magnetite has been replaced by gold-bearing iron sulphides; and (c) it has not been possible to relate the geometry of the altered rocks, or the geochemical gradients in the alteration, to the type of footwall alteration that may be expected in a deposit emplaced at surface. Consequently, many of the units that were previously considered to be gold-bearing sediments are being re-evaluated.

The more obvious and well-established features of gold deposits point unequivocally to emplacement at depth. These include: the distribution of the deposits and the associated alteration in tabular deformation zones of regional dimensions; the association of the veins with progressive deformation in these zones; and evidence (admittedly less well documented) that the geochemistry of alteration assemblages of deposits may reflect the P-T conditions of regional metamorphism. It is evident that the deposits are related to regional-scale geological features, which implies that ore genesis is related to regional-scale processes.

The generation of the fluids by decarbonation and dehydration reactions at the base of the greenstone sequence in the amphibolite facies is compatible with the relationship of the deposits to regional geological features (Kerrich, 1983; Phillips and Groves, 1983; Groves *et al.*, 1984). According to the model, the deformation zones serve as conduits, and the rise of the fluids to lower P-T conditions bring them into disequilibrium with some of their host rocks. Thus emplacement would be controlled in part by the transition across metamorphic boundaries (Groves *et al.*, 1984).

Principally because of the association of felsic porphyries with gold deposits, a magmatic hydrothermal source for the ore fluids was probably the most popular theory among Canadian geologists until the late 1960s. Recently, as more evidence has indicated that CO_2-rich fluids may be generated by magmatic processes, the theory has received renewed attention. Macdonald and Hodgson (1986) have pointed out that geochemical features of gold deposits, such as the addition of alkalis to the altered rocks, and the associated elements B, Mo, W, As and Sb, are typical of deposits generally accepted to be of magmatic hydrothermal origin. In their study of deposits in the Temagami district, Fyon and Crocket (1986) describe the zoning of occurrences within a shear zone as being spatially related to late granitoid intrusions; Mo-Cu veins occur closest to the contact of the intrusion, and Cu-Zn-Au and Au zones occur progressively away from the intrusion. Macdonald (1984) reports a similar Mo, Cu, Au zonation related to large felsic intrusions in the Geraldton district. Mason and Melnik (1986) describe similar zoning, but around a comparatively small porphyry body, in the Hollinger-McIntyre vein sytem in Timmins. Mason and Melnik (1986) propose a porphyry-type model in which the gold vein mineralization is peripheral to a porphyry-type hydrothermal system centered on a pipe-like zone of copper stockwork orebodies within the Pearl Lake porphyry. They also conclude that mineralization preceded the regional deformation. In contrast, Wood *et al.* (1986) and Burrows and Spooner (1986) in their study of the same deposit, concluded that mineralization post-dated regional deformation, and that the hydrothermal system is not related to the porphyry bodies. They propose a more remote, magmatic source: the "domal tonalite gneiss-granodiorite-quartz-monzonite-type material which intrudes the lower part of Archean greenstone belts" (Wood *et al.*, 1986, p. 78).

It is apparent that the geological evidence concerning the timing of gold deposits in relationship to the various intrusive igneous events and deformation may be ambiguous or incomplete in some environments. These ambiguities may be appreciated in the light of the conclusions of Andrews *et al.* (1986), from their study of the Red Lake district, that contact metamorphism, shear deformation and alteration were all broadly coeval and linked to batholith emplacement. Thus the emplacement of gold deposits is linked to the regional structural, metamorphic and intrusive igneous history of the greenstone belt. This is similar to the conclusion reached by Helmstaedt and Padgham (1986) regarding the relationship of gold deposits to the igneous, structural and metamorphic development of the Yellowknife district, Northwest Territories. Andrews *et al.* (1986, p. 21) point out that the ore-forming fluids "could have had their source in metamorphic dehydration and decarbonation ahead of the rising diapirs, and also magmatic degassing from the diapirs themselves".

Acknowledgements

An earlier version of the manuscript was critically reviewed by Patricia Sheahan of Konsult International Inc., Soussan Marmont of the Ontario Geological Survey, and Lee Barker of Lacana Mining. I am grateful to Michael Gray, President of Queenston Gold Mines and Dean Rogers of Dome Mines for allowing me to include previously unpublished information in this paper. The structural relationships at the Upper Canada Mine were first pointed out to me by Terry Bottrill, Consulting Geologist, and subsequently described in a B.Sc. thesis by Paul Stewart of the University of Waterloo. I acknowledge financial support from the Ontario Geological Survey Geoscience Research Grant Program and grants from Energy, Mines and Resources, Canada.

References

Andrews, A.J., Hugon, H., Durocher, M., Corfu, F. and Lavigne, M.J., 1986, The anatomy of a gold-bearing greenstone belt: Red Lake, Northwestern Ontario, *in* Macdonald, A.J., ed., Proceedings of Gold '86, an International Symposium on the Geology of Gold: Toronto, 1986, p. 3-22.

Andrews, A.J. and Wallace, H., 1983, Alteration, metamorphism and structural patterns associated with Archean gold deposits — preliminary observations in the Red Lake area, *in* Colvine, A.C., ed., The Geology of Gold in Ontario: Ontario Geological Survey, Miscellaneous Paper 110, p. 111-122.

Anhauesser, C.R., Fritz, K., Fyfe, W.S. and Gill, R.C.O., 1975, Gold in "primitive" Archean volcanics: Chemical Geology, v. 16, p. 129-135.

Archambault, G., 1985, Archean wrench fault tectonics and structural evolution of the Blake River Group, Abitibi Belt: discussion: Canadian Journal of Earth Sciences, v. 22, p. 943-945.

Baragwanath, W., 1953, Ballarat goldfields, *in* Edwards, A.B., ed., Geology of Australian Ore Deposits: Australasian Institute of Mining and Metallurgy, Melbourne, p. 986-1002.

Barnett, E.S., Hutchinson, R.W., Adamcik, A. and Barnett, R., 1982, Geology of the Agnico Eagle gold mine, Quebec, *in* Hutchinson, R.W., Spence, C.D. and Franklin, J.M., eds., Precambrian Sulphide Deposits: Geological Association of Canada, Special Paper 25, p. 403-426.

Bavinton, O.A. and Keays, R.R., 1978, Precious metal values from interflow sedimentary rocks from the komatiite sequence at Kambalda, Western Australia: Geochimica et Cosmochimica Acta, v. 42, p. 1151-1163.

Boyle, R.W., 1961, The geology, geochemistry and origin of the gold deposits of the Yellowknife district: Geological Survey of Canada, Memoir 310, 193 p.

Boyle, R.W., 1979, The geochemistry of gold and its deposits: Geological Survey of Canada, Bulletin 280, 584 p.

Burrows, D.R. and Spooner, E.T.C., 1986, The McIntyre Cu-Au deposit, Timmins, Ontario, Canada, *in* Macdonald, A.J., ed., Proceedings of Gold '86, an International Symposium on the Geology of Gold: Toronto, 1986, p. 23-39.

Burrows, D.R., Wood, P.C. and Spooner, E.T.C., 1985, Carbon isotope evidence for a magmatic origin for Archean gold-quartz vein ore deposits: Nature, v. 321, p. 851-854.

Burrows, D. R. and Spooner, E.T.C., 1985, Generation of an Archean H_2O-CO_2 fluid enriched in Au, W and Mo by fractional crystallization in the Mink Lake intrusion, N.W. Ontario: Geological Society of America, Abstracts with Program, v. 17, p. 536.

Cameron, E.M. and Hattori, K., 1985, The Hemlo gold deposit, Ontario: a geochemical and isotopic study: Geochimica et Cosmochimica Acta, v. 49, p. 2041-2050.

Charlewood, G.H., 1964, Geology of Deep Developments on the Main Ore Zone at Kirkland Lake, NTS 42A/1, Timiskaming District, Ontario: Ontario Department of Mines, Geological Circular 11, 49 p.

Cherry, M.E., 1983, Association of gold and felsic intrusions - examples from the Abitibi belt, *in* Colvine, A.C., ed., The Geology of Gold in Ontario: Ontario Geological Survey, Miscellaneous Paper 110, p. 48-55.

Colvine, A.C., Andrews, A.J., Cherry, M.E., Durocher, M.E., Fyon, A.J., Lavigne, M.J., Macdonald, A.J., Marmont, S., Poulsen, K.H., Springer, J.S. and Troop, D.G., 1984, An integrated model for the origin of Archean lode gold deposits: Ontario Geological Survey, Open File Report 5524, 99 p.

Cooke, D.L. and Moorhouse, W.W., 1969, Timiskaming volcanism in the Kirkland Lake Area, Ontario, Canada: Canadian Journal of Earth Sciences, v. 6, p. 117-132.

Crawford, M.L., 1981, Fluid inclusions in metamorphic rocks - low and medium grade, *in* Hollister, L.S. and Crawford, M.L., eds., Fluid Inclusions: applications to petrology: Mineralogical Association of Canada, Short Course Handbook, v. 6, p. 157-181.

Crocket, J.H. and Lavigne, M.J., 1984, Sulphur sources in the Dickenson gold mine as suggested by sulphur isotopes, *in* Foster, R.P., ed., Gold '82: Zimbabwe Geological Society, Special Publication No. 1, p. 417-434.

Crocket, J.H., Cowan, P. and Kusmirski, R.T.M., 1981, Gold content of volcanic-hosted interflow sedimentary rocks in the Red Lake area: implications on ore genesis at the Dickenson mine, *in* Pye, E.G. and Roberts, R.G., eds., Genesis of Archean, volcanic hosted gold deposits: Ontario Geological Survey, Miscellaneous Paper 97, p. 128-143.

Deines, P., 1970, The carbon and oxygen isotopic composition of carbonates from the Oka carbonatite complex, Quebec, Canada: Geochimica et Cosmochimica Acta, v. 34, p. 1199-1225.

Deines, P. and Gold, D.P., 1973, The isotopic composition of carbonatite and kimberlite carbonates and their bearing on the isotopic composition of deep-seated carbon: Geochimica Cosmochica Acta, v. 37, p. 1709-1733.

Dimroth, E., Imreh, L., Rocheleau, M. and Goulet, N., 1982, Evolution of the south-central part of the Archean Abitibi Belt, Quebec. Part I: Stratigraphy and paleogeographic model: Canadian Journal of Earth Sciences, v. 19, p. 1729-1758.

Donnelly, T.H., Lambert, I., Oehler, D.Z., Halleberg, J., Hudson, D.R., Smith, J.W., Bavinton, O.A. and Golding, L., 1977, A reconnaissance study of stable isotope ratios in Archean rocks from the Yilgarn Block, Western Australia: Journal of the Geological Society of Australia, v. 24, p. 409-420.

Fehlberg, B. and Giles, C.W., 1984, Archean volcanic exhalative gold mineralization at Spargoville, Western Australia, *in* Foster, R.P., ed., Gold '82: Zimbabwe Geological Society, Special Publication No. 1, p. 285-304.

Fripp, R.E.P., 1976, Stratabound gold deposits in Archean banded iron formation, Rhodesia: Economic Geology, v. 71, p. 58-75.

Fryer, B.J. and Hutchinson, R.W., 1976, Generation of metal deposits on the sea floor: Canadian Journal of Earth Sciences, v. 13, p. 126-135.

Fryer, B.J., Kerrich, R., Hutchinson, R.W., Pierce, M.G. and Rogers, D.S., 1979, Archean precious metal hydrothermal systems, Dome Mine, Abitibi greenstone belt. I. Patterns of alteration and metal distribution: Canadian Journal of Earth Sciences, v. 16, p. 421-439.

Fyon, J.A. and Crocket, J.H., 1982, Gold exploration in the Timmins district using field and lithogeochemical characteristics of carbonate alteration zones, *in* Hodder, R.W. and Petruk, W., eds., Geology of Canadian Gold Deposits: Canadian Institute of Mining and Metallurgy, Special Paper 24, p. 113-129.

Fyon, A.J. and Crocket, J.H., 1986, Gold mineralization in Strathy Township, Temagami area, Ontario, *in* Chater, A.M., ed., Gold '86, an International Symposium on the Geology of Gold: Toronto, 1986, Poster paper abstracts, p. 44.

Fyon, J.A., Crocket, J.H. and Schwarcz, H.P., 1983a, Application of stable isotope studies to gold metallogeny in the Timmins-Porcupine camp, NTS 42A, Cochrane District: Ontario Geological Survey, Open File Report 5464, 182 p.

Fyon, J.A., Crocket, J.H. and Schwarcz, H.P., 1983b, The Carshaw and Malga iron-formation hosted gold deposits of the Timmins area, *in* Colvine, A.C., ed., The Geology of Gold in Ontario: Ontario Geological Survey, Miscellaneous Paper 110, p. 98-110.

Gibbins, W., 1981, Gold and Precambrian iron formation in the Northwest Territories, *in* Morton, R.D., ed., Proceedings of the Gold Workshop: Yellowknife Gold Workshop Committee, Yellowknife, Northwest Territories, p. 258-287.

Gibson, I.L., Roberts, R.G. and Maloney, J.A., 1982, Felsic volcanism and the origin of gold deposits at Timmins and Kirkland Lake: Geological Association of Canada - Mineralogical Association of Canada, Program with abstracts, v. 7, p. 52.

Golding, S.D. and Wilson, A.F., 1983, Geochemical and stable isotope studies of the No. 4 lode, Kalgoorlie, Western Australia: Economic Geology, v. 78, p. 438-450.

Goodwin, A.M. and Ridler, R.H., 1970, The Abitibi orogenic belt, *in* Baer, A.J., ed., Symposium on Basins and Geosynclines of the Canadian Shield: Geological Survey of Canada, Paper 70-40, p. 1-24.

Graves, M.C. and Zentilli, M., 1982, A review of the geology of gold in Nova Scotia, *in* Hodder, R.W. and Petruk, W., ed., Geology of Canadian Gold Deposits: Canadian Institute of Mining and Metallurgy, Special Paper 24, p. 233-242.

Groves, D.I., Phillips, G.N., Ho, S.E., Henderson, C.A., Clark, M.E. and Wood, G.M., 1984, Controls on distribution of Archean hydrothermal gold deposits in Western Australia, *in* Foster, R.P., ed., Gold'82: Geological Society of Zimbabwe, Special Publication No. 1, p. 689-712.

Guha, J., Archambault, G. and Leroy, J., 1983, A correlation between the evolution of mineralizing fluids and the geomechanical development of a shear zone as illustrated by the Henderson Z Mine, Quebec: Economic Geology, v. 78, p. 1605-1618.

Guha, J., Gauthier, A., Vallee, M., Descarreaux, J. and Lange-Brard, F., 1982, Gold mineralization patterns at the Doyon mine (Silverstack), Bousquet, Quebec, *in* Hodder, R.W. and Petruk, W., eds., Geology of Canadian Gold Deposits: Canadian Institute of Mining and Metallurgy, Special Paper 24, p. 50-57.

Helmstaedt, H. and Padgham, W.A., 1986, A new look at the stratigraphy of the Yellowknife Supergroup at Yellowknife, N.W.T. - implications for the age of gold-bearing shear zones and Archean basin evolution: Canadian Journal of Earth Sciences, v. 23, p. 454-475.

Hemley, J.J. and Jones, W.P., 1964, Chemical aspects of hydrothermal alteration with emphasis on hydrogen metasomatism: Economic Geology, v. 59, p. 538-569.

Henley, R.W., 1973, Solubility of gold in hydrothermal chloride solutions: Chemical Geology, v. 11, p. 73-87.

Higgins, N.C., 1980, Fluid inclusion evidence for the transport of tungsten by carbonate complexes in hydrothermal solutions: Canadian Journal of Earth Sciences, v. 17, p. 823-830.

Hodgson, C.J., 1982, Gold deposits of the Abitibi Belt, Ontario, *in* Wood, J., White, O.L., Barlow, R.B. and Colvine, A.C., eds., Summary of field work, 1982, Ontario Geological Survey, Miscellaneous Paper 106, p. 192-197.

Hodgson, C.J., 1983, Preliminary report on a computer file of gold deposits of the Abitibi Belt, Ontario, *in* Covine, A.C., ed., The Geology of Gold in Ontario: Ontario Geological Survey, Miscellaneous Paper 110, p. 11-37.

Hodgson, C.J. and MacGeehan, P.J., 1982, A review of the geological characteristics of "gold only" deposits in the Superior Province of the Canadian Shield, *in* Hodder, R.W. and Petruk, W., eds., Geology of Canadian Gold Deposits: Canadian Institute of Mining and Metallurgy, Special Volume 24, p. 211-228.

Hoefs, J., 1980, Stable isotope geochemistry: Springer-Verlag, New York, 208 p.

Hugon, H. and Schwerdtner, W.M., 1985, Structural signature and tectonic history of deformed gold-bearing rocks in N.W. Ontario: Ontario Geological Survey, Geoscience Research Seminar and Open House, abstracts, p. 2.

Hutchinson, R.W., 1976, Lode gold deposits: the case for volcanogenic derivation, *in* Proceedings Volume, Pacific Northwest Mining and Metals Conference, Portland, 1975: Oregon Department of Geology and Industrial Minerals, p. 64-105.

Hutchinson, R.W. and Burlington, J.L., 1984, Some broad characteristics of greenstone belt gold lodes, *in* Foster, R.P., ed., Gold '82: Geological Society of Zimbabwe, Special Publication No. 1, p. 339-372.

Jensen, L.S., 1985, Stratigraphy and petrogenesis of Archean metavolcanic sequences, southwestern Abitibi Subprovince, Ontario, *in* Ayres, L.D., Thurston, P.C., Card, K.D. and Weber, W., eds., Evolution of Archean Supracrustal Sequences: Geological Association of Canada, Special Paper 28, p. 65-88.

Karvinen, W.O., 1981, Geology and evolution of gold deposits, Timmins area, Ontario, *in* Pye, E.G. and Roberts, R.G., eds., Genesis of Archean, volcanic hosted gold deposits: Ontario Geological Survey, Miscellaneous Paper 97, p. 29-46.

Keays, R.R., 1984, Archean gold deposits and their source rocks: the upper mantle connection, *in* Foster, R.P., ed., Gold '82: Geological Society of Zimbabwe, Special Publication Number 1, p. 17-52.

Keays, R.R. and Scott, R.B., 1976, Precious metals in ocean ridge basalts: implications for basalts as source rocks for gold mineralization: Economic Geology, v. 71, p. 705-720.

Kerrich, R., 1983, Geochemistry of gold deposits in the Abitibi greenstone belt: Canadian Institute of Mining and Metallurgy, Special Volume 27, 75 p.

Kerrich, R. and Watson, G.P., 1984, The Macassa Mine Archean lode gold deposit, Kirkland Lake, Ontario: patterns of alteration and hydrothermal regimes: Economic Geology, v. 79, p. 104-1130.

Kerrich, R., Fryer, B.J., Milner, K.J. and Pierce, M.G., 1981, The geochemistry of gold-bearing chemical sediments, Dickenson mine, Ontario: a reconnaissance study: Canadian Journal of Earth Sciences, v. 18, p. 624-637.

Kerswell, J.A., 1984, The Lupin deposit, Contwoyto Lake area, N.W.T.: styles of gold distribution and possible genetic models: Geological Association of Canada - Mineralogical Association of Canada, Program with abstracts, v. 9, p. 78.

Kreulen, R., 1980, CO_2-rich fluids during regional metamorphism on Naxos (Greece): carbon isotopes and fluid inclusions: American Journal of Science, v. 280, p. 745-771.

Kuhns, R.J., Sawkins, F.J. and Ito, E., 1986, Alteration associated with the Golden Giant deposit, Hemlo, Ontario: Geological Association of Canada - Mineralogical Association of Canada, Annual Meeting, Program with abstracts, v. 11, p. 91.

Kwong, Y.T.J. and Crocket, J.H., 1978, Background and anomalous gold in rocks of an Archean greenstone assemblage, Kakagi Lake area, northwestern Ontario: Economic Geology, v. 73, p. 50-63.

Lambert, I.B., Phillips, G.N. and Groves, D.L., 1984, Sulphur isotope compositions and genesis of Archean gold mineralization, Australia and Zimbabwe, *in* Foster, R.P., ed., Gold '82: Geological Society of Zimbabwe, Special Publication No. 1, p. 389-416.

Latulippe, M., 1982, An overview of the geology of gold occurrences and developments in northwestern Quebec, *in* Hodder, R.W. and Petruk, W., eds., Geology of Canadian Gold Deposits: Canadian Institute of Mining and Metallurgy, Special Volume 24, p. 9-14.

Lavigne, M.J. and Crocket, J.H., 1983, Geology of the East South "C" ore zone, Dickenson mine, Red Lake, *in* Colvine, A.C., ed., The Geology of Gold in Ontario: Ontario Geological Survey, Miscellaneous Paper 110, p. 141-158.

Macdonald, A.J., 1984, Gold mineralization in Ontario, I: The role of banded iron formation, *in* Guha, J. and Chown, E.H., eds., Chibougamau - stratigraphy and mineralization: Canadian Institute of Mining and Metallurgy, Special Volume 34, p. 412-430.

Macdonald, A.J. and Hodgson, C.J., 1986, The case for the magmatic-hydrothermal origin of Archean lode gold deposits, *in* Chater, A.M., ed., Gold '86, an International Symposium on the Geology of Gold: Toronto, 1986, Poster paper abstracts, p. 98.

MacGeehan, P.J. and Hodgson, C.J., 1982, Environments of gold mineralization in the Campbell Red Lake and Dickenson Mines, Red Lake District, Ontario, *in* Hodder, R.W. and Petruk, W., eds., Geology of Canadian Gold Deposits: Canadian Institute of Mining and Metallurgy, Special Paper 24, p. 184-210.

Mason, R. and Melnik, N., 1986, The anatomy of an Archean gold system - the McIntyre-Hollinger Complex at Timmins, Ontario, Canada, *in* Macdonald, A.J., ed., Proceedings of Gold '86, an International Symposium on the Geology of Gold: Toronto, 1986, p. 40-55.

Matthews, A. and Beckinsale, R.D., 1979, Oxygen isotope equilibration systematics between quartz and water: American Mineralogist, v. 64, p. 232-240.

Mathieson, N.A. and Hodgson, C.J., 1984, Alteration, mineralization, and metamorphism in the area of the East South "C" ore zone, 24th level of the Dickenson mine, Red Lake, northwestern Ontario: Canadian Journal of Earth Sciences, v. 21, p. 35-52.

Melling, D.R., Watkinson, D.H., Poulsen, K.H., Chorlton, L.B. and Hunter, A.D., 1986, The Cameron Lake gold deposit, Northwestern Ontario, Canada: geological setting, structure and alteration, *in* Macdonald, A.J., ed., Proceedings of Gold '86, an International Symposium on the Geology of Gold: Toronto, 1986, p. 149-169.

MERQ-OGS, 1983, Lithostratigraphic map of the Abitibi Subprovince: Ministère de l'Energie et des Ressources, Québec — Ontario Geological Survey, scale 1:500,00; catalogued as "Map 284" in Ontario and "DV-83-16" in Quebec.

Michie, H., 1967, Hollinger mine, *in* Northwest Quebec and Northern Ontario: Canadian Institute of Mining and Metallurgy, Centennial Field Excursion, p. 111-117.

Narayanaswami, S., Ziauddin, M. and Ramachandra, A.V., 1960, Structural control and localization of gold-bearing lodes, Kolar gold field, India: Economic Geology, v. 55, p. 1429-1459.

Ohmoto, H. and Rye, R.O., 1979, Isotopes of sulphur and carbon, *in* Barnes, H.L., ed., Geochemistry of Hydrothermal Ore Deposits, Second Edition: Wiley, New York, p. 509-567.

Padgham, W.A., 1981, Gold deposits of the Northwest Territories, *in* Morton, R.D., ed., Proceedings of the Gold Workshop: Yellowknife Gold Workshop Committee, Yellowknife, Northwest Territories, p. 174-201.

Page, C.E. and Roberts, R.G., 1984. The Cullaton Lake B-Zone deposit, Northwest Territories: geology and geochemistry of a gold-bearing iron formation: Geological Association of Canada - Mineralogical Association of Canada, Program with abstracts, v. 9, p. 94.

Pattison, E.F., Auerbrei, J.A., Hannila, J.J. and Church, J.F., 1986, Gold mineralization in the Casa-Beradi area, Quebec, Canada, *in* Macdonald, A.J., ed., Proceedings of Gold '86, an International Symposium on the Geology of Gold: Toronto, 1986, p. 170-183.

Phillips, G.N., 1986, Geology and alteration in the Golden Mile, Kalgoorlie: Economic Geology, v. 81, p. 779-808.

Phillips, G.N. and Groves, D.I., 1983, The nature of Archean gold fluids as deduced from gold deposits of Western Australia: Journal of the Geological Society of Australia, v. 30, p. 25-39.

Phillips, G.N. and Groves, D.I., 1984, Fluid access and fluid-wall rock reactions in the genesis of the Archean gold-quartz vein deposit at the Hunt mine, Kambalda, Western Australia, *in* Foster, R.P., ed., Gold '82: the geology, geochemistry and genesis of gold deposits: Geological Society of Zimbabwe, Special Publication No. 1, p. 389-416.

Phillips, G.N., Groves, D.I. and Martyn, J.E., 1984, An epigenetic origin for Archean banded iron-formation hosted gold deposits: Economic Geology, v. 79, p. 162-171.

Pineau, F., Javoy, M. and Bottinga, Y., 1976, $^{13}C/^{14}C$ ratios of rocks and inclusions in popping rocks of the Mid-Atlantic Ridge and their bearing on the problem of isotopic composition of deep-seated carbon: Earth and Planetary Science Letters, v. 29, p. 413-421.

Pirie, J., 1981, Regional geological setting of gold deposits in the Red Lake Area, northwestern Ontario, *in* Pye, E.G. and Roberts, R.G., eds., Genesis of Archean, volcanic hosted gold deposits: Ontario Geological Survey, Miscellaneous Paper 97, p. 71-93.

Platt, J.P., 1983, Progressive refolding in ductile shear zones: Journal of Structural Geology, v. 5, p. 619-622.

Poulsen, K.H., 1983, Structural setting of vein-type gold mineralization in the Mine Centre-Fort Frances area: implication for the Wabigoon subprovince, *in* Colvine, A.C., ed., The Geology of Gold in Ontario: Ontario Geological Survey, Miscellaneous Paper 110, P. 174-180.

Poulsen, K.H., Ames, D.E. and Galley, A.G., 1986, Gold mineralization in the Star Lake pluton, La Ronge belt, Saskatchewan: a preliminary report: Geological Survey of Canada, Current Research, Paper 86-1A, p. 205-212.

Poulsen, K.H. and Franklin, J.M., 1981, Copper and gold mineralization in an Archean trondhjemitic intrusion, Sturgeon Lake, Ontario: Geological Survey of Canada, Paper 81-1A, p. 9-14.

Pyke, D.R., 1976, On the relationship of gold mineralization and ultramafic volcanic rocks in the Timmins Area, northeastern Ontario: Canadian Institute of Mining and Metallurgy, Bulletin, v. 69, p. 79-87.

Pyke, D.R., 1982, Geology of the Timmins Area, NTS 42A/6, Cochrane District, Ontario: Ontario Geological Survey, Report 219, 141 p.

Ramsay, J.G., 1980, Shear geometry: a review: Journal of Structural Geology, v. 2, p. 83-99.

Rigg, D.M. and Helmstaedt, H., 1981, Relations between structures and gold mineralization in Campbell Red Lake and Dickenson mines, Red Lake area, Ontario, *in* Pye, E.G. and Roberts, R.G., eds., Genesis of Archean, volcanic hosted gold deposits: Ontario Geological Survey, Miscellaneous Paper 97, p. 111-127.

Robert, F. and Brown, A.C., 1986, Archean gold-bearing quartz veins at the Sigma mine, Abitibi greenstone belt, Quebec, Part II: vein paragenesis and hydrothermal alteration: Economic Geology, v. 81, p. 595-616.

Robert, F., Brown, A.C. and Audet, A.J., 1983, Structural control of gold mineralization at the Sigma Mine, Val d'Or, Quebec: Canadian Institute of Mining and Metallurgy, Bulletin, v. 76, p. 72-80.

Roberts, R.G., 1981, The volcanic-tectonic setting of gold deposits in the Timmins area, *in* Pye, E.G. and Roberts, R.G., eds., Genesis of Archean, volcanic hosted gold deposits, Ontario Geological Survey, Miscellaneous Paper 97, p. 1-28.

Roberts, R.G. and Reading, D.J., 1981, The volcanic-tectonic setting of gold deposits in the Timmins district - carbonate-bearing rocks at the Dome mine, *in* Pye, E.G., ed., Geoscience Research Grant Program, Summary of Research 1980-1981: Ontario Geological Survey, Miscellaneous Paper 98, p. 222-228.

Rye, D.M. and Rye, R.O., 1974, Homestake gold mine, South Dakota: I. Stable isotope studies: Economic Geology, v. 69, p. 293-317.

Saager, R. and Meyer, M., 1984, Gold distribution in Archean granitoids and supracrustal rocks from southern Africa; a comparison, *in* Foster, R.P., ed., Gold '82: Geological Society of Zimbabwe, Special Publication Number 1, p. 53-70.

Saager, R., Meyer, M. and Muff, R., 1982, Gold distribution in supracrustal rocks in Archean greenstone belts in Southern Africa and from Paleozoic ultramafic complexes of the European Alps: metallogenic and geochemical implications: Economic Geology, v. 77, p. 1-24.

Sanborn, M. and Schwerdtner, W.M., 1986, The role of brittle-ductile shear in the formation of gold-bearing quartz-carbonate veins in the west carbonate zone of the Cochenour-Willans mine, Red Lake, Ontario: Geological Association of Canada - Mineralogical Association of Canada, Program with abstracts, v. 11, p. 123.

Schwerdtner, W.M., Stone, D., Osadetz, K., Morgan, J. and Stott, G.M, 1979, Granitoid complexes and the Archean tectonic record in the southern part of northwestern Ontario: Canadian Journal of Earth Sciences, v. 19, p. 1965-1977.

Seward, T.M., 1973, Thio-complexes of gold and the transport of gold in hydrothermal ore solutions: Geochimica et Cosmochimica Acta, v. 37, p. 379-399.

Seward, T.M., 1984, The transport and deposition of gold in hydrothermal systems, *in* Foster, R.P., ed., Gold '86: Geological Society of Zimbabwe, Special Publication No. 1, p. 165-182.

Sheppard, S.M.F. and Schwarcz, H.P., 1970, Fractionation of carbon and oxygen isotopes and Mg between co-existing and metamorphic calcite and dolomite: Contributions to Mineralogy and Petrology, v. 20, p. 161-198.

Simpson, C., 1983, Displacement and strain patterns from naturally occurring shear zone terminations: Journal of Structural Geology, v. 5, p. 497-506.

Smith, P.M., 1986, Duport, a structurally controlled gold deposit in Northwestern Ontario, Canada, *in* Macdonald, A.J., ed., Proceedings of Gold '86, an International Symposium on the Geology of Gold: Toronto, 1986, p. 197-212.

Smith, T.J. and Kesler, S.E., 1985, Relation of fluid inclusion geochemistry to wall rock alteration and lithogeochemical zonation at the Hollinger-McIntyre gold deposit, Timmins, Ontario: Canadian Institute of Mining and Metallurgy, Bulletin, v. 78, p. 35-46.

Taylor, H.P., 1979, Oxygen and hydrogen isotope relationships, *in* Barnes, H.L., ed., Geochemistry of Hydrothermal Ore Deposits, Second Edition: Wiley, New York: p. 236-272.

Taylor, H.P., Frechen, J. and Degens, E.T., 1967, Oxygen and carbon isotope studies of carbonatites from the Laacher See district, West Germany and the Alno district, Sweden: Geochimica et Cosmochimica Acta, v. 31, p. 407-430.

Tchalenko, J.S., 1968, The evolution of kink-bands and the development of compressive textures in sheared clays: Tectonophysics, v. 6, p. 159-174.

Thode, H.G. and Goodwin, A.M., 1983, Further sulphur and carbon isotope studies of late Archean iron-formation of the Canadian Shield and the rise of sulphate reducing bacteria: Precambrian Research, v. 20, p. 337-356.

Thompson, A.B., 1971, p_{CO_2} in low grade metamorphism; zeolite, carbonates, clay mineral, prehnite relations in the system $CaO-Al_2O_3-SiO$: Contributions to Mineralogy and Petrology, v. 33, p. 145-161.

Tilling, R.I., Gottfried, D. and Rowe, J.J., 1973, Gold abundance in igneous rocks and its bearing on gold mineralization: Economic Geology, v. 68, p. 168-186.

Touret, J., 1981, Fluid inclusions in high grade metamorphic rocks, *in* Hollister, L.S. and Crawford, M.L., eds., Short course in fluid inclusions: applications in petrology: Mineralogical Association of Canada, Short Course Handbook, v. 6, p. 182-208.

Valliant, R.I., Mongeau, C. and Doucet, R., 1982, The Bousquet pyritic gold deposits, Bousquet Region, Quebec: descriptive geology and preliminary interpretations of genesis, *in* Hodder, R.W. and Petruk, W., eds., Geology of Canadian Gold Deposits: Canadian Institute of Mining and Metallurgy, Special Volume 24, p. 41-49.

Viljoen, M.J., 1984, Archean gold mineralization and komatiites in southern Africa, *in* Foster, R.P., ed., Gold '82: Geological Society of Zimbabwe, Special Publication No. 1, p. 595-628.

Walsh, J.F., Haynes, F.M. and Kesler, S.E., 1984, Fluid inclusion analyses of auriferous vein quartz from the Porcupine district, Ontario: Geological Association of Canada - Mineralogical Association of Canada, Program with abstracts, v. 9, p. 114.

Wanless, R.K., Boyle, R.W. and Lowdon, J.A., 1960, Sulphur isotope investigations of the gold-quartz deposits of the Yellowknife District: Economic Geology, v. 55, p. 1591-1621.

Ward, J.H.W., 1984, Barberton rocks and exhalative gold, *in* Pearton, T.N., ed., Archean gold, Barberton centenary symposium, (abstract), p. 26.

Wood, P.C., Burrows, D.R., Thomas, A.V. and Spooner, E.T.C., 1986, The Hollinger-McIntyre Au-Quartz vein system, Timmins, Ontario, Canada; geologic characteristics, fluid properties and light stable isotopes, *in* Macdonald, A.J., ed., Proceedings of Gold '86, an International Symposium on the Geology of Gold: Toronto, 1986, p. 56-80.

Wood, P.C., Thomas, A.V., Burrows, D.R., Macdonald, A.J., Noble, S.R. and Spooner, E.T.C., 1984, CO_2-bearing, low-moderate salinity fluids in Archean gold-quartz-carbonate-(W-Mo) vein deposits and magmatically derived Mo, W, Ta, and Sn mineralization: Geological Society of America, Abstracts with Program, v. 16, p. 700.

Accepted, as revised, 12 February 1987.
Originally published in
Geoscience Canada v. 14 Number 1
(March 1987)

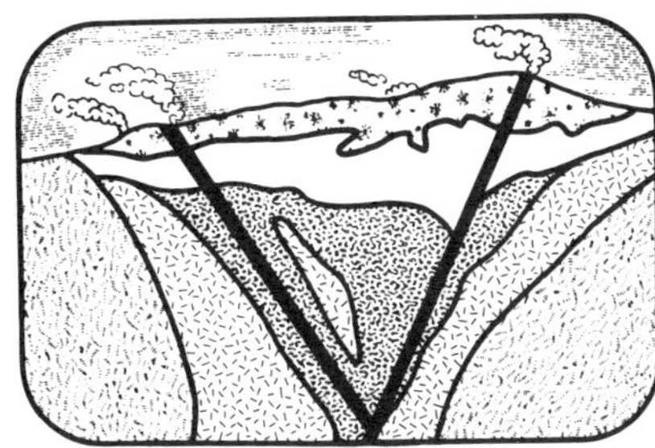

Disseminated Gold Deposits

S.B. Romberger
Department of Geology
Colorado School of Mines
Golden, Colorado 80401

Introduction

Disseminated gold deposits have been attractive exploration targets since the early 1960s when the Carlin deposit in northeastern Nevada was discovered. Before that time, such deposits were known, however they were considered to be too low grade or too small to be mined profitably. The low gold price and the lack of technology for recovering fine-grained gold from low-grade materials contributed to the unattractiveness of these deposits. However, locally small-scale mining did occur from veins and disseminations where gold became concentrated enough to warrant such activity. In fact, certain disseminated gold deposits have been discovered by reinvestigating historically active lode districts.

For the purpose of this paper, disseminated gold deposits are defined as those deposits originating from hydrothermal processes which contain economic amounts of gold finely dispersed in host rocks of variable composition where little or no fabric control on mineralization is apparent, at least at the hand specimen scale. Beyond this, the variabilities observed in size, grade, textures, mineral associations, nature and degree of structural control, composition of host rocks, association with igneous rocks, nature and degree of alteration, and other factors are such that it is difficult to define these deposits more specifically. The gold is most commonly sub-microscopic or microscopic but locally may become visible to the naked eye. The deposits occur in rocks of all ages, however, the mineralization usually can be related to some Tertiary event.

The variabilities observed make it difficult to develop a general model that can be applied to all deposits. In addition, because of the fine-grained nature of the gold and associated mineralization and the difficulties in recognizing distinctive hydrothermal features such as wall-rock alteration, these deposits are difficult to study both in the field and laboratory. However, there are a number of features common to many of these deposits that must be considered in their genesis: (1) the common occurrence of silicification, expressed as jasperoids in most of the sedimentary rock-hosted deposits; (2) the close association between gold and pyrite in primary ores; (3) the anomalously low abundance of common base metals such as copper, lead and zinc; and (4) the occurrence of significant amounts of arsenic, antimony, mercury and thallium in the ores. These factors will be incorporated into the model discussed below along with other characteristics commonly thought of as typical for the deposits, such as the association with carbonaceous material. As will become clear, the latter is not a universal occurrence but is limited to certain sedimentary rock-host deposits.

Before developing a genetic model for disseminated gold deposits, it is necessary to review the characteristics exhibited by these deposits. In this review, emphasis will be placed on the deposits located in the western United States. Specific deposits will be used in the discussion because of their unique or outstanding characteristics which are important in developing a genetic model, availability of literature, and familiarity to the author.

Distribution and Size

Figure 1 shows the location of selected disseminated gold deposits in the western United States. These deposits are listed in Table 1 along with the estimated reserves and grade. Other important deposits are known; however, they are relatively recent discoveries and the available information on these deposits is limited. However, including them here would not seriously affect, either in a positive or negative way, the genetic model.

Host Rocks

Commonly, the host rocks for hydrothermal deposits are considered important in the ore-forming process because they serve as chemical sinks for material dissolved in the solutions. Evidence that exchanges of material occur between rocks and solutions lay in the occurrence of wall-rock alteration envelopes

Table 1 Size and grade for selected disseminated gold deposits of Nevada and Utah.

Deposit	Size * *(Millions of tons)*	Grade *(oz/ton)*	Source
Alligator Ridge	5.0	0.12	Klessig, 1984b
Battle Mountain			Blake *et al.*, 1984
Minnie-Tomboy	3.9	0.09	
Fortitude	16.0	0.15	
Borealis	2.5	0.08	Wilkins, 1984
Carlin	22.0	0.30	Wilkins, 1984
Getchell	4.4	0.28	Wilkins, 1984
Jerritt Canyon	11.9	0.22	Wilkins, 1984
Mercur	10.1	0.15	Wilkins, 1984
Northumberland	8.0	0.08	Wilkins, 1984
Pinson	3.2	0.16	Kretschmer, 1984
Round Mountain	200.0	0.05	Mills, 1984

* Calculated from historical production and estimated reserves

around hydrothermal veins. However, the variety in the composition of the host rocks for disseminated gold deposits precludes these having a unique genetic role in the precipitation of the gold. The host rocks for most of these deposits range from early Paleozoic sedimentary units to late Tertiary volcanic rocks.

At the Carlin deposit (Figure 1), most of the gold mineralization occurs in the upper 250 m of the Silurian to Devonian Roberts Mountains Formation. In the mine area, this formation consists of 550-600 m of dark grey thin-bedded laminated silty calcareous dolomite and limestone. Ore occurs as replacements of carbonate minerals, primarily calcite, in the thin-bedded argillaceous arenaceous dolomitic units, whereas pelloidal wackestones within the same stratigraphic horizon appear unfavourable for mineralization (Radtke *et al.*, 1980). The main ore host at Jerritt Canyon is a middle unit of the Ordovician to Silurian Hanson Creek Formation (Birak and Hawkins, 1984). This favourable interval consists of more than 90 m of alternating carbonaceous micritic limestone beds and laminated carbonaceous dolomitic limestone beds. Gold mineralization favours the more permeable laminated units. Chert lenses, less than 10 cm long and 2 cm thick, occur sporadically in the lower parts of this unit (Birak and Hawkins, 1984). At Jerritt Canyon, mineralization also occurs in the lower part of the Roberts Mountains Formation which overlies the Hanson Creek Formation.

At the Getchell deposit (Figure 1), the main ore hosts are limestone units within the Cambrian Preble Formation, which consists of intercalated thin arenaceous limestone and limy carbonaceous shale beds (Berger, 1980). At the Pinson Mine, about eight kilometres south of the Getchell deposit, gold mineralization is hosted by the Cambrian to Ordovician Comus Formation which directly overlies the Preble. The former unit consists of thin-bedded carbonate and shale which are locally rhythmically interbedded and laminated (Kretschmer, 1984). Carbonate units also host disseminated gold at the Mercur Mine in the southern Oquirrh Mountains of Utah (Kornze *et al.*, 1984). Here the mineralization occurs in the Mercur Mine series within the upper half of the Lower Great Blue Formation of Mississippian age. This series consists of interbedded lime wackestones, packstones, micritic limestones, mudstones, grainstones, fine-grained sandstones and chert. Most of the gold occurs in the Mercur bed which consists of fossiliferous mudstones and grainstones.

At the Alligator Ridge deposit, 110 km northwest of Ely, Nevada (Figure 1), the most important ore host is the lower 100 m section of the 135 m thick Mississippian Pilot Shale. This formation consists of interbedded thin-bedded calcareous, carbonaceous siltstones and claystones. The ore horizon is dominated by dark grey to greyish-black siltstones with local thin lenses of limestone. This lower silty zone is separated from the more clay-rich upper beds by a discontinuous zone of interbedded dark grey, lenticular cherts and light-colored clays (Klessig, 1984a,b). At Copper Canyon, Nevada, gold deposits are hosted by siliceous and calcareous conglomerates in the basal 30 m of the lower member of the Middle Pennsylvanian Battle Formation. The conglomerate consists of sub-angular to sub-rounded clasts of chert, quartzite and rare limestone up to nearly 70 cm in diameter. The matrix originally consisted of calcareous medium-grained sandstone, however, metamorphism has resulted in the formation of tremolite (Blake and Kretschmer, 1983; Black *et al.*, 1984).

Disseminated gold deposits are hosted by volcanic rocks as well as sedimentary units. At Round Mountain, Nevada (Figure 1), substantial amounts of gold occur in Miocene intracaldera, poorly to densely welded, rhyolite tuff (Mills, 1982, 1984). Disseminated gold occurs in Miocene andesitic volcanic rocks at the Borealis Mine near Hawthorne, Nevada (Reid, 1984). Many more occurrences contain gold hosted by volcanic units, however, most show a strong structural control and therefore have not been considered disseminated deposits. In terms of the geochemical processes responsible for mineralization, there may be little difference between the typical vein and disseminated deposits.

Structural Control

The structural control of disseminated gold deposits can be considered on both regional and local scales. Both are important because of the necessity to recognize features responsible for focussing geological processes which result in economic concentrations of gold, and also the need for plumbing

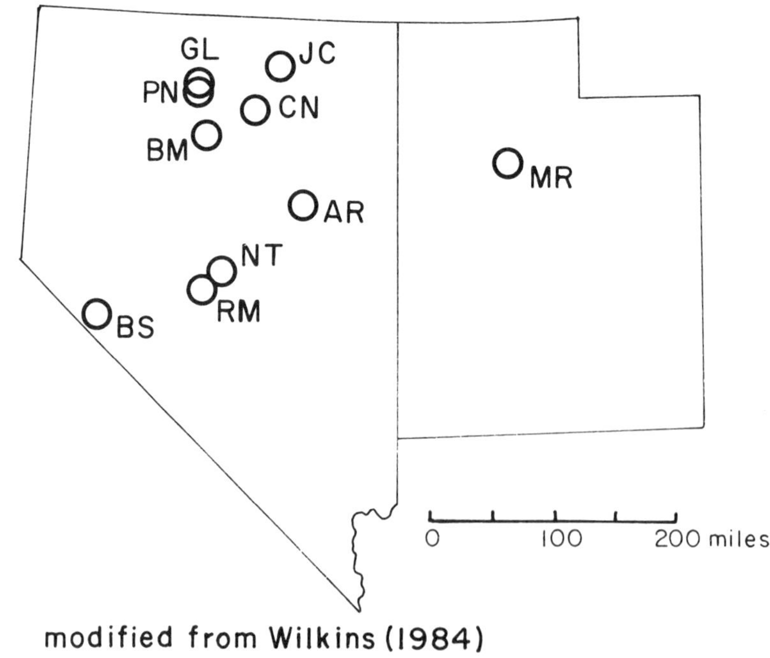

Figure 1 *Selected disseminated gold deposits of Nevada and Utah.* ***AR**, Alligator Ridge;* ***BM**, Battle Mountain;* ***BS**, Borealis;* ***CN**, Carlin;* ***GL**, Getchell;* ***JC**, Jerritt Canyon;* ***MR**, Mercur;* ***NT**, Northumberland;* ***PN**, Pinson;* ***RM**, Round Mountain.*

systems to transmit the mineralizing solutions. Yet, there is no structural feature or features which appear to be unique to disseminated deposits. Regional structures may play the role of localizing igneous activity that supplies the energy to drive the hydrothermal system as well as the stress required for mechanical ground preparation. Post-mineralization tectonic adjustments along major belts also may be responsible for exposing the deposits to discovery and exploitation.

On a more local scale, ground preparation through fracturing is important to produce not only the permeability necessary for solution migration but also open spaces for mineral precipitation. The formation of a disseminated deposit, as opposed to a vein, may depend only on the degree and scale at which the mineralizing solutions can penetrate the host rocks, which in turn depends on the nature of fracturing. The two types of occurrences are found together in many districts and deposits, where structures occupied by veins may have served as feeder systems for the disseminated mineralization.

The style of fracturing will be host-rock dependent as well. Argillaceous sedimentary rocks will yield, as well as fracture, at depth along a few planes in response to directed stresses. In contrast, volcanic rocks emplaced at or near the surface experience brittle failure resulting in many well-developed open fractures. Fracturing may be in response to stresses produced by the igneous activity (caldera collapse, for example) and permeable volcanic breccias are a common result. Therefore, it is not surprising that disseminated deposits are hosted most commonly by sedimentary rocks and vein deposits by volcanic units.

Roberts (1960, 1966) suggested that mining districts in northern Nevada were aligned along northwest-trending belts. Much of his work was done before discovery of the large disseminated gold deposits in this area, although minor vein occurrences in proximity to the latter were known. One belt extends for over 100 km from the Getchell and Pinson Mines in the northwest to the Eureka district in the southwest. Bonham (1984) described the Carlin gold belt, a northwest-trending zone containing a number of important disseminated and vein deposits. This belt can be extended southeast along trend to include Alligator Ridge and the deposits around Ely, Nevada. Many of the disseminated gold deposits of this belt are exposed in windows eroded through the upper plate of the shallow-dipping Roberts Mountains thrust. This is a major zone of faulting with significant west-to-east displacement which occurred during the Devonian to Mississippian Antler orogeny. There may be a genetic relationship between the events producing the uplifts in the window areas and the gold mineralization so that occurrence of the deposits in these windows is not fortuitous.

The Mercur deposit lies in a major west-trending belt of igneous and hydrothermal activity which includes the Bingham porphyry copper deposit and the Park City base and precious metal deposits. This belt is the western extension of the Uinta Arch. In all these belts, middle to late Tertiary normal faulting and igneous activity are superimposed on Paleozoic geosynclinal sedimentary sequences and thrust plates with local occurrences of Mesozoic rocks.

The occurrence of disseminated gold deposits in distinct belts in western Nevada is less clear. Most of these deposits appear to be related to centres of Tertiary volcanic activity, which, in themselves, suggest zones of crustal weakness. Many of these centres developed on basements consisting of metamorphosed late Paleozoic to Mesozoic sedimentary and volcanic assemblages and igneous intrusions.

On a more local scale, even though most of the gold is hosted by sedimentary, and to a lesser extent volcanic, units, the mineralization shows a strong spatial relationship to fractures in all deposits. Most fracture systems show normal displacement and developed during the Tertiary; however, some may have originated during the Mesozoic and experienced later recurrent movement.

At the Carlin deposit, gold mineralization occurs in the Lynn window in the rocks constituting the lower plate of the Roberts Mountains thrust. Ore occurs in and adjacent to high-angle normal faults of variable strikes and up to a few hundred metres of displacement. This normal faulting had the result of producing extensive areas of shattered rock in the upper part of the Roberts Mountains Formation just below the thrust fault. The normal faults are interpreted as the conduits which fed hydrothermal solutions to the structurally prepared host rock (Radtke *et al.*, 1980). Particularly favourable structural traps may have been produced by a combination of the movements along the Roberts Mountains thrust and the intersecting normal faults. Locally, mineralization also occurs in the hanging wall of the thrust. A similar structural framework occurs at the Jerritt Canyon deposit although the principal host rocks are different (Birak and Hawkins, 1984).

At the Getchell and Pinson deposits structural control of gold mineralization is much more pronounced. The major structural features in this area are a series of anastomosing high-angle normal faults which form the eastern margin of the Osgood Mountain block. Ore occurs in breccias and gouge along strands of this major fault zone; individual ore widths range up to 70 m (Berger, 1980; Kretschmer, 1984). At Battle Mountain, a series of northwest-trending, northeast-dipping mineralized normal faults served as channels for the solutions to penetrate the adjacent conglomerates (Blake and Kretschmer, 1983; Blake *et al.*, 1984).

At the Alligator Ridge deposit, the relationship of mineralization to structure is less clear. The rocks in the area have been folded into a series of gentle north-trending folds and subsequently cut by a series of high-angle normal faults. Mineralization is post-normal faulting, however, the spatial and genetic relationships of ore to fractures have not been clearly defined. Klessig (1984a, b) states that mineralizing solutions migrated upward along unspecified conduits and believes that gold deposition is relatively young in age because of the existence of geothermal waters in the area.

Structural control of gold deposits in Tertiary volcanic centres is usually relatively easy to discern. At Round Mountain, the mineralization is localized in the Jefferson Caldera in the southern Toquima Range. This caldera is one of four recognized in the area and developed in response to the eruption of large volumes of rhyolitic and rhyodacitic tuffs on a basement of Paleozoic metasedimentary and Cretaceous plutonic rocks (Mills, 1982, 1984). Mineralization appears to be localized along a segment of the structural margin of the caldera where late resurgence of igneous activity occurred. A major northwest-trending shear zone and associated dilatant fractures are overprinted by a system of radial and concentric fractures associated with a central breccia. Vein, stockwork and disseminated mineralization appear to be associated with these zones of intense fracturing. At Borealis, such structural control has not been as clearly defined. According to Reid (1984), ore occurs in a breccia adjacent to a partially mineralized northeast-trending fault zone.

Association With Igneous Activity

Igneous activity and its products may serve as sources of metals, transporting solutions and/or heat to drive the hydrothermal systems. The fracture systems required for the migration of the transporting solutions also may be a product of such activity. Magmas are important sources of all three components in some mineralizing systems such as porphyry copper deposits. However, epithermal precious metal mineralizing systems, of which disseminated gold deposits represent a subgroup, appear to be dominated by heated meteoric water (Taylor, 1979; Radtke *et al.*, 1980). Under these circumstances metals must be derived from any material through which the solutions may pass, including the source of the latter, the host rocks, or emplaced igneous material. Alternatively, the latter may suppy only the heat to drive the mineralizing system.

All disseminated gold deposits have some plutonic or volcanic rocks occurring in their vicinity, however, the genetic connection between the two has been a matter of significant debate. Deposits hosted by volcanic rocks are usually genetically related to their hosts even though the volcanic history may be quite complex and gold deposition may occur a significant time after emplacement of the hosts.

The genetic importance of igneous rocks in the sedimentary rock-hosted deposits is much less clear. There appears to be no unique association which sets these systems apart from typical metalliferous hydrothermal deposits. On the contrary, these deposits occur in a number of different types of igneous environments with a variety of compositions. The gold deposits at Battle Mountain and Mercur occur in the peripheral zone of major porphyry systems, although the latter deposit lies 20 km south of the centre of the Bingham district and may not be directly related to the copper deposits. At Mercur, fine-grained porphyritic rhyolite intrudes the host rocks adjacent to the gold deposits, but this rock is only weakly altered and contains no gold and may be post-mineralization (Kornze *et al.*, 1984). At Battle Mountain, the gold deposits lie approximately 1500 m south of the main Copper Canyon copper deposits which are associated with a middle Tertiary granodiorite porphyry intrusion. The disseminated gold mineralization occurs within the metamorphic aureole of the pluton and a dyke of the latter occurs in at least one of the deposits (Blake and Kretschmer, 1983; Blake *et al.*, 1984).

At Getchell and Pinson Mines, the host Paleozoic sedimentary rocks have been intruded by large masses of Cretaceous granodiorite (Berger, 1980; Kretschmer, 1984). At both deposits, portions of the main pluton and cross-cutting dykes are variably altered to sericite, chlorite and clay. At Getchell, alteration appears to be strongest in the vicinity of the ore. Both plutons are surrounded by contact metamorphic aureoles and the gold deposits lie within these zones. Based on spatial relationships and age determinations on alteration minerals, Silberman *et al.* (1974) concluded that gold mineralization was Cretaceous in age. However, Berger (1980) reported a major Miocene thermal event in the area based on fission-track studies on apatite from an ore zone. At the Pinson deposit, argillized and sericitized andesitic or dacitic porphyry dykes containing low amounts of gold near the ore have been reported (Kretschmer, 1984). These dykes are younger than the granodiorite, but their absolute age is unknown.

Mesozoic plutonic rocks underlie the regions around both the Carlin and Jerritt Canyon districts. However, they are considered to be older than the gold mineralization even though dykes of this material occupy normal faults at the Carlin deposit (Radtke *et al.*, 1980). These latter dykes are altered and faulted along younger fractures. Tertiary volcanic rocks with an age of 14 Ma are found in the Carlin district, but are not exposed in the mine. Based on overall geologic relationships, this igneous event is considered the heat source responsible for driving the hydrothermal system. Middle Tertiary andesitic and rhyolitic volcanism occurred in the Jerritt Canyon region as well, however, these units are only exposed in the extreme north-eastern section of the district (Birak and Hawkins, 1984). Tertiary rhyolite tuffs and basalts occur in the vicinity of the Alligator Ridge deposits and some of these units appear very young, younger than normal faulting. However, the genetic relationship between the volcanic rocks and mineralization is unknown. Klessig (1984a, b) believes the mineralization to be quite young because of the existence of an active geothermal cell in the mine area.

In the western Great Basin most, if not all, precious metal deposits are associated with various volcanic centres (Albers and Kleinhampl, 1970). McKee (1979) reported a detailed study of the relationship of mineral deposits to volcanism and noted that the majority of these deposits were associated with andesitic volcanism. Most of these deposits are strongly structurally controlled and usually classified as bonanza vein-type ores. However, a few, such as Round Mountain and Borealis, contain significant amounts of disseminated mineralization. The volcanic host rocks for these deposits have been described above.

Ore Mineralogy and Petrography

Similarities between the various disseminated deposits begin to appear when studied at the hand specimen or microscopic scale. These observations are important in developing a genetic model that is generally applicable to most of these deposits. The close association of iron sulphides with the gold in primary ores is a key element in the formation of these deposits. The features these deposits have in common will become clearer in the discussion of the various deposits below.

The most detailed studies of ores in sedimentary hosts have been carried out at the Carlin Mine (Radtke *et al.*, 1980). However, the ores from other carbonate-hosted deposits are remarkably similar to those at Carlin. At the latter deposit, the primary ores have been classified into five gradational types based on the relative abundance of various components: (1) normal, (2) siliceous, (3) pyritic,

(4) carbonaceous, and (5) arsenical. An important step in the formation of these ores is the removal of up to 50% of the original calcite in the carbonate host. This ground preparation is thought to have been carried out by the hydrothermal solutions early in their evolutionary history and is important in producing the permeability necessary for the penetration of the ore solutions. However, close inspection of the rocks from this and other deposits reveals many small closely spaced fractures which give a brecciated appearance.

Over half of the ore consists of the normal type where the gold occurs with mercury, antimony and arsenic as surface coatings on, and fracture fillings in, pyrite grains. In all the primary ore types, gold occurs mostly as coatings on the pyrite. As the carbon, silica and arsenic contents increase to produce the carbonaceous, siliceous and arsenical ores, respectively, small amounts of gold may be found with these other components. Organic carbon contents in the normal ore are similar to, but slightly higher than, those of unmineralized host rock. It increases from about 0.3% in the normal ore to up to 0.9% in the pyritic ore to over 5% in the carbonaceous ore. Carbon occurs as dispersed grains of amorphous material, hydrocarbons and organic material and in small veinlets and seams of hydrocarbons. Pyrite is the most common sulphide, and increases to 5-10% of the primary ore in the pyritic type. Realgar and stibnite occur in minor amounts in all ore types but increase in the carbonaceous and arsenical ores. The latter ores are paragenetically late and contain gold associated with carbonaceous material and in realgar veins as well as with pyrite. Arsenic contents range from 0.5% to over 10%. The arsenical ores contain unusually high amounts of mercury, antimony and thallium in rare sulphides and sulphosalts. The siliceous ore represents a transition between normal ore and jasperoid as the amount of introduced silica increases. In this type, small amounts of fine-grained gold may be found included in quartz grains.

Radtke *et al.* (1980) describe both an acid-leached zone and an oxidized zone at Carlin. They attribute the former zone to hypogene acid oxidizing solutions produced by oxidation of H_2S to sulphuric acid. The H_2S was released from the hydrothermal solutions during boiling at depth at a temperature of approximately 275°C. This represents a significant increase over the 200°C for the deposition of the sulphide stages in which no evidence of boiling was found. In the formation of the leached zone, barite, anhydrite, quartz and kaolinite were formed and dolomite, sulphides and organic carbon were removed. During the boiling episode, the salinities of the solutions increased from approximately 3 wt.% NaCl equivalent to almost 15 wt.%. Carbon dioxide probably was an important component in the vapour phase because calcite veins were formed above the leached zone during this stage. Finally, the ores were oxidized during normal weathering processes upon exposure during erosion.

The primary ores at Jerritt Canyon, Mercur, Getchell and other carbonate-hosted deposits are very similar to those at Carlin. At Jerritt Canyon, gold occurs in carbonaceous and pyritic silty grainstones or calcareous siltstones (Hawkins, 1984). However, the intimate association between gold and pyrite has not been reported. Birak and Hawkins (1984) state that the most reliable mineralogical indicators of gold are realgar and orpiment; the former is the most abundant arsenic mineral found in the ores. These two minerals occur in carbonaceous rocks in veins with or without calcite and as disseminated grains with remobilized carbon along fractures. Minor arsenopyrite has been detected with X-ray diffraction. Other accessory minerals found are cinnabar, stibnite and barite. In their geologic history of the Jerritt Canyon deposits, Birak and Hawkins (1984) imply that a period of oxidation occurred along with argillization after gold deposition and predating the precipitation of stibnite, barite and the arsenides. Finally, the primary ores have been oxidizied by supergene solutions, producing a variety of antimony and arsenic oxides.

At the Getchell deposit, gold occurs as sub-microscopic grains associated with carbonaceous material, within sulphide grains, and between quartz and clay grains (Berger, 1980). The carbonaceous material is a mixture of amorphous carbon, organic carbon complexes and graphite, and occurs as thin laminae parallel to quartz layers in the bedding. The pyrite is intergrown with the quartz and commonly contains small blebs and rims of arsenopyrite. Visible gold has been reported in association with pyrite, arsenopyrite and carbonaceous material. However, overall there appears to be a closer association of gold with the sulphides and quartz than with the carbonaceous material. Realgar and orpiment are late in the paragenetic sequence and occur interstitial to ore and gangue minerals along fractures, veins and/or bedding planes. They are also enclosed in masses of late-stage remobilized carbonaceous material. Cinnabar, stibnite, minor base metal sulphides, barite and rare sulphosalts have been reported as gangue minerals.

At the Pinson deposit, the ore consists of a dense jasperoid containing gold, iron oxides and scattered remnants of iron sulphides. Gold has been reported as micron-size inclusions in arsenian pyrite (Kretschmer, 1984).

In the unoxidized ore at Mercur, gold occurs with pyrite, realgar, orpiment, marcasite and cinnabar in relative order of abundance (Kornze *et al.*, 1984). Organic carbon occurs as thin irregular veinlets, thin films coating fossil fragments or as disseminated amorphous material. Some carbonaceous material has been remobilized into carbon-rich pods. Even though there is always some organic carbon present with the gold, the converse is not true. Similarly, gold is always accompanied by sulphides but not *vice versa*. Barite and calcite occur in late stage veins. Kornze *et al.* (1984) recognize a period of late stage oxidation produced by hypogene hydrothermal solutions. This resulted in the destruction of sulphides and carbonaceous material and the formation of iron sulphates and anhydrite, now gypsum. Some supergene oxidation has occurred, however, most is interpreted as being hypogene.

At Alligator Ridge, most of the ore is of the oxidized type and consists of gold, specular hematite, jarosite, stibiconite, goethite, quartz, barite, calcite, gypsum, alunite and kaolinite (Klessig, 1984a, b). The unoxidized ore contains gold in carbonaceous material along with stibnite, pyrite, orpiment, realgar and calcite. The oxidized ore is considered to be hypogene hydrothermal in origin.

The gold mineralization at Battle Mountain contrasts with those so far described in that it occurs in the calcareous matrix of a conglomerate unit with no carbonaceous material present and arsenide minerals are not reported. Sulphide minerals present are pyrite, pyrrhotite and minor amounts of sphalerite, galena, marcasite and chalcopyrite. These occur as disseminations and replacements in the tremolite-bearing matrix. Outside the ore zone, pyrite is more abundant than pyrrhotite and total sulphide content is less than 2%. Within the ore zone, pyrrhotite exceeds pyrite, and total sulphide contents range from 10% to 50%. Gold is associated with the sulphides in the disseminations, replacements and fracture fillings (Blake and Kretschmer, 1983; Blake *et al.*, 1984).

The mineralization hosted by volcanic rocks is quite different in composition from that in sedimentary rocks yet there are some important similarities, notably the association of gold with sulphides in primary ores. At Round Mountain, gold occurs as immiscible blebs in pyrite in both veins and disseminations. In some high-grade veins, free gold occurs in intimate association with quartz and pyrite. Minor accessory minerals also found in the veins are fluorite, realgar, calcite, adularia and alunite (Mills, 1982, 1984). At Borealis, the ores have been acid leached and oxidized, however, evidence for the preoxidation occurrence of pyrite has been reported (Reid, 1984). Acid leaching has resulted in the formation of quartz, barite, jarosite, alunite and various iron oxides.

Alteration

The most important type of alteration which has occurred in disseminated gold deposits is silicification. In the sedimentary rock-hosted deposits, with the exception of Battle Mountain, this has resulted in the extensive replacement of carbonate units by silica and the formation of jasperoid bodies. These jasperoids are structurally controlled and sometimes change into zones of quartz veining at depth. The existence of jasperoids is a very good indication that gold mineralization is present, however, they vary in gold content from being quite barren to containing several grams per tonne. Because of their resistance to weathering, the jasperoids often form ridges, and as such, have served as very good guides to ore. At Jerrit Canyon, approximately 40% of the rocks in the deposit are jasperoids but only 10% represent ore-grade material (Birak and Hawkins, 1984). The deposits in volcanic rocks also exhibit extensive silicification. Round Mountain contains a zone of intense silicification located centrally and vertically above the mineralized area. Silicification was particularly intense in the tuffaceous sedimentary rocks. However, this alteration was not accompanied by gold or pyrite even though late silica veins in the silicified zone contain these minerals (Mills, 1982, 1984). At Borealis, silicification was intense also where zones in the host andesite were completely converted to a rock consisting of quartz and kaolinite.

Other types of alteration recognized in the disseminated gold deposits are decalcification, argillization and oxidation. The early removal of calcite is recognized as being an important process in the preparation of the host rock for later gold mineralization at Carlin, Mercur, Getchell and Alligator Ridge (Radtke *et al.*, 1980; Berger, 1980; Kornze *et al.*, 1984; Klessig, 1984a, b). Early stages of jasperoid are reported for Jerritt Canyon and Pinson (Birak and Hawkins, 1984; Kretschmer, 1984) and supposedly this required removal of carbonate. This decalcification suggests the early hydrothermal solutions were acid, however, they were pre-ore and may be different in composition than the actual mineralizing solutions. In contrast, the formation of calcite veins occurs in the shallow parts of the mineralizing systems in association with the acid leaching at depth.

There are two periods of argillic alteration which are recognized in many disseminated gold deposits. The first is closely associated with the main periods of mineralization and is represented by the formation of various proportions, but small amounts, of kaolinite and sericite. This argillization is best recognized in impure carbonate hosts, volcanic hosts and igneous bodies within mineralized zones. At Battle Mountain in the conglomerate hosts, this alteration is represented by the replacement of tremolite and other metamorphic minerals by epidote, chlorite and clay minerals (Blake and Kretschmer, 1983; Blake *et al.*, 1984). The second and more important episode of argillic alteration occurs during acid leaching in the shallow portions of the hydrothermal system. The mineralogical changes observed include the destruction of sulphides and organic matter, the removal of dolomite in carbonate hosts, and the formation of kaolinite and various sulphates such as iron sulphates, barite and anhydrite. This type of alteration may be accompanied by the precipitation of large amounts of quartz. Because of the buffering capacity of carbonate, this argillic alteration is developed only locally in limestone and dolomite hosts, and may not be recognized at all. In volcanic environments, this argillization may result in the complete destruction of the host rock and the replacement of it with an assemblage of kaolinite and quartz with smaller amounts of alunite and other sulphates.

Oxidation is important in the process of argillic alteration described above. Supergene oxidation is widely recognized in disseminated gold deposits and occurs usually in a zone spatially related to a present or paleoerosion surface. During this process, the sulphides and organic carbon are removed, and a variety of iron, arsenic and/or antimony oxides are formed. Some recrystallization of gold may occur as visible gold becomes more common and its grain size may increase by an order of magnitude or more. Because of the greater permeability along fractures, supergene oxidation may be detected to greater depths along veins. This creates confusion between supergene and hypogene oxidation and has led to significant debate concerning the relative importance of each in the formation of gold deposits.

Based on spatial and geochemical data, hypogene oxidizing solutions have been postulated in the formation of many disseminated gold deposits. Whether or not these solutions actually transported gold, or were superimposed on an earlier depositional event is still debatable. In addition, the oxidizing event may be occurring at one point in the system as gold was deposited elsewhere. The oxidized zone at Carlin has been described already (Radtke *et al.*, 1980). The spatial distribution of oxidized rock at Alligator Ridge suggests that hypogene oxidizing solutions permeated the ore zones at some stage in their development (Klessig, 1984a, b). Hypogene oxidizing solutions are described for Mercur (Kornze *et al.*, 1984) and Jerritt Canyon (Birak and Hawkins, 1984). The mineralogical changes occurring are similar to those for supergene oxidation. However, remnants of sulphide commonly occur within the oxidized zone and the grain size of the gold increases. These facts may lead to the conclusion that the oxidizing episode was superimposed upon an earlier mineralizing event involving more reduced solutions.

The Model

Most of the models proposed for epithermal precious metal deposits have been centered around the bonanza vein-type deposits, with the exception of the model for the Carlin deposit by Radtke *et al.* (1980). In their development of genetic models for epithermal deposits, Buchanan (1981) and Berger (1982) assumed mineralization to have occurred in fossil geothermal areas where the hydrothermal solutions were chemically evolved, heated ground waters (Taylor, 1979). In addition, Henley and Ellis (1883) discussed the similarities between present-day geothermal systems and epithermal gold deposits, and implied the importance of ground water in the formation of the latter. Buchanan (1981) and Roedder (1984) summarized the compositions of solutions from which gold precipitated in hydrothermal systems, obtaining data from fluid inclusions and natural ore, gangue and alteration mineral assemblages. The temperatures of formation for most epithermal deposits were in the 200-300°C range. Many deposits exhibit evidence that the mineralizing solutions boiled at least in some stage of the deposition of gold. The solutions had salinities averaging around 3 wt.% NaCl equivalent and rarely exceeded 10 wt.%. Available data suggest sodium chloride is the most abundant dissolved component with smaller amounts of calcium and potassium also present. Carbon dioxide may be present in significant quantities in hydrothermal systems responsible for the sedimentary rock-hosted deposits.

The genetic models developed for most epithermal vein-type deposits can be applied only in a general way to the sedimentary rock-hosted disseminated deposits. These models assume a sub-vertical open fracture system along which hot solutions rise and precipitation occurs as a result of some physico-chemical changes occurring in the system. The result is a mineral deposit with a relatively simple and predictable symmetry and consistent zoning around individual fractures or fracture zones. These models also assume that boiling is an important process in the genesis of the deposits and there is no doubt that this is an effective mechanism for gold deposition (Romberger, 1982, 1983, 1984;

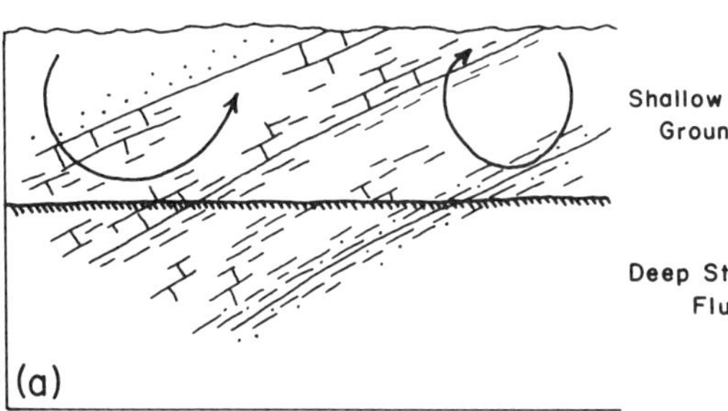

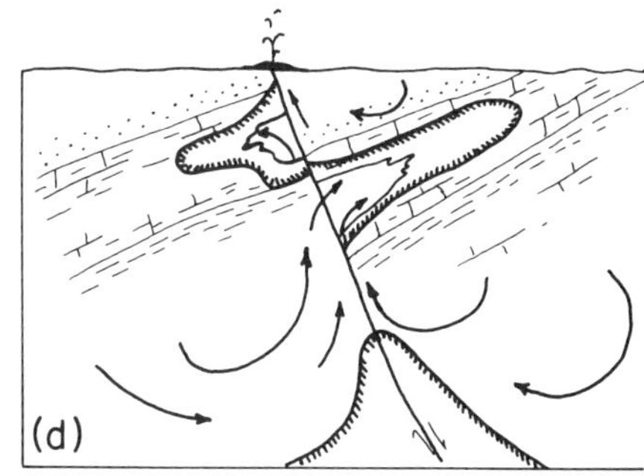

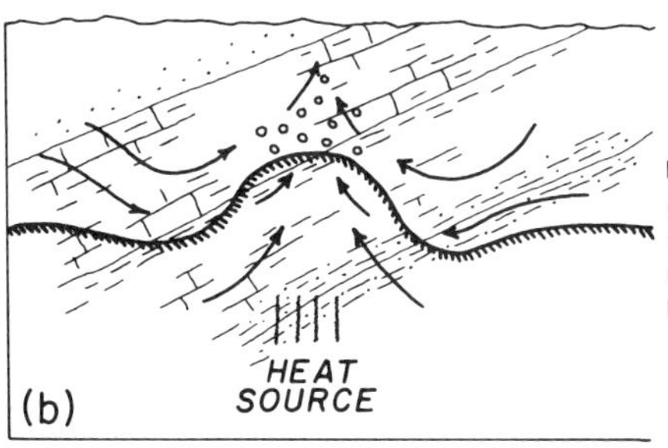

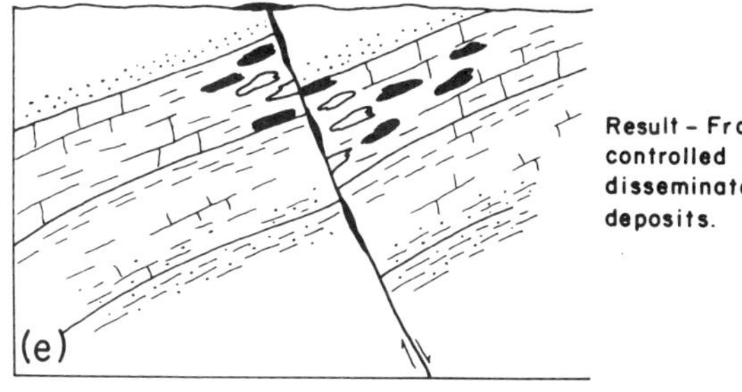

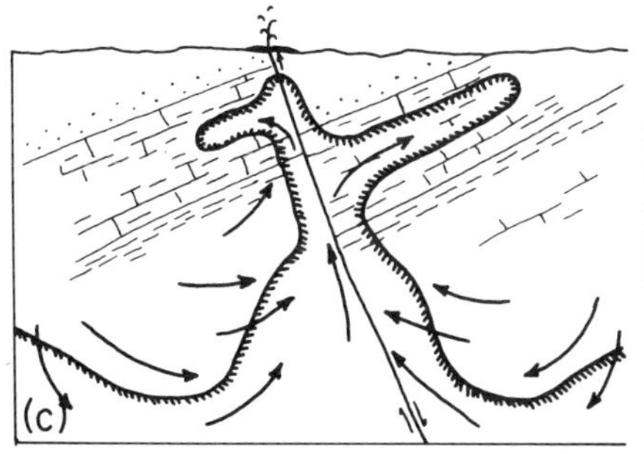

Figure 2 *Generalized and schematic model for the origin of disseminated gold deposits:*
(a) *hydrologic and geochemical equilibrium state before increase in geothermal gradient;*
(b) *increase in geothermal gradient and migration of volatiles before fracturing;*
(c) *increase in geothermal gradient accompanied by fracturing showing migration of fluids and recharge from shallow ground waters;*
(d) *cutoff of deep source by encroachment of oxygenated ground waters; and,*
(e) *resultant fracture-controlled partially oxidized disseminated gold deposit. See text for discussion.*

Drummond and Ohmoto, 1985). However, evidence for boiling in the sedimentary rock-hosted deposits is sparse. Even in the well-studied Carlin deposit, evidence for boiling was found only during the stage of acid leaching (Radtke *et al.*, 1980).

The hydrothermal system must be open to the surface for boiling to occur. Even though much of the mineralization in disseminated deposits is fracture controlled, portions of the ore are stratigraphically and structurally situated such that open conditions could not have existed locally. Rather than a simple upward flow, solutions probably migrated in a variety of directions, depending on the distribution of hydrologic potentials. The factors influencing the flow of the hydrothermal solutions actually may originate well outside the site of ore deposition; the latter may represent a local geochemical anomaly in a much larger flow regime.

Based on the temperature data so far collected on these deposits, it seems likely that increases in geothermal gradient, either regionally or on a local scale, are the driving force behind solution migration during mineralization. Before this thermal perturbation the ground water and deeper pore fluids were close to a state of hydrologic equilibrium where permeabilities and porosities decreased with depth, depending on the nature of the aquifers (Figure 2a). Under such conditions, the deeper pore waters would have a significant residence time, especially in impermeable shales, while the shallower ground waters experienced reasonable rates of recharge, again, depending on the stratigraphy. The change from a shallow recharging system to a deeper static regime may be gradational with a transitional mixing zone between. Under these conditions, the deeper waters will approach chemical equilibrium with their reservoirs while the shallower waters maintain a certain degree of oxygenation due to recharge from the atmosphere. The pore fluids in the deep zones become more and more reducing, particularly if the reservoir rocks contain significant amounts of sulphides and/or organic carbon. During this chemical evolution these fluids will leach various components from the rocks which are compatible with their developing reduced composition. Thus, we have two contrasting ground-water environments, a deep, reduced solution overlain by a relatively more oxygenated one.

As the geothermal gradient begins to increase from below, bulges will appear at the interface between the two zones (Figure 2b). These irregularities will develop at sites where the rocks have increased permeabilities such as along fractures. As more heat is added, these plumes of reduced pore fluids expand up along the fractures within the oxidized zone. The shapes of the plumes will be controlled by the distribution of permeabilities, expanding out from the fractures in permeable hosts and narrowing where claystones and other aquitards are encountered (Figure 2c). During the rise in the geothermal gradient and the doming of the ground-water surface, heating of the deep pore fluids may result in the migration of volatile components such as CO_2 and H_2S upward into the oxygenated ground-water environment. These components will dissolve and the H_2S will oxidize to produce locally acid solutions. These acid waters may be responsible for local solution of carbonate rocks, an important process in the preparation of the host rocks for subsequent gold mineralization. As the reduced pore fluids continue to migrate, the source rocks must be recharged from below, laterally or above, depending on overall hydrologic regime. Migration of the solutions will be slow at first because the geothermal gradient will be increasing from below rocks of low permeability. Recharge may be slow enough to maintain the overall zoning in the composition of the waters. The rate at which the system develops may play an important role in controlling the size and grade of the disseminated deposits; the slower the development, the more gold is transported out of the reduced zone up into the host rocks. Eventually, the convection cell will develop to the point that mixing occurs between zones, recharge of the source rocks breaks down because the ground waters will not migrate up the thermal gradient and/or overwhelming recharge from the oxidized zone will cut off the source at depth (Figure 2d). The latter will result in the superposition of an oxidized assemblage on an earlier reduced one. This entire process may wax and wane with the growth and decline of the geothermal gradient, and over the hundreds of thousands or even millions of years that these deposits have to develop, several episodes of gold remobilization and deposition may occur. At the same time tectonic activity, erosion, sedimentation and/or volcanic activity can change the structural frame-work, further complicating the mineralizing history. This model can be applied equally as well to the volcanic environment except the geothermal gradient would be steeper, events would occur over a shorter period of time, the host rocks would be fractured and brecciated volcanic units and shallow ground water would be much more involved.

It is now necessary to build into this model the unique geochemical nature of the deposits. The common occurrence of silicification is due primarily to the transporting solutions moving down significant thermal gradients. In most solutions, the primary factor controlling the solubility of silica is temperature, particularly at the depths and pressures typical of disseminated gold mineralization. Quartz has a prograde solubility so that cooling is an adequate mechanism for deposition. In contrast, calcium carbonate exhibits a retrograde solubility. The combination of these two contrasting chemistries will result in the replacement of limestone by quartz by a cooling solution saturated with silica or the formation of jasperoid. Cooling may be caused by loss of heat to the wall rocks or mixing with cooler ground waters. Silica is one of the easiest of the rock-forming components to dissolve in a hydrothermal system, and the deposition of quartz or other forms of silica is not unusual to such environments. The presence of jasperoid or quartz veins need not be accompanied by gold or other metals. Their presence does, however, indicate that the hydrothermal processes necessary for the formation of gold deposits had occurred.

The association of gold with elements such as mercury, arsenic, antimony and thallium and the lack of occurrence of the common base metals such as copper, lead and zinc can both be explained by processes taking place at the source. As the deep pore solutions approach chemical equilibrium with their enclosing rocks, they become more and more reducing, particularly if the aquifers contain significant amounts of organic carbon and sulphide. As these solutions leach the rocks, the only components that will go into solution are those that are soluble at very low oxygen activities in the presence of sulphur. Such elements are those that form soluble sulphide complexes such as gold, mercury, arsenic and perhaps antimony and thallium. Barium may be included here because it does not form an insoluble sulphide but is insoluble as a sulphate. Thus, barium will be mobile in the reducing environment and immobile under oxidizing conditions, as its presence in the oxidized zones as barite indicates. In contrast, those elements which form insoluble sulphides and are soluble as chloride complexes in an oxidizing environment such as the base metals will not go into solution in the reduced pore fluids. Thus a separation of the gold "group" elements from base metals will occur at the source. Any base metals that are dissolved will become more soluble as the solutions penetrate a more oxidized environment and, therefore, will not be deposited. In contrast,

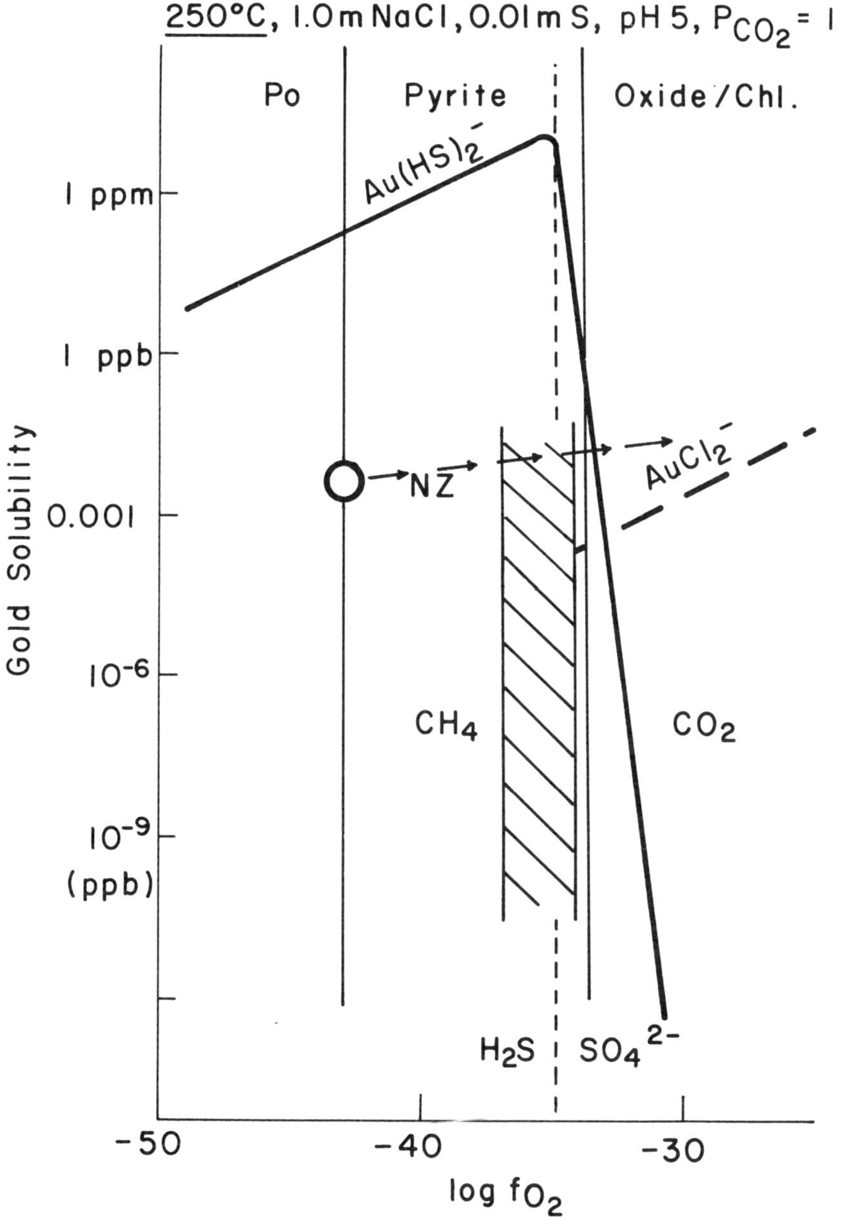

Figure 3 *Calculated gold solubility versus oxygen activity at 250°C, pH 5, and CO_2 (CH_4) pressure of 1 atm showing the solution path during oxidation of a solution starting with 0.004 ppb Au buffered by the mineral pair pyrite-pyrrhotite. See text for discussion.*

Romberger (1982, 1983, 1984) has shown that oxidation is a very efficient mechanism for the deposition of gold transported as bisulphide complexes (Figure 3) and would apply equally well to the other metals in solution as one sort of sulphide complex. In addition, the same mechanism would result in the deposition of barite as the activity of sulphate increased during oxidation of sulphide. Figure 3 shows the solubility of gold as a bisulphide complex in the heavy line as a function of the activity of oxygen, increasing to the right, for a solution at 250°C and pH of 5. Neutral pH at this temperature is 5.5, and the relationships will not change significantly by assuming a higher pH. The other assumed solution parameters are given. Also shown are the ranges of oxidation state over which pyrrhotite (Po), pyrite and iron oxides or chlorite would be stable. The field of stability of graphite is shown as the hatchured area, as are oxidation conditions under which methane, carbon dioxide, reduced sulphide (H_2S) and sulphate would predominate. The diagram shows that once a solution passes from the sulphide field to that of sulphate, the solubility of gold bisulphide complex decreases quite rapidly. Therefore, gold would be deposited as a result of oxidation of solutions saturated with this metal. The diagram shows the path (labelled NZ) a solution would take which was originally in equilibrium with the mineral pair pyrrhotite and pyrite and had a gold content of 0.004 ppb (Weissberg *et al.*, 1979). This solution would be undersaturated with respect to gold originally, however, upon oxidation, its composition would pierce the solubility surface and gold would precipitate, as shown. The solubility of gold as the chloride complex is shown, for comparison, as the heavy dashed line. It is interesting to note that the oxidation of a similar solution containing methane would result in the deposition of carbon.

Romberger (1984) also argued that most hydrothermal solutions in equilibrium with pyrite never reach saturation with respect to gold and the latter coprecipitates with sulphides. He based his arguments on the high solubility of gold in bisulphide solutions (Seward, 1973) and the intimate association between gold and pyrite in most primary ores (Figure 3). The gold ends up either in solid solution in the pyrite or other sulphides or as blebs in, or coatings on, the grains. As hydrothermal activity proceeds, recrystallization will result in the gold forming discrete grains but still in association with the sulphides. Later oxidation may remove the sulphides, converting them to oxides, however, the gold will remain because of its limited solubility in oxidized solutions.

Summary

In summary, using the physical, hydrologic and geochemical model proposed above, it is possible to explain both the structural variations and geochemical similarities exhibited by these deposits. Minor unexplained features do occur and subsequent major geologic events may mask the original environment. In addition, the behaviour of trace metals is presently only conjectural and needs to be strengthened by more work. However, the model is strong enough to withstand a few unknowns. It is believed that further studies on these deposits and the contained material will serve to strengthen it even more.

References

Albers, J.P. and Kleinhampl, F.J., 1970, Spatial relation of mineral deposits to Tertiary volcanic centers in Nevada: United States Geological Survey, Professional Paper 700-C, p. C1-C10.

Berger, B.R., 1980, Geological and geochemical relationships at the Getchell mine and vicinity, Humboldt Country, Nevada: Society of Economic Geologists, field conference on epithermal gold deposits, Reno, Nevada, p. 111-135.

Berger, B.R., 1982, The geological attributes of Au-Ag-base metal epithermal deposits: United States Geological Survey, Open-File Report 82-795, p. 119-126.

Birak, D.J. and Hawkins, R.B., 1984, The geology of the Enfield Bell mine and the Jerritt Canyon district, Elko County, Nevada: Society of Economic Geologists, Field guide to sediment-hosted precious metal deposits, northern Nevada.

Blake, D.W. and Kretschmer, E.L., 1983, Gold deposits at Copper Canyon, Lander County, Nevada: Nevada Bureau of Mines and Geology, Report 36, p. 3-10.

Blake, D.W., Wotruba, P.R. and Theodore, T.G., 1984, Zonation in the skarn environment at the Minnie-Tomboy gold deposits, *in* Wilkins, J., ed., Gold and Silver Deposits of the Basin and Range Province Western USA: Arizona Geological Society, Digest, v. XV, p. 67-72.

Bonham, H.F., 1984, General Geology of the Carlin gold belt: Society of Economic Geologists, Field guide to sediment-hosted precious metal deposits, northern Nevada.

Buchanan, L.J., 1981, Precious metal deposits associated with volcanic environments in the southwest, *in* Relations of Tectonics to Ore Deposits: Arizona Geological Society, Digest, v. XIV, p. 237-262.

Drummond, S.E. and Ohmoto, H., 1985, Chemical evolution and mineral deposition in boiling hydrothermal systems: Economic Geology, v. 80, p. 126-147.

Hawkins, R.B., 1984, Discovery of the Bell gold mine — Jerritt Canyon district, Elko County, Nevada: Arizona Geological Society, Digest, v. 15, p. 53-58.

Henley, R.W. and Ellis, A.J., 1983, Geothermal systems ancient and modern: a geochemical review: Earth Science Reviews, v. 19, p. 1-50.

Klessig, P.J., 1984a, History and geology of the Alligator Ridge gold mine, White Pine County, Nevada: Society of Economic Geologists, Field guide to sediment-hosted precious metal deposits, northern Nevada.

Klessig, P.J., 1984b, History and geology of the Alligator Ridge gold mine, White Pine County, Nevada, *in* Wilkins, J., ed., Gold and Silver Deposits of the Basin and Range Province Western USA: Arizona Geological Society, Digest, v. XV, p. 77-88.

Kornze, L.D., Faddies, T.B., Goodwin, J.C. and Bryant, M.A., 1984, Geology and geostatistics applied to grade control at the Mercur gold mine, Mercur, Utah: American Institute of Mining and Metallurgical Engineers, Preprint No. 84-442, 21 p.

Kretschmer, E.L., 1984, Guidebook geology of the Pinson mine, Humboldt County, Nevada: Society of Economic Geologists, Field guide to sediment-hosted precious metal deposits, northern Nevada.

McKee, E.H., 1979, Ash-flow sheets and calderas: their genetic relationship to ore deposits in Nevada, *in* Chapin, C.E. and Elston, W.E., eds., Ash-Flow Tuffs: Geological Society of America, Special Paper 180, p. 205-211.

Mills, B.A., 1982, Geology of the Round Mountain gold deposit: Nye County, Nevada: American Mining Congress, Las Vegas, Nevada.

Mills, B.A., 1984, Geology of the Round Mountain gold deposit, Nye County, Nevada, *in* Wilkins, J., ed., Gold and Silver Deposits of the Basin and Range Province Western USA: Arizona Geological Society, Digest, v. XV, p. 89-99.

Radtke, A.S., Rye, R.O. and Dickson, F.W., 1980, Geology and stable isotope studies of the Carlin gold deposit, Nevada: Economic Geology, v. 75, p. 641-62.

Reid, R.F., 1984, The geology of the Borealis deposit: Geological Society of America, Abstracts with Program, Reno, Nevada, p. 632.

Roberts, R.J., 1960, Alignment of mining districts in north central Nevada: United States Geological Survey, Professional Paper 400-B, p. B17-B19.

Roberts, R.J., 1966, Metallogenic provinces and mineral belts in Nevada: Nevada Bureau of Mines, Report 13, pt. A, p. 47-72.

Roedder, E., 1984, Fluid inclusion evidence bearing on the environments of gold deposition, *in* Foster, R.P., ed., Gold '82: The Geology, Geochemistry and Genesis of Gold Deposits: A.A. Balkema, Rotterdam, p. 129-164.

Romberger, S.B., 1982, Transport and deposition of gold in hydrothermal systems at temperatures up to 300°C: Geological Society of America, Abstracts with Program.

Romberger, S.B., 1983, Transport and deposition of gold in hydrothermal systems at temperatures up to 250°C with geologic implications: Geological Association of Canada — Mineralogical Association of Canada, Program with Abstracts, v. 8, p. A59.

Romberger, S.B., 1984, Mechanisms of deposition of gold in low temperature hydrothermal systems: Association of Exploration Geochemists, Program with Abstracts.

Seward, T.M., 1973, Thiocomplexes of gold and the transport of gold in hydrothermal ore solutions: Geochimica et Cosmochimica Acta, v. 37, p. 379-399.

Silberman, M.L., Berger, B.R. and Koski, R.A., 1974, K-Ar age relations of grandiorite emplacement and tungsten and gold mineralization near the Getchell mine, Humboldt County, Nevada: Economic Geology, v. 69, p. 646-656.

Taylor, H.P., 1979, Oxygen and hydrogen isotope relationships in hydrothermal mineral deposits, *in* Barnes, H.L., ed., Geochemistry of Hydrothermal Ore Deposits, Second Edition: Wiley, New York, p. 236-277.

Weissberg, N.H., Browne, P.R.L. and Seward, T.M., 1979, Ore metals in active geothermal systems, *in* Barnes, H.L., ed., Geochemistry of Hydrothermal Ore Deposits, Second Edition: Wiley, New York, p. 738-780.

Wilkins, J., 1984, The distribution of gold- and silver-bearing deposits in the Basin and Range Province, western United States, *in* Wilkins, J., ed., Gold and Silver Deposits of the Basin and Range Province Western USA: Arizona Geological Society, Digest, v. XV, p. 1-27.

Accepted, as revised, 7 November 1985.
Originally published in
Geoscience Canada v. 13 Number 1
(March 1986)

A Canadian Cordilleran Model for Epithermal Gold-Silver Deposits

Andrejs Panteleyev
Senior Project Geologist
Geological Branch
Mineral Resources Division
British Columbia Ministry of Energy, Mines and Petroleum Resources
Victoria, British Columbia V8V 1X4

Introduction

Discovery of lode gold in 1851 and placer gold in 1857 initiated the first large-scale economic activity in British Columbia and led to a major influx of miners during the gold rushes of the mid- and late 1800s. This explosion of non-native population necessitated colonial expansion, the creation of new settlements, developments of infrastructure, and provided the economic base for the fledgling colony.

Total gold production to date from the Canadian Cordillera (British Columbia and Yukon) is close to 1,182 tonnes (38 million ounces), approximately one-third of the amount produced by Ontario. About 60% of Cordilleran gold comes from lode deposits; the remainder has been won from placer workings. Placer gold production peaked in 1900; lode gold production peaked in 1939 when 18.3 tonnes of gold were produced in British Columbia. Thereafter, output declined steadily until the early 1970s when by-product gold, mainly from porphyry copper mines as well as massive sulphide and skarn deposits, became the dominant source. These accounted for up to 80% of the annual 3-4 tonnes of gold production.

Gold was liberated in international markets in 1968 from the fixed price of $35US per ounce set in 1934. The resulting price increases, culminating in spectacular price peaks in late 1979, renewed interest in known deposits and enlivened the quest for new supplies of both gold and silver. The search for gold has been the main focus during the 1980s of the mining exploration community, a $100 million annual enterprise in the Canadian Cordillera. One of the primary targets has been gold-silver deposits of the "epithermal type", also known variously as "bonanza ores", "Tertiary type", "precious metal deposits of volcanic association" or "fossil hot spring type" (Figure 1 and Table 1).

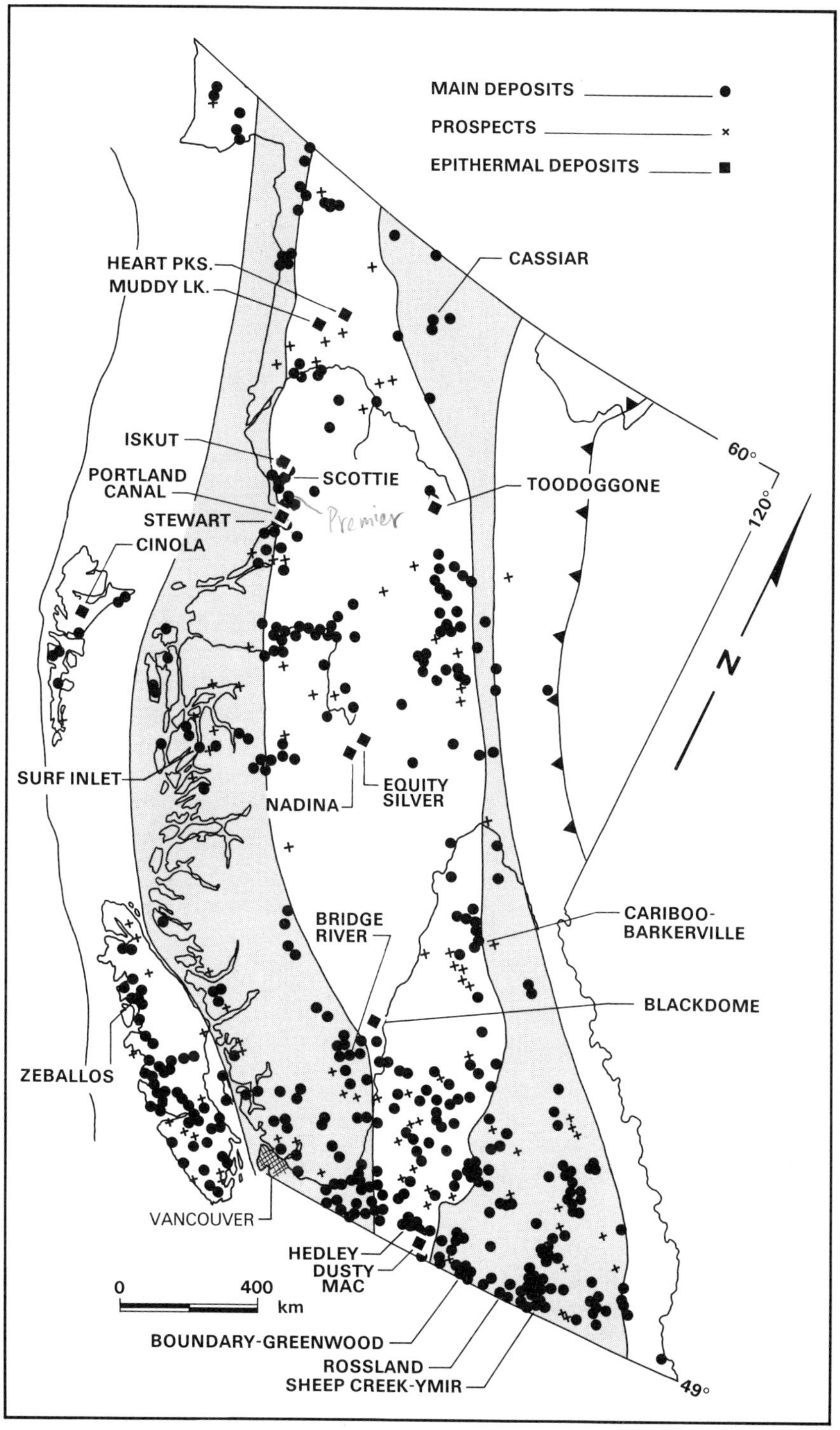

Figure 1 *Distribution of gold deposits in British Columbia showing major camps, individual deposits and areas of recent exploration activity. Lines indicate major tectono-physiographic boundaries. Crystalline-metamorphic terranes of the Coast Plutonic Belt in the west and Omineca Belt in the east are shown by the hachured pattern.*

Epithermal precious metal deposits are attractive, especially when base metal prices are depressed, because they have high unit values of precious metals with generally low or no base metal content. The deposits commonly occur as small vein systems (less than a million tonnes in size), but they tend to have good grades, and many contain high-grade ore shoots. They provide quick payback at high rates of return on modest amounts of invested capital. Consequently, epithermal deposits are particularly attractive to smaller companies that seek financing primarily from public sources. In addition, major mining companies, with long-term investment strategies and the ability to support major capital outlays, seek larger epithermal prospects suitable for open-pit operations. An example of the type of superior epithermal deposit that explorationists strive to find is the El Indio Mine, Chile, which began limited production in 1979. In its first year of operation, 12,800 kilograms of gold were produced from 50,000 tonnes of direct shipping ore. There remained 70,000 tonnes of similar material containing 277 grams gold per tonne as well as main reserves of 3.2 million tonnes with 12.3 grams gold per tonne, 141 grams silver per tonne, and 4.0% copper (Walthier *et al.*, 1982).

Characteristics of Epithermal Deposits

The term "epithermal" was introduced by Waldemar Lindgren in 1933. It is part of a genetic classification of ore deposits that describes hydrothermal fluid sources and depth zones in such terms as "epithermal", "mesothermal", and "hypothermal". Epithermal deposits were considered (Lindgren, 1933, p. 212) to be formed "by hot ascending waters of uncertain origin, but charged with igneous emanations. Deposition and concentration (of ore minerals occurs) at slight depth". He considered temperatures to range from 50-200°C under conditions of "moderate" pressure. Lindgren also noted that epithermal deposits have "striking analogies to those products of the hot springs". Other workers, such as Buddington (1935), recognized that temperatures greater than those suggested by Lindgren were possible in near surface hydrothermal environments. Soon the upper temperature limit for epithermal deposits was extended to at least 300°C. Use of the term "epithermal" became well entrenched in North America and elsewhere during the 1940s and 1950s. Many descriptions of epithermal-type deposits and districts accumulated and document the geometry, structural controls and mineralogical variations of these deposits.

Review articles describing "epithermal" deposits abound; each period of renewed interest in gold produced updated summaries. Notable descriptions include: Schmitt (1950a), Wisser (1966), Sillitoe (1977), Berger (1982), Berger and Eimon (1982), Silberman (1982), and Heald-Wetlaufer *et al.* (1983). They emphasize the following characteristics:

(1) The deposits form near the surface. Mineralization takes place from surface to a maximum depth of about 1,000 m. Ore can be developed over a considerable strike length, but is restricted in vertical extent to intervals varying from 100 to 1,000 m. Average vertical range of ore is about 350 m; it rarely exceeds 600 m. Ore zones (ore shoots) bottom in either barren rock or pass downward into subeconomic zones containing base metal sulphides.

(2) Veins are the most common ore host; they tend to branch or flare upward into complicated, wedge-like or cone-like features. Breccia zones, stockworks, and fine-grained bedding replacement zones also occur; larger zones of these types may extend to tens of millions of tonnes in size.

(3) Deposits form in extensional tectonic settings, in areas with well-developed tension fracture systems and normal faults. The fracture systems are commonly, but not necessarily, associated with large-scale volcanic collapse structures.

Table 1 Reserves and Production, British Columbia Epithermal Deposits

Deposit	Production (million tonnes)		Au (kg)	Ag (kg)	Reserves
Silbak Premier	1918-1976	4.2	56,000	1,271,000	~ mt; 2.43 g/t Au; 110.4 g/t Ag
Equity Silver (Sam Goosly)	1980-present	~7	3,300	696,000	~21 mt; 0.35% Cu; 0.85 g/t Au; 109 g/t Ag
Big Missouri	1927-1942	0.8	1,800	1,600	~2 mt; 3.36 g/t gold-equivalent
Baker (Chappelle)	1980-1983	0.08	1,200	23,000	—
Dusty Mac	1969-1976	0.06	600	10,500	—
Nadina	1972-1973	0.2	<10	13,700	577,600 t; 3.1 g/t Au; 213 g/t Ag; some Pb, Zn, Cu
Cinola	—				~34 mt; 2.06 g/t Au
Lawyers	—				>1 mt; 7.27 g/t Au; 254.2 g/t Ag
Blackdome	—				195,000 t; 27 g/t Au; 110 g/t Ag
Mt. Johnny (Iskut R. area)	—				554,000 t; 21.3 g/t Au; some Ag, Pb, Zn, Cu

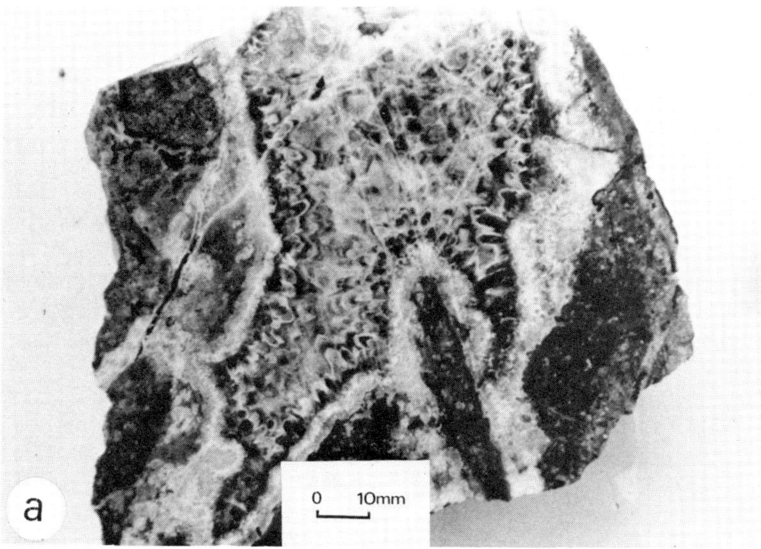

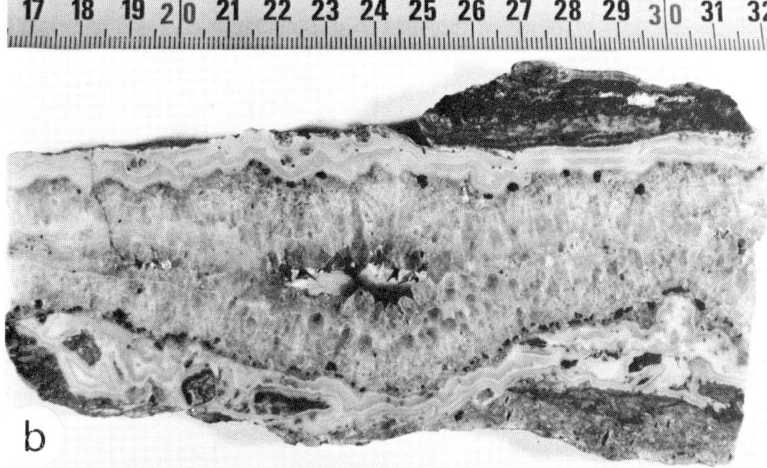

Figure 2 *Textures of epithermal ores:* **(a)** *crystalline amethystine quartz in an irregular vein. Fine-grained quartz marks the vein margin and there is an adularia selvage at the wall-rock contact, Lawyers deposit, Toodoggone area, British Columbia;* **(b)** *banded chalcedonic quartz forms the vein walls; crystals of galena and sphalerite (dark grains), and fine-grained fluorite occur at the contact with inner zone amethystine quartz, Creede, Colorado;* **(c)** *vuggy breccia with fragments rimmed by chalcedonic quartz and fine-grained crystalline quartz overgrowths, Lawyers deposit, Toodoggone area, British Columbia.*

(4) Mineralization commonly occurs in volcanic terranes with well-differentiated, subaerial pyroclastic rocks, and numerous small sub-volcanic intrusions. Hot spring deposits and fumarolic volcanic phenomena are sometimes evident where centres of hydrothermal discharge have not been deeply eroded.

(5) Ore and associated minerals are deposited dominantly as open space filling with banded, crustiform, vuggy, drusy, colloform, and cockscomb textures. Repeated cycles of mineral deposition are evident. Ore minerals are generally fine-grained but commonly have coarse-grained, well-crystallized overgrowths of gangue minerals. Some replacement textures are evident; pseudomorphs of quartz after calcite are characteristic. (See Figure 2).

(6) Gold and silver are the main economic metals, and occur along with enhanced amounts of Hg, As, Sb and rarely Tl, Se and Te. Gold to silver ratios range widely; silver is typically more abundant than gold. Main ore minerals are native gold and silver, electrum, acanthite (argentite), and silver-bearing arsenic-antimony sulphosalts. Tellurides are locally important. In addition, galena and sphalerite are common; copper occurs generally as chalcopyrite but in some deposits forms enargite. Cinnabar, stibnite, tetrahedrite and selenides are important in some deposits.

(7) Gangue minerals are mainly quartz and calcite with lesser fluorite, barite and pyrite. Chlorite, hematite, dolomite, rhodonite and rhodochrosite are less common. Silica occurs in many varieties, most commonly as quartz or amethystine quartz, but also as opal, chalcedony and cristobalite.

(8) Hydrothermal alteration is pronounced. Precious metal mineralization is frequently associated with silicification. Zones of silicification can be flanked by zones of illite-sericite and clay alteration, all occurring within larger zones of propylitic alteration. At depth, vein structures contain adularia; near the surface, broad argillic zones, some containing alunite, can predominate. Some deposits have aluminous, advanced argillic alteration assemblages containing: kaolinite/dickite, sericite, pyrophyllite and andalusite with accessory diaspore, corundum, topaz, zunyite, lazulite or scorzalite, dumortierite, and rutile or anatase.

Genesis of Epithermal Deposits

Studies of recent and active geothermal sytems, such as those by Henley and Ellis (1983), Weissberg (1969), Weissberg *et al.* (1979), Ewers and Keays (1977) and White (1981), have done much to demonstrate the relationship between hot springs and epithermal deposits, a relationship noted by Schmitt (1950b) and strongly promoted by White as early as 1955. The concept that ascending magmatic-source hydrothermal fluids are important has been largely eliminated, mainly by fluid inclusion and stable isotope studies. Epithermal deposits are now considered to form from relatively dilute, near-neutral to weakly-alkaline chloride water (<5 wt.% NaCl equivalent) that undergo boiling or effervescent degassing, fluid mixing, and oxidation at temperatures generally between 200-300°C and most commonly between 230-260°C. Boiling or mixing of fluids as they ascend or migrate laterally appear to be the two most important cooling mechanisms. Downward migration of fluids has been documented in one locality, Creede, Colorado, where hydrothermal fluids in at least part of the hydrothermal system have mixed with denser, cooler brines (Bethke and Rye, 1979).

Recent detailed studies of a number of Tertiary epithermal deposits in caldera settings, principally by geologists from the United States Geological Survey, thoroughly document the geologic settings and present perceptive, well-researched explanations of the origins of these deposits (Steven and Eaton, 1975; Lipman and Steven, 1976; Lipman *et al.*, 1976; Casadevall and Ohmoto, 1977; Slack, 1980; and others). The authoritative descriptions and genetic interpretations of these deposits has led to extensive comparisons with other deposits and areas. Unfortunately, many new workers and explorationists place undue emphasis on the caldera or resurgent caldera setting of the deposits. Consequently, there is a widely held notion that calderas are a requisite for the development of epithermal deposits. This is not the case. In fact, calderas, as described by Smith and Bailey (1968), are simply a type of very specialized volcano; they do not inherently contain any mineralization. In Nevada, only 2 out of 31 recognized calderas are known to contain ore (McKee, 1979), and in the western United States, only 14 out of 125 known calderas have any associated ore (Rytuba, 1981). If ore occurs, it is because calderas produce large fracture systems, but any major fracture system that channels hydrothermal fluids can localize mineralization.

Radiometric data from epithermal deposits in Tertiary volcanic areas of the southwestern United States (Silberman and McKee, 1974; and others) shows that ores are 2-17 million years younger than the caldera-forming volcanism. Thus it seems that hydrothermal activity is not genetically related to caldera volcanism, but is related to younger, sub-volcanic magmatic activity in structurally disturbed rocks at the caldera margins and in the surrounding rocks.

Association of epithermal ores with felsic volcanic rocks (rhyolite and dacite flows, domes, ash flow sheets and tuffs) has also been overemphasized. Epithermal ores occur in all rock types, particularly those that sustain large, open-fracture systems over extended periods of time during hydrothermal activity. In Nevada, McKee (1979) noted that in 98 mining districts with economically significant ore production, only 5 districts were in siliceous tuffs; the

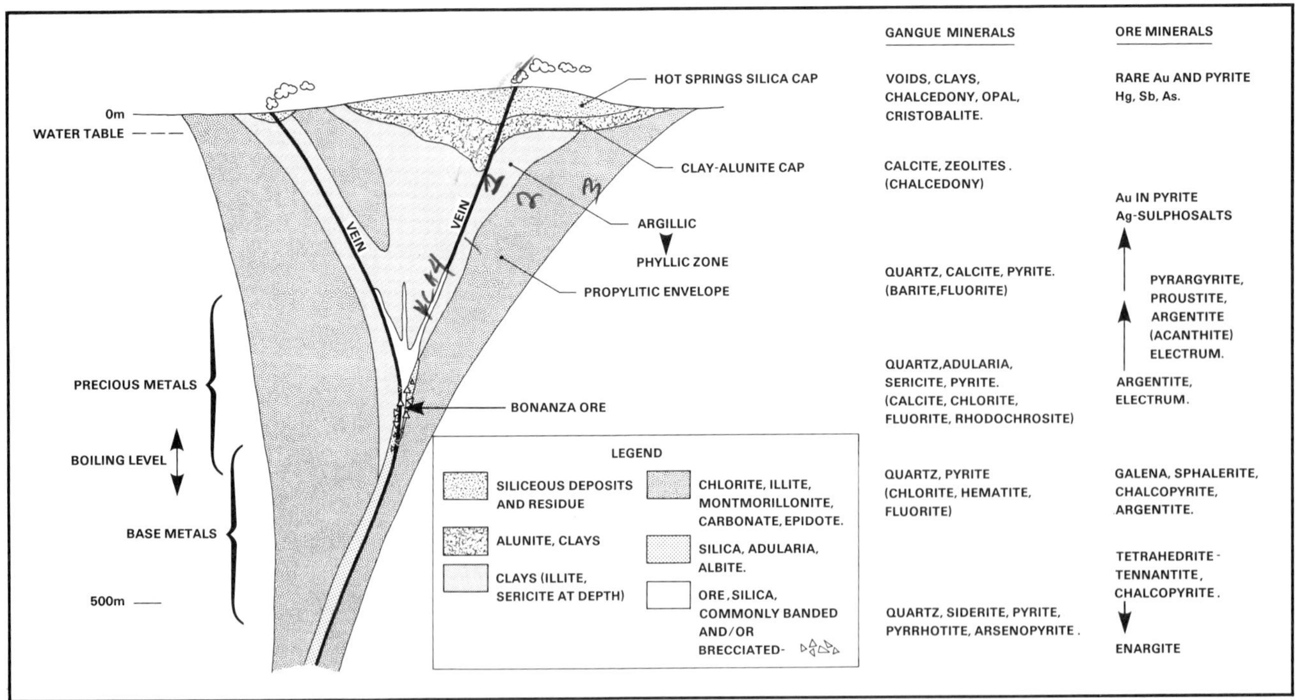

Figure 3 *Idealized section of a bonanza epithermal deposit. (After Buchanan, 1981). Real systems are commonly more complex because this single-stage model is overprinted by several stages of mineralization related to migration of fluid boiling or degassing levels.*

majority of mineral deposits were in andesitic hypabyssal and extrusive volcanic rocks. Andesitic pyroclastic rocks or flow breccias appear to preferentially maintain zones of primary high permeability during hydrothermal activity. In addition, andesites can sustain fault and fracture-related dilatent structures and openings, such as cymoid structures or cymoid loops, over long periods of mineralization.

In summary, hydrothermal activity is only rarely related to caldera development or resurgence, but the ore-controlling structures may be. The heat for hydrothermal activity does not appear to be the latent heat of volcanism, but is more likely derived from structurally controlled sub-volcanic intrusions or deeper plutons. Any rock type that maintains primary or structurally induced permeability and permits focussed hydrothermal fluid flow can provide sites for ore deposition.

Zoning of Hydrothermal Alteration—The Key Exploration Guide

Passage of hydrothermal fluids through fractured rocks produces structurally controlled zones of hydrothermal alteration. In most epithermal districts at least some hydrothermal alteration is evident as readily visible zones of bleached rock. Larger alteration zones can be many kilometres in dimension; the ore zones are a few metres or tens of metres, at best. The challenge to the explorationist is to properly assess the alteration zones for economic potential and to locate and define structural features that provided a focus for sustained hydrothermal fluid flow.

Propylitic alteration (chlorite, calcite, pyrite, epidote, zeolite) is an early-developed, widespread and district-wide alteration in many epithermal districts. The term "propylite" was introduced by von Richthofen in the late 1800s to describe distinctive rocks in the famous Comstock Lode epithermal camp, Virginia City, Nevada. Within the broad areas of propylitic alteration are more restricted zones of sericitic alteration or recessive weathering clay alteration (illite-kaolinite-montmorillonite). These surround central zones of silicification or quartz veining, some portions of which may be mineralized. The silicified zones are more resistant and commonly form local heights of land, many of which, in old mining areas, are sites of mine headframes or other workings.

The Buchanan "Boiling" Model. Buchanan (1981) summarized and tabulated data for many of the western US epithermal deposits and presented a model, now widely circulated and utilized by explorationists in the Cordillera, that effectively illustrates the geometric arrangement of ore and alteration zones in "typical" epithermal veins that are hosted by volcanic rocks (Figure 3). The model describes a zone of mineralization that occurs along a dominant sub-vertical fracture system from a depth of about 500 metres to the surface; close to surface, it splits into a series of subsidiary structures. The centre of ore deposition, placed by Buchanan at a depth of about 350 metres, has a discrete top and bottom. Above the ore zone, quartz veining persists, but diminishes progressively in abundance upward, as do precious and base metal amounts. Similarly, quartz becomes progressively finer-grained upward and becomes opaline silica or chalcedony in the upper part of the zone. If the silica-rich hydrothermal fluids discharge at surface as hot springs, mushroom-shaped caps of siliceous sinter are deposited. Alternatively, in veins with less dynamic fluid flow regimes, surface expressions of deeper hydrothermal activity may be nothing more than thin calcite veins or clay-altered zones in wall rocks adjoining faults or fractures. At the base of the ore zone, Buchanan describes two types of ore terminations. In one type, quantities of ore diminish downward and quartz veins that continue to depth are barren or contain only minor chalcopyrite and pyrrhotite. In the second more common type, precious metals, galena, and sphalerite occur in sub-economic amounts at depth together with minor pyrite and chalcopyrite. In both cases, calcite and adularia contents decrease with depth.

Buchanan (1981) concluded that ore deposition and attendant wall-rock alteration resulted from boiling and oxidation of the ascending hydrothermal fluids. In the model, repeated self-sealing is followed by episodic refracturing and brecciation along the ore structures. The attendant pressure drops cause boiling that results in mineral precipitation. This produces multiple stages of ore deposition, and the distinctive, layered and symmetrically banded ore textures so common in epithermal deposits. The boiling level is controlled by the temperature and salinity of the fluids; hydrostatic fluid pressures are considered to prevail in this type of shallow, open-fracture system. Boiling causes a decrease in fluid temperatures and vapour loss that increases pH in the ore fluids. At the start of boiling, after a slight loss of volatiles, solutions become neutral to slightly alkaline; with a 20% vapour loss and more alkaline conditions, silver minerals deposit along with adularia, minor sericite, and abundant silica. At the top of the boiling zone, vigorous boiling leads to faster vapour loss with attendant rapid cooling. At higher structural levels, the released vapours may condense into oxidized, acidic fluids. In this environment, the various gold complexes (bisulphate, chloride, thiocomplexes or others) destabilize and gold precipitates along with copious amounts of silica (Seward, 1973). At or near surface, any remaining volatiles that are released condense to form a highly oxidized, acidic environment. The strong acid solutions formed can be diluted by the neutral ground waters but generally are only weakly buffered chemically. Extensive and intense sericitic or argillic alteration develops in this zone. This so-called "low pH or acid capping" characterizes many epithermal deposits. The clay alteration zones are generally not of ore grade but commonly have anomalous amounts of Au, As, Pb and less commonly Hg, Sb, W, Mo, B and Ag; some contain alunite ($KAl_3(SO_4)_2(OH)_6$). The size of sericite-clay zones tends to be proportional to that of related orebodies — the larger the alteration zone, the larger the zone of mineralization.

Buchanan's model is frequently criticized. Despite many detailed studies of epithermal deposits and despite theoretical studies (Drummond and Ohmoto, 1985), only a few deposits show evidence of boiling. In fluid inclusions, boiling is indicated by widespread entrapment of various amounts of vapour in ore and gangue minerals, and the presence of remnant concentrated brine. An alternative explanation, based on the dilute nature of fluids in epithermal deposits and the isotopic signatures of vein minerals (O'Neil and Silberman, 1974; Radtke *et al.*, 1980), is that the hydrothermal systems are free-flowing fluids derived from, and recharged by, meteoric waters that undergo little isotopic exchange with wall rocks. Models that postulate mixing of fluids as a cooling and oxidizing mechanism have been discussed by Henley and McNabb (1978), Henley and Ellis (1983), Hedenquist and Henley (1985) and others. In their models, developed from studies of active geothermal fields, hydrothermal fluids, which are primarily deeply circulating meteoric waters, are heated, rise in a buoyant thermal plume, and then mix with cooler, oxygenated, neutral to acidic surface waters.

Buchanan's major contribution is his clear description of ore and alteration-mineral zoning and the spatial relationship between mineralizing solutions and the paleosurface. Undoubtedly most ore fluids have complex histories, some including both boiling (Drummond and Ohmoto, 1985) and fluid mixing (Casadevall and Ohmoto, 1977). Uncertainty about the mechanism of ore deposition in epithermal systems does not lessen the usefulness of Buchanan's empirical model. Once the position of the paleosurface has been recognized or postulated from geological field data, the model provides a useful guide for estimating depths to mineralization.

A Hot Spring Model — Acid-Sulphate Siliceous Alteration. White (1955) described a number of active springs with associated epithermal mercury and gold-silver deposits. In 1981, he stated, "The correlation of fossil geothermal systems (ore deposits) with present-day active systems provides insights into possible origins of various constituents of ore-generating systems". One of the best-studied examples is at Steamboat Springs, near Reno, Nevada, where active hot springs and areas of steaming ground have been extensively studied, and nearby mineral deposits, genetically related to extinct hot springs, have been exposed by shallow pits and tested to depth by boreholes.

Modern siliceous hot spring deposits and their partially exposed underlying altered rocks represent the "silica cap" and "low pH acid cap" described by Buchanan (1981) as the surface products of an epithermal system. The siliceous sinter deposited is porous at surface but compacted and cemented at depth; it consists of opaline silica, chalcedony, and cristobalite. The deposits at Steamboat Springs were formed by hot springs related to underlying, 1-3 million-year-old rhyolite domes. Maximum water temperature measured in boreholes is 186°C. The sinters are enriched in Au, Ag, Sb, Hg, As, Tl and B (White, 1981); locally, black, siliceous muds contain crystalline stibnite, or metastibnite and cinnabar. Uphill from the active hot springs, fossil ponds are now seen as small lenses of chalcedony, some of which contain up to 0.1% Hg in the form of cinnabar or metacinnabar.

Altered rocks underlying the hot spring deposits, which were originally basaltic andesite and granodiorite, are now exposed in the nearby "Silica Pit" as zones of bleached, white clay-alunite-silica rock. The alteration is caused by sulphuric acid fumes that still issue from fractures in the walls and floor of the pit. The vapours deposited enough sulphur and mercury to permit recovery of crystalline native sulphur and earthy cinnabar from another, nearby excavation. Visible cinnabar is restricted to a zone within 15 metres of the present ground surface, and analytical detection of mercury is possible to 26 metres (White, 1981). Stibnite along with arsenic values are found somewhat deeper, to a maximum depth of 46 metres. Values of up to 15 ppm Au, 150 ppm Ag, 3.9% Sb, and 0.1% Hg are obtained from siliceous muds in the sinter; native sulphur and possibly some cinnabar are deposited above the ground-water table as sublimates from vapours. The position of the ground-water table marks the boundary between hot waters and the vapour-dominanted fluids (White *et al.*, 1971). Above this boundary, where H_2S is oxidized to H_2SO_4 and condenses to aqueous sulphuric acid, descending refluxing solutions cause the most severe acid-leaching. Schoen *et al.* (1974) have shown that the acid capping (solfateric alteration) at Steamboat Springs consists of the following zones from surface to depth: opal, cristobalite, and some anatase (presumably from the titanium-rich basaltic country rocks); opal, alunite, quartz, and minor pyrite; and kaolinite, alunite, montmorillonite and pyrite. Below the clay-alunite zone, narrow zones of montmorillonite and illite alteration are restricted to the walls of the hydrothermal channelways.

The Marysvale Replacement Model — Hydrothermal Alunite Deposits. Alunite-silica deposits associated with hydrothermal activity in volcanic areas can be large and distinctly zoned. In these deposits, central, silica-rich core zones or cappings are flanked and underlain by alunite-silica zones that pass laterally into argillic zones within broad areas of propylitic alteration (Figure 4). The deposits form bleached, locally iron-stained, resistant outcrops with rare native sulphur and cinnabar mineralization. They are of interest to exploration geologists because they indicate sulphur-rich hydrothermal activity and the potential for nearby or deeper precious metal deposits. The alunite-bearing hydrothermal systems represent another type of fossil hot spring deposit and constitute part of the low pH capping in Buchanan's model. Alunite mineralization forms deeper parts of the acid-sulphate alteration zones found at Steamboat Springs and other active hot springs.

Cunningham *et al.* (1984) derived a model for the replacement and vein alunite deposits (Figure 4) from long-term studies near Marysvale, Utah. Their model clearly describes the setting of the deposits, illustrates the alteration zoning, and explains their origins. At Marysvale, Miocene alunite deposits formed in lava flows, under highly oxidizing conditions, at the tops of convecting hydrothermal plumes spaced at three to four kilometre intervals around a central monzonite stock. A series of circular altered areas formed that are zoned laterally and vertically. Central siliceous alunitic cores are surrounded by kaolinized rocks within a pyritic propylitic zone of regional extent. Vertical zoning produces a layered or "stacked" sequence in which a basal zone of alunite replacement is overlain, in sequence, by zones of silica-jarosite, silica-hematite, and a flooded silica capping. This mineralization apparently forms above the ground-water table, where ascending hydrothermal fluids "flash" into wet steam resulting in some condensation of vapours and a reflux of aqueous sulphuric acid into the system (Cunningham *et al.*, 1984). This type of alteration takes place above the water table at depths of generally less than 50 metres. This is a vital fact for explorationists because it establishes the position of the paleosurface, the most important datum plane in all depth-zoning models.

Cupriferous Pyritic Gold-Silver Deposit Model — Acid-Sulphate Aluminous Alteration. Arsenic and antimony-rich cupriferous pyritic gold-silver deposits, referred to as the "hot and/or deep acid-sulphate type", form a distinctive group of mineral occurrences associated with acid-sulphate advanced argillic alteration (Knight, 1977). These deposits are characterized by aluminous alteration minerals, native sulphur, and abundant pyrite, enargite or other As, Sb minerals. Ashley (1982) classified them as enargite-gold deposits and Sillitoe (1983) described them as enargite-bearing massive sulphide deposits that form at high levels in porphyry copper systems.

The mineralization occurs over a wide vertical range, from shallow, near-surface volcanic environments with abundant breccia-related mineralization, to deeper, replacement deposits and porphyry copper stockwork systems in proximity to porphyritic intrusive rocks. Hydrothermal alteration is primarily argillic with abundant quartz, alunite, kaolinite, sericite-illite and montmorillonite. Somewhat restricted advanced argillic zones contain the aluminous minerals pyrophyllite, kaolinite and andalusite, as well as the related alteration minerals quartz, dumortierite, scorzalite, lazulite, corundum, zunyite, pyrite, hematite, topaz and rutile (Wojdak and Sinclair, 1984). Advanced argillic alteration zones generally underlie or occur within the broader argillic alteration zones. Ore mineralization occurs within the advanced argillic zones, but is more commonly present at the transition from argillic to advanced argillic aluminous alteration, as at Goldfield, Nevada (Ashley, 1974). Preliminary fluid inclusion data from Goldfield (Ashley, 1984), and the widespread presence of pyrophyllite and some diaspore, suggest that the advanced argillic aluminous alteration assemblages form above the hydrothermal stability limit of kaolinite at temperatures of at least 300°C. The ore minerals, which generally line or fill open spaces in breccia zones, formed later, probably as the hydrothermal systems cooled. Mineralized structures may contain abundant pyrite; zones with up to 15%, or more, pyrite are common. Local small massive sulphide lenses may also be present. Enargite-luzonite, tetra-

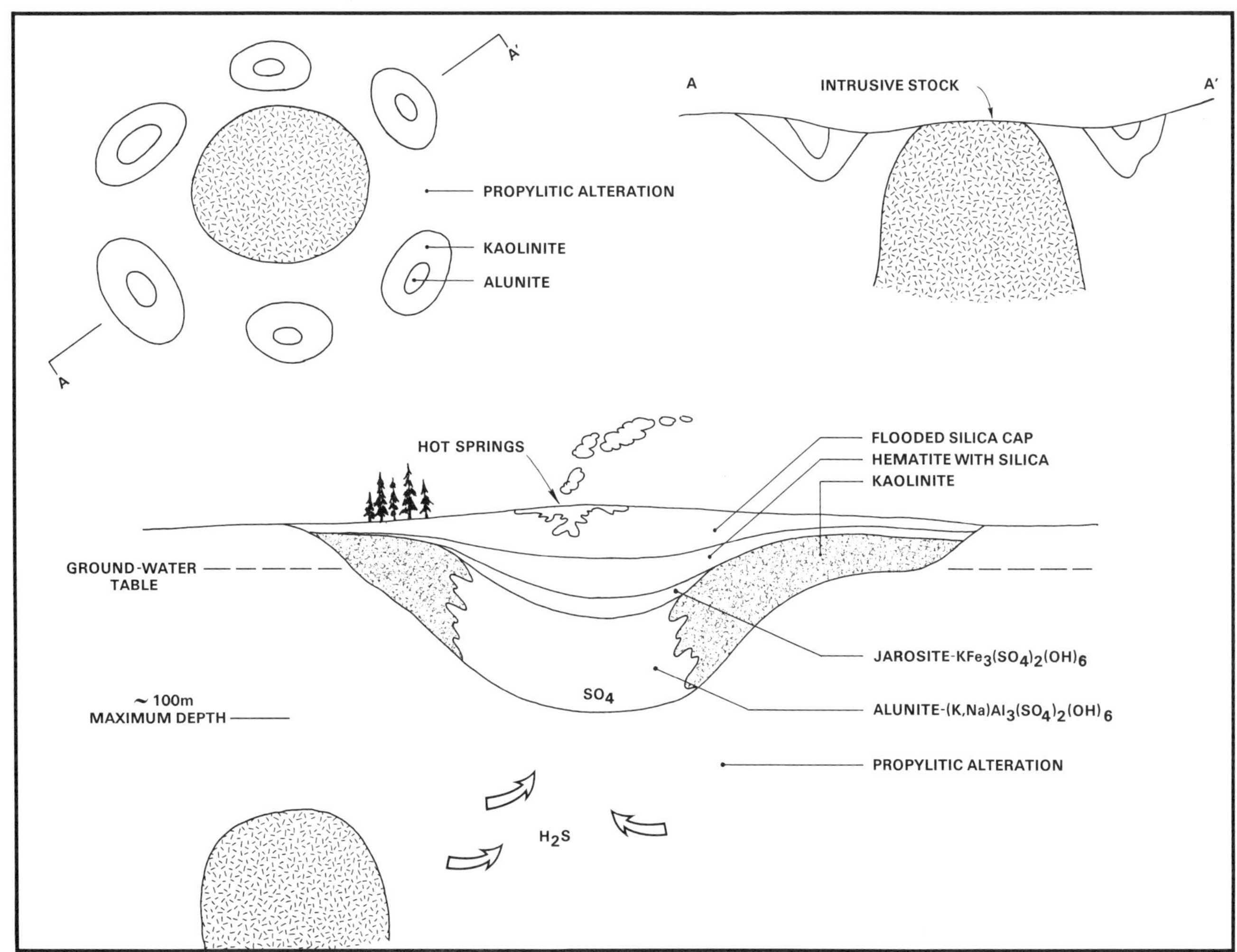

Figure 4 *Diagrammatic representation of the argillic-alunite alteration model described by Cunningham* et al. *(1984). A central intrusive body is surrounded by hydrothermal convection cells that cause zoned areas of alteration to form. Distinct vertical and some lateral mineral zoning in individual alteration zones (lower part of figure) is related both to near surface boiling and oxidation of fluids at the ground-water table.*

hedrite-tennantite, silver sulphosalts and minor amounts of base metal sulphides are the main ore minerals; tellurides occur in some deposits.

The deposits originated from hot fluids emplaced at depths ranging from slight to far below any hot spring discharge sites. The near-surface deposits are, therefore "telescoped', as defined by Spurr (1923, p. 292-308) in his book *The Ore Magmas*. The definition implies that deposition took place at unusually high temperatures for such a shallow setting, and also that thermal gradients were very steep near the surface. Preliminary isotopic studies indicate a magmatic source for sulphur, but the other fluid components are meteoric in origin (Ashley, 1982). Alunite and native sulphur are dominantly hypogene in these deposits. Supergene alunite that forms in some acid-sulphate alteration zones and leached cappings can be distinguished by stable isotope analysis and radiometric dating.

A Canadian Cordilleran Epithermal Model

A General Model. The western part of the Canadian Cordillera is a dominantly eugeoclinal region. Allochthonous volcanic terranes, flanking sedimentary basins, and their metamorphosed and intruded equivalents were accreted during Mesozoic time against the ancestral continental margin of North America (Monger *et al.*, 1982). Most of the accreted terranes are only locally metamorphosed and large regions are only moderately deformed with little disruption of stratigraphic continuity. Within these terranes are numerous areas of subaerial rocks with related or younger, structurally controlled, high-level plutons. These provide geological settings suitable for formation of epithermal deposits. The oldest known favourable host rocks are subaerial andesitic rocks that were deposited near the end of Early Jurassic island arc volcanism. Extensive Cretaceous and Tertiary to Recent continental volcanic rocks were deposited following Jurassic to Cretaceous accretion and consolidation of the Cordillera in a predominantly extensional tectonic regime with much major strike-slip faulting. Epithermal deposits in these rocks bear remarkable similarity to many Tertiary epithermal deposits in the southwestern United States.

Much of the Cordillera was extensively glaciated and is deeply dissected and eroded. Nevertheless, some sites of high-level hydrothermal activity and fossil hot spring deposits have survived. In the Toodoggone area, in northern British Coumbia, for example, the recognition of 190 Ma alunite replacement deposits (Schroeter, 1982) and related epithermal deposits in rocks of Early Jurassic age (204-183 Ma) demonstrates the presence of pre-Tertiary epithermal deposits in the Canadian Cordillera. Recognition of paleotopographic surfaces by means of attendant preserved hot spring deposits has important implications for exploration using depth-zoning models.

In a model developed from British Columbia deposits (Figure 5), there are three main hydrodynamic components that represent hydrothermal flow regimes in epithermal systems as illustrated by Berger and Eimon (1982).

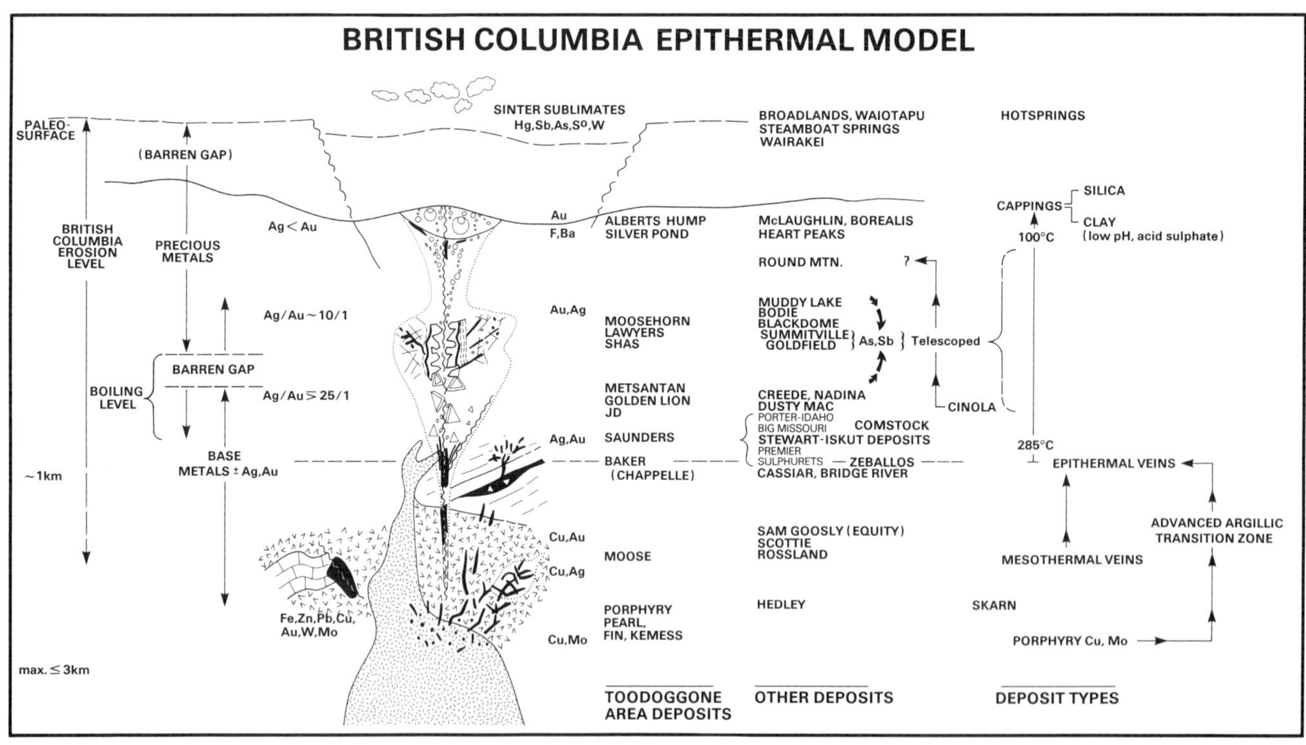

Figure 5 *British Columbia epithermal model. The model is based on studies of epithermal deposits in the Toodoggone area by T.G. Schroeter and A. Panteleyev, and comparisons with deposits elsewhere. The model infers a continuum from porphyry copper and skarn through transitional deposits, to epithermal veins, and hot spring discharge deposits.*

These are: (1) the near surface to hot spring discharge component; (2) the ascending, free-flowing, hydrothermal component that is open to surface (a physically "open" but chemically "closed" system); and (3) the intermittently sealed or constrained fluid flow component in which there are stacked hydrothermal cells and some lateral flow. The hot spring component with surface deposits of siliceous sinter, acid-sulphate and clay alteration zones and underlying replacement alunite deposits has been described above. The ascending open hydrothermal component is essentially described by Buchanan's (1981) model, in which boiling and oxidation cause mineral deposition from ascending hydrothermal fluids in steep structurally controlled conduits that remain open and provide unrestricted travel for fluids to the surface. The stacked hydrothermal cell component, described in part by Berger and Eimon (1982), is more complicated. In this situation, hydrothermal fluids that encounter impermeable cap rocks, which restrict their upward flow, tend to migrate laterally. Similarly, in permeable zones, periodic constriction and self-sealing of hydrothermal conduits by precipitation of vein minerals, fault movements, dyke emplacement or other mechanisms, restricts passage of fluids. When the system is sealed, fluid pressure builds until it exceeds the strength of the confining rocks and fracturing takes place as fluids break through to the surface. The resulting interplay of fluid pressures — from hydrostatic to those approaching lithostatic conditions — produces migrating boiling levels and hydrofracturing along the conduits. Rapid depressurization by hydrofracturing or faulting can be accompanied by extensive breccia development and may lead to phreatic explosions at surface. In this type of flow regime, lateral, lithologically controlled fluid flow can dominate when the system is sealed. The deposition sequence in veins is complicated by overlapping, periodically repeated mineralization, and dissolution and replacement of earlier minerals. Older vein material is commonly broken and disrupted during later fracturing and brecciation. Wall-rock interaction with fluids is more pronounced than in open, free-flowing systems, resulting in broader, more pervasive alteration envelopes around veins with both prograde and retrograde multicyclic alteration overprints.

The ore mineral zoning in epithermal deposits illustrated by Buchanan (1981) is part of a larger zoning pattern recognized long ago and described in a "reconstructed vein system" model by Emmons (1924). Ore zoning models, whether related to boiling levels or to progressive cooling away from magmatic fluid sources as proposed in older theories, predict that deeper vein deposits will contain base metals with some precious metals, and silver will increase in abundance upward. Above will be a zone with both gold and silver, and it will be capped by a zone containing minor amounts of gold in the form of electrum near the surface. At surface, porous rocks and sinter will contain a little cinnabar, arsenic, and antimony, and possibly some barite and fluorite. Rarely, these siliceous sinters contain gold in economic amounts.

The generalizations about precious metal zoning, as defined by silver to gold ratios in mine workings, rarely stand close scrutiny in epithermal districts, in individual deposits, or in orebodies. Zoning patterns are commonly inconsistent between epithermal deposits in the same district. Ore zones in these deposits tend to be erratic in distribution and grade, a feature that makes ore reserve calculations frustrating and difficult. For example, at the Silbak Premier Mine, one of the largest past producers in the Canadian Cordillera, the average silver to gold ratio within the same orebody of 23:1, varied from 112:1 to 5:1 over a vertical range of 400 metres (Grove, 1971). Gold content consistently increased with depth relative to silver, contrary to traditional expectations. Perhaps the only valid generalization about metal distributions in epithermal deposits is that base metals tend to increase with depth and silver is more abundant than gold. Silver to gold ratios are generally greater than one, and most are commonly 10:1 to 25:1, or greater.

Another zone invariably discovered during the course of mining operations or encountered during exploration is the "barren gap", one of which is shown on Figure 5. A barren gap, as shown by Emmons (1924), occurs below the bonanza gold and gold-silver deposits and above the deeper silver-rich and base metal zones. As Emmons (1924, p. 986) stated, "so many of the precious-metal deposits of the Tertiary (epithermal) type pass downward abruptly into worthless gangue that this has come to be regarded as an outstanding characteristic". The explanation for barren gaps is not certain. They might represent the zones in which the deeper, hot, mildly reducing and alkaline, dilute chloride brines pass through the transition into oxidized, neutral to acid solutions; possibly no ore minerals are deposited in this environment. More likely, the barren gap is a zone where earlier deposited minerals are dissolved and remobilized during hypogene leaching (Brimhall, 1980). Periodic shifts between precipitation and dissolution can be expected as pH, redox, acidity and fluid pressure conditions fluctuate in response to migrating boiling levels, self-sealing of conduits and mixing of fluids.

British Columbia Epithermal Deposits

The model for epithermal deposits shown in Figure 5 is based in large part on study of the Jurassic deposits in the Toodoggone area of northern British Columbia. In this area, much topographic relief and block faulting allow reconstruction of a large vertical range in hydrothermal systems from their intrusive source areas to hot spring discharge sites. The highest structural levels, and sites of hot spring discharge are represented by the Alberts Hump and Silver Pond deposits. These are siliceous alunite-clay cappings that are analogous to fossil hot spring deposits and acid-sulphate alteration zones. At Alberts Hump, a number of these alunite-clay zones occur within an area of hydrothermal alteration about 6 kilometres in diameter, in which many small, fracture-controlled zones of high-grade gold-barite mineralization have been discovered recently.

A detailed study of Clark (1983) describes the vertical ordering of alteration zones from surface to depth as: alunite-quartz, clay-quartz-barite, clay-quartz, quartz-hematite and quartz-pyrite (in propylitic rocks). This sequence, with the exception of abundant barite, bears a remarkable similarity to the Marysvale replacement alunite model of Cunningham *et al.* (1984).

Similar alteration is evident in the Heart Peaks silver prospect, British Columbia, which contains amethystine quartz stockworks in Pliocene, or younger, rhyolite (Schroeter, 1985). These deposits are also similar to the Borealis Mine, Nevada, where a silicified wedge of ore with a capping of siliceous sinter is mined, and the McLaughlin deposit in California, where a Plio/Pleistocene hot spring has deposited 18 million tonnes of siliceous material containing 5.5 grams gold per tonne. This ore is adjacent to pits previously mined for mercury.

No large, low-grade disseminated hot spring deposit has been discovered in British Columbia, with the possible exception of the Cinola Mines (Babe) deposit. Cinola is a prospect in Miocene conglomerate and sandstone beds containing more than 30 million tonnes with about 2 grams gold per tonne. The mineralization occurs in a silicified zone, surrounded by argillic rocks and is localized along a major fault that has been intruded by a Miocene rhyolite dyke (Champigny and Sinclair, 1982). Cinola poses an intriguing problem in depth classification; it appears to have hot spring affinity but detailed examination of fluid inclusion, stratigraphic and age data suggest a depth of formation of 1.1-1.8 km (Shen *et al.*, 1982).

The main Toodoggone epithermal deposits are divided into two groups — those with amethystine quartz and little base metal content (Lawyers, Moosehorn, Shas) and the base metal-bearing quartz-carbonate vein deposits (Metsantan, Golden Lion, JD). These deposits resemble many other epithermal prospects or past producers in British Columbia such as Engineer, Blackdome, Nadina, Dusty Mac and a number of occurrences in the Stewart-Iskut area. The Stewart area deposits, including the Silbak Premier Mine and 12 other large, former producing mines and numerous smaller mines and prospects, are especially noteworthy for their variety of mineralization and regional zoning patterns. Most of the mine production came from gold-silver vein deposits that occur with andesitic volcanic and dyke rocks and are genetically related to Middle Jurassic felsic volcanism (Alldrick, 1985). All these British Columbia deposits have some features that resemble those of the well-known "classic" epithermal deposits in the southwestern United States, such as Round Mountain, Nevada; Bodie, California; Creede, Colorado; and the famous Comstock Lode, Nevada.

Deeper levels of hydrothermal activity in the Toodoggone area are represented by the pyritic, sericite-altered and molybdenite-bearing Saunders prospect and by Baker Mine (Chappelle). Mineralization at Baker Mine is in quartz veins in faults in Triassic volcanic rocks that underlie the subaerial Jurassic rocks. The Moose, Porphyry Pearl, Fin, and Kemess deposits are spatially associated with intrusive rocks. Mineralization in these deposits is characteristic of the porphyry copper-molybdenum type, but with enhanced amounts of gold and silver; large, low-grade zones of fracture-controlled galena, sphalerite and gypsum veinlets suggest lower temperature, epithermal affinities. Elsewhere, some other epithermal deposits, such as those at Zeballos on Vancouver Island, also show a close spatial relationship with intrusive stocks (Sinclair and Hansen, 1983).

The other British Columbia deposits on Figure 5 that are shown to occur at deeper structural levels, can be considered to be "mesothermal" type. They are generally compact, crystalline white quartz or ribboned, graphitic quartz veins containing ankeritic carbonate minerals and some native gold, arsenopyrite, pyrite, pyrrhotite, base metal sulphides and minor scheelite. Host rocks are commonly greenstone in strongly deformed terranes that have undergone greenschist-grade metamorphism. The veins tend to have relatively thin, carbonate-rich alteration envelopes containing quartz-sericite or quartz-pyrite-fuchsite-albite-chlorite-talc. Some mesothermal auriferous veins are devoid of quartz and contain massive, fine-grained intergrowths of pyrite, pyrrhotite, and minor base metal sulphides. These mesothermal-type veins are similar to the California Mother Lode veins (Nesbitt *et al.*, 1986) or Archean lode gold deposits (Colvine *et al.*, 1984). Gold-bearing veins in the Cassiar and Bridge River areas are examples of mesothermal quartz, quartz-carbonate, and ribboned quartz-graphite veins associated with carbonate-sericite-fuchsite alteration in basic to ultrabasic host rocks. The veins in the Rossland area and the Scottie deposits are deeper-seated, auriferous, mesothermal-type pyrite-pyrrhotite chalcopyrite veins with related silicate alteration. Hedley is an arsenopyrite-rich skarn or tactite gold deposit.

Examples of the telescoped, acid-sulphate enargite-alunite-bearing orebodies are not known to occur in British Columbia. However, Sam Goosly deposit (Equity Silver Mine), an important silver, gold, copper, antimony, arsenic deposit with initial reserves of 28 million tonnes with 0.38% Cu, 106 g Ag per tonne, 0.96 g Au per tonne and 0.085% Sb (Cyr *et al.*, 1984), is somewhat similar. The deposit was discovered in 1967 and can be regarded (Schroeter and Panteleyev, 1985) as a "transitional deposit" linking mineralization of the porphyry copper and epithermal types. The porphyry copper-epithermal transition deposits are Cu, Ag, Au ores containing abundant pyrite and arsenic-antimony minerals such as enargite and/or tetrahedrite. They are akin to hot acid-sulphate aluminous epithermal deposits such as Goldfield, Nevada, and Summitville, Colorado (Sillitoe, 1983). However, they are deeper-seated, more restricted "closed" systems spatially associated with intrusive rocks, and not sufficiently oxidized to produce extensive alunite or native sulphur. The transitional deposits contain higher temperature aluminous alteration minerals such as andalusite, pyrophyllite, diaspore, and commonly have peripheral or overprinted zones of sericite and kaolinite alteration. Fluid inclusion homogenization temperatures approach those of porphyry copper deposits but show a larger temperature range. Fluids are generally more saline than those of epithermal deposits, but display a broad range from relatively concentrated to dilute. Shen and Sinclair (1982) and Wojdak and Sinclair (1984) described temperatures of mineralization at Equity Silver that range from over 400°C to about 200°C, and salinities in ore veins up to 22 wt.% equivalent NaCl. A number of the enargite deposits described by Sillitoe (1983) could be considered to be transitional types, as are parts of the Butte, Montana deposit.

Summary
Figure 5 presents the comparative position of epithermal deposits relative to other major genetic types of deposits. Mineral deposits in a volcanic intrusive setting are shown to form a continuum from intrusive-related porphyry copper and skarn to hot spring discharge deposits. Clearly there will be gaps between deposits and neither all the ore fluids nor all the metals will be derived from the magmas. Further from intrusive stocks, meteoric fluids will probably dominate the ore-forming solutions and magmatic activity will simply provide heat to drive the convecting hydrothermal systems. Epithermal deposits form at high structural levels, at some distance from intrusions, except perhaps minor dykes. Temperatures would be less than 285°C, within the thermal stability limit of kaolinite. Higher temperatures occur in the deeper, transitional deposits and the telescoped, aluminous, acid-sulphate epithermal deposits. Epithermal hydrothermal systems produce extensive high-level argillic zones and solfateric near-surface deposits of Hg, Sb, As, barite, fluorite and native sulphur. These mineral occurrences are indicators of potential bonanza gold and gold-silver ores at depth, ideally at depths of less than 1,000 metres.

Future Trends
The search for epithermal-type hydrothermal systems will be extended from the Mesozoic and younger, primarily subaerial volcanic terranes to other volcanic terranes, areas with sub-volcanic and high-level plutonic intrusions, and sedimentary terranes of the Cordillera. The sedimentary terranes, including miogeoclinal rocks of the eastern tectonic belts, will be explored for sediment-hosted, fine-grained disseminated gold deposits in thin-bedded calcareous rocks, shales and phyllites. Deposits in these rocks, similar to the Carlin, Nevada and other deposits in the southwestern US and the manto-type deposits in Mexico, are considered by many to have formed from epithermal-type hydrothermal systems (Bonham and Giles, 1983).

In addition to epithermal veins, the search will continue for large tonnage deposits suitable for open-pit mining. For example, Westmin Resources Limited recently announced reserves of over 4 million tonnes grading 2.43 grams gold per tonne and 110.4 grams silver per tonne in the Old Glory Hole area at Silbak Premier Mine in the Stewart area, where over 56 tonnes of gold and 1,270 tonnes of silver were recovered in the period 1918-1976, mainly from underground workings.

Innovative planning and research to assess the potential of heap or dump leaching might also lead to economic success. The attraction of leaching processes is that diffuse epithermal mineralization, low-grade stockpiles, or old waste dumps that are suitable for bulk mining and leaching extraction methods, might become economically viable with gold concentrations of only 1-2 ppm. In the Canadian Cordillera, the temperate climate and high rainfall might restrict leaching to seasonal operations or batch leaching, possibly under cover. The economic success of leach extraction of gold and silver from ores with 0.7-1.4 ppm gold at elevations of 1,675 metres near Zortman-Landusky in the Little Rocky Mountains of Montana provides an important example and an inspiration for possibilities in the Canadian Cordillera (Chamberlain and Pojar, 1984). Possibly the biggest inhibitor to viable operations in the Canadian Cordillera will not be climate, but rather, lack of well-developed zones of weathering oxidation in the extensively glaciated terrane.

Acknowledgements
This report is an outgrowth of metallogenic and mineral deposits investigations, educational travel opportunities and research engaged in the British Columbia Ministry of Energy, Mines and Petroleum Resources. I am grateful to my colleague, Tom G. Schroeter, District Geologist, Smithers, British Columbia, for his enthusiasm and inspiration during our mutual field projects, and numerous field trips and discussions. Study of the Toodoggone deposits by Schroeter and the writer has benefitted greatly from the dedication of a number of seasonal Ministry staff, most notably Larry J. Diakow. In the field, the cooperation of the exploration geologists working in the Toodoggone area is gratefully acknowledged. Discussions with many geologists in the British Columbia exploration community have filled gaps in my knowledge of epithermal deposits; the errors or deficiencies in the above discussions are my own.

This paper has benefited from the comments and editorial reviews by W.J. McMillan, L. Barker, T.L. Bottrill and P. Sheahan. Draughting was done by J. Armitage; photography by R. Player, and manuscript typing by D. Bulinckx, all of the Geological Branch. Permission to publish was granted by the Chief Geologist, Geological Branch, Ministry of Energy, Mines and Petroleum Resources.

References

Alldrick, D.J., 1985, Stratigraphy and petrology of the Stewart mining camp, *in* Geological Fieldwork, 1984: British Columbia Ministry of Energy, Mines and Petroleum Resources, Paper 85-1, p. 316-342.

Ashley, R.P., 1974, Goldfield mining district, *in* Guidebook to the geology of four Tertiary volcanic centres in central Nevada: Nevada Bureau of Mines and Geology, Report 19, p. 49-66.

Ashley, R.P., 1982, Occurrence model for enargite-gold deposits *in* Erickson, R.L., compiler, Characteristics of Mineral Deposit Occurrences: United States Geological Survey, Open-File Report 82-795, p. 144-147.

Ashley, R.P., 1984, Goldfield, Nevada: Unpublished field trip guidebook, Society of Economic Geologists, 14 p.

Barr, D.A., 1980, Gold in the Canadian Cordillera: Canadian Institute of Mining and Metallurgy, Bulletin, v. 73, n. 818, p. 59-76.

Berger, B.R., 1982, The geological attributes of Au-Ag-base metal epithermal deposits, *in* Erickson, R.L., compiler, Characteristics of Mineral Deposit Occurrences: United States Geological Survey, Open-File Report 82-795, p. 119-126.

Berger, B.R. and Eimon, P.I., 1982, Comparative models of epithermal silver-gold deposits: American Institute of Mining and Metallurgical Engineers, Preprint 82-13, 25 p.

Bethke, P.M. and Rye, R.O., 1979, Environment of ore deposition in the Creede mining district, San Juan Mountains, Colorado: Part IV, Sources of fluids from oxygen, hydrogen, and carbon isotope studies: Economic Geology, v. 74, p. 1832-1851.

Bonham, H.F., Jr. and Giles, D.L., 1983, Epithermal gold/silver deposits: the geothermal connection: Geothermal Resources Council, Special Report, n. 13, p. 257-262.

Brimhall, G.H., Jr., 1980, Deep hypogene oxidation of porphyry copper potassium silicate protore at Butte, Montana: a theoretical evaluation of the copper remobilization hypothesis: Economic Geology, v. 75, p. 384-409.

Buchanan, L.J., 1981, Precious metal deposits associated with volcanic environments in the southwest, *in* Dickinson, W.R. and Payne, W.D., eds., Relations of tectonics to ore deposits in the southern Cordillera: Arizona Geological Society, Digest, v. XIV, p. 237-262.

Buddington, A.F., 1935, High-temperature mineral associations at shallow to moderate depths: Economic Geology, v. 30, p. 205-222.

Casadevall, T. and Ohmoto, H., 1977, Sunnyside mine, Eureka mining district, San Juan County, Colorado: geochemistry of gold and base metal ore deposition in a volcanic environment: Economic Geology, v. 72, p. 1285-1320.

Chamberlain, P.G. and Pojar, M.G., 1984, Gold and silver leaching practices in the United States: United States Bureau of Mines, Report IC-8969, p. 54.

Champigny, N. and Sinclair, A.J., 1982, Cinola gold deposit, Queen Charlotte Islands, British Columbia: A Canadian Carlin-type deposit, *in* Hodder, R.W. and Petruk, W., eds., Geology of Canadian Gold Deposits: Canadian Institute of Mining and Metallurgy, Special Volume 24, p. 243-254.

Clark, J.R., 1983, Thermobarageochemical definition of epithermal Au-Ag system, Toodoggone district, British Columbia: (Abstract) *in* Mineral Exploration Research Institute, Note de Recherche 83-1, p. 7-8.

Colvine, A.C., Andrews, A.J., Cherry, M.E., Durocher, M.E., Fyon, A.J., Lavigne, M.J., Jr., Macdonald, A.J., Marmont, Soussan, Poulsen, K.H., Springer, J.S. and Troop, D.G., 1984, An integrated model for the origin of Archean lode gold deposits: Ontario Geological Survey, Open File Report 5524, 98 p.

Cunningham, C.G., Rye, R.O., Steven, T.A. and Mehnert, H.H., 1984, Origin and exploration significance of replacement and vein-type alunite deposits in the Marysvale volcanic field, west central Utah: Economic Geology, v. 79, p. 50-71.

Cyr, J.B., Pease, R.B. and Schroeter, T.G., 1984, Geology and mineralization at Equity silver mine: Economic Geology, v. 79, p. 947-968.

Drummond, S.E. and Ohmoto, H., 1985, Chemical evolution and mineral deposition in boiling hydrothermal systems: Economic Geology, v. 80, p. 126-147.

Emmons, W.H., 1924, Primary downward changes in ore deposits: American Institute of Mining and Metallurgical Engineers, Transactions, v. 70, p. 964-992.

Ewers, G.R. and Keays, R.R., 1977, Volatile and precious metal zoning in the Broadlands geothermal field, New Zealand: Economic Geology, v. 72, p. 1337-1354.

Grove, E.W., 1971, Geology and mineral deposits of the Stewart area: British Columbia Ministry of Energy, Mines and Petroleum Resources, Bulletin 58, 219 p.

Hall, R.B., 1982, Model from hydrothermal alunite deposits, *in* Erickson, R.L., compiler, Characteristics of Minsteral Deposit Occurrences: United States Geological Survey, Open-File Report 82-795, p. 148-154.

Heald-Wetlaufer, P., Hayba, D.O., Foley, N.K. and Goss, J.A., 1983, Comparative anatomy of epithermal precious- and base-metal districts hosted by volcanic rocks: United States Geological Survey, Open-File Report 83-710, 16 p.

Hedenquist, J.W. and Jenley, R.W., 1985, Hydraulic fracturing, hydrothermal eruptions and mineralization in the Waiotapu thermal system, New Zealand: Their origin, associated breccias and relation to precious metal mineralization: Economic Geology, v. 80, p. 1640-1668.

Henley, R.W. and Ellis, A.J., 1983, Geothermal systems, ancient and modern: a geochemical review: Earth Science Reviews, v. 19, p. 1-50.

Henley, R.W. and McNabb, A., 1978, Magmatic vapor plumes and ground water interaction in porphyry copper emplacement: Economic Geology, v. 73, p. 1-20.

Knight, J.E., 1977, A thermochemical study of alunite, enargite, luzonite, and tennantite deposits: Economic Geology, v. 72, p. 1321-1336.

Lindgren, W., 1933, Mineral deposits: McGraw-Hill Book Company, New York, 930 p.

Lipman, P.W., Fisher, F.S., Mehnert, H.H., Naerser, C.W., Luedke, R.G. and Steven, T.A., 1976, Multiple ages for mid-Tertiary mineralization and alteration in the western San Juan Mountains, Colorado: Economic Geology, v. 71, p. 571-588.

Lipman, P.W. and Steven, T.A., 1976, Calderas of the San Juan volcanic field, southwestern Colorado: United States Geological Survey, Professional Paper 958, 35 p.

McKee, E.H., 1979, Ash-flow sheets and calderas: their genetic relationship to ore deposits in Nevada, *in* Chapin, C.E. and Elston, W.E., eds., Ash-Flow Tuffs: Geological Society of America, Special Paper 180, p. 205-211.

Monger, J.W.H., Price, R.A. and Tempelman-Kluit, D.J., 1982, Tectonic accretion and the origin of the two major metamorphic and plutonic welts in the Canadian Cordillera: Geology, v. 10, p. 70-75.

Nesbitt, B.E., Murowchick, J.B. and Muelenbachs, K., 1986, Dual origins of lode gold deposits in the Canadian Cordillera, Geology, v. 14, p. 506-509.

O'Neil, J.R. and Silberman, M.L., 1974, Stable isotope relations in epithermal Au-Ag deposits: Economic Geology, v. 69, p. 902-909.

Radtke, A.S., 1981, Geology of the Carlin gold deposit, Nevada: United States Geological Survey, Open-File Report 81-97, p. 154.

Radtke, A.S., Rye, R.O. and Dickson, F.W., 1980, Geology and stable isotope studies of the Carlin gold deposit, Nevada: Economic Geology, v. 75, p. 641-672.

Prytuba, J.J, 1981, Relation of calderas to ore deposits in the western United States, *in* Dickinson, W.R. and Payne, W.D., eds., Relations of tectonics to ore deposits in the southern Cordillera: Arziona Geological Society, Digest, v. XIV, p. 227-236.

Schmitt, H., 1950a, Origin of the "epithermal" mineral deposits: Economic Geology, v. 45, p. 191-200.

Schmitt, H., 1950b, The fumarolic-hot spring and "epithermal" mineral deposit environment: Colorado School of Mines, Quarterly, v. 45, n. 18, p. 209-229.

Schoen, R., White, D.E. and Henley, J.J., 1974, Argillization by descending acid at Steamboat Springs, Nevada: Clays and Clay Mineralogy, v. 22, p. 1-22.

Schroeter, T.G., 1982, Toodoggone River, *in* Geological Fieldwork, 1981: British Columbia Ministry of Energy, Mines and Petroleum Resources, Paper 82-1, p. 122-133.

Schroeter, T.G., 1985, Heart Peaks prospect, *in* Geological Fieldwork, 1984: British Columbia Ministry of Energy, Mines and Petroleum Resources, Paper 85-1, p. 359-364.

Schroeter, T.G. and Panteleyev, A., 1985, Lode gold-silver deposits in Northwestern British Columbia, *in* Morin, J.A., ed., Mineral Deposits of the Northern Cordillera: Canadian Institute of Mining and Metallurgy, Special Volume 37, p. 178-190.

Seward, T.M., 1973, Thiocomplexes of gold and the transport of gold in hydrothermal ore solutions: Geochimica et Cosmochimica Acta, v. 37, p. 379-399.

Shen, Kun, Champigny, N. and Sinclair, A.J., 1982, Fluid inclusion and sulphur isotope data in relation to genesis of the Cinola gold deposit, Queen Charlotte Islands, British Columbia, *in* Hodder, R.W. and Petruk, W., eds., Geology of Canadian Gold Deposits: Canadian Institute of Mining and Metallurgy, Special Volume 24, p. 255-260.

Shen, Kun, and Sinclair, A.J., 1982, Preliminary results of a fluid inclusion study of Sam Goosly deposit, Equity Silver Mines Ltd., Houston, British Columbia, *in* Geological Fieldwork, 1981: British Columbia Ministry of Energy, Mines and Petroleum Resources, Paper 82-1, p. 229-233.

Silberman, M.L., 1982, Hot-spring type, large tonnage, low-grade gold deposits, *in* Characteristics of mineral deposit occurrences: United States Geological Survey, Open-File Report 82-795, p. 116-125.

Silberman, M.L. and McKee, E.H., 1974, Ages of Tertiary volcanic rocks and hydrothermal precious-metal deposits in central and western Nevada: Nevada Bureau of Mines, Geology Report 19, p. 67-72.

Sillitoe, R.H., 1977, Metallic mineralization affiliated to subaerial volcanism: a review, *in* Gass, I.G., ed., Volcanic Processes in Ore Genesis: Geological Society of London, Special Publication, Number 7, p. 99-116.

Sillitoe, R.H., 1983, Enargite-bearing massive sulphide deposits high in porphyry copper systems: Economic Geology, v. 78, p. 348-352.

Sinclair, A.J. and Hansen, M.C., 1983, Resource assessment of gold-quartz veins, Zeballos mining camp, Vancouver Island, *in* Geological Fieldwork, 1982: British Columbia Ministry of Energy, Mines and Petroleum Resources, Paper 83-1, p. 291-304.

Slack, J.F., 1980, Multistage vein ores of the Lake City district, western San Juan Mountains, Colorado: Economic Geology, v. 75, p. 963-991.

Smith, R.L. and Bailey, R.A., 1968, Resurgent cauldrons: Geological Society of America, Memoir 116, p. 613-662.

Spurr, J.E., 1923, The Ore Magmas, a series of essays on ore deposition: McGraw-Hill Book Company, New York, 588 p.

Steven, T.A. and Eaton, G.P., 1975, Environment of ore deposition in the Crede mining district, San Juan Mountains, Colorado: Part I. Geologic, hydrologic, and geophysical setting: Economic Geology, v. 70, p. 1023-1037.

Walthier, T.N., Araneda, G. and Crawford, J.W., 1982, The El Indio gold, silver, copper deposit, region of Coquimbo, Chile, unpublished preliminary report, 13 p.

Weissberg, B.G., 1969, Gold-silver ore-grade precipitated from New Zealand thermal waters: Economic Geology, v. 64, p. 95-108.

Weissberg, B.G., Browne, P.R.L. and Seward, T.M., 1979, Ore metals in active geothermal systems, *in* Barnes, H.L., ed., Geochemistry of Hydrothermal Ore Deposits, Second Edition: Wiley, New York, p. 738-780.

White, D.E., 1955, Thermal springs and epithermal ore deposits: Economic Geology, 50th Anniversary Volume, p. 99-154.

White, D.E., 1981, Active geothermal systems and hydrothermal ore deposits: Economic Geology, 75th Anniversary Volume, p. 392-423.

White, D.E., Muffler, L.J.P. and Truesdell, A.H., 1971, Vapor-dominated hydrothermal systems compared with hot water systems: Economic Geology, v. 66, p. 75-97.

Wisser, E., 1966, The epithermal precious-metal province of northwest Mexico: Nevada Bureau of Mines, Report 13, Part C, p. 63-92.

Wojdak, P.J. and Sinclair, A.J., 1984, Equity Silver silver-copper-gold deposit: alteration and fluid inclusion studies: Economic Geology, v. 79, p. 969-990.

Accepted, as revised, 7 November 1985
Originally published in
Geoscience Canada v. 13 Number 2
(June 1986)

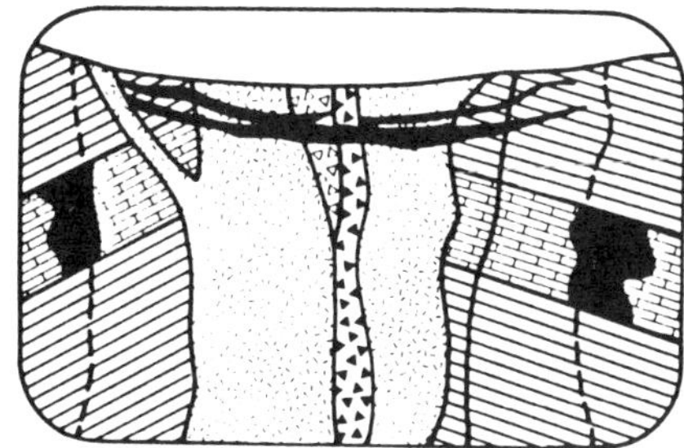

Porphyry Copper Deposits

W.J. McMillan and Andrejs Panteleyev
Geological Branch
British Columbia Ministry of Energy, Mines and Petroleum Resources
Victoria, British Columbia V8V 1X4

Introduction

In this review, we discuss porphyry copper deposits in which the products are copper, copper and molybdenum, or copper and gold. In the first, descriptive part, we outline the definition, history, distribution, and geologic characteristics of porphyry copper deposits; in the second part, a model is presented that incorporates known characteristics with genetic concepts. Most references cited are recent summary papers, which can be used as sources for more extensive research. Our discussion reflects most directly on the Canadian Cordillera, although we anticipate that it may be applicable elsewhere. Porphyry molybdenum deposits, though similar to porphyry coppers, are sufficiently different to warrant a separate paper and are not discussed here.

Originally, the term porphyry copper was applied to mineral deposits with widely dispersed copper mineralization in acid porphyritic rocks. Now the term combines engineering considerations with geologic features and refers to large, relatively low-grade, epigenetic, intrusion-related copper deposits that can be mined using mass mining techniques.

The generalized geological characteristics of porphyry copper deposits are as follows: they are spatially and genetically related to igneous intrusions; the intrusions are generally felsic but range widely in composition; the intrusions are epizonal and invariably porphyritic; multiple intrusive events, dyke swarms, intrusive breccias, and pebble dykes are characteristic; hosts for the intrusions can be any rock type, and range from unrelated country rocks to co-magmatic extrusive equivalents; the intrusions and surrounding rocks are intensely fractured; mineralization and alteration form large zones that exhibit lateral zoning; supergene alteration can produce vertical zoning that results in leached cappings and zones of secondary mineralization that can be critical to the economics of mining.

The large size of intrusive-related porphyry copper hydrothermal systems is possibly their most impressive feature. Lowell (1974) suggests that a deposit should have at least 20 million tonnes containing a minimum of 0.1% copper to be called a porphyry copper. The world's largest porphyry copper deposits have reserves of 1.5 to 3 billion tonnes of 0.8 to 2% copper (Table 1). A typical giant — say, 2 billion tonnes at 1.5% — might eventually produce 30 million tonnes of copper metal. Based on 1978 figures, such a mine could supply Canadian consumption for 130 years or world consumption for more than three years. The largest porphyry copper of the Canadian Cordillera is approximately one billion tonnes with grades just under 0.5% copper; most are much smaller. At the present time, approximately half the world's copper reserves, 60% of Canadian copper resources, and 90% of British Columbia's reserves are contained in porphyry deposits.

History

Large, low-grade supergene copper deposits were discovered during the 19th-century in the southwestern United States, Chile and Peru. The grade of these deposits, enriched by secondary processes, was about 2% copper but they remained uneconomic until the development of a method of mass mining, at Bingham Canyon, Utah, in 1906. The concomitant development of froth flotation techniques, which allowed selective separation of copper sulphides, was a key factor in making the enterprise profitable. Soon after, similar deposits were brought into production at Ely in Nevada, Santa Rita in New Mexico, Globe-Miami in Arizona, and El Teniente and

Table 1 Some giants of the porphyry copper world. (Modified after Sutulov, 1974 and other sources). For locations see Figure 2.

Country	Name	Size (million tonnes)	Copper (percent)	Molybdenite (percent)
			(grade approximate)	
United States	(1) Bingham Canyon	1,400	0.71	0.05
	(2) Butte *	large	?	?
	(3) San Manuel	1,000	0.75	0.015
	(4) Twin Buttes	800	0.74	0.017
	(5) Safford	2,000	0.40	—
Mexico	(6) La Caridad	600	0.75	0.016
	(7) Cananea	1,700	0.79	—
Panama	(8) Cerro Colorado	3,000	0.8	present
Chile	(9) Chuquicamata	>2,000	1.3	0.04
	(10) El Teniente	3,250	0.87	0.03
	(11) El Abra	1,500	0.80	—
New Guinea	(12) Bougainville	750	0.47	—
Philippines	(13) Biga	700	0.5	—
Iran	(14) Sar Cheshmeh	450	1.13	0.03
Canada	(15) Valley Copper	> 750	0.48	—
	(16) Lornex	400	0.41	0.014

* No reserve or grade figures available; from 1880-1964, 326 million tonnes with 2.45% Cu mined.

Figure 1 *Highland Valley, south-central British Columbia in June 1976. Viewed southwesterly: Bethlehem Copper (foreground) and Lornex open pits. 1A to F show aspects of development and mining from:* ***(A)*** *exploration (Berg deposit);* ***(B)*** *preparation for production (Afton);* ***(C) to (F)*** *mining, milling, grinding, flotation (Highland Valley), respectively.*

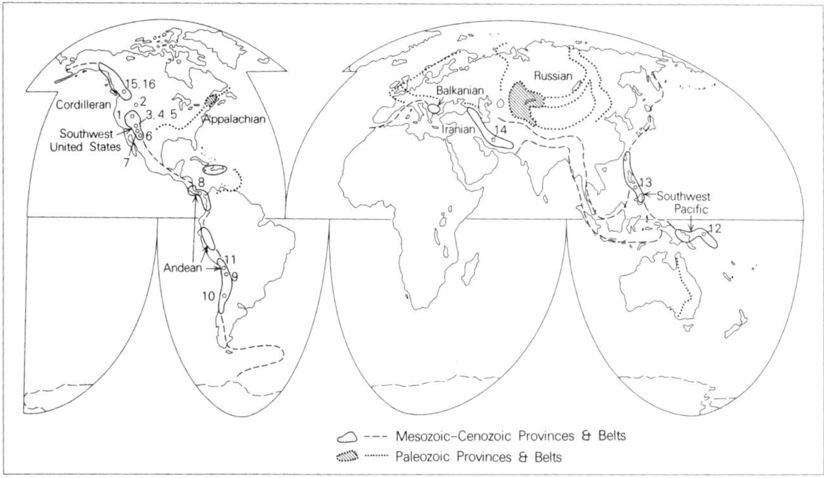

Figure 2 *World-wide distribution of porphyry provinces. Numbers refer to deposits described in Table 1.*

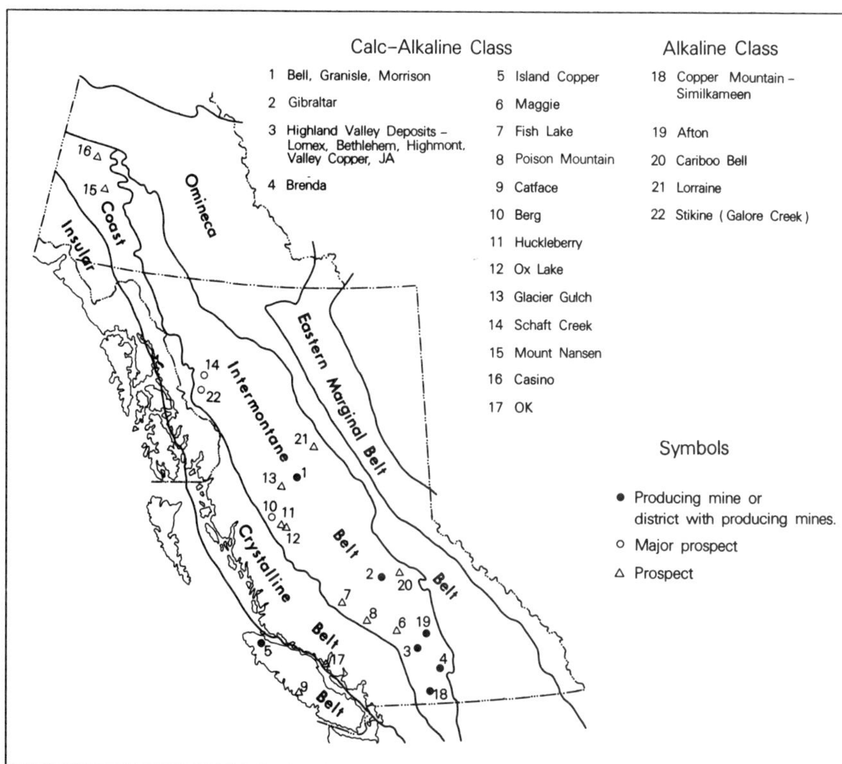

Figure 3 *Cordilleran porphyry mines and prospects and their tectonic settings.*

Chuquicamata in Chile. In all these deposits, mining began in secondarily enriched (supergene) ore. Consequently, it took several decades for the relationship between the supergene deposits and unweathered, generally uneconomic, primary mineralization (protore) to be understood. Similarly, the significance of associated porphyry intrusions was slow to be recognized (Emmons, 1927). The genesis of these supergene deposits became evident only when improved mining techniques and advanced in equipment design lowered cutoff grades and enabled mining of the protore at depth.

During World Wars I and II copper demand and production increased dramatically; after each war, the market for copper softened and there was little reason to prospect for new porphyry deposits. Later, demand linked to the Korean War rejuvenated porphyry copper exploration in the southwest United States. In the mid-1950s, exploration was extended into the Canadian Cordillera, South America, the southwest Pacific and other regions.

Development of these large, low-grade deposits depended and still depends on advances in engineering and ore dressing techniques, on world price and demand, and on taxation policies.

Distribution and Age

Porphyry copper provinces seem to coincide, worldwide, with orogenic belts (Figures 2 and 3). This remarkable association is clearest in Circum-Pacific Mesozoic to Cenozoic deposits but is also apparent in North American, Australian and Soviet Paleozoic deposits within the orogenic belts. Porphyry deposits occur in two main settings within the orogenic belts; in island arcs and at continental margins. Deposits of Cenozoic and, to a lesser extent, Mesozoic age predominate. Those of Paleozoic age are uncommon and only a few Precambrian deposits with characteristics similar to porphyry coppers have been described (Kirkham, 1972; Gaál and Isohanni, 1979). Deformation and metamorphism of the older deposits commonly obscured primary features, hence they are difficult to recognize (Griffis, 1979).

Porphyry Copper Classification

Porphyry copper deposits comprise three broad types: plutonic, volcanic, and those we will call "classic". The general characteristics of each are presented in Table 2 and illustrated in Figure 4. Plutonic porphyry copper deposits occur in batholithic settings with mineralization principally occurring in one or more phases of plutonic host rock. Volcanic types occur in the roots of volcanoes, with mineralization both in the volcanic rocks and in associated co-magmatic plutons. Classic types occur with high-level, post-orogenic stocks that intrude unrelated host rocks; mineralization may occur entirely within the stock, entirely in the country rock, or in both. The earliest mined deposits, as well as the majority of Cenozoic porphyry copper deposits, are of the classic type. Their characteristics, particularly for deposits in the southwest United States, have been extensively described (Titley and Hicks, 1966; Lowell and Guilbert, 1970). The term "classic" has been applied to them because of their historical significance, because of the role they played in development of genetic models, and because no other term currently in the literature adequately describes them. Deposits of this type have variously been labelled simple, cylindrical, phallic (Sutherland Brown, 1976) and hypabyssal.

In the Canadian Cordillera, Mesozoic deposits are of the volcanic or plutonic types (Figures 6 and 7) and are associated with calc-alkalic or alkalic plutons which commonly intrude and mineralize co-magmatic volcanic piles. Cenozoic deposits are generally of the classic type (Figure 5). To date, the majority of Canadian Cordilleran porphyry deposits occur in the Intermontane Belt, although a few have been discovered in the Insular and Coast Crystalline Belts (Figure 3).

Intrusions Associated with Porphyry Copper Deposits

Intrusions associated with porphyry copper deposits are diverse but generally felsic and differentiated. Those in island arc settings have primitive strontium isotopic ratios ($^{87}Sr/^{86}Sr$ of 0.702 to 0.705) and, therefore, are derived either from upper mantle material or recycled oceanic crust. In contrast, ratios from intrusions associated with deposits in continental settings are generally

Table 2 Characteristics of the three types of porphyry copper deposits.

	Classic (Stock-related)	Volcanic	Plutonic
Setting	Associated with post-orogenic stocks intruding unrelated host rocks; co-magmatic volcanic piles rarely preserved. Cordilleran deposits are of Late Mesozoic to Tertiary age.	In basic to intermediate volcanic piles intruded by comagmatic calc-alkalic or alkalic (dioritic or shoshonitic suite) plutons; magmatism produces consanguineous and intimately associated intrusive/extrusive assemblages. Cordilleran deposits are of Mesozoic age.	In large calc-alkalic plutons emplaced in or near comagmatic volcanic rocks; plutons typically have mafic borders and are moderately to strongly differentiated. Cordilleran deposits are of Mesozoic age.
Plutons	Multiple phases emplaced as successive, small (0.5 to 2 km^2), cylindrical porphyritic intrusions; numerous pre-, intra-, and post-mineral porphyry dykes emplaced at shallow depth.	Calc-alkalic – very small to small sheets, dykes and plugs (0.2 to 10 km^2), with much textural variety; subvolcanic emplacement. Alkalic – high level sheets, dykes, plugs associated with underlying differentiated mesozonal pluton or small batholith.	Batholithic rocks (>100 km^2) immobilized at relatively deep levels (2 to 4 km). Phaneritic coarse-grained to porphyritic rocks with local swarms of pre- to post-ore porphyritic dykes.
Structural Control of Intrusions	Passive, structure need not be significant; many stocks localized by intersections of regional faults.	Calc-alkalic – emplacement in volcanic vents, fault zones, radial fractures. Alkalic – intrusive centres localized by regional structures. High level intrusive rocks invade volcanic vents and fault zones.	Diapiric emplacement; magmatic pulses and differentiation cause sharp to gradational internal phase boundaries.
Breccias	Abundant and characteristic; post-ore argillic diatremes are common. Other types present include collapse breccias, intrusive breccias, and carapace or stoping breccias. Early breccias can be mineralized.	Calc-alkalic – common and diverse; include primary pyroclastic tephra, alteration pseudo-breccia, vent agglomerate, shatter and igneous breccias. Mineralized breccias are characteristic; some contain magnetite or tourmaline. Alkalic – intrusive and volcanic breccias common and generally mineralized, as in calc-alkalic types.	Common in association with late-stage porphyry dyke swarms. Breccias pre-, intra-, and post-ore, some contain specularite or tourmaline.
Alteration	Potassic, phyllic, and propylitic universally developed as annular shells around intrusions; argillic of varying importance. Early developed biotite (EDB) can be part of an isochemical hornfels and has often been misidentified as part of the potassic zone.	Calc-alkalic – propylitic is widespread; potassic is more restricted but can be intense; alteration centred on zones of high permeability. Similar to classic-type deposits with small core zones of potassic and local phyllic and/or argillic shells. Alkalic – local intense to pneumatolytic potassic alteration; early hydrothermal biotite overprinted by propylitic, then by sodic and/or potassic (albite-K-feldspar) and rarely scapolite alteration.	Phyllic, phyllic-argillic, and propylitic types are best developed; local potassic alteration. Fracture controlled to pervasive, commonly as alteration envelopes on multistage fractures and veins. Centred on orebodies but patterns of zoning complicated by overprinting.
Orebodies	In margins and adjacent to porphyry intrusion(s) as annular ore shells, or as domal cappings; pronounced lateral zoning. Pyrite is found throughout; the weakly mineralized core is surrounded by zones dominated by molybdenite, then chalcopyrite, and, finally, a pyritic halo.	Calc-alkalic – generally Cu-Mo deposits intimately associated with breccias and intensely altered rocks; orebodies lensoid and irregular, with some preferential bedding control. Most ore contains chalcopyrite with rare bornite or molybdenite as 'dry' fracture fillings. Alkalic – generally Cu-Au deposits in intrusive breccia or in highly fractured country rock; some replace porous country rock. Locally magnetite-apatite of magmatic origin present as vein or breccia infillings; zoning is from chalcopyrite ±magnetite and bornite outward to a pyrite halo.	Large and diffuse vein stockworks; some breccia control, some faults mineralized; sulphides relatively sparse. Zoning is evident with iron content increasing outward from bornite to chalcopyrite to pyrite rich zones; Mo distribution is variable. Some deposits have low-grade quartz-rich core zones.

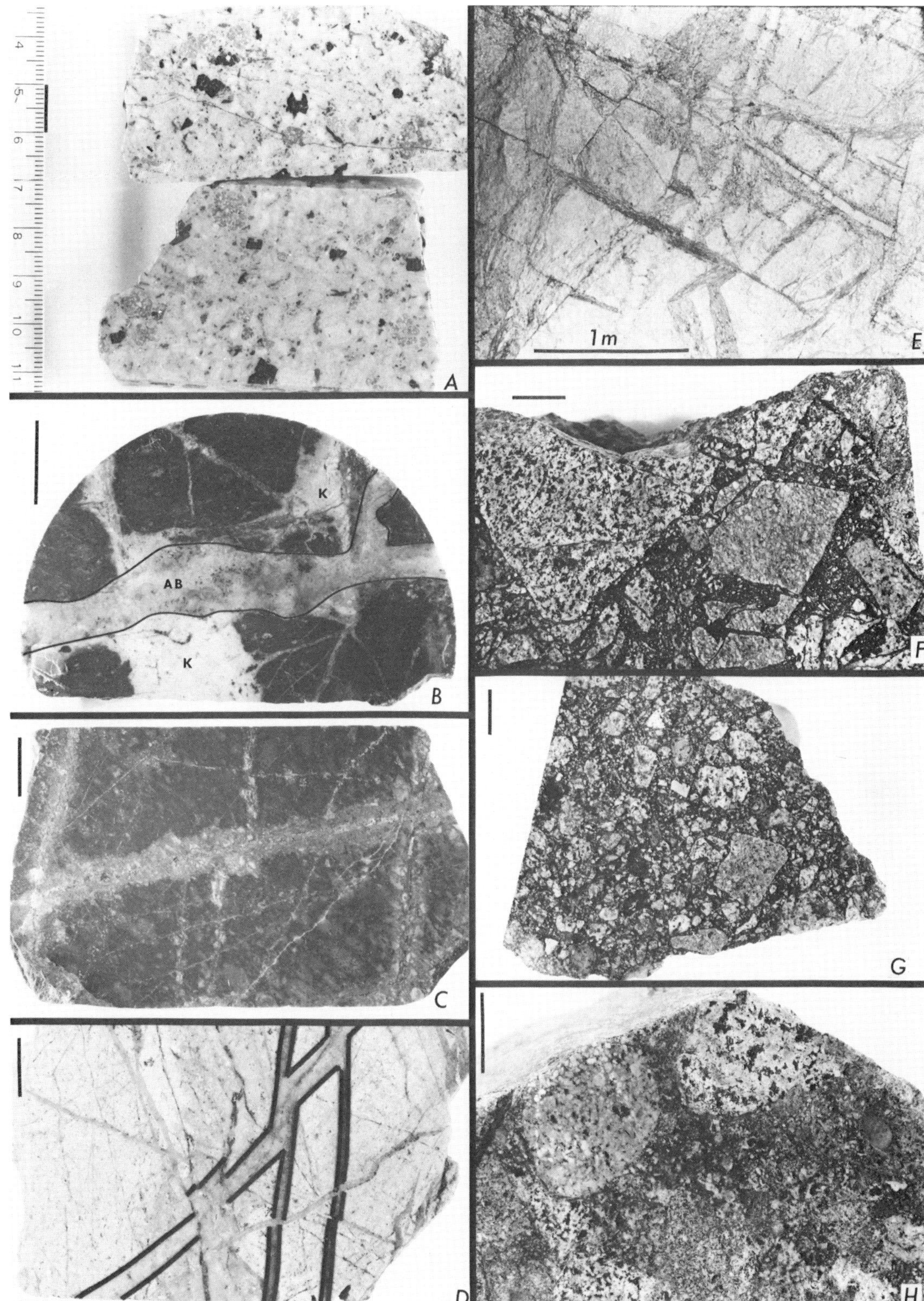

Figure 4 *Lithologic and alteration types in porphyry deposits. Scale bar is 1 cm long except where noted.*

(A) *biotite quartz feldspar porphyry* ***(QFP)****;*

(B) *biotite hornfels cut by K-feldspar veins* ***(K)*** *which are cut by an anhydrite-biotite vein* ***(AB)****; the hydrothermal fluids were in equilibrium with country rock;*

(C) *biotite hornfels with pale lapilli cut by quartz-pyrite veins with alteration envelopes; hydrothermal fluids were not in equilibrium with the country rock;*

(D) *multistage veins in phyllic tuff;*

(E) *quartz-chalcopyrite veins with flakey sericite-quartz envelopes in shattered quartz monzonite porphyry;*

(F) to (H) *breccias showing a progession from incipient, to angular transported, to rounded transported fragments.*

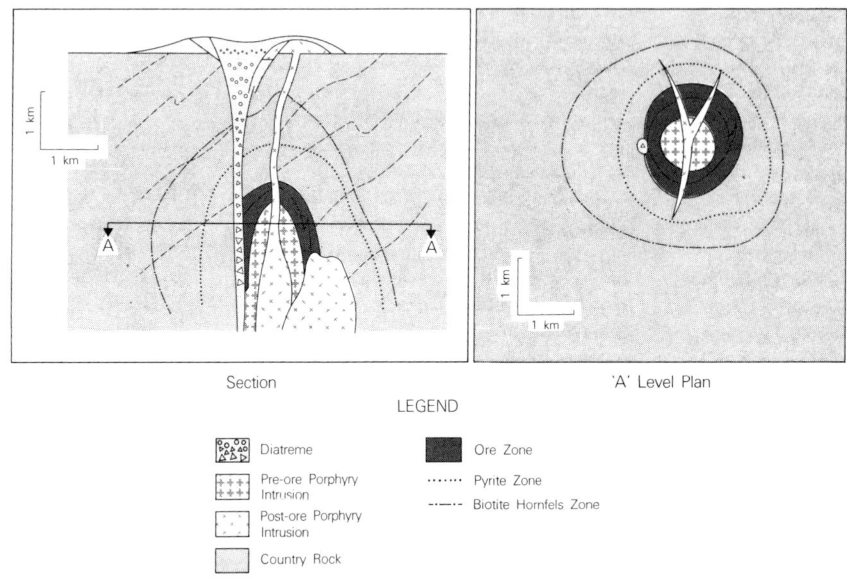

Figure 5 *Model of classic-type porphyry copper deposits. (After Sutherland Brown, 1976).*

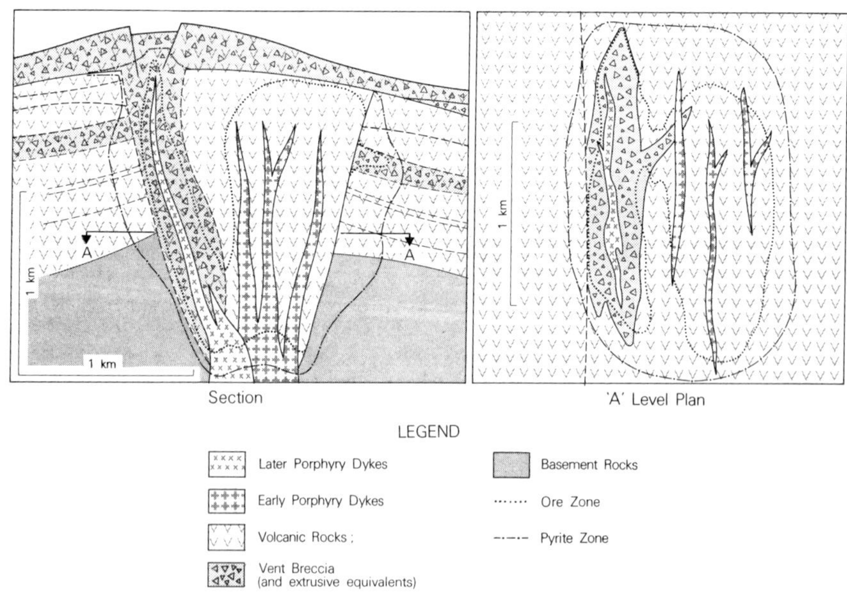

Figure 6 *Model of volcanic-type porphyry copper deposits. (After Sutherland Brown, 1976).*

higher, and are either derived from or, more likely, contaminated by crustal material. Compositions generally range from quartz diorite to quartz monzonite or granite in calc-alkalic suites; and from diorite to syenomonzonite or syenite in alkalic to shoshonitic (sometimes called dioritic) suites. Multiple intrusive events are characteristic of porphyry districts and many deposits are related to intrusions that are among the most differentiated of those present. In some deposits, however, mineralizing and nonmineralizing intrusions are practically identical. Differentiation alone does not result in the formation of porphyry copper deposits.

Magma composition might influence the behaviour of ore constituents. For example, copper is partitioned into octahedral sites in a residual melt, and the ratio of octahedral to tetrahedral sites is high when aluminum is abundant relative to total alkalis (Feiss, 1978). Therefore potassium-poor island arc suites, which usually have high aluminum-alkali ratios, are likely to produce copper-rich hydrothermal fluids. On the other hand, copper enrichment in potassium-rich continental suites may be accounted for by high oxygen fugacity and high water pressure in the magma (Mason and Feiss, 1979). Thus, conditions leading to residual metal and volatile concentration, not just chemistry, determine whether a magma will have associated mineralization.

Porphyry copper deposits associated with volcanism generally form during an intrusive phase late in the volcanic cycle and mineralization usually follows one or more pulses of magma emplacement. At Ray, Arizona, for example, early quartz diorite was intruded at 70 Ma, a porphyritic phase at 63 Ma, and a mineralized porphyry at 61 Ma (Cornwall and Banks, 1977). Similarly, at El Salvador and at OK Tedi, the onset of mineralization occurred 1 to 3 million years after initial magma emplacement (Gustafson and Hunt, 1975; Page, 1975, respectively).

Intrusions associated with porphyry copper deposits were generally emplaced as crystal-liquid mixtures at less than four kilometres depth; most were emplaced at only one to two kilometres. Almost invariably they are porphyritic, reflecting "sudden" crystallization due to rapid chilling or to concentration and subsequent release of a volatile phase. Porphyry dykes are nearly ubiquitous and many breccia bodies associated with porphyry copper systems reflect the sometimes explosive escape of volatiles. Some breccias comprise post-ore diatremes but those that predate or form during mineralization can be important hosts for ore. Several periods of brecciation commonly occur and many mechanisms

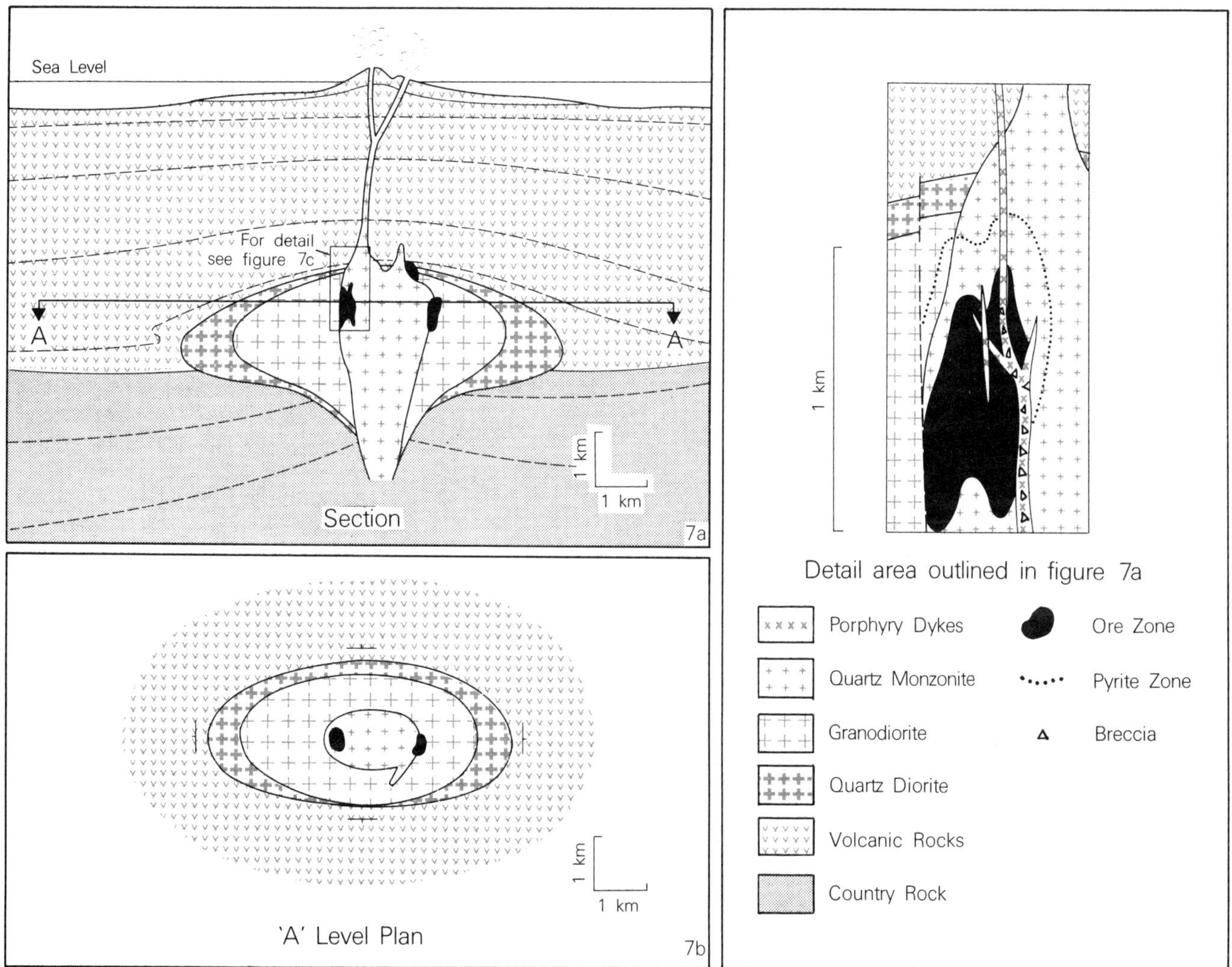

Figure 7 *Model of plutonic-type porphyry copper deposits.*

operate to cause brecciation. These include: explosive release of volatiles, fluidization of fault breccia, solution along fractures, chemical brecciation, roof collapse, shrinkage during crystallization, and others (Kents, 1961; Bryner, 1961). The breccias are typified by transported fragments that range from large, angular, rotated blocks to rounded, milled fragments in a finely comminuted matrix. Like the breccias, porphyry dykes can be pre-, intra- or post-ore in age (Kirkham, 1971). Commonly, intense hydrothermal alteration accompanies and affects the breccias and dykes. In such cases, it can be difficult to distinguish porphyry dykes from similar host rocks and, on occasion, even to recognize breccia bodies.

Structural Features

Faults localized magma emplacement in many porphyry copper districts. Fault intersections and strongly fractured zones are particularly important controls. In some areas, plutons seem to be localized by regional basement structures (Schmitt, 1966: Seraphim and Hollister, 1976; and many others) or large-scale, circular, cauldron subsidence (?) structures (Eggers, 1979).

Ground preparation in the deposits themselves is generally complex. The intrusions are often fractured by rejuvenation of regional faults along which they were emplaced. Furthermore, dyke emplacement, formation of breccias and hydrofracturing in response to hydrothermal activity also enhance permeability and help create the "plumbing sytems" followed by later ore-bearing hydrothermal fluids. Multiple episodes of healing and refracturing typically occur as is shown by cross-cutting relationships in veins, mineralized fractures and faults.

Alteration

In general, strong alteration zones develop in and around intrusions associated with porphyry copper deposits. Hydrothermal fluids derived from both the magma and heated groundwaters cause the alteration reactions and lead to formation of stable mineral assemblages analogous to metamorphic facies. Alteration is typically a base leaching process that is controlled by the metal cation to hydrogen ion ratio in the altering solution (Hemley and Jones, 1964). If the alkali to hydrogen ratio is low, feldspars, micas, and other silicates become unstable and hydrolysis occurs, releasing cations and driving the hydrothermal system toward equilibrium. Reactions are controlled, for the most part, by temperature and pressure, but also by the abundance, composition, and dynamic behaviour of the fluids, and the amount of wall-rock interaction.

Four alteration types are common: propylitic, argillic, phyllic, and potassic. Under conditions of weak hydrolysis, quartz and alkali feldspar are stable, but plagioclase and mafic minerals react with fluid to form the propylitic assemblage of albitized plagioclase, chlorite, epidote, carbonate and montmorillonite (with or without hydromica) or, less commonly, tremolite/actinolite. More intense hydrolysis produces argillic or phyllic alteration. Argillic assemblages, which are characterized by quartz, kaolinite and chlorite, with lesser montmorillonite, appear to be transitional into phyllic alteration assemblages. Phyllic assemblages are characterized by quartz and sericite, commonly accompanied by pyrite. Intense hydrolysis, at elevated temperature, produces advanced argillic assemblages consisting of quartz, pyrophyllite, kaolinite or dickite, and, in some cases, andalusite. Under conditions of very intense hydrolysis, the end product of alteration could be a porous mass of quartz. Potassic alteration takes place at high temperature in the presence of concentrated hydrothermal fluids. Conditions are equivalent to those in a late magmatic environment and, except where the country rock is granite or quartz monzonite, all constituents of the rock are unstable. Alteration assemblages consist typically of quartz (commonly as resorbed grains), K-feldspar, biotite, intermediate plagioclase (oligoclase to andesine) and rare anhydrite.

In a generalized model, alteration assemblages form distinct zones around the mineralized intrusion. They form shells with a core of potassic alteration grading outward through phyllic, argillic and propylitic alteration zones into unaltered country rock (Lowell and Guilbert, 1970). In reality, the complete alteration sequence is rarely developed or preserved, and assemblages are strongly influenced by the composition of the host rocks (Guilbert and Lowell, 1974). For example, potassic alteration might produce secondary K-feldspar and sericite in rhyolite, but biotite in andesite. Furthermore, pressure, temperature and permeability, conditions that determine lateral and vertical alteration zoning, change during the course of mineralization. These changes with time result in superimposed and cross-cutting stages of pervasive and vein-related alteration. In strongly fractured or otherwise permeable rocks, alteration tends to be pervasive and younger assemblages may completely mask older ones. In less permeable rocks, where alteration is fracture and vein controlled, changes in temperature, pressure and fluid composition can be inferred both from various suites of alteration minerals and from fluid inclusion data. Commonly adjacent fractures or veins have different alteration assemblages.

Invariably stockworks of veins with many cross-cutting relationships are present in porphyry copper deposits. These veins demonstrate that multiple episodes of fracturing and healing occur and that each stage may have hydrothermal fluids of different character. In general, the age sequence of alteration types is similar but not identical to the lateral zoning sequence; from oldest to youngest, vein alteration types are commonly potassic and propylitic, then phyllic, and finally argillic.

Hypogene Mineralization and Zoning

Hydrogene mineralization consists of varying amounts of pyrite, chalcopyrite, bornite and molybdenite as disseminations and fracture fillings, and in quartz veinlets. Zoning in porphyry coppers differs, not only between individual deposits, but also between classes of deposits (Table 2). In deposits of the classic type, a typical pattern would be as follows: a weakly mineralized or barren core zone centered on the intrusion has minor chalcopyrite and molybdenite and rare bornite; pyrite is generally less than 2%. Surrounding ore shells have enrichment in first molybdenite, then chalcopyrite; pyrite abundance increases outward in the ore shells. A peripheral pyrite-rich halo with 10-15% pyrite but only minor amounts of chalcopyrite and molybdenite encloses the ore shells. Base metal veins with gold and silver values are usually found in radial fracture zones peripheral to the pyrite halo. Overall, pyrite is the most abundant and widespread sulphide mineral in porphyry copper deposits.

The zoning discussed above adequately describes classic-type deposits in the southwestern United States and Tertiary deposits in the Canadian Cordillera. However, Mesozoic deposits in the Cordillera are of the volcanic and plutonic types, and differ from classic type (Table 2). Volcanic types usually have poorly defined metal zoning, in which central, weakly pyritized ore zones, containing chalcopyrite, bornite and magnetite are flanked by barren pyritic zones. Mineral zoning in plutonic types generally proceeds from bornite in the core through a chalcopyrite zone into poorly developed pyritic halos; some have a poorly mineralized siliceous core zone; molybdenite zones are irregularly distributed.

Ore Fluids and Sulphur Sources

Studies of the compositions and variations in composition of hydrothermal fluids, and of temperature and pressure conditions, reveal much about porphyry copper systems. Both fluid inclusion and isotopic studies have provided methods for evaluating the nature of ore-forming fluids (Nash, 1976; Sheppard, 1977). The fluids involved in alteration and ore formation are metal- and salt-rich brines containing both magmatic and meteoric components. Proportions of each may change at any stage in the hydrothermal process and may vary from place to place in the porphyry system.

Homogenization temperatures from various deposits range from 250°C to over 750°C. At Cerro Verde, Peru (LeBel, 1979a, b), for example, fluid inclusions homogenized at between 380°C and 410°C. Temperatures derived from study of sulphur isotopes, sulphide-sulphate ratios, ^{13}C and the composition of sericite (actually phengitic muscovite) substantiate the homogenization temperatures.

Sulphur, hydrogen and oxygen isotope studies shed light on the sources of sulphur and water in the ore deposits. At Cerro Verde, sulphur isotopes from pyrite and chalcopyrite display magmatic or mantle values, whereas those from sulphate minerals have meteoric values. Detailed study revealed that sulphates and carbonates began crystallizing under magmatic conditions but were subsequently modified by meteoric waters. Similarly, in the United States, studies of hydrogen and oxygen isotopes (Sheppard *et al.*, 1971) and of fluid inclusions from Bingham Canyon and Butte (Roedder, 1971) indicate that heated meteoric water is involved in porphyry copper formation. At Valley Copper, in British Columbia, as at Cerro Verde, the ore fluid was apparently a mixture (Jones, 1975). Magmatic water comprised roughly 75% of the ore fluid during main-stage mineralization; later, the system was quenched by an influx of meteoric or seawater.

Geological and geochemical evidence in porphyry deposits invariably suggest formation depths of less than four kilometres; most formed at one or two kilometres depth (Sillitoe, 1973). At Cerro Verde and Valley Copper, for example, LeBel and Jones, respectively, inferred pressures of 200 to 300 bars, equivalent to a depth of one to two kilometres.

Sources of Metals in Porphyry Copper Deposits

The close liaison between porphyry belts and orogenic belts suggests that the fundamental control of porphyry belts is tectonic. Isotopic evidence indicates that sulphur in the deposits is largely derived from the upper mantle or remelted oceanic crust, although meteoric waters play an important role in alteration and metal deposition in the porphyry environment.

The origin of the metals in the deposits is more speculative. Metals and sulphur in hydrothermal fluids may be concentrated as by-products of magmatic crystallization. However, Noble (1970) and, more recently, Banks and Page (1977) argued that magmas are incapable of transporting sufficient quantities of metals and sulphur to produce porphyry copper deposits. They concluded that hydrothermal fluids originate independently from magmas but in the same source area. In this theory, porphyry intrusions are associated with the deposits only because magma and later hydrothermal solutions followed the same access routes. Another hypothesis is that metals and/or sulphur are derived from the country rock. In this view, metal is scavenged from the country rock by convecting fluids in cells driven by the heat of the associated magma.

Post-Depositional Effects

Metamorphism and deformation are rarely significant in Cordilleran porphyry copper deposits. One exception is Gibraltar, a plutonic porphyry deposit in central British Columbia, in which contemporaneous mineralization and deformation occurred (Drummond *et al.*, 1976).

In older terranes, metamorphism and deformation may mask original alteration types and zoning through retrograde or prograde reactions and fabric readjustments. Alteration assemblages most resistant to change in low-grade metamorphic terranes wil be propylitic, phyllic and argillic, whereas only phyllic and potassic alteration will survive in higher grade terranes. Aluminum silicates derived from argillic (aluminous) assemblages may signal earlier hydrothermal activity, particularly in granitoid rocks.

Supergene effects have received little attention in this review because only a few porphyry deposits in the Canadian Cordillera contain significant zones of supergene mineralization. Some of these deposits show grade enrichment, but often the supergene zones present metallurgical problems which result in poor recovery and low-grade concentrates. At Afton, however, although the supergene zone is not enriched, the natural beneficiation converted sulphide ore into native copper and oxide ore, which simplified milling and smelting. Nevertheless, an understanding of supergene effects and processes in the porphyry environment is necessary, especially at the exploration stage, to interpret the weathered outcrops and leached cappings that constitute many of the surface showings in the Canadian Cordillera. For a thorough account of supergene effects, see Ney *et al.* (1976).

Relationship with Plate Tectonics

Porphyry copper deposits are found mainly in island arcs and near continental margins; both represent destructive boundaries of lithospheric plates (Mitchell and Garson, 1972). In this setting, a genetic relationship between subduction, magmatism and porphyry deposits is stated or implied and is generally accepted (Sillitoe, 1972; Creasey, 1977). Beyond this generalization, the relationship is often difficult to substantiate, even in the youngest Cenozoic orogenic belts (Gustafson, 1978), let alone in older terranes (Sangster, 1979).

For example, at OK Tedi, the youngest known porphyry copper deposit, mineralization is 1.1-1.2 Ma old (Page and McDougall, 1972) but subduction apparently took place some 30 million years earlier. Furthermore, some porphyry deposits lie in continental settings. Mesozoic to Cenozoic porphyry deposits in the southwestern United States, for example, are hundreds of kilometres from the continental margin and 200 km inland from the western edge of the Precambrian craton (Rogers *et al.,* 1974). This is much too far inland to be related to a typical subduction zone and debate continues whether there are any genetic ties between this mineralization and subduction (Lowell, 1974; Sillitoe, 1975).

Models for Porphyry Copper Deposits
No single model can adequately portray the alteration and mineralization processes that have produced the wide variety of porphyry copper deposits. However, volatile-enriched magmas emplaced in highly permeable terranes are ore-forming regimes that can be described in a series of models that represent successive stages in an evolving process. End-member models of hydrothermal regimes (Figure 8) attempt to show contrasting conditions for systems dominated by magmatic and meteoric waters, respectively. Both are depicted after enough time has elapsed following magmatic emplacement for convection cells to become established in the country rock in response to the magmatic heat source. The convecting fluids transfer mass and heat from the magma into the country rock and redistribute elements in the convective system. In intrusive settings, where these hydrothermal regimes operate, temperatures range from magmatic at depth, to ambient at the surface (>800°C to 20°C). At depth, fluid pressure is lithostatic and probably equivalent to a maximum load of four to five kilometres; near the surface, it approaches hydrostatic. At depth, the main cooling is accomplished through conduction; near the surface, cooling results from convective fluid movement.

The two models shown in Figure 8 represent end members of a continuum. The fundamental difference between them is the source and flowpath of the hydrothermal fluids. In the traditional orthomagmatic end member, volatiles and metals are concentrated during crystallization of the magma then break through the crystallized carapace, as hydrothermal fluids, in the post-magmatic stage. The initial wave of escaping fluids fractures the country rock, creating crackle zones and a primeval plumbing system that controls the travel paths of subsequent hydrothermal fluids and localizes alteration and mineralization (Burnham, 1967; Holland, 1972; Whitney, 1975; Henley and McNabb, 1978). Further crackling results from magmatic pressures, boiling and hydrofracturing (Phillips, 1973). In the convective end member, the fluid is mostly ground water whose source is meteoric, connate or seawater (Cathles, 1977; Norton and Knapp, 1977; Norton, 1978). In this model, thermally driven convective cells are initiated by emplacement of

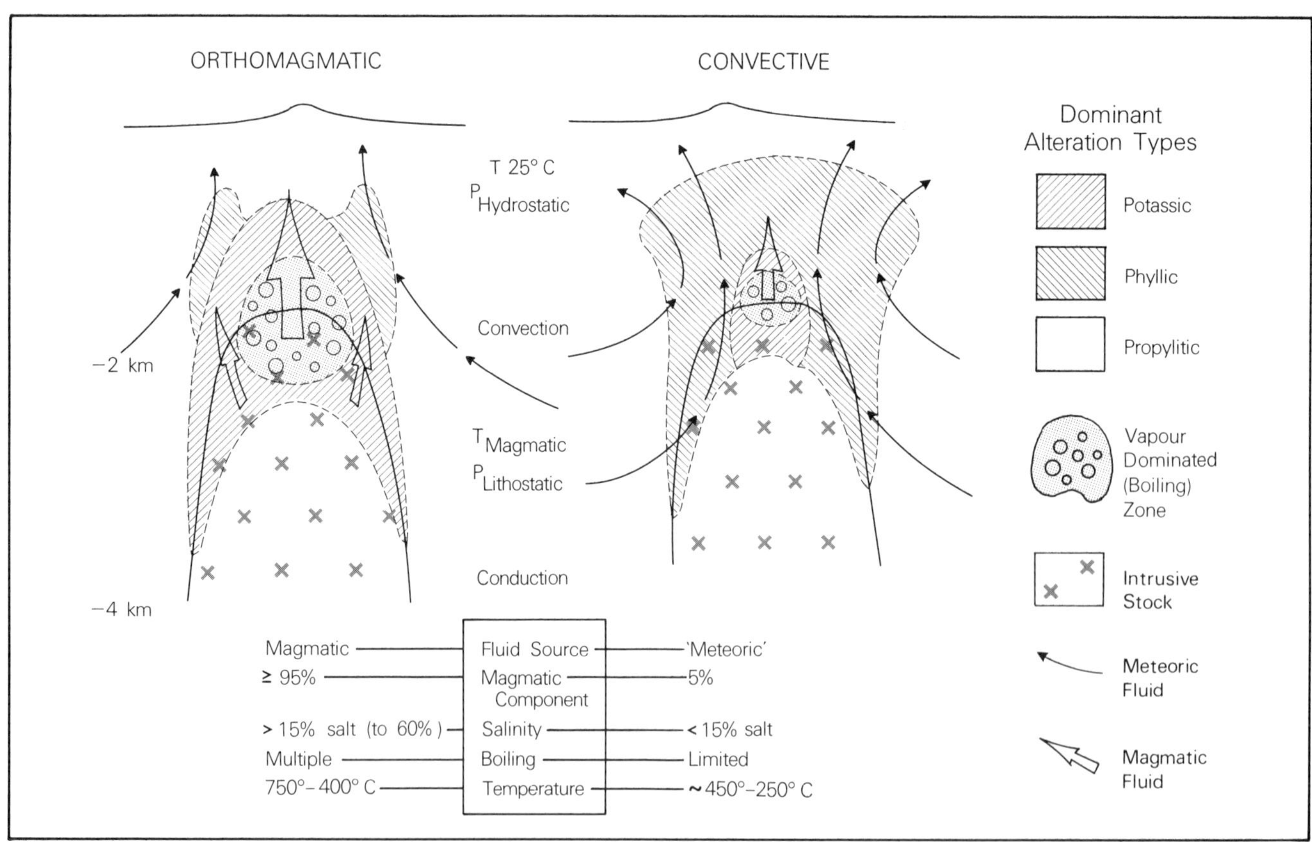

Figure 8 *Model of hydrothermal systems with contrasting orthomagmatic and convective fluid flow patterns. For explanation, see text.*

the magma. The permeability of the country rock, enhanced by the intrusive events, is sufficient to allow convective circulation to begin. Convection thoroughly redistributes fluids and concentrates ore and gangue constituents in and near the intrusion.

In the evolving dynamic systems that produce porphyry copper deposits, proportions of magmatic and meteoric fluids can be expected to fluctuate, both in time and space. Application of fluid inclusion studies, stable isotope and fluid dynamic simulation studies to the end-member models yields the following observations:

(1) In the orthomagmatic model, the cooling stock generates an ascending hydrothermal plume. There is some peripheral entrainment of meteoric water. In the convective model, permeable country rocks are the primary source of fluids. Ground water flows into the convective cells from as much as 2 km above and 5 km lateral to the stock.

(2) The magmatic component constitutes up to 95% of the hydrothermal fluid in the orthomagmatic system and as little as 5% in the convective system.

(3) Usually, salinity is relatively high in ore zones. In orthomagmatic sytems, saline fluids with greater than 15 wt.%, and ranging as high as 70 wt.%, NaCl equivalent can be found. In convective systems, overall salinity is low to moderate, generally less than 15 wt.% NaCl equivalent, though boiling might cause local areas of high salinity.

(4) Highly saline fluid inclusions, with co-existing gas and fluid-rich inclusions, provide the best evidence that boiling occurred. In orthomagmatic systems, there is widespread evidence of boiling or high-temperature entrapment of supercritical fluid. Often, second or multiple episodes of boiling occurred as fluid pressure fluctuated between lithostatic and hydrostatic. These rapid changes in hydraulic pressure seem to have been caused by throttling and repeated self-sealing and refracturing of the rocks. In the convective systems, boiling appears to have been local and of limited duration.

(5) In orthomagmatic systems, fluid temperatures range from magmatic down to 400°C; seemingly, high temperatures persisted for a protracted period of time. In convective systems, heat transfer efficiency is greater, and although temperatures briefly reach 450°C or more, they quickly drop to about 250°C. These lower temperatures are evidently maintained for a considerable length of time.

(6) The following alteration patterns emerge. Orthomagmatic systems are dominated by potassic and propylitic alteration, with narrow zones of phyllic alteration in the area of interaction between magmatic and meteoric fluids. As a consequence, pervasive alteration and mineralization form a series of shells around the core of the intrusion. Convective systems are dominated by phyllic alteration, with peripheral propylitic alteration around restricted, locally obliterated, potassic core zones. Alteration and mineralization are both pervasive and fracture controlled.

(7) Sulphide distribution patterns can be identical in the two settings, however, there is a fundamental diference in the sources of ore constituents. In the orthomagmatic system, metals and sulphur are derived from the magma and are concentrated in residual fluids. In the

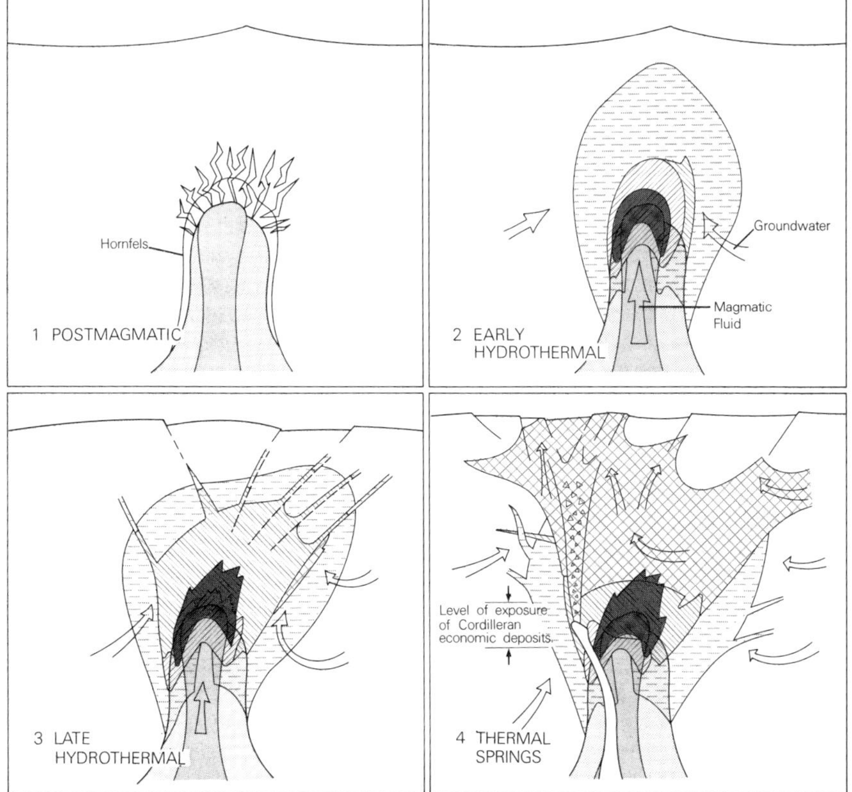

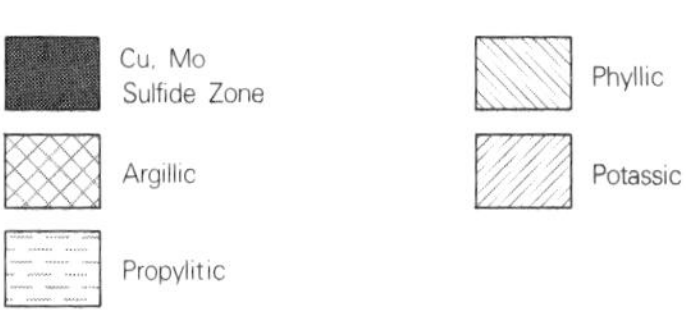

STAGED ALTERATION MODEL
After Gustafson & Hunt 1975

Figure 9 *Model showing four sequential stages of alteration mineralization. For explanation, see text. (After Gustafson and Hunt, 1975).*

convective system, metals and sulphur are scavenged from the enclosing rocks by convecting heated ground waters.

A few porphyry copper deposits, for example Granisle and Bell in British Columbia (Wilson *et al.*, 1980), closely resemble the orthomagmatic and convective end members, respectively. Most deposits combined elements of both, commonly with evidence for early orthomagmatic and later convective alteration/mineralization. Problems in identifying all the events and their sequence arise because younger, superimposed episodes can mask older ones completely. These complications make static, end-member models, such as are shown in Figure 8, inadequate to describe actual porphyry systems; staged models which incorporate changes with time are more realistic. In Figure 9, the four *main* stages of mineralization/alteration that typically occur in porphyry copper systems are illustrated. Figure 9 is patterned after the description of the El Salvador deposit in Chile by Gustafson and Hunt (1975).

Intrusion of the magma causes thermal metamorphism due to conductive heat flow (Figure 9, stage 1). This produces biotite hornfels, often referred to as early developed biotite (EDB). Later, upward and outward flow of fluids increases the rate of cooling of the pluton, and causes additional fracturing and attendant mass transfer (Figure 9, stage 2). Heating during stage 1, abetted by the heat and mass transfer of stage 2, leads to initial mineralization and produces a potassic core, peripheral propylitic alteration zones, and possibly minor phyllic zones. Stages 1 and 2 constitute the orthomagmatic end-member model outlined above. At least in the potassic zone, stage 2 hydrothermal processes take place at high temperature and lithostatic pressure. For brief periods, however, fluid pressures may exceed lithostatic pressure and become an important control in fracture and breccia propagation. Alteration reactions in this stage trend toward pervasive prograde chemical equilibrium in the altered rocks. Alteration and mineralization processes may terminate at the end of stage 2, but generally continue and evolve.

All three types of porphyry copper deposits in the Canadian Cordilleran display elements of stages 1 and 2. In some, such as the classic-type Granisle deposit (Wilson *et al.*, 1980) and several volcanic-type deposits, such as Copper Mountain, Stikine Copper and Schaft Creek, much of the ore was deposited during these orthomagmatic stages. In plutonic deposits, because the country rock is granitic, thermal metamorphic effects (stage 1) are difficult to recognize. Stage 2 potassic alteration is erratically distributed; either it was not originally widespread or it has been overprinted by later alteration. In the plutonic deposits, main-stage mineralization has a predominant orthomagmatic component.

Later alteration and mineralization are controlled by convective hydrothermal circulation involving both magmatic and meteoric fluids, but dominated by the latter (Figure 9, stage 3). Ground water flows toward and through the crackled intrusion, resulting in widespread phyllic overprinting of earlier alteration types. So long as permeability is maintained, pressure is hydrostatic and heat loss is rapid. This cooler, more acidic, hydrothermal regime produces K-feldspar and biotite-destructive alteration. Pervasive alteration may result, or retrograde margins or envelopes develop on veins and fractures as is common in the plutonic deposits of Highland Valley, British Columbia. Extensive remobilization and enrichment of early formed copper sulphides by means of hypogene leaching can take place at this stage (Gustafson and Hunt, 1975; Brimhall, 1979).

As the system cools, hydrothermal activity wanes, and the convective cell begins to collapse inward and downward (stage 4). The result is a relatively low-temperature, dilute-acid hot spring environment that causes argillic overprinting. At the same time, interaction of post-ore porphyry intrusions with the cool ground water may propagate pebble breccia pipes or diatremes. This stage is rare in classic-type Cordilleran deposits but is well developed in at least one volcanic-type deposit (Island Copper) and several plutonic-type deposits (Highland Valley).

Conclusion

The spectrum of characteristics of a porphyry copper deposit reflects the various influences of each of the four main and many transient stages in the evolution of the porphyry hydrothermal system. Not all stages develop fully, nor are all the stages of equal importance. Various factors, such as magma type, volatile content, the number, size, timing and depth of emplacement of mineralizing porphyry plutons, variations in country rock composition and fracturing, all combine to ensure a wide variety of detail. As well, the rate of fluid mixing, density contrasts in the fluids, and pressure and temperature gradients influence the end result. Different depths of erosion alone can produce a wide range in appearances even in the same deposit. The search for porphyry copper deposits, especially buried ones, must be founded on detailed knowledge of their tectonic setting, geology, alteration patterns, and geochemistry. Sophisticated genetic models incorporating these features will be used to design and control future exploration programs.

Acknowledgements

This paper was solicited by J.M. Allen on behalf of the Mineral Deposits Division of the Geological Association of Canada and published in 1980. We thank our colleagues in the Geological Division, A. Sutherland Brown, N.C. Carter, T. Höy, and V.A. Preto for their discussion and comments on the manuscript and are grateful for the editorial scrutiny of Richard Butler. We thank R.W. Hodder and T.J. Bottrill for their thoughtful reviews of the manuscript.

Draughting was done by J.P. St. Gelais and the manuscripts typed by D. Bulinckx and J. Patenaude. Permission to publish was granted by the Chief Geologist, Mineral Resources Branch, British Columbia Ministry of Energy, Mines and Petroleum Resources.

References

(* denotes classic or key paper)

Banks, N.A. and Page, N.J., 1977, Some observations that bear on the origin of porphyry copper deposits: United States Geological Survey, Open-File Report 77-127, 14 p.

Beane, R.E. and Titley, S.R., 1981, Porphyry copper deposits, Part II. Hydrothermal alteration and mineralization: Economic Geology, 75th Anniversary Volume, p. 235-269.

Brimhall, G.H., Jr., 1979, Lithologic determination of mass transfer mechanisms of multiple stage porphyry copper mineralization at Butte, Montana: Vein formation by hypogene leaching and enrichment of potassium-silicate protore: Economic Geology, v. 74, p. 556-589.

*Bryner, L., 1961, Breccia and pebble columns associated with epigenetic ore deposits: Economic Geology, v. 56, p. 488-508.

Burnham, C.W., 1967, Hydrothermal fluids at the magmatic stage, *in* Barnes, H.L., ed., Geochemistry of Hydrothermal Ore Deposits, First Edition: Wiley, New York, p. 34-76.

Cathles, L.M., 1977, An analysis of the cooling of intrusives by groundwater convection which includes boiling: Economic Geology, v. 72, p. 804-826.

Cornwall, H.R. and Banks, N.G., 1977, Igneous rocks and copper mineralization in the Porphyry Copper District, Arizona: United States Geological Survey, Open-File Report 77-255, 10 p.

Cox, D.P., 1982, A generalized empirical model for porphyry copper deposits, *in* Erikson, R.L., ed., Characteristics of Mineral Occurrences: United States Geological Survey, Open-File Report 82-795, p. 27-32.

Creasey, S.C., 1977, Intrusives associated with porphyry copper deposits, *in* Relations between Granitoids and Associated Ore Deposits of the Circum-Pacific Region: Geological Society of Malaysia, Bulletin 9, p. 51-66.

Drummond, A.D., Sutherland Brown, A., Young, R.J. and Tennant, S.J., 1976, Gibraltar — regional metamorphism, mineralization, hydrothermal alteration and structural development, *in* Sutherland Brown, A., ed., Porphyry Deposits of the Canadian Cordillera: Canadian Institute of Mining and Metallurgy, Special Volume 15, p. 195-205.

Eggers, A.J., 1979, Large-scale circular features in North Westland and West Nelson, New Zealand: Possible structural control for porphyry molybdenum-copper mineralization: Economic Geology, v. 74, p. 1490-1494.

Emmons, W.J., 1927, Relations of the disseminated copper ores in porphyry to igneous intrusives: American Institute of Mining and Metallurgical Engineers, Transactions, v. 75, p. 797-815.

Feiss, P.G., 1978, Magmatic sources of copper in porphyry copper deposits: Economic Geology, v. 73, p. 397-404.

Gaál, G. and Isohanni, M., 1979, Characteristics of igneous intrusions and various wall rocks in some Precambrian porphyry copper-molybdenum deposits in Pohjinamaa, Finland: Economic Geology, v. 74, p. 1198-1210.

Griffis, A.T., 1979, An Archean "porphyry-type" disseminated copper deposit, Timmins, Ontario — A discussion: Economic Geology, v. 74, p. 695-696.

*Guilbert, J.M. and Lowell, J.D., 1974, Variations in zoning patterns in porphyry ore deposits: Canadian Institute of Mining and Metallurgy, Bulletin, v. 67, p. 99-109.

Gustafson, L.B., 1978, Some major factors of porphyry copper genesis: Economic Geology, v. 73, p. 600-607.

*Gustafson, L.B. and Hunt, J.P., 1975, The porphyry copper deposit at El Salvador, Chile: Economic Geology, v. 70, p. 857-912.

*Gustafson, L.B. and Titley, S.R., 1978, Porphyry copper deposits of the southwestern Pacific Islands and Australia: Economic Geology, v. 73, p. 597-985.

*Hemley, J.J. and Jones, W.R., 1964, Chemical aspects of hydrothermal alteration with emphasis on hydrogen metasomatism: Economic Geology, v. 59, p.538-569.

Henley, R.W. and McNabb, A., 1978, Magmatic vapour plumes and ground-water interaction: Economic Geology, v. 73, p. 1-20.

Holland, H.D., 1972, Granites, solutions, and base metal deposits: Economic Geology, v. 67, p. 281-301.

Hollister, V.F., 1974, Regional characteristics of porphyry copper deposits of South America: American Institute of Mining and Metallurgical Engineers, Transactions, v. 255, p. 45-53.

Hollister, V.F., Potter, R.R. and Barker, A.L., 1974, Porphyry-type deposits of the Appalachian Orogen: Economic Geology, v. 69, p. 618-630.

Jones, M.B., 1975, Hydrothermal alteration and mineralization of the Valley copper deposit, Highland Valley, B.C., Unpublished Ph.D. Thesis, Oregon State University.

Kents, P., 1961, Brief outline of a possible origin of copper porphyry breccias: Economic Geology, v. 56, p. 1465-1471.

*Kents, P., 1964, Special breccias associated with hydrothermal developments in the Andes: Economic Geology, v. 59, p. 1551-1563.

Kesler, S.E., Jones, L.M. and Walker, R.L., 1975, Intrusive rocks associated with porphyry copper mineralization in island arc areas: Economic Geology, v. 70, p. 515-526.

Kirkham, R.V., 1971, Intermineral intrusions and their bearing on the origin of porphyry copper and molybdenum deposits: Economic Geology, v. 66, p. 1244-1250.

Kirkham, R.V., 1972, Geology of Copper and Molybdenum Deposits: Geological Survey of Canada, Paper 72-1A, p. 82-87.

LeBel, L., 1979a, Christallochimie des Micas et son Application à la Metallogenie des Porphyres Cuprifères: Résumé des Principaux Resultats Scientifiques et Techniques du Service Géologique National pour 1978: BRGM, Paris, France, p. 88-89.

LeBel, L., 1979b, Micas Magmatiques et Hydrothermaux dans l'Environnement du Porphyre Cuprifère de Cerro Verde — Santa Rosa, Pérou: Bulletin de Minéralogie, v. 102. p. 35-41.

Lowell, J.D., 1974, Regional characteristics of porphyry copper deposits of the southwest: Economic Geology, v. 69, p. 601-617.

*Lowell, J.D. and Guilbert, J.M., 1970, Lateral and vertical alteration-mineralization zoning in porphyry ore deposits: Economic Geology, v. 65, p. 373-408.

Mason, D.R. and Feiss, P.G., 1979, On the relationship between whole rock chemistry and porphyry copper mineralization: Economic Geology, v. 74, p. 1506-1510.

McMillan, W.J., 1985, Geology and ore deposits of the Highland Valley camp: Geological Association of Canada, Mineral Deposits Division, Field Guide and Reference Manual No. 1, 121 p.

Meyer, C. and Hemley, J.J., 1967, Wall rock alteration, *in* Barnes, H.L., ed., Geochemistry of Hydrothermal Ore Deposits, First Edition: Wiley, New York, p. 166-235.

Mitchell, A.H.G. and Garson, M.S., 1972, Relationship of porphyry copper and Circum-Pacific tin deposits to Paleo Benioff Zones: Institution of Mining and Metallurgy, Transactions, v. 81. p. B10-B25.

*Nash, J.T., 1976, Fluid-inclusion petrology — data from porphyry copper deposits and applications to exploration: United States Geological Survey, Professional Paper 907-D, 16 p.

Ney, C.S., Cathro, R.J., Panteleyev, A. and Rotherham, D.C., 1976, Supergene copper mineralization, *in* Sutherland Brown, A., ed., Porphyry Deposits of the Canadian Cordillera: Canadian Institute of Mining and Metallurgy, Special Volume 15, p. 72-78.

Noble, J.A., 1970, Metal provinces of the western United States: Geological Society of America, Bulletin, v. 81, p. 1607-1624.

Norton, D., 1978, Source lines, source regions and pathlines for fluids in hydrothermal systems related to cooling plutons: Economic Geology, v. 73, p. 21-28.

Norton, D. and Knapp, R., 1977, Transport phenomena in hydrothermal systems: Nature of Porosity: American Journal of Science, v. 277, p. 913-936.

Page, R.W., 1975, Geochronology of Late Tertiary and Quaternary mineralized intrusive porphyries in the Star Mountains of Papua, New Guinea and Irian Jaya: Economic Geology, v. 70, p. 928-936.

Page, R.W. and McDougall, I., 1972, Ages of mineralization of gold and porphyry copper deposits in the New Guinea Highlands: Economic Geology, v. 67, p. 1034-1048.

*Perry, V.D., 1961, The significance of mineralized breccia pipes: Mining Engineering, v. 13, p. 367-376.

Phillips, W.J., 1973, Mechanical effects of retrograde boiling and its importance in the formation of some porphyry ore deposits: Institution of Mining and Metallurgy, Transactions, v. 82, p. B90-B98.

Roedder, E., 1971, Fluid inclusion studies on the porphyry-type ore deposits at Bingham, Utah, Butte, Montana and Climax, Colorado: Economic Geology, v. 66, p. 98-120.

*Roedder, E., 1977, Fluid inclusions as tools in mineral exploration: Economic Geology, v. 72, p. 503-525.

Rogers, J.J.W., Burchfiel, B.C., Abbott, E.W., Anepohl, J.K., Ewing, A.H., Koehnken, P.J., Novitsky-Evans, J.M. and Talukdars, S.C., 1974, Paleozoic and Lower Mesozoic Volcanism and Continental Growth in the Western United States: Geological Society of America, Bulletin, v. 85, p. 1913-1924.

Sangster, D.F., 1979, Plate tectonics and mineral deposits: Geoscience Canada, v. 6, p. 185-188.

Schmitt, H.A., 1966, The porphyry copper deposits in their regional settting, *in* Titley S.R. and Hicks, C.L., eds., Geology of the Porphyry Copper Deposits, Southwestern North America: University of Arizona Press, Tucson, Arizona, p. 17-34.

Seraphim, R.W. and Hollister, V.F., 1976, Structural settings, *in* Sutherland Brown, A., ed., Porphyry Deposits of the Canadian Cordillera: Canadian Institute of Mining and Metallurgy, Special Volume 15, p. 30-43.

Sheppard, S.M.F., 1977, Identification of the origin of ore-forming solutions by the use of stable isotopes, *in* Gass, I.G., ed., Volcanic Processes in Ore Genesis: Institution of Mining and Metallurgy, Special Publication 7, p. 25-41.

Sheppard, S.M.F., Nielsen, R.L. and Taylor, H.P., Jr., 1971, Hydrogen and oxygen isotope ratios in minerals from porphyry copper deposits: Economic Geology, v. 66, p. 515-542.

*Sillitoe, R.H., 1972, A plate tectonic model for the origin of porphyry copper deposits: Economic Geology, v. 67, p. 184-197.

Sillitoe, R.H., 1973, The tops and bottoms of porphyry copper deposits: Economic Geology, v. 68, p. 799-815.

Sillitoe, R.H., 1975, Subduction and porphyry copper deposits in southwestern North America - a reply to recent objections: Economic Geology, v. 70, p. 1474-1477.

Sillitoe, R.H., 1977, Andean mineralization: A model for the metallogeny of convergent plate margins, *in* Strong, D.F., ed., Metallogeny and Plate Tectonics: Geological Association of Canada, Special Paper 14, p. 59-100.

Sutherland Brown, A., 1976, Morphology and classification, *in* Sutherland Brown, A., ed., Porphyry Deposits of the Canadian Cordillera: Canadian Institute of Mining and Metallurgy, Special Volume 15, p. 44-51.

*Sutherland Brown, A., 1976, ed., Porphyry Deposits of the Canadian Cordillera: Canadian Institute of Mining and Metallurgy, Special Volume 15, 510 p.
A survey of all known porphyry occurrences in British Columbia and Yukon. This includes reviews of world and Cordilleran distribution, structural setting, classification, hypogene mineralization/alteration, supergene mineralization and age; also four regional reviews, descriptions of 33 individual deposits, and a summary of recent research.

Sutulov, Alexander, 1974, Copper porphyries: University of Utah Printing Services, Salt Lake City, Utah, 200 p.
A synopsis of world-wide porphyry copper deposits.

Titley, S.R., 1982, Advances in Geology of the Porphyry Copper Deposits Southwestern North America: The University of Arizona Press, Tucson, Arizona, 560 p.
Updated version of Titley and Hicks (1966).

Titley, S.R. and Beane, R.E., 1981, Porphyry copper deposits, Part I. Geologic settings, petrology and tectogenesis: Economic Geology, 75th Anniversary Volume, p. 214-234.

*Titley, S.R. and Hicks, C.L., 1966, eds., Geology of the Porphyry Copper Deposits, southwestern North America: University of Arizona Press, Tucson, Arizona, 287 p.

White, D.E., 1974, Diverse origins of hydrothermal ore fluids: Economic Geology, v. 69, p. 954-973.

*White, D.E., Muffler, L.B.P. and Truesdell, A.H., 1971, Vapor-dominated hydrothermal systems, compared with hot-water systems: Economic Geology, v. 66, p. 75-97.

*Whitney, J.A., 1975, Vapor generation in a quartz monzonite magma: a synthetic model with application to porphyry copper deposits: Economic Geology, v. 70, p. 346-359.

Wilson, J.W.J., Kesler, S.E., Cloke, P.L. and Kelly, W.C., 1980, Fluid inclusion geochemistry of the Granisle and Bell Porphyry Copper Deposits, British Columbia: Economic Geology, v. 75, p. 45-61.

Accepted, as revised, 17 March 1980
Originally published in
Geoscience Canada v. 7 Number 2
(June 1980)
Revised August 1985

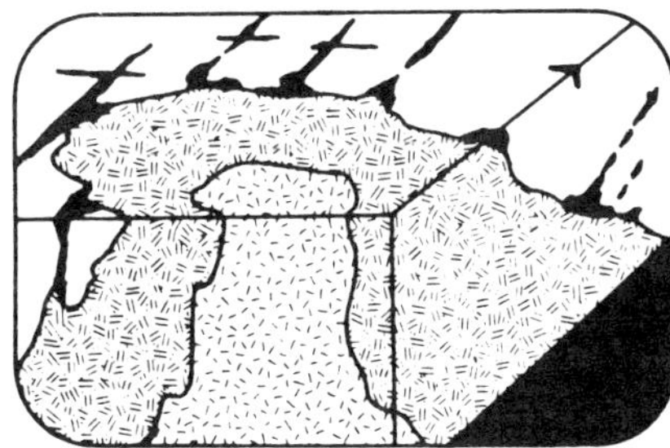

A Model for Granophile Mineral Deposits

D.F. Strong
Department of Earth Sciences
Memorial University of Newfoundland
St. John's, Newfoundland A1B 3X5

Introduction

The aim of this paper is to outline a petrological-geochemical framework or "model" in which to consider mineral deposits that are associated with granitoid rocks. (I use the term "granitoid" rocks as only a textural term for the broad group of granite-textured or phaneritic intrusive igneous rocks which might have compositions from diorite to granite.) Although tin deposits are emphasized, the model should apply equally well to deposits of many other elements that are concentrated by the same processes. The paper does not aim to provide either detailed information on specific deposits nor any broad-scale metallogenic assessment, since this has been provided in a number of recent works (*e.g.*, Ahlfeld and Schneider-Scherbina, 1964; Hutchinson and Taylor, 1978; Hosking, 1968; Ishihara and Takenouchi, 1980; Gass, 1977; Stemprok *et al.*, 1978; Strong, 1976; Taylor, 1979).

Mineral deposits associated with granitoid rocks may be conveniently grouped into two broad categories on the basis of a wide range of geological and geochemical features, some of which are summarized in Figure 1. Those of the first category, *porphyry-type deposits*, have been described in a previous paper in this series (McMillan and Panteleyev, 1980, 1988-this volume), and are characterized by Cu and Mo mineralization, in granitoid rocks of mainly intermediate composition, which were emplaced at shallow depths. The second category I term *granophile* deposits as they are typically hosted by quartz-rich leucocratic granitoids enriched in the so-called "granophile" elements. I would subdivide these granophile elements into three groups, *i.e.*, the large, highly charged cations Sn^{4+}, W^{6+}, U^{4+}, Mo^{6+}, the small, variably charged cations Be^{2+}, B^{3+}, Li^{+} and P^{5+}, and the anions or anionic complexes CO_3^{2-}, Cl^-, F^-, which might be readily recalled under the respective acronyms SWUM, BEBLIP and CCF. The SWUM group contains the economically important elements and may have others (*e.g.*, Nb, Ta, Bi, Ag) concentrated with them. The BEBLIP group are concentrated in the mineral assemblages characteristic of greisens, produce geochemical anomalies useful in exploration, and may have associated concentrations of Na, Rb, Cs, REE, *etc*. The CCF group, possibly including sulphide and hydroxide complexes, are important in forming soluble complexes for transport of the SWUM and BEBLIP groups. Both the BEBLIP and CCF groups play a critical role in controlling the genesis of magmas and their subsequent solidification behaviour. The SWUM group, on the other hand, is passively controlled by these processes of magma generation and solidification.

Deposits of tin and other granophile elements have been classified using a variety of criteria such as tectonic setting or associated mineral assemblages (see Taylor, 1979), and a reasonably comprehensive and useful approach to classification is that of Varlamoff (1978) which relates deposit characteristics to depth of emplacement. This is shown in Figure 2, modified to include a number of different examples, and to expand the depth range which Varlamoff considered. Needless to say, no single deposit (including the examples given) will exhibit the full range of characteristics of a given category.

Porphyry Cu-Mo deposits, as reviewed by McMillan and Panteleyev (1980, 1988-this volume), are characterized by hydrothermal alteration assemblages which result mainly from variable degrees of leaching by fluids consisting essentially of water. While granophile deposits may also display such effects as kaolinization or sericitization, they

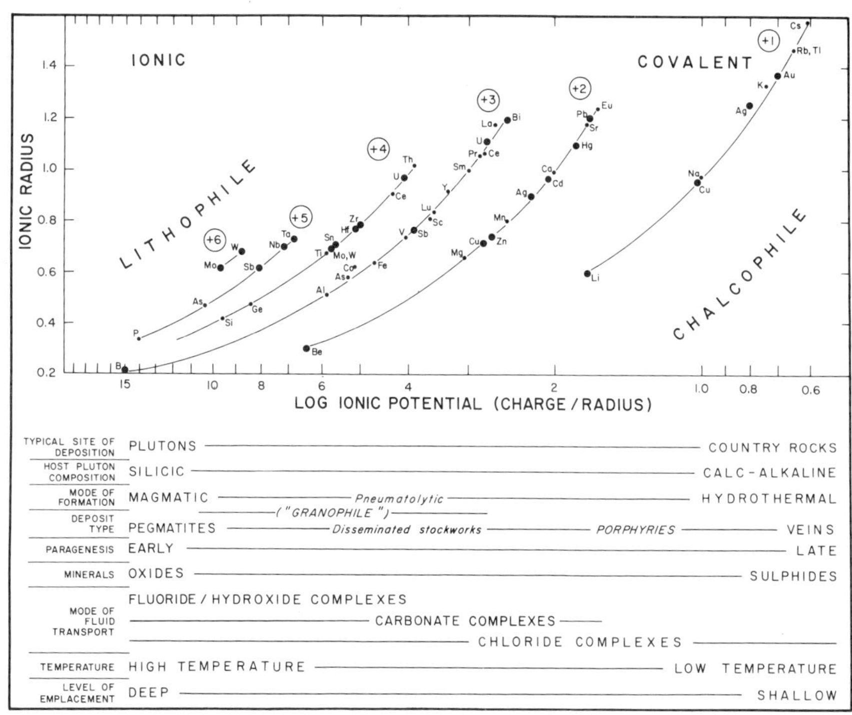

Figure 1 *Schematic outline of the effect of two of the many variables which influence the geochemical behaviour of the elements which make up different types of granitoid mineral deposits. Some geological conditions controlling the formation of such deposits are crudely summarized in the lower half of the diagram, indicating the common patterns of occurrence from the more lithophile or ionically bonded elements on the left to the more chalcophile or covalently bonded elements on the right. (After Strong, 1980).*

are more characteristically associated with greisens. *Greisen* can be described as the secondary mineral assemblage formed by alteration of either the granitic or country rock by the addition of elements such as Li, Be, B, F and Si and H_2O to produce assemblages with varied combinations of white mica (commonly Li-rich), quartz, topaz, tourmaline and fluorite. Greisen alteration patterns may be zoned, as shown in Figure 3, but they tend to be more irregular than those of porphyry Cu-Mo deposits, and may be telescoped, superimposed or asymmetrical. There is often evidence of feldspathization preceding greisenization, and kaolinization following it (*e.g.*, Badham *et al.*, 1976), reflecting changing fluid-rock equilibria with falling temperatures.

Granophile deposits tend to concentrate toward the contact zones of related granitoids, and occur as disseminations or pegmatites in the pluton (endocontact) and as veins and stockworks developed upward or outward from it (exocontact). There is typically a zonation of elements, with Sn, W, As and U passing outward through (U, Ni, Co) to Cu to Pb-Zn-Ag to Fe and Sb sulphides. This pattern, shown in Figure 4, is well established in the Hercynian granites of western Europe, and is perhaps best known for those of Cornwall. Such element zonation is known to occur on a very broad scale and can serve as a useful exploration guide in distinguishing between plutons of high and low mineralization potential (*cf.* Davenport, 1982).

Granitoid Petrogenesis

There is no doubt that water is the single most important compositional variable in controlling both the physical and chemical behaviour of granitoid melts, and that it ultimately determines the characteristics of any related granophile mineral deposits, a theme most elegantly developed by Burham (1967, 1979).

As shown in Figure 5, a range of rock types could melt at comparable temperatures under conditions of water saturation over a range of pressures, and the melt would be essentially granitic. In the absence of free water vapour (*i.e.*, under-saturation), melting would be governed by the breakdown of the hydrous phases, *i.e.*, muscovite, biotite or amphibole, at successively higher

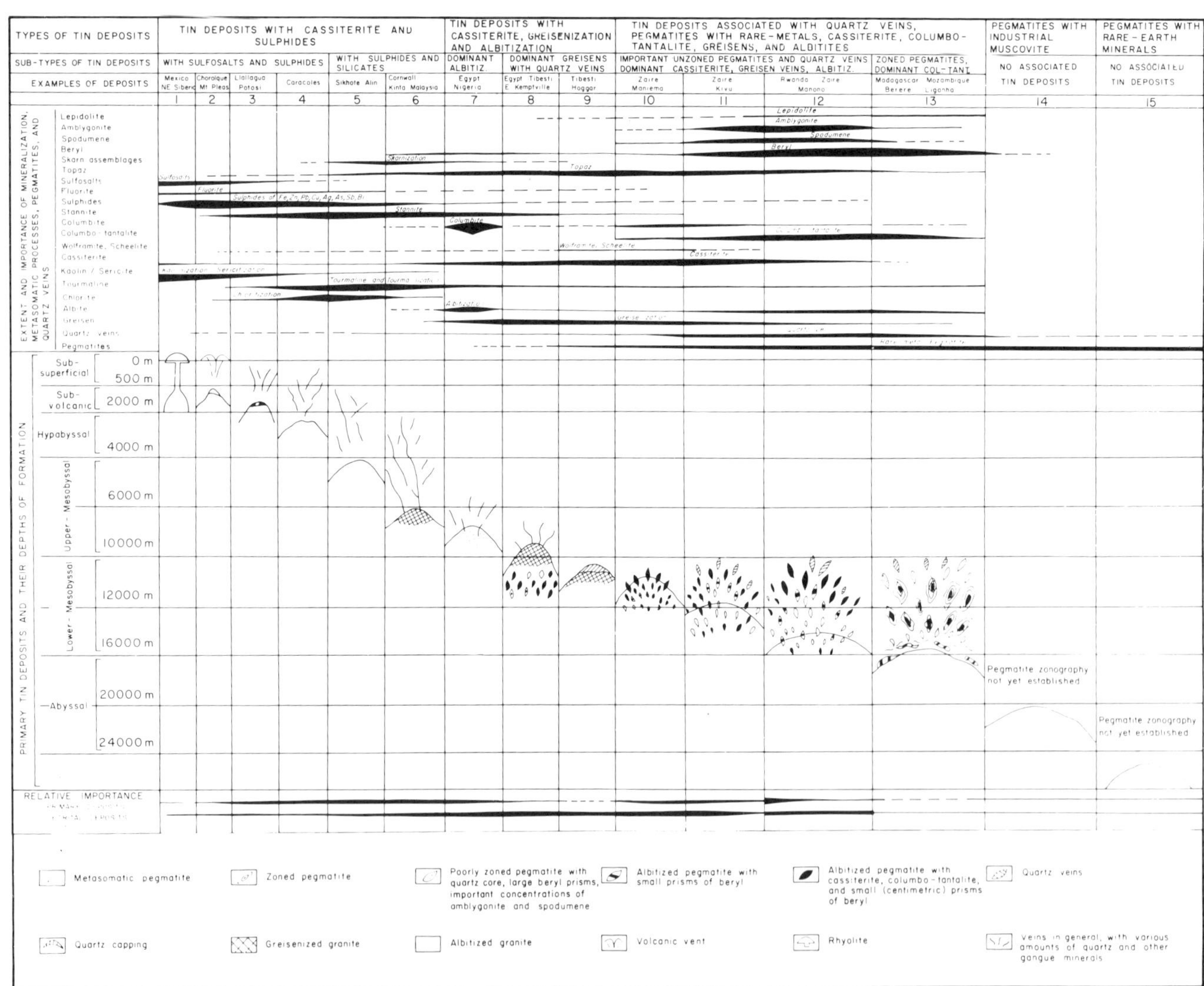

Figure 2 *Classification of tin deposits relating general features of alteration and associated mineral assemblages to approximate depths of formation. (Modified from Varlamoff, 1978).*

temperatures. Three important consequences of source-rock melting being controlled by the breakdown of hydrous minerals are:

(1) The amount of water contained in the melt would be highest for muscovite breakdown and lowest for amphibole, *i.e.*, more "primary" magmatic water would be available for mineralizing processes in muscovite-bearing granitoids or those derived from muscovite-rich source rocks such as shale.

(2) More advanced melting of most source rocks at progressively higher temperatures and pressures would produce progressively less silicic "drier" melts (Brown and Fyfe, 1970).

(3) These drier higher-temperature melts, *i.e.*, biotite- or hornblende-bearing, would be capable of rising and intruding to shallower levels in the crust before they would reach their solidus curves and solidify, whereas primary muscovite-bearing magmas would tend to freeze at depths of about 12-16 km.

These predictions accord with the general observation that high-level, subvolcanic plutons tend to have calc-alkaline intermediate compositions, *i.e.*, those associated with porphyry-type deposits. In contrast, muscovite-bearing leucogranites, which generally host the SWUM deposits, are typical of metamorphic terranes and seldom break the surface. Thus, we have a natural link between the petrogenetic and metallogenetic history of granitoid rocks suggesting two extremes, *i.e.*, large volumes of magmatic water, deeper levels of emplacement, primary muscovite-bearing silicic minimum melts and granophile deposits, *versus* water-poor intermediate composition, amphibole- or biotite-bearing sub-volcanic (and therefore with access to groundwater circulation) intrusions with porphyry-type deposits. These two extremes provide us with a framework in which to focus on the similarities and differences between the range of deposits associated with granitoid rocks.

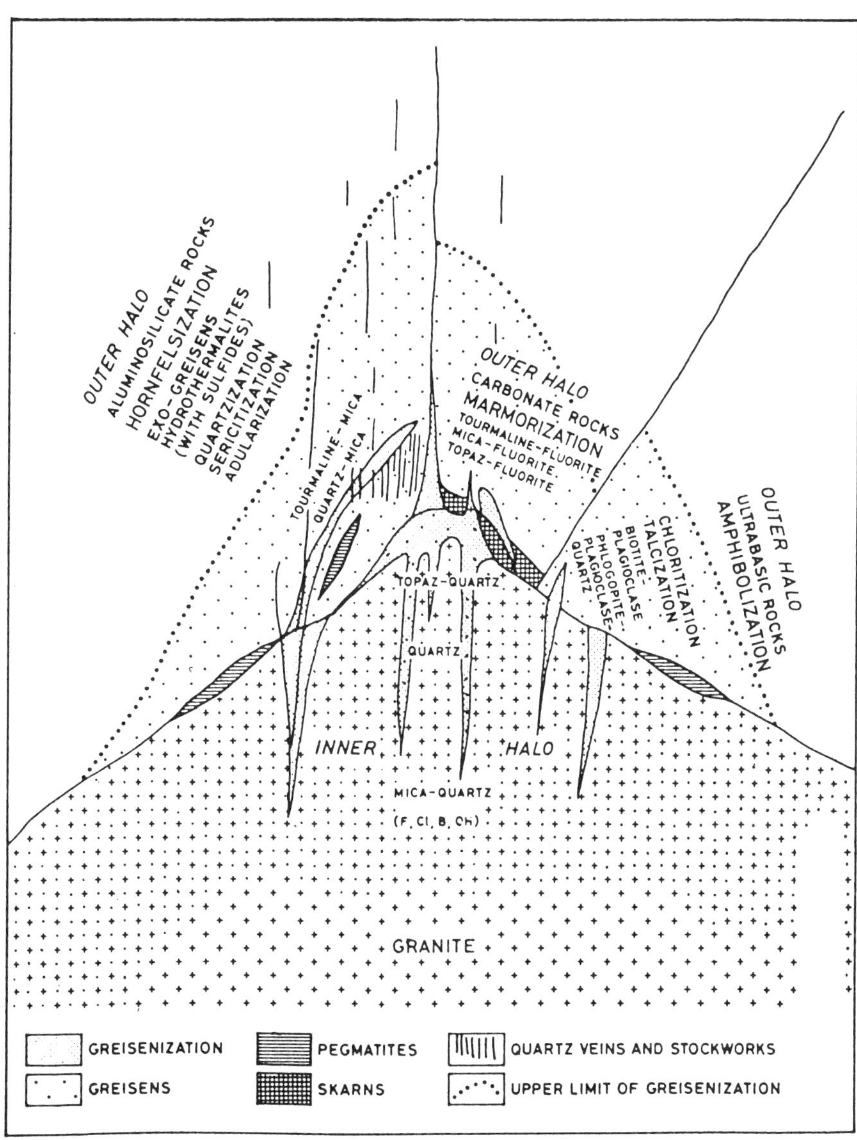

Figure 3 *Alteration-zonation patterns of greisens in aluminosilicate, carbonate and ultramafic environments. (After Taylor, 1979, redrawn from Scherba, 1970a,b).*

Granophile Metallogenesis

Figure 1 shows two of the properties which control geochemical behaviour of the elements of interest, *i.e.*, ionic radius and "ionic potential" (a rough measure of the tendency to form ionic bonds — Taylor, 1965), which illustrates the grouping described above. Thus, the strongly lithophile group tend to form tetrahedral complexes ($(SnO_4)^{4-}$, $(WO_4)^{2-}$, $(NbO_4)^{3-}$, $(TaO_4)^{3-}$, *etc.*) and be strongly concentrated in silicic melts formed through either crystal fractionation or as initial partial melts. Those elements with very small ionic radii (Li, Be) may also form such complexes, but can also be concentrated because of their exclusion from silicate structures because of their small radii. The strongly chalcophile Cu can either enter silicate lattices or form sulphides, and thus tends to be randomly distributed during cooling of a magma or partial melting. The other "lithochalcophile" elements, although comparable in ionic radii to silicate-forming ions such as Ca^{2+}, have strongly covalent bonds with oxygen which exclude them from silicate lattices. They would thus be concentrated in differentiated liquids, unless sulphur activity were such that sulphides of certain elements like Mo can form and cause their early removal from magmas.

When aqueous fluids co-exist with silicate magmas, the behaviour of the elements is critically dependent upon their fluid-melt partition coefficients, which depend upon a range of variables, particularly composition. As can be seen from Figure 6, an important effect of our BEBLIP and CCF element groups is to lower or raise the melting temperature,

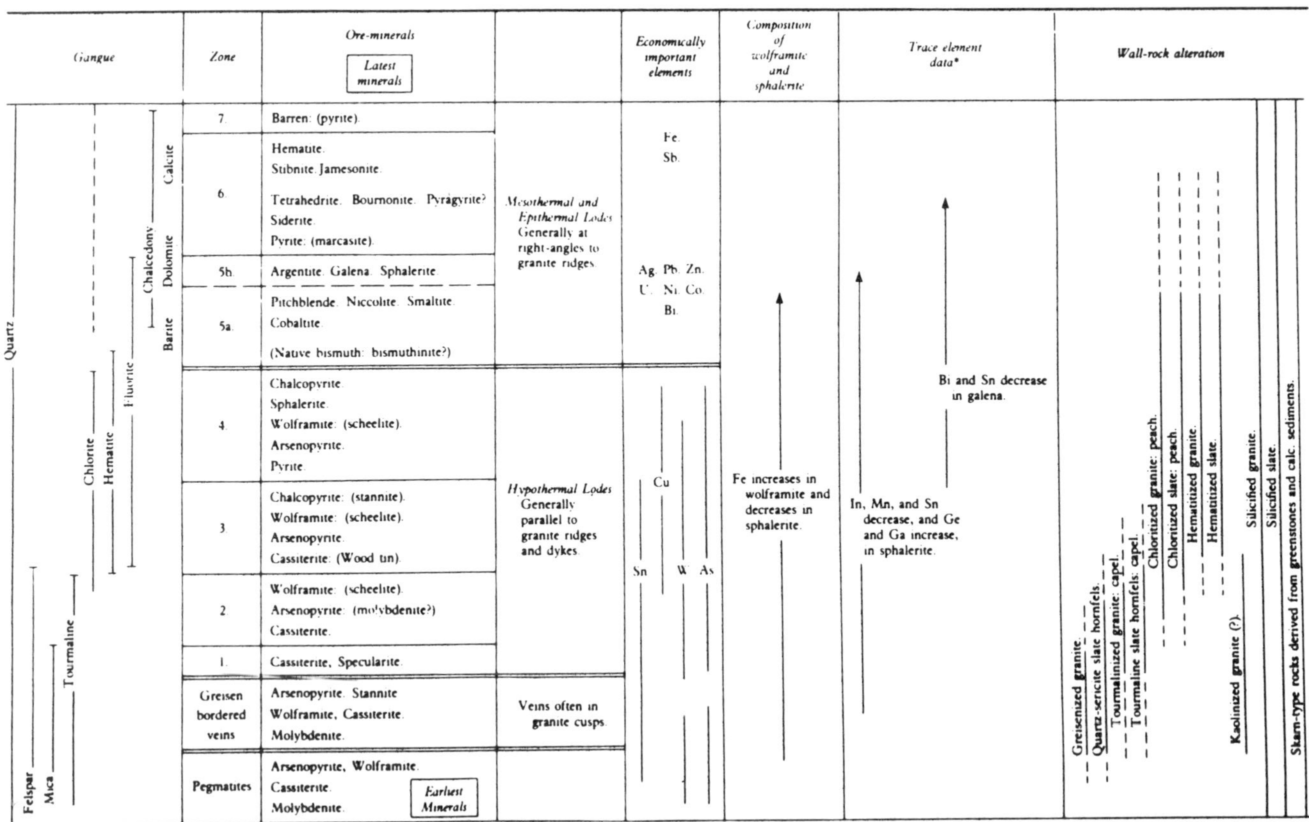

Figure 4 *General order of mineral deposition and wall-rock alteration associated with the mineral deposits of southwest England. (After Hosking, 1964).*

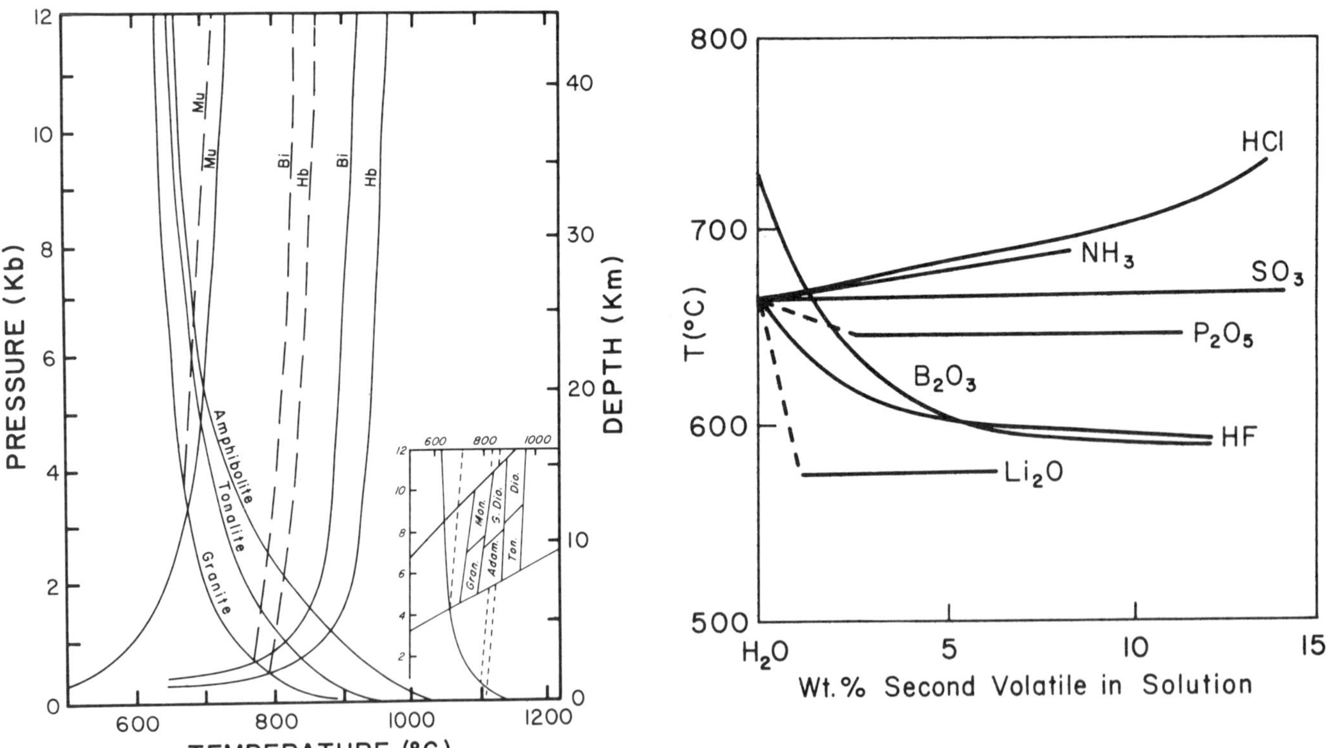

Figure 5 *(left) Pressure-temperature projection of some melting and reaction relations relevant to the genesis of granitoid rocks. The solid lines labelled "granite", "tonalite" and "amphibolite" are the solidi (beginning of melting) of these compositions as shown by Wyllie (1977). The solidus (dashed) and liquidus (solid) curves for muscovite (Mu), biotite (Bi), and hornblende (Hb), are after Burnham (1979). Inset shows the approximate melt compositions formed under different P-T conditions (after Brown and Fyfe, 1970).*

Figure 6 *(right) The effects of different elements or compounds on the melting temperature of granitic compositions at 2.75 kbar. Modified from Luth (1976) to include the data for boron at 1 kbar from Chorlton and Martin (1978).*

reflecting their relative partitioning into the melt or aqueous solution, *e.g.*, F or Cl. Thus, F, P, Li or B would tend to stay with the concentrate in residual melts, lowering their viscosity and freezing temperature and prolonging their differentiation. Alternatively, they may lower the melting temperature of protoliths and promote magma generation in otherwise unfavourable situations. These effects would also permit the intrusion of muscovite-bearing granites to significantly shallower depths, as illustrated in Figure 7. Although the chemical processes are not well understood, there tends to be strong correlation between these "solidus-lowering" elements and the SWUM group in both granites (Taylor, 1979) and silicic volcanic rocks (Hildreth, 1979), suggesting a genetic relation between them. The "solidus-raising" elements (Cl, CO_2) would, on the other hand, tend to concentrate in the aqueous phase and separate from the melt before significant differentiation, and not enhance the concentration of any elements in residual or early melts. Because chloride and possibly carbonate (Higgins, 1980) solutions are transition metal solvents, their separation and circulation would govern the nature of the base metal porphyry-type and other vein-type deposits.

In general, the formation of granitoid mineral deposits depends upon whether the element of interest is dispersed or concentrated during crystallization and partial melting. With dispersal, *e.g.*, Cu in general or Mo in S-bearing melts, a stage of hydrothermal leaching is necessary to form economic concentrations. With magmatic concentration, hydrothermal vein-type deposits may be formed during cooling of a water-saturated melt if the element is preferentially partitioned into the aqueous phase, *e.g.*, the lithochalcophile elements, thus not requiring particularly strong fractionation (*i.e.*, not evolved granitoid compositions). With fractionation of a water-saturated magma, or if the element is preferentially partitioned into the silicate melt as are the SWUM and BEBLIP element groups, highly evolved magmas (and generally high SiO_2, *etc.*) may be necessary to produce economic concentrations.

The Model

An attempt to summarize and integrate the above observations is shown in Figure 7. Figure 7a is an expanded and inverted version of Figure 5, to illustrate possible conditions of melt generation and behaviour, and Figure 7b illustrates the intrusive and mineralization phenomena which might be observed. Melts which formed by the breakdown of hornblende, biotite or muscovite at the arbitrarily chosen depths of regions I, II and III, would have the respective compositions of diorite, granodiorite and granite (Brown and Fyfe, 1970) and water contents of about 2.7, 3.3 and 8.4 wt.%, respectively (Burnham and Ohmoto, 1980). Alternatively, the more silicic melts might form by differentiation from the more mafic, with similar resulting water contents. These melts could ascend to a level where they encounter their respective solidi (*e.g.*, points D, C, and B), with the more mafic (driest) reaching shallower depths. Thus, muscovite-bearing melts would solidify at about 4 kb (12-16 km), whereas hornblende- and biotite-bearing melts would rise to very close to the surface. If the melts were sufficiently enriched in the BEBLIP group or F, either by differentiation or during partial melting, the solidus would be significantly depressed, allowing much shallower ascent, say to point A.

The line XYZ shows the H_2O-NaCl critical curve (Sourirajan and Kennedy, 1962), which intersects the water-saturated granite solidus at about 700°C and 1250 bars (point Z), and the BEBLIP-saturated solidus at about 650°C and 1100 bars (point Y). Assuming that XYZ applies for more complex aqueous systems containing the BEBLIP group elements, this line gives an indication of the maximum depth (about 5.5 km) at which the aqueous fluid (which has already boiled off the silicate melt — "resurgent" or "second" boiling) will boil again ("third" boiling).

It is generally assumed that the expansive evolution of gas during resurgent boiling is responsible for brecciation of the intrusion and host rocks associated with high-level porphyry deposits. The "third" boiling would account for the preponderance of highly saline fluids and possibly induce further hydraulic fracturing during hydrothermal alteration. Note that the "third" boiling might occur if XYZ or its equivalent is encountered by either pressure-release (*e.g.*, BEBLIP- or CO_2-rich fluid rising from point A) or temperature-drop (*e.g.*, a Cl-rich fluid cooling from D or C). Although the physical effects of each may be similar in each case, it should be emphasized that the former depends on the prior enrichment of the melt in BEBLIP elements, results in alteration phenomena such as tourmalinization and greisenization, and is associated with SWUM deposits. The latter case results simply from post-magmatic cooling of a Cl-rich fluid which scavenges the intrusive or host rocks, with typical hydrothermal alteration and typical porphyry Cu-Mo deposits. Because of these basic differences, I term the former case "pseudoporphyries" to distinguish them from the true porphyries, recognizing, alas, that there are certainly gradations between these extremes.

Another implication of the model is that granitoid rocks intruded at depths between B and C (Figure 7a) would tend to be low in primary magmatic water, unlike those deeper than B, and would be unlikely to encounter ground-water systems, unlike those shallower than C. Hence, they would have little opportunity for, and low potential for, mineralization. I suggest that these are the large plutons of microcline-megacrystic biotite granite such as those which form most of the large Appalachian batholiths — and unfortunately tend to be barren. Of course, exceptions to this generalization may occur in the cases where such plutons can establish convection systems, *e.g.*, in the rare cases where they intruded deep sedimentary sequences containing formation waters (*cf.* Plant *et al.*, 1980), or where cooling conditions allow efficient concentration of small amounts of magmatic water.

At this point, the opportunity should be taken to point out the close correspondence between Figure 7a and the classic ideas of Fersman (1931), and to revive his very appropriate definitions of common terminology. He suggested that the order of mineralization in a cooling magmatic

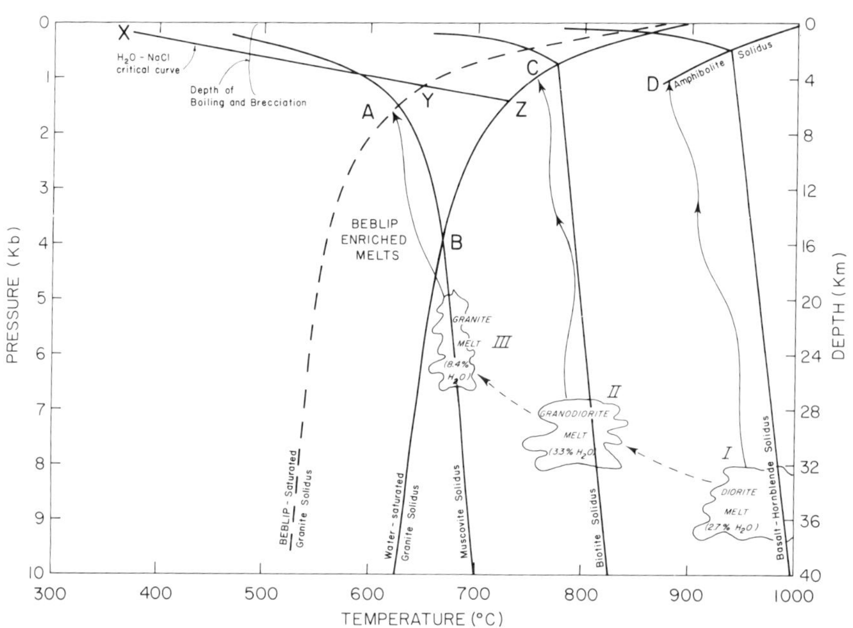

Figure 7 (A) (upper right) Approximate conditions of generation of granitoid melts and mineral deposits. The water-saturated granite solidus and mineral solidi are taken from Figure 5 or Burnham and Ohmoto (1980). The "BEBLIP-saturated granite solidus" is interferred from Figure 6, and the H_2O-NaCl critical curve from Sourirajan and Kennedy (1962). The compositions of melts forming at the arbitrarily chosen depths of I, II and III are inferred from the inset in Figure 5 (Brown and Fyfe, 1970), and the minimum water contents from Burnham and Ohmoto (1980). The solid arrows approximate paths of ascent, and dashed arrows represent differentiation of one magma from a more mafic one. See text for further discussion.

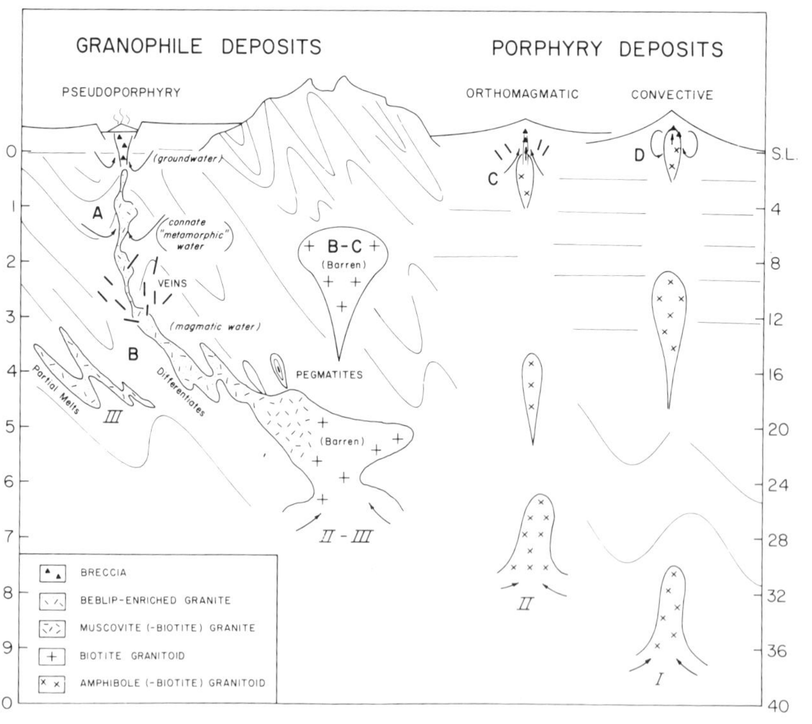

Figure 7 (B) (lower right) Schematic representation of features to be expected with types of deposits associated with different types of granitoid rocks formed as in Figure 7a. The more mafic calc-alkaline diorite and granodiorite melts formed at I or II could rise to shallow depths and discordantly intrude their volcanic carapace (letters correspond to those of Figure 7a). Low initial water contents in I may require augmenting by ground water to generate a large porphyry-type deposit, e.g., the "convective" variety of McMillan and Panteleyev (1980, 1988-this volume). Larger initial water contents in II may be sufficient to form a large porphyry-type deposit, e.g., the "orthomagmatic" version of McMillan and Panteleyev (1980, 1988-this volume).

Intermediate composition granitoids formed between II and III and intruded between depths B and C (Figure 7a) are thought to represent the biotite-megacrystic granites-monzonites which form large discordant to concordant unmineralized batholiths of orogenic belts. They may be barren because of low initial water contents or lack of interaction with ground water, lack of brecciation, etc., because of intrusion at too great a depth. Similar lithologies may form by differentiation of melts formed at sites I and II.

Muscovite-bearing granites may form as anatectic melts at site III or as differentiates of the above, and may be concordant to discordant. They might be enriched in the BEBLIP group elements, typically in cupolas at the tops of larger plutons, which prolongs their time of crystallization and allows for greater SWUM-enrichment and shallower intrusion than for a normal water-rich muscovite-granite. This shallower intrusion permits second boiling and brecciation, forming "pseudo-porphyries" with many physical features like those of Cu and Mo porphyry deposits. They may, however, form granophile deposits at deeper levels due to high initial H_2O content, or even set up convection cells with connate or metamorphic fluids at intermediate depths (as might biotite granites, but apparently very rarely). Pegmatites tend to be more abundant with these "type III" granites, as do greisens, with variations as illustrated in Figures 2 and 3. Dispersion of fluids and contained elements through the plutons and country rocks might result in veins with the characteristic zonation patterns shown in Figure 4.

system could be described in the following successive stages:

(1) *Magmatic stage,* at which equilibrium is maintained between silicate melt and crystalline phases, *i.e.*, at temperatures above the solidi of Figure 7a.

(2) *Pegmatitic stage,* throughout which the melt, crystalline phases and gas phases co-exist, *e.g.*, along the water-saturated or BEBLIP-saturated solidi of Figure 7a.

(3) *Pneumatolytic stage,* characterized by equilibrium between crystals and supercritical fluid, *i.e.*, between the solidi and line XYZ of Figure 7a.

(4) *Hydrothermal stage,* in which equilibrium is maintained between crystals, aqueous solutions, and aqueous gas, *i.e.*, after the supercritical fluid has intersected XYZ.

Further observations regarding Figure 7b are given in the caption.

It might be appropriate to conclude this discussion with the following statement from Hosking (1968) who wrote:

> "As far as possible the writer has refrained from discussing those questions of relationship which require consideration of the sources of granitic magmas, granitising agents and ore-forming ones, and theories of the nature of these agents and the chemistry of ore genesis. He has adopted this line of action because he believes that a good exploration programme must be based essentially on facts, and that a programme founded largely on theoretical concepts, however sophisticated the latter might appear to be, is little better than one which requires only that a blind-folded person should stick a pin in a map in order to find a tin deposit."

Although there have been a number of recent studies which suggest approaches to tin exploration (*e.g.*, Smith and Turek, 1976; Badham, 1980; Ivanov and Narnov, 1970; Flinter *et al.*, 1972; Tauson and Kozlov, 1973), one cannot go wrong by reading Hosking's numerous "factual" treatments of the search for tin (*e.g.*, 1963a,b,c, 1964, 1968, 1970). Meanwhile, I hope that the preceding more speculative model will provide a step toward understanding tin and other granophile deposits in their geochemical-petrological context, and lead to eventual removal of the blindfold.

Acknowledgements

I am grateful to S.J.P. Burry, W.F. Marsh and C. Neary for assistance in preparation of the manuscript, to R.H. Flood, W.J. McMillan, A. Panteleyev, R.P. Taylor, D.H.C. Wilton and R.F. Cormier for criticism of it, and the Natural Sciences and Engineering Research Council for financial support in the form of an operating grant.

References

Ahlfeld, F.E. and Schneider-Scherbina, A., 1964, Los yacimientos minerales y de hidrocarboros de Bolivia: Bolv. Dep. Nac. Geol. No. 5 (Especial), 338 p.

Badham, J.P.N., 1980, Late magmatic phenomena in the Cornish batholith — useful field guides for tin mineralization: Ussher Society, Proceedings, p. 44-53.

Badham, J.P.N., Stanworth, C.W. and Lindsay, R.P., 1976, Post-emplacement events in the Cornubian batholith: Economic Geology, v. 71, p. 534-539.

Brown, G.C. and Fyfe, W.S., 1970, The production of granitic melts during ultrametamorphism: Contributions to Mineralogy and Petrology, v. 28, p. 310-318.

Burnham, C.W., 1967, Hydrothermal fluids at the magmatic stage, *in* Barnes, H.L., ed., Geochemistry of Hydrothermal Ore Deposits, First Edition: Wiley, New York.

Burnham, C.W., 1979, Magmas and Hydrothermal Fluids, *in* Barnes, H.L., ed., Geochemistry of Hydrothermal Ore Deposits, Second Edition: John Wiley Interscience, New York.

Burnham, C.W. and Ohmoto, H., 1980, Late stage processes of felsic magmatism, *in* Ishihara, S. and Takenouchi, S., eds., Granitic Magmatism and Related Mineralization: Society of Mining Geologists of Japan, Mining Geology, Special Issue No. 8, p. 1-12.

Chorlton, L.B. and Martin, R.F., 1978, The effect of boron on the granite solidus: Canadian Mineralogist, v. 16, p. 239-244.

Davenport, P.H., 1982, The Identification of Mineralized Granitoid Plutons from Ore Element Distribution Patterns in Regional Lake Sediment: Geochemical Data: Canadian Institute of Mining and Metallurgy, Bulletin, v. 75, no. 840, p. 79-90.

Fersman, A.E., 1931, Les pegmatites, leur importance scientifique et pratique: Academy of Sciences, USSR, Leningrad. French translation by J. Thoreau, Louvain, 3 vols., 1951.

Flinter, B.H., Hesp, W.R. and Rigby, D., 1972, Selected geochemical, mineralogical and petrological features of granitoid rocks of the New England Complex, Australia, and their relation to Sn, W, Mo and Cu mineralization: Economic Geology, v. 67, p. 1241-1262.

Gass, I.G., 1977, ed., Volcanic Processes in Ore Genesis: Institution of Mining and Metallurgy and Geological Society of London, Special Publication 7, 188 p.

Higgins, N.C., 1980, Fluid inclusion evidence for the transport of tungsten by carbonate complexes in hydrothermal solutions: Canadian Journal of Earth Sciences, v. 17, p. 823-830.

Hildreth, W., 1979, The Bishop Tuff: Evidence for the origin of compositional zonation in silicic magma chambers, *in* Chapin, C.E. and Elston, W.E., eds., Ash-Flow Tuffs: Geological Society of America, Special Paper 180, p. 43-75.

Hosking, K.F.G., 1963a, The search for tin: Mining Magazine, v. 113, no. 4, p. 261-273.

Hosking, K.F.G., 1963b, The search for tin: Mining Magazine, v. 113, no. 5, p. 368-383.

Hosking, K.F.G., 1963c, The search for tin: Mining Magazine, v. 113, no. 6, p. 449-461.

Hosking, K.F.G., 1964, Permo-Carboniferous and later primary mineralization of Cornwall and southwest Devon, *in* Hosking, K.F.G. and Shrimpton, G.J., eds., Present Views of Some Aspects of the Geology of Cornwall and Devon: Royal Geological Society of Cornwall, 330 p.

Hosking, K.F.G., 1968, The relationship between primary tin deposits and granitic rocks, *in* Technical Conference on Tin 1967: London International Tin Council, p. 269-306.

Hutchinson, C.S. and Taylor, D., 1978, Metallogenesis in SE Asia: Geological Society of London, Journal, v. 135, p. 407-429.

Ishihara, S. and Takenouchi, S., 1980, eds., Granitic Magmatism and Related Mineralization: Society of Mining Geologists of Japan, Mining Geology, Special Issue No. 8, 247 p.

Ivanov, V.S. and Narnov, G.A., 1970, On the behaviour of tin in the granitoid intrusives of northeastern USSR: Geokhimiya, No. 5, p. 601-609.

Luth, W.C., 1976, Granitic Rocks, *in* Bailey, D.K. and MacDonald, I., eds., The Evolution of the Crystalline Rocks: Academic Press, New York, p. 335-417.

McMillan, W.J. and Panteleyev, A., 1980, Porphyry Copper Deposits: Geoscience Canada, v. 7, p. 52-63.

McMillan, W.J. and Panteleyev, A., 1988, Porphyry Copper Deposits, *in* Roberts, R.G. and Sheahan, P.A., eds., Ore Deposit Models: Geological Association of Canada, Geoscience Canada Reprint Series 3p. 45-58.

Plant, J., Brown, G.C., Simpson, P.R. and Smith, R.T., 1980, Signatures of metalliferous granites in the Scottish Caledonides: Institution of Mining and Metallurgy, Transactions, v. 89, p. B182-B197.

Scherba, G.N., 1970a, Greisens: International Geology Review, v. 12, p. 114-151.

Scherba, G.N., 1970b, Greisens: International Geology Review, v. 12, p. 239-254.

Smith, T.E. and Turek, A., 1976, Tin-bearing potential of some Devonian Granitic rocks in S.W. Nova Scotia: Mineralium Deposita, v. 11, p. 234-245.

Sourirajan, S. and Kennedy, G.C., 1962, The system H_2O-NaCl at elevated temperatures and pressures: American Journal of Science, v. 260, p. 115-141.

Stemprok, M., Burnol, L. and Tischendorf, G., 1978, eds., Metallization Associated with Acid Magmatism: Ustredni Ustav Geologicky, Praha, v. I (1974), 410 p.; v. II (1977), 166 p.; and v. III (1978), 446 p.

Strong, D.F., 1976, ed., Metallogeny and Plate Tectonics: Geological Association of Canada, Special Paper 14, 660 p.

Strong, D.F., 1980, Granitoid rocks and associated mineral deposits of eastern Canada and western Europe, *in* Strangway, D.W., ed., The Continental Crust and Its Mineral Deposits: Geological Association of Canada, Special Paper 20, p. 741-769.

Tauson, L.V. and Kozlov, V.D., 1973, Distribution functions and ratios of trace-element concentrations as estimators of the ore-bearing potential of granites, *in* Jones, M.J., ed., Geochemical Exploration, 1971: Institution of Mining and Metallurgy, London, p. 37-44.

Taylor, R.G., 1979, Geology of Tin Deposits: Elsevier, New York, 543 p.

Taylor, S.R., 1965, The application of trace elements data to problems in petrology, *in* Physics and Chemistry of the Earth, v. 6, p. 133-213.

Varlamoff, N., 1978, Classification and spatial-temporal distribution of tin and associated mineral deposits, *in* Stemprok, M., Burnol, L., and Tischendorf, G., eds., Metallization Associated with Acid Magmatism, v. 3: Ustredni Ustav, Geologicky, Praha, p. 139-158.

Wyllie, P.J., 1977, Crustal anatexis: an experimental review: Tectonophysics, v. 43, p. 41-71.

Accepted, as revised, 24 September 1981
Originally published in
Geoscience Canada v. 8 Number 4
(December 1981)

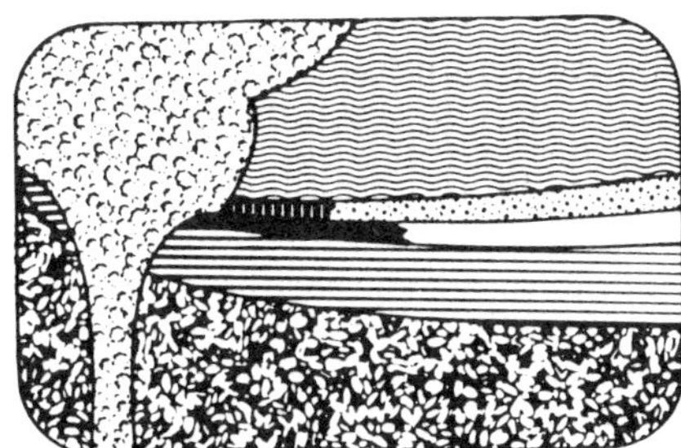

Sedimentary-Type Stratiform Ore Deposits: Some Models and A New Classification

J.M. Morganti
Placer Dome Inc.
Box 350, Suite 3500, IBM Tower
Toronto Dominion Centre
Toronto, Ontario M5K 1N3

Introduction

Many mineral deposits containing zinc, lead, copper, barium and/or precious metals are stratiform in that their general morphology is similar to sedimentary strata (Stanton, 1972, p. 498-503). Some of these deposits occur in predominantly clastic sedimentary sequences where volcanic rocks are not demonstrably related to ore formation. These are herein referred to as sedimentary-type stratiform deposits (Figure 1). Major examples are McArthur River, Sullivan, Meggen, Mufulira and XY. Typically, these deposits consist of stratiform sulphide bodies that internally contain at least some bedded sulphides suggesting that deposition of the sulphides occurred before lithification. Furthermore, many deposits, when specifically grouped, occur in one major sedimentary basin, although individual deposits may occur within separate, second-order or sub-basins. Examples of major basins or first-order basins are found in the Zambian Copperbelt where Mufulira, Muliashi and Chambishi subbasins represent remaining roots of a very extensive basin in which Katanga sediments were deposited (Fleischer *et al.*, 1976). The Kupferschiefer deposits of central Europe are contained within the Permian Zechstein Basin, and all the Howards Pass deposits occur within the Selwyn Basin of the Northern Cordillera.

As a group, the deposits are loosely associated with carbonaceous sedimentary rocks, but individually, some occur within specific associated lithologies. The stratiform nature of the deposits and their clastic sedimentary rock association help separate these deposits from the strata-bound Mississippi Valley type deposits (Anderson, 1978) which are generally associated

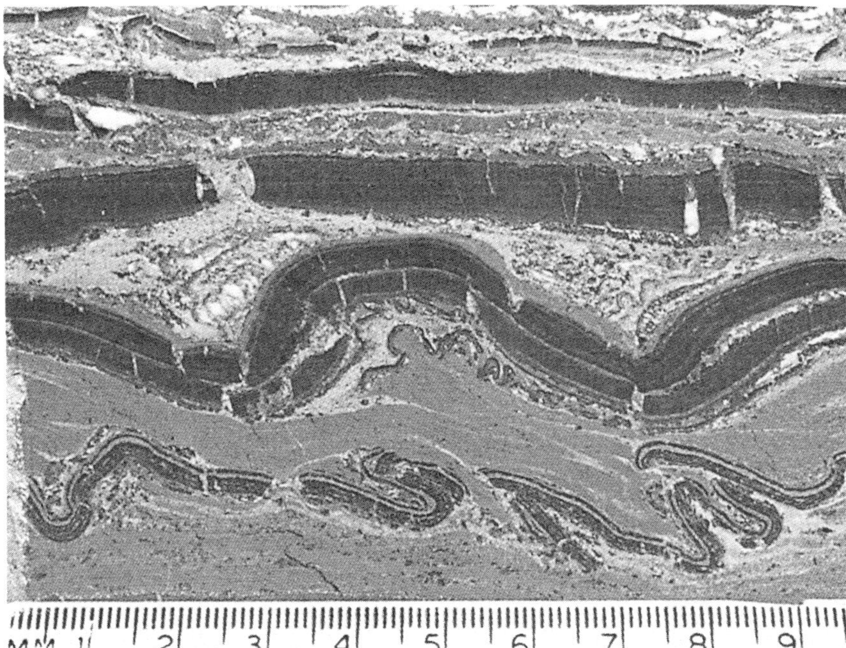

Figure 1 *Specimens of laminated ore from sedimentary-type stratiform ore deposits:*
(a) *laminated sulphide and chert, from Mt. Isa, Australia;*
(b) *laminated argillite with sulphides, from the Sullivan deposit, Canada;*
(c) *laminated sulphide and darker chert, from the Howards Pass deposits, Canada.*

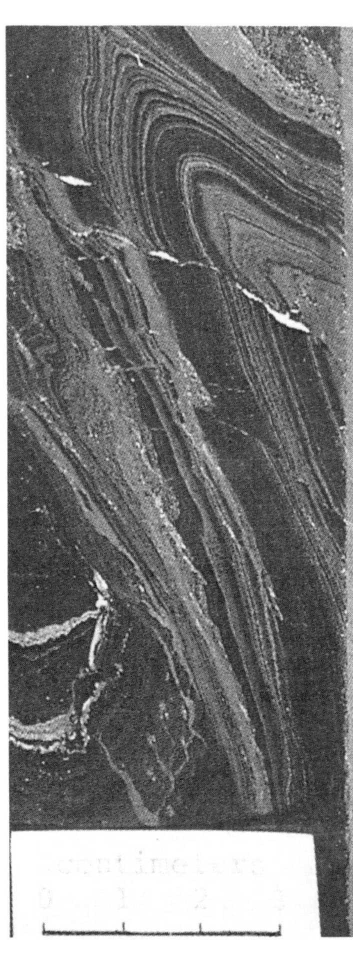

with carbonate rock sequences. The separation of the sedimentary-type deposits and distal volcanogenic stratiform deposits (Plimer, 1978; Large, 1979; Jambor, 1979) is gradational, since few sedimentary basins are totally lacking in a volcanic component (Conybeare, 1979).

The sedimentary-type stratiform deposits constitute substantial reserves of zinc, lead, copper and barium. For example, the McArthur River deposit in Australia contains 200 million tonnes grading 10% Zn, 4% Pb and 45 g/tonne Ag (Murray, 1975). The areal extent of this class of deposits is exemplified by the Zambian Copperbelt where an area 7500 km^2 contains most of the deposits, or the Howards Pass area where potential mineral deposits occur over 130 km of regional strike length. Thicknesses of the deposits are highly variable, ranging from 15 cm at Creta (Argall, 1975) to over 650 m at Mt. Isa (Mathias and Clark, 1975). Locations of deposits discussed in this paper are shown in Figure 2, and represent the major examples of the class of deposits termed sedimentary-type stratiform deposits.

This paper is a review of some of the models proposed for the formation of sedimentary-type stratiform mineral deposits; and proposes a three-fold classification based on the type of sedimentary basin of deposition for the associated sediments. This classification aids in rationalizing the location of these deposits and allows for comparison of deposits. The combination of this classification with the possible behaviour of ore-forming fluids allows for an appreciation of the classes' diversity.

Most sedimentary-type stratiform ore deposits show evidence of similar processes operating during formation. Thus, models proposed by many workers have several features in common. For the purpose of model construction, the generation of an ore deposit involving a hydrous fluid is considered to have four critical aspects: (1) a source for the ore constituents; (2) solution of the ore constituents, at least in part, in a hydrous fluid; (3) migration of the fluids after acquiring their metal content, in directions controlled by pressure and/or chemical differentials; and, (4) formation of the ore deposits by selective precipitation of certain constituents in response to physical and/or chemical changes as the fluids migrate into new environments. Therefore, models must focus on source, solution of elements, migration and precipitation.

The present paper emphasizes the migration and deposition of metals. Four possibilities exist in the sedimentary environment: (1) metal and sulphur are both truly sedimentary; (2) metal is truly sedimentary while sulphur is imported and fixed in the sediment during diagenesis; (3) sulphur is truly sedimentary while metal is imported and fixed in the sediment during diagenesis; or, (4) metal and sulphur are both imported and fixed in the sediment during diagenesis. These four models of formation may operate either individually or in combination to produce any particular deposit.

Tectono-Sedimentary Framework Classification

For discussion and for mineral exploration purposes, sedimentary-type stratiform deposits may be divided into three sub-classes based on gross sedimentation related to major tectonostratigraphic environments: (1) intracratonic basin sulphide deposits in shallow water shales, silt and sandstones associated with carbonates and evaporites; (2) flysch basin sulphide and barite deposits in turbidites and associated lithologies; or (3) platform-marginal basin sulphide deposits in carbonaceous laminites associated with deep-water mudrocks and cherts, outboard of cratons or platforms. Idealized stratigraphic sections associated with the three sub-classes are shown in Figure 3.

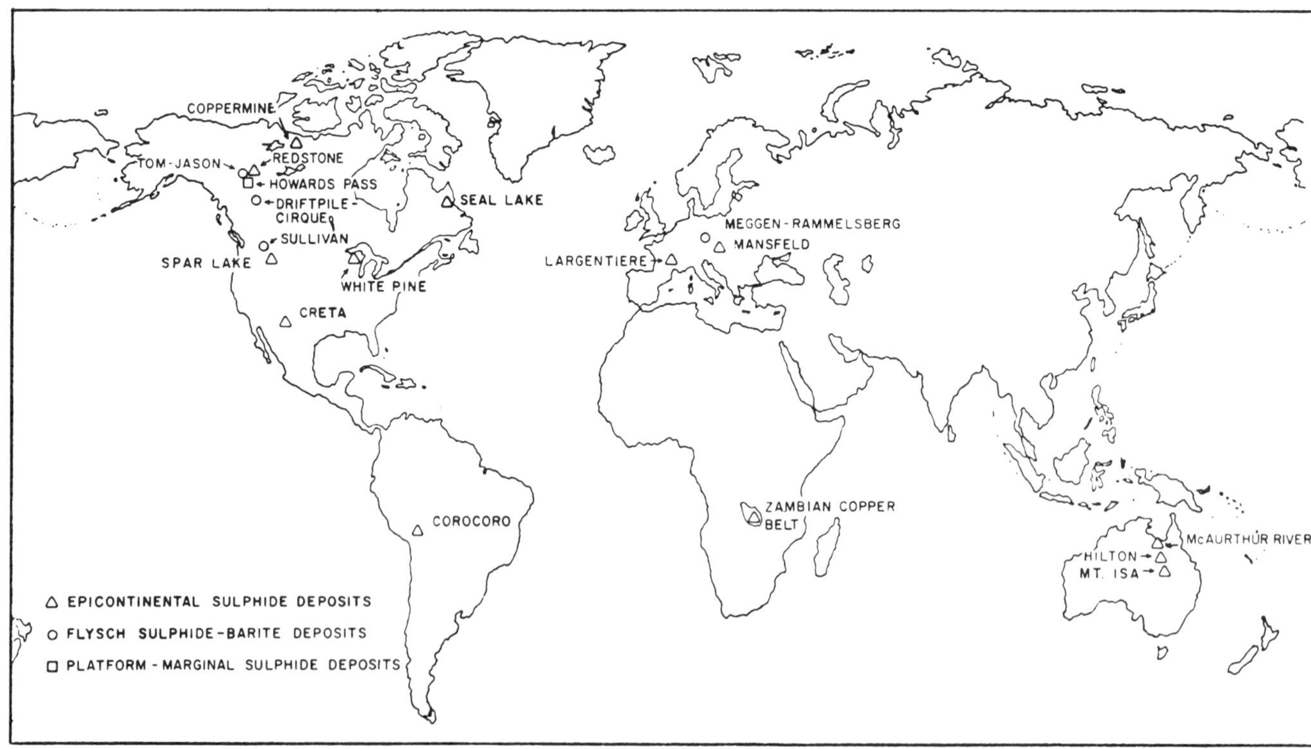

Figure 2 *World location map of sedimentary-type stratiform ore deposits. Symbols identify the types of deposit as proposed in this paper.*

Intracratonic Basin Deposits

Intracratonic basins occur on a continental shelf or within a craton. Facies models (Walker, 1979) included in this type of basin are coarse alluvial, fluvial, deltas, barrier island, shallow marine and supratidal systems. Examples of ore deposits associated with intracratonic basins are those of the Kupferschiefer (intracratonic basin), those of the Zambian Copperbelt (marine marginal intracratonic basin), McArthur River (intracratonic trough) and Largentiere (alluvial fan complex).

Five general characteristics are common to deposits of the intracratonic subclass. (1) The deposits are associated with poorly sorted sandstones, siltstones, silty limestones and dolomites; locally, conglomerate may be associated with or underlie the deposits. Examples of this association are the Copper Harbour Conglomerate underlying the White Pine deposit, (Ensign *et al.*, 1968), the calcareous siltstones and silty limestones of the Copper Cap Formation in the Redstone deposits (Helmstaedt *et al.*, 1979), the footwall conglomerate and sandstone in the Zambian Copperbelt (Annels, 1979). (2) Typically many of the clastics in the associated sequence are coloured red by the presence of hematite. Crosscutting red "Rote Fäule" associated with the Mansfeld deposit (Jung and Knitzschke, 1976), the Copper Harbour Conglomerate and Freda Sandstone near the White Pine deposit (Ensign *et al.*, 1968), and the W-fold shale underlying the McArthur River deposits (Williams, 1978) exemplify the red bed association. (3) Most of the deposits of this sub-class occur in, or very near, mudstones or shales that were reducing in nature. Examples are carbonaceous shales containing the Mansfeld deposit (Rentzsch, 1974), the carbonaceous dolomitic mudrocks, containing the McArthur River deposits (Croxford and Jephcott, 1972), carbonaceous mudrock with pyrite containing the White Pine deposit (Brown, 1971), algal mat-related organic matter in some of the Zambian Copperbelt deposits (Renfro, 1974), and green, possibly methane-reduced zones associated with the Belt-Purcell copper-silver deposits. (4) Gypsum is present or inferred to be associated with the deposits. Examples include, the Zambian Copperbelt (Annels, 1979), Kupferschiefer (Renfro, 1974) and West Texas-Oklahoma areas (Johnson, 1976). (5) Lateral chemical zoning may be evident in many of the deposits. Examples include the Mansfeld deposit where the zoning of copper, lead, zinc and pyrite occurs away from the "Rote Fäule" sediments (Deans, 1948), the Roan (Zambian Copperbelt) where a basinward zoning of chalcocite, bornite, chalcopyrite, pyrite is evident (Garlick, 1961), and McArthur River where copper occurs near the Emu Fault and the (Zn + Pb)/(Cu + Zn + Pb) ratio increases basinward (Wiliams, 1978) (Figure 4).

There are many possible sources of base metals in these deposits. For example, in the Zambian Copperbelt and the Kupferschiefer, it has been proposed that copper was released initially by weathering of basement rocks which, in the case of the Copperbelt, contain porphyry copper-type mineralization (Wakefield, 1978). At White Pine, Coppermine River and Seal Lake, copper may have originated from underlying cupriferous basalts. For example, late Proterozoic mafic flows containing native copper in amygdaloidal and fragmental flow tops underlie the White Pine deposit (Brown, 1974).

The association of red beds and evaporites may be important in the transport of base metals. The formation of strong complexes between cuprous ion (Cu^+) and chloride ion is recorded in the chemical literature. If chloride solutions are responsible for dissolution and/or transport of copper, then deposits formed from these solutions should be associated with sources of chloride such as evaporites (Rose, 1976). In many cases, the porous nature of the red beds may also provide the medium for brine migration.

EPICRATONIC DEPOSITS

SILTY LIMESTONES
LOCALLY EVAPORITES
Cu (Ag,Zn,Pb) CARBONACEOUS SHALE
CONGLOMERATES, SANDSTONES (LOCALLY HEMATITIC AND Cu RICH)
BASALTIC FLOWS (Cu RICH)

FLYSCH DEPOSITS

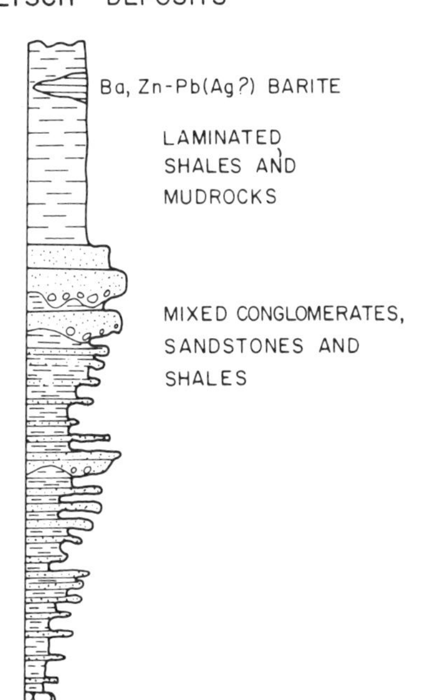

PLATFORM MARGINAL DEPOSITS

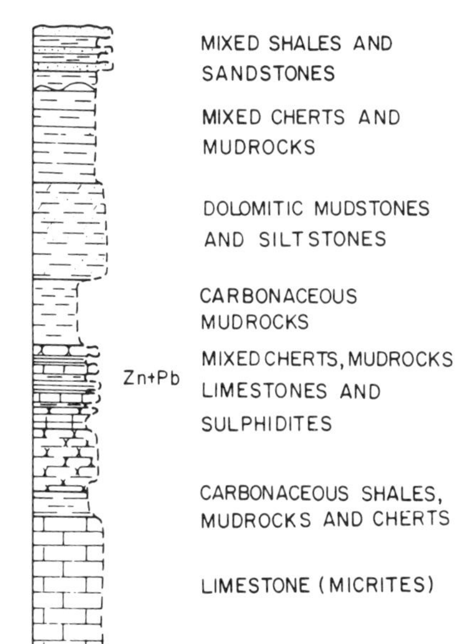

Figure 3 *Comparative generalized stratigraphic columns for the three sub-classes of sedimentary-type stratiform ore deposits. Columns not to scale.*

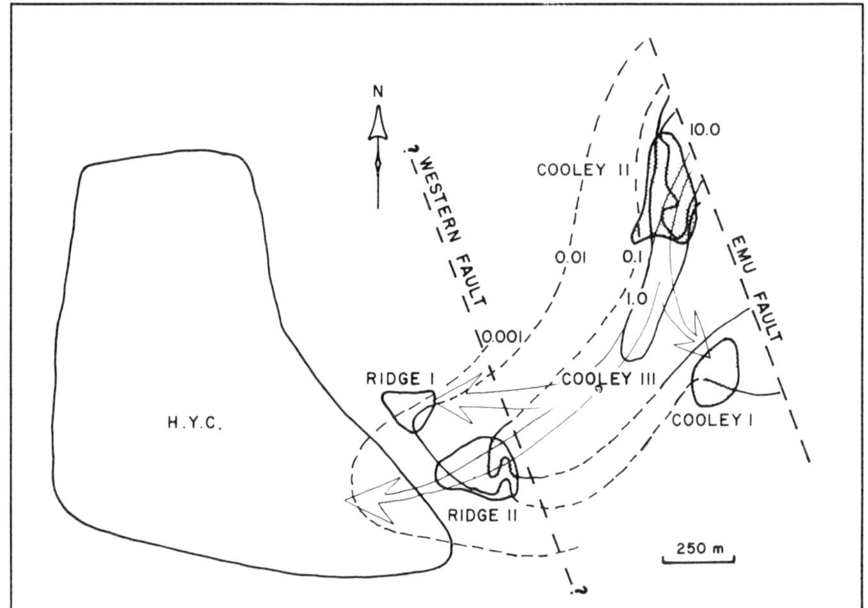

Figure 4 *(upper)* *Generalized plan of the sulphide deposits in the McArthur River area. Numbers refer to Cu/(Zn + Pb) ratios and suggest a zoning of elements related to the Emu Fault. The arrows show the inferred direction of ore fluid migration. Shaded deposits are epigenetic; unshaded deposits are bedded. This suggests that fluids migrated within the carbonates to the east of the western fault and subsequently surfaced near that fault and flowed into the H.Y.C. sub-basin. (Modified from Williams, 1978).*

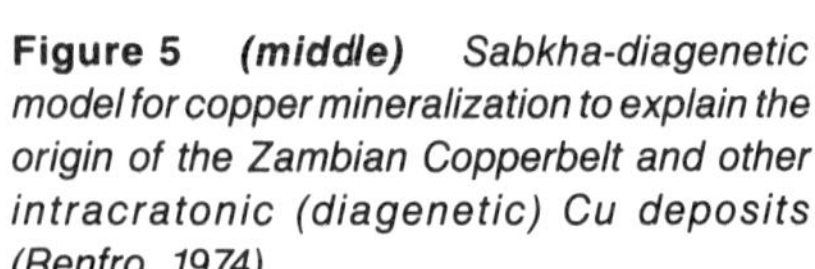

Figure 5 *(middle)* *Sabkha-diagenetic model for copper mineralization to explain the origin of the Zambian Copperbelt and other intracratonic (diagenetic) Cu deposits (Renfro, 1974).*

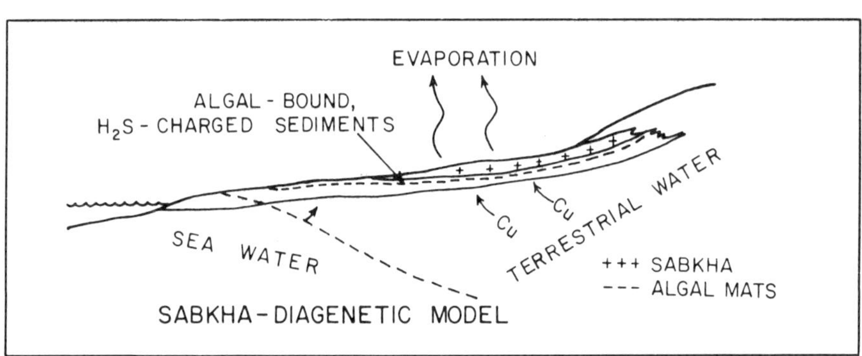

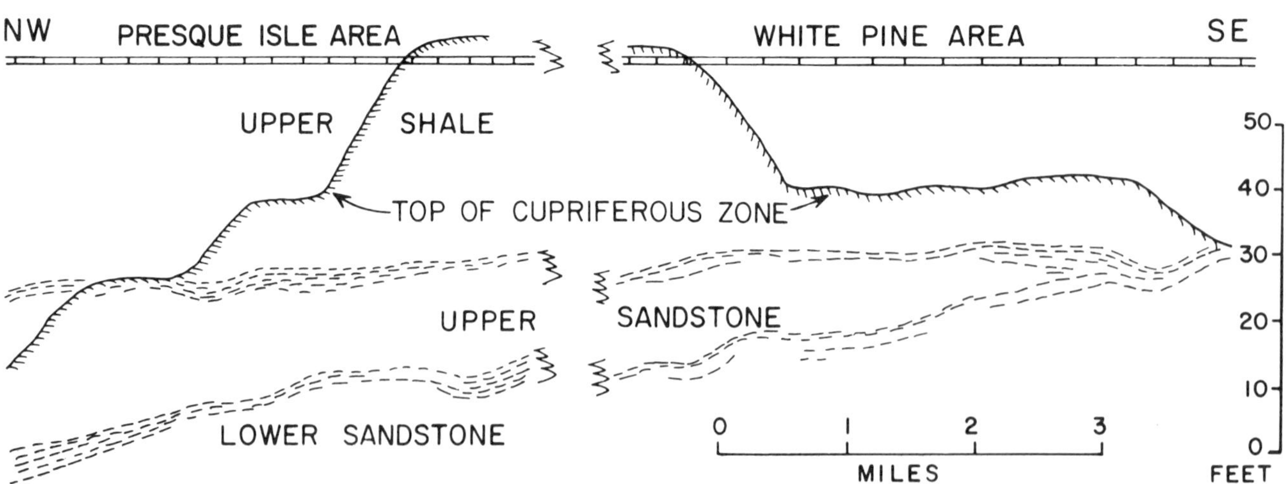

Figure 6 *Schematic cross-section showing the transgressive nature of the cupriferous zone at White Pine, Michigan. (Modified from Ensign* et al., *1968).*

The association of reductants with the intracratonic deposits indicates that sulphides were formed at the site of deposition and not transported in solution. Reduced sulphur in low temperature environments may be formed by bacterial sulphate reduction, or at higher temperature (>80°C) by chemical reduction (Orr, 1977) possibly related to biogenic methane generation. Reduced sulphur may also be formed by the destruction of other sulphides such as pyrite (White and Wright, 1966).

Several genetic models have been proposed for intracratonic ore deposits. The sabkha model (Renfro, 1974) attributes the formation of evaporite-associated stratiform metalliferous deposits to diagenetic processes of coastal sabkhas. This model is applicable to stratigraphic sequences such as that shown in Figure 5. Coastal sabkhas form in a hot, arid climate with a large evaporation debit. Regression causes the evaporite-encrusted sabkhas to prograde basinward across the landward-thinning wedge of strongly reducing, organic-rich intertidal-lagoonal sediment. The trailing edge of the sabkha is nourished by sub-surface flow of metal-bearing, oxygenated, terrestrial water. This dilute, metalliferous solution must pass upward from its oxygenated source beds through the overlying, hydrogen sulphide-charged algal mats in order to reach the area of evaporative discharge. The hydrogen sulphide-laden algal mats act as a reduction membrane that causes the trace metals in the ascending water to be precipitated as sulphide minerals (Figure 5). In the White Pine district where copper-rich basalts underlie the deposits, migration of fluids up through the oxygenated red beds moved metals upward through the reductant shale (Figure 6). Although both of the above models infer that the base metals are diagenetic, they differ in proposed depth of burial and temperature of formation.

Most workers who have studied the geology of the McArthur River deposit have concluded that it formed on the sea floor from metalliferous exhalations (*e.g.*, Murray, 1975; Lambert, 1976; Croxford and Jephcott, 1972). This is based on the conformable nature of the ore, its sedimentary and early diagenetic structures and its association with tuffaceous sediments. An epigenetic origin has also been proposed for the deposits (Williams, 1978), but the relatively minor amounts of sulphides which must have formed after deposition (Lambert, 1976) support the sedimentary origin for the HYC deposit. A general model for McArthur River includes exhalation into the sub-basins from associated synsedimentary faults. Such exhalation may not necessarily be related to volcanism, but may simply be caused by abnormally high heat flow or late-stage, deep compaction. In this context, the distinction between "formation waters" and "volcanic exhalation" may be of great importance.

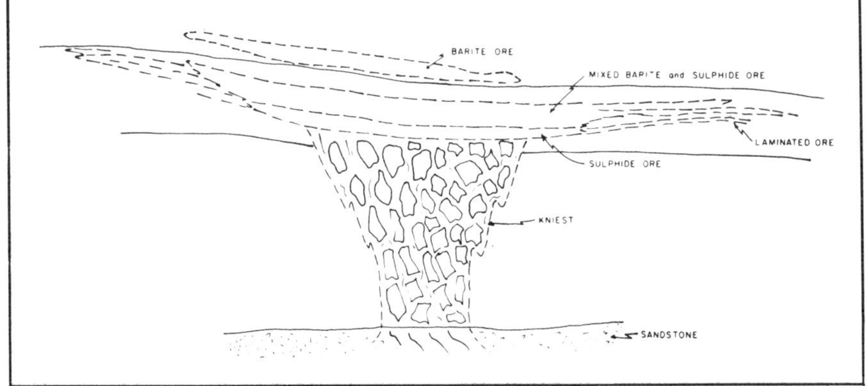

Figure 7 *Reconstructed cross-section of the Rammelsberg deposit showing the relationships between the underlying kniest, laminated sulphides and overlying barite. (Modified from Gunzert, 1969).*

Flysch Basin Deposits

Flysch basin sulphide and related barite deposits occur in thick turbidite sequences. Typical rocks associated with these are deep-water greywackes, siltstones, conglomerates and mudrocks (Walker, 1976).

Three general characteristics are common to the flysch basin deposits. (1) Barite is a major constituent with most Phanerozoic deposits and is present in some Proterozoic examples. Barite occurs with and above the sulphide ore at Rammelsberg (Gunzert, 1969) (Figure 7). The Tom deposit is essentially a Pb-Zn-Ag-Ba deposit within intercalated carbonaceous mudrocks, sulphides and laminated barite. In contrast, the Proterozoic example of this subclass, the Sullivan, does contain minor barite, but not nearly in the proportions associated with the younger deposits. In the northern Cordillera, laminated barite deposits are abundant at approximately the same stratigraphic position as stratiform sulphide deposits, but separated laterally from them. The barite deposits consist of laminated lenticles of barite, some of which contain over one million tonnes grading over 50% $BaSO_4$. The main differences between these barite deposits and the Pb-Zn-Ag rich deposits is simply the lack of sulphide in the former. (2) Many flysch basin deposits have a related alteration-feeder zone underlying or adjacent to them. The Rammelsberg deposit is underlain by kniest ore, a hard, partly brecciated, mineralized silicified rock (Gunzert, 1969) in which sulphides have filled veins and fractures (Figure 7). The Sullivan ore body is underlain by breccia and extensive tourmalinization (Freeze, 1966)(Figure 8). The Tom deposit is underlain by a siderite alteration zone (Carne, 1979) with minor veins of chalcopyrite and tetrahedrite (Figure 9). (3) Many of the deposits are contained in sub-basins, related to synsedimentary grabens. The lenticular shape of the Rammelsberg and Sullivan deposits and the rapid thickening of the sulphide-barite horizon at the Tom and other similar barite deposits, suggest that the sub-basins may be fault bounded. However, reactivation of the faults may have obscured movement penecontemporaneous with sedimentation. The MacMillan Pass graben (Smith, 1978) contains at least two sulphide-barite deposits, the Tom and Jason (Figure 10). Faults associated with the Sullivan deposit may also represent a much smaller graben.

The possible sources of the base metals in these deposits are most likely the underlying sediments. For example, the Tom deposit and other regionally extensive barite deposits in the Earn Group of Devonian-Mississippian age are associated with carbonaceous mudrocks having high barium, zinc and lead background levels. Thus, it is reasonable to consider that underlying sediments provided the metals for the deposits. Turbidite lithologies can potentially provide large amounts of connate fluids during compaction. The sediments are rapidly deposited, and as a consequence, interstitial fluids are trapped within the sedimentary sequence. As the weight of the overlying sediments increases, internal overpressures may be generated. Sediments which show such overpressures are also undercompacted because the weight of the overlying sediments is in part supported by the interstitial fluid rather than by the contained grains (Chapman, 1972). Faults, common in flysch environments, cutting the overpressured sediments, bring about release of the pressure and removal of the fluids. The resulting collapse of the sediments would produce grabens and provide a brine source. Igneous activity such as the intrusion of sills or the extrusion of tuffs can produce similar features or complement them, causing high heat flow and higher temperature ore fluids. Excessively rapid loading by deposition of conglomerates associated with submarine fan progradation could also initiate rapid fluid migration. Although highly simplified, this model fits the general geology of these deposits.

The deposition of sulphide and barite in the flysch deposits reflects the proximity to the vents within the sub-basins. Sulphur isotope studies (Anger *et al.*, 1966) suggest that the cause of fractionation between metal sulphides and barite in a hydrothermal precipitate is largely a process of oxidation of the hydrothermal solutions (Lydon *et al.*, 1979). The ore solutions undergo progressive oxidation from the sulphide to the barite zones. At Rammelsberg, the majority of the barite samples have a sulphur isotope composition similar to that of contemporaneous seawater. It is this mixing of oxygenated seawater with the ore solutions that is the major cause of the oxidation process. Thus, there are two sources of sulphate: contemporaneous seawater and the oxidation of hydrothermal-reduced sulphur species. High rates of discharge within the sub-basin provide optimum conditions for displacement of marine water from the vent area and tend to insulate the immediate vent area. At lower rates of discharge, or the opening of the sub-basin, seawater sulphate would be more important resulting in more sulphate deposition. If this is true, then the regionally associated barite deposits differ from sulphide-rich barite deposits because of a lack of sulphide-rich or metal-rich discharge, low discharge rate, lack of a closed basin, or combinations of the above parameters.

An alternative theory that also fits the data is that of a low-sulphur brine entering an isolated sub-basin. This model envisions the alternate isolation and opening of the sub-basin. In this case, reducing brine would reduce seawater sulphate to form sulphide when the basin is isolated, but if the brine influx is

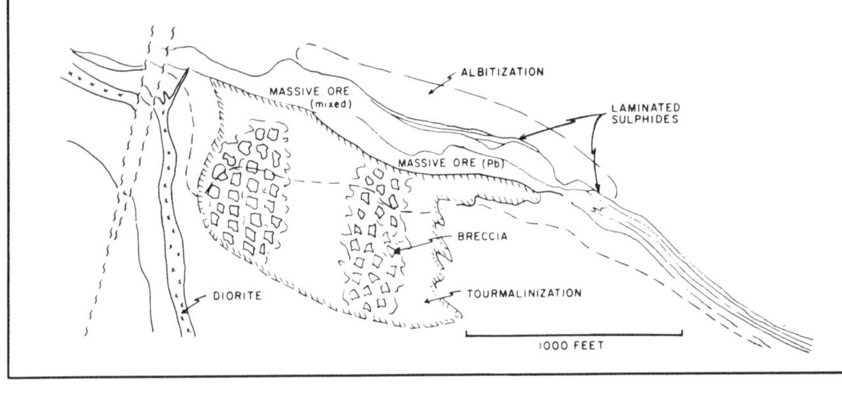

Figure 8 *Longitudinal section of the Sullivan deposit showing the marginal, laminated sulphides and the underlying tourmaline alteration and breccia. (Modified from Freeze, 1966).*

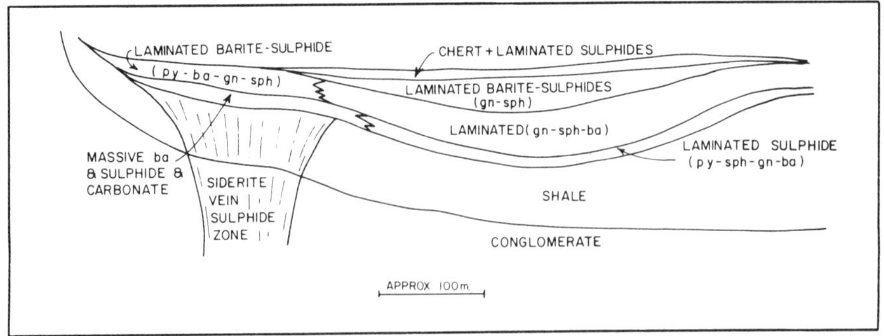

Figure 9 *Reconstructed cross-section of the Tom deposit. Note the sub-basin containing most of the laminated sulphides and the underlying feeder zone. (Modified after Carne, 1979).*

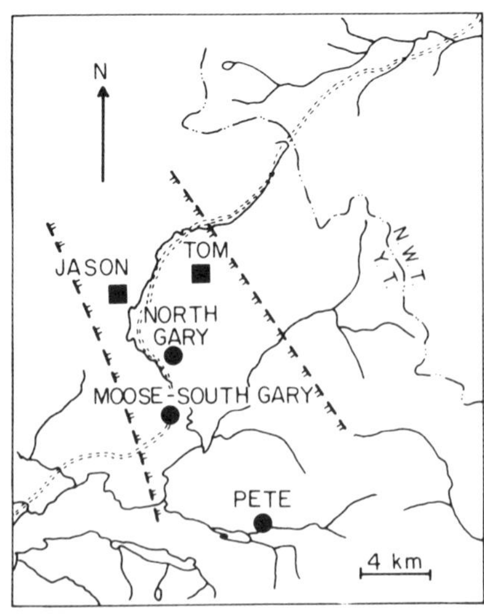

Figure 10 *Spatial relationships between Ba (circles), Ba-Zn-Pb-Ag (squares) deposits and the McMillan Pass Graben. (Modified from Smith, 1978).*

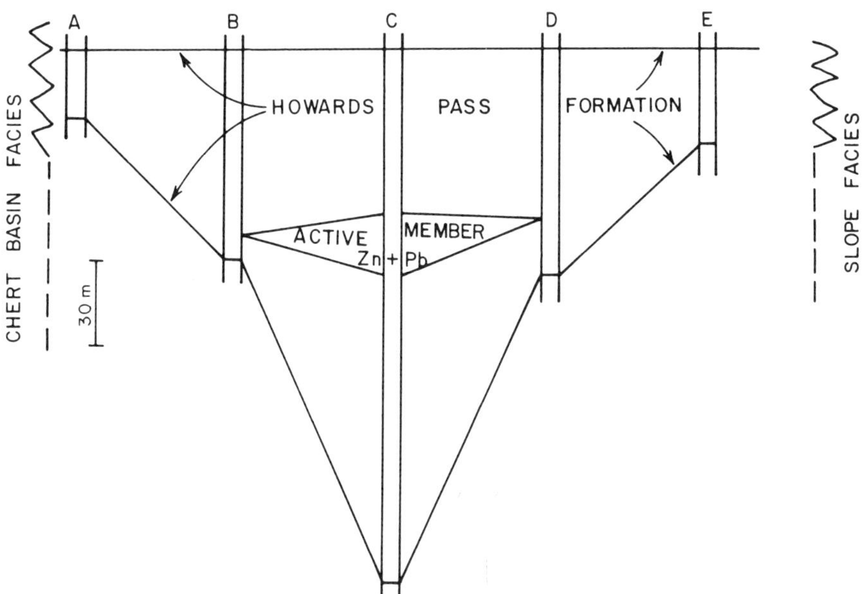

Figure 11 *Composite stratigraphic sections of the Howards Pass Formation across the base of slope facies (see Figure 12). These sections show a general thickening of the Formation and the presence of the Zn-Pb containing active member in the sub-basins. Distance between sections A and E is approximately 8 km.*

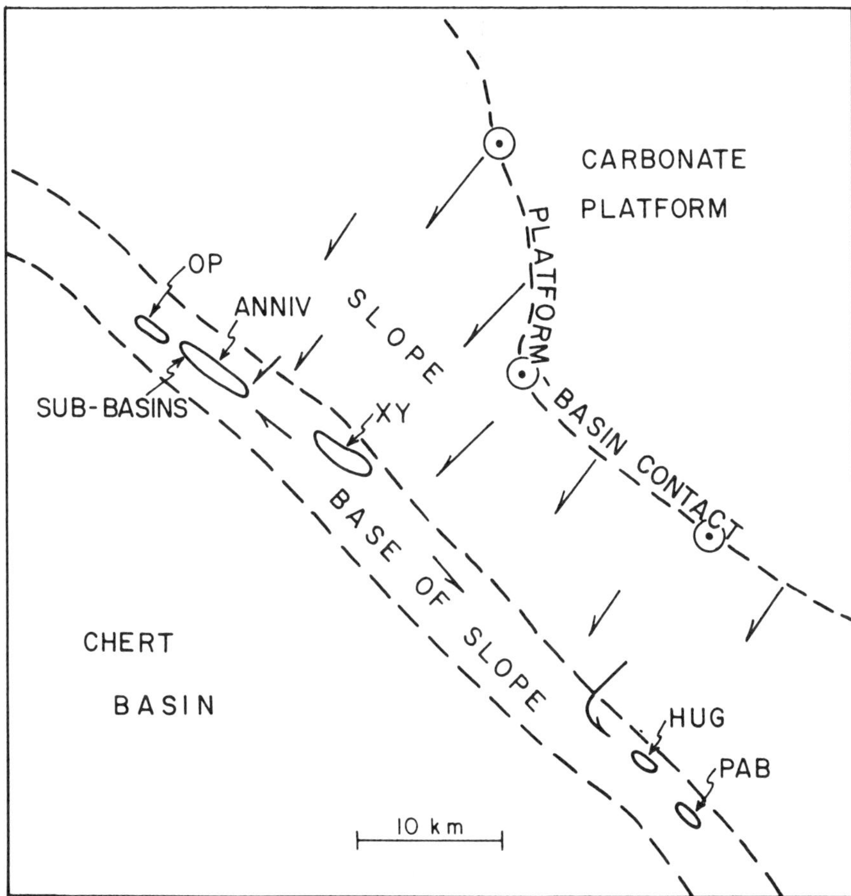

Figure 12 *General model for the formation of the platform-marginal deposits showing the geometry of the platform-slope-base of slope and chert basin facies. Arrows show surface movement of ore fluids which migrate up to the sediment-water interface, down the basin slope and are trapped in the sub-basins at the base of the slope.*

diminished or the basin is open to seawater sulphate, barite becomes stable. If this model is accepted, then the regionally extensive barite deposits differ from the sulphide-rich barite deposits because of low discharge rate or lack of sub-basin isolation.

Platform-Marginal Deposits

Platform-marginal sedimentary-type stratiform sulphide deposits occur in basins seaward of major platforms or shelves associated with cratons. These basins differ from the intracratonic basins in that a large portion of the basin is deep water. In contrast, local deeps within a generally shallow water environment are associated with intracratonic basins (*e.g.*, the McArthur River area). Deposition rates within these deep-water, starved basins are low compared with the flysch environment. Although the only deposits grouped into this subclass are the XY, ANNIV and OP (*i.e.*, Howards Pass) deposits of the Selwyn Basin, sufficient work has been completed to demonstrate that these deposits define a third sub-class (Morganti, 1979). Five characteristics are exhibited by all of the deposits. (1) The sulphide mineralogy is simple: predominantly sphalerite and galena. The XY, ANNIV and OP deposits, for example, show galena, sphalerite and pyrite to be the only sulphides, except for a few grains of chalcopyrite noted in the XY deposit. (2) The pyrite content is low compared with other stratiform sulphide deposits. For example, the XY deposit contains less than 5% pyrite, even where sphalerite and galena constitute 70% of the rock. Furthermore, the pyrite content within the Howards Pass Formation is almost constant throughout, and the main difference between the pyrite of the deposit and that of the rest of the stratigraphic section is textural, in that framboidal pyrite occurs in the deposit and nodular pyrite in the rest of the section. (3) The Ba content is low compared with other Paleozoic stratiform sulphide deposits, typically less than 2000 ppm; furthermore, there are no barite deposits directly associated with the sulphide bodies. (4) There are no copper zones associated with these deposits such as those found at Mt. Isa or Meggen. Of the few hundred copper analyses completed on material from the Howards Pass deposits, the highest copper value obtained was 150 ppm.

(5) The deposits are associated with anomalously thick sedimentary sequences (*i.e.*, sub-basins, Figure 11), but lack evidence for rapidly formed graben structures similar to the flysch deposits.

With the Selwyn Basin, platform-marginal deposits occur in graptolitic carbonaceous mudrocks, cherts and limestone of the Howards Pass Formation (Morganti, 1977). This unit contains slope, base of slope and basin floor facies which developed west of the MacKenzie Platform (Figure 12). Within the base of slope (rise?) facies, sub-basins occur as sea-floor depressions (Figure 13) in which cherts, mudrocks, limestones and sulphides were deposited. The laminated, sulphide-rich beds occur in a rhythmic sequence limited to these sub-basins within the active member of the Howards Pass Formation. A general trend up-section of limestone-mudrock-chert (Figure 14) also occurs as individual cycles within the major cycle. A sequence such as this could be the result of the increasing isolation of a sub-basin, accompanied by formation of limestone as a by-product of sulphate to sulphide reduction. A decrease in pH and resultant change in the stability of carbonate relative to amorphous silica could have been brought about by the influx of low pH, metalliferous brine into the high pH, reducing sub-basins. The lack of associated feeder zones underlying the deposits of this sub-class, their low copper and silver contents, and the large lateral dimensions of the deposits all suggest that the source of the brine was not nearby, such as is evident in flysch deposits. In the case of the eastern Selwyn Basin, tuffs associated with basin-platform transition suggest that anomalous heat flow and possible brine exhalation may have occurred and provided a brine which subsequently migrated down-slope.

Ore-Forming Fluids for Sedimentary-Type Stratiform Deposits

The fluids which transport the metals to their site of deposition constitute the second parameter to be considered here. These brines deposited their metals within the lithifying sediment or at the sediment-water interface.

Within the sediment, migration of ore-forming brine may be somewhat analogous to oil migration. Thus, primary

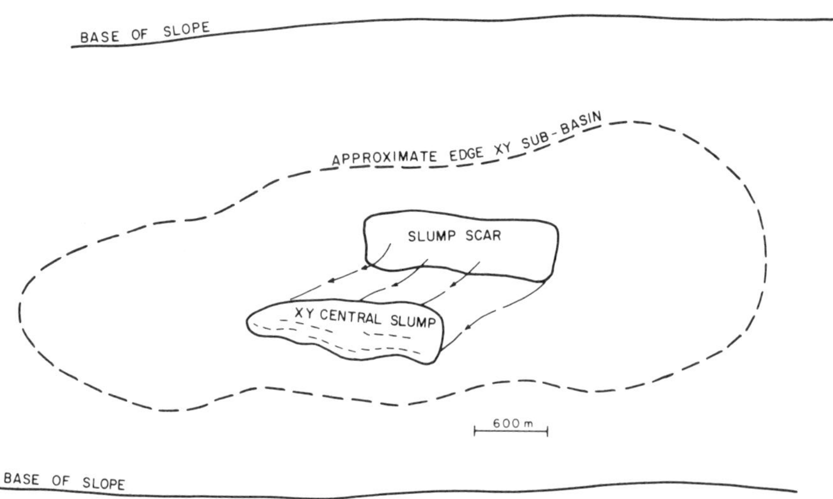

Figure 13 *Generalized plan view of reconstructed XY sub-basin, Howards Pass. The sub-basin axis of elongation parallels that of the base of slope. Slumping has produced a high-grade "plumb" within the larger XY deposit. (Arrows indicate direction of movement of slumped sediment.)*

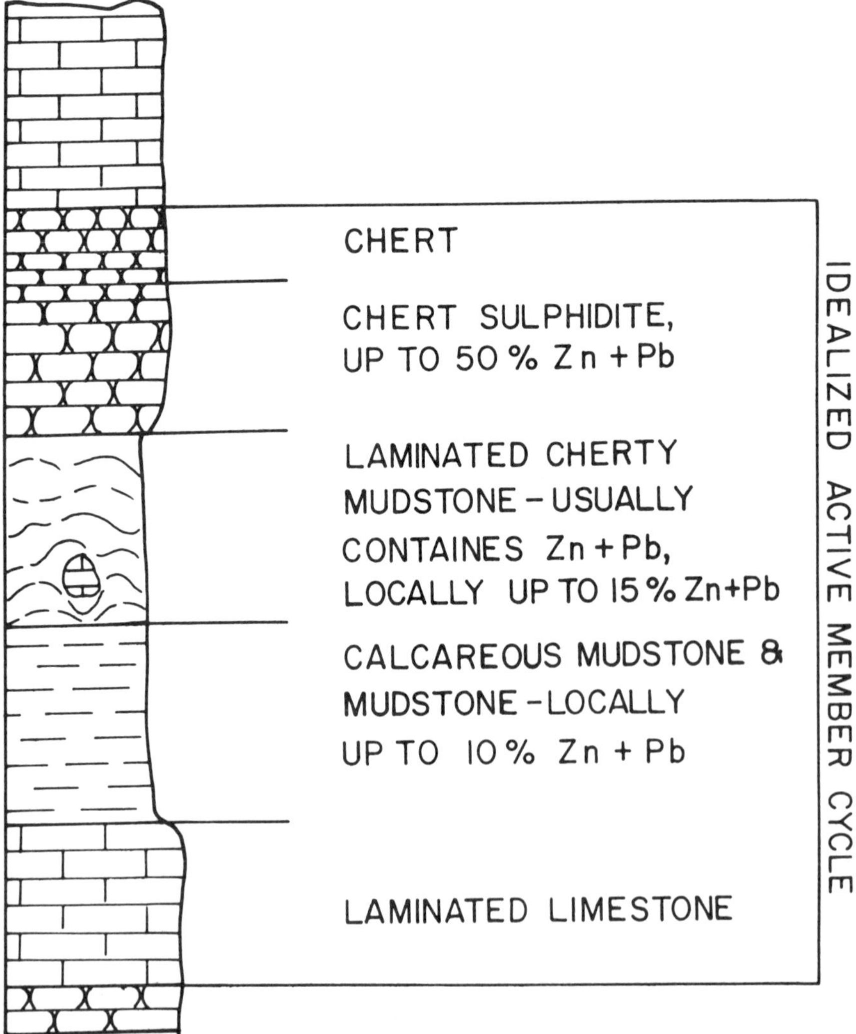

Figure 14 *Generalized stratigraphic section of an idealized active member showing a typical sequence of facies. Note that there is a decrease in carbonate and an increase in chert upward.*

migration includes the release of metals from source beds and their transport within and through the capillaries. Metal complexes expelled from a source bed pass through the pores of more permeable rock units (Tissot and Welte, 1978). Migration of brines above the sediment-water interface is here also considered as secondary migration of brine. Secondary migration in the sub-surface occurs through primary or secondary porosity. Examples of deposits which show associated sub-surface deposition in secondary porosity are the Cu deposits in the Mt. Isa and McArthur River areas (Flinlow-Bates and Stumpfl, 1979; Williams, 1976) and the feeder zones underlying the flysch deposits such as the Tom and the Sullivan. Sub-surface brines responsible for White Pine and Mufulira appear to have migrated through primary porosity (Brown, 1971; Annels, 1979).

Little is known about the nature of these brines, but recent studies on fluid inclusions (Roedder, 1976, 1979) and stability relationships from Mississippi Valley-type deposits (Anderson, 1973, 1978), and recent sub-surface brines (Carpenter *et al.*, 1974) suggest that the brines were dense (> 1.1 g·cm^{-3}), moved slowly (few m·yr^{-1}), were of a low temperature (k100-160°C), high salinity (> 15 wt.% NaCl equivalent), and had a low sulphur content.

Upon exhalation onto the sea floor, a brine may behave variably depending on its physical properties. Such low temperature brines occurring in seawater may be classified on the basis of physical behaviour (Sato, 1972). Subsequent model experiments (Turner and Gustafson, 1978) have supported this classification. The density of brines with various concentrations of NaCl at varying temperatures can be compared with seawater density (Figure 15), resulting in three major brine types, two of which are considered here.

Type I brines are low-temperature, relatively high-salinity brines which, upon exhalation, flow down slope due to a higher density than the surrounding seawater. Mixing with seawater during down-flowing will take place only to a very limited extent because the fresh, ascending solution forms a stable bottom layer. Because of the high brine density, the resultant deposits are strongly controlled by sea-floor topography. Alteration feeder zones are poorly developed because of the low temperature of the brine, and they may be spatially removed from the stratiform mineralization because of the high density of the brine. Homogenization due to internal mixing may occur during transport, but density stratification is characteristic of Type I brines after sub-basin containment. As a result, there is very little lateral metal zoning in the resultant deposits. Examples which display many of the characteristics of a deposit formed by a Type I brine are the Howards Pass deposits and McArthur River. The control of submarine topography, a lack of distinct mineral zoning and a large distance between proposed feeder zones and stratiform sulphides are characteristic of these deposits.

Type II brines are moderate-to high-temperature solutions of moderate salinity. Their relatively high temperatures produce increased leaching capacity in the aquifer, and an obvious, associated feeder zone. Type II brines are subdivided into two sub-types based on a reversal of density trend during their evolution. Brines of Type IIa are heavier than seawater. Only minor mixing with seawater takes place, and therefore, the geometry of the deposit is strongly influenced by sea-floor topography. Alteration pipes or stockwork stringer mineralization should underlie the deposit because of the association with graben structures. Both the Sullivan and Tom deposits exhibit many of the characteristics expected in a deposit formed from such a brine. Related feeder zones are characteristic of these two examples.

Type IIb brines are slightly higher temperature and less saline than Type IIa. Upon exhalation into seawater, these brines are less dense than seawater, but convective mixing above the vent (Turner and Gustafson, 1978) causes cooling which, in turn, causes the brines to become more dense than seawater. Yet, because of the initially less dense nature of the brine, sea-floor topography does not exhibit as important a control on the distribution of stratiform mineralization as brine Types I and IIa. Because of the high temperature, copper may be deposited near the vent. Mixing of brine and seawater away from the vent would produce an intimate association of layered sulphide and barite, and furthermore, a strong zonation within the deposits. The Rammelsberg deposit is strongly zoned and underlain by a well-developed feeder zone suggesting it is a deposit formed by a Type IIb brine.

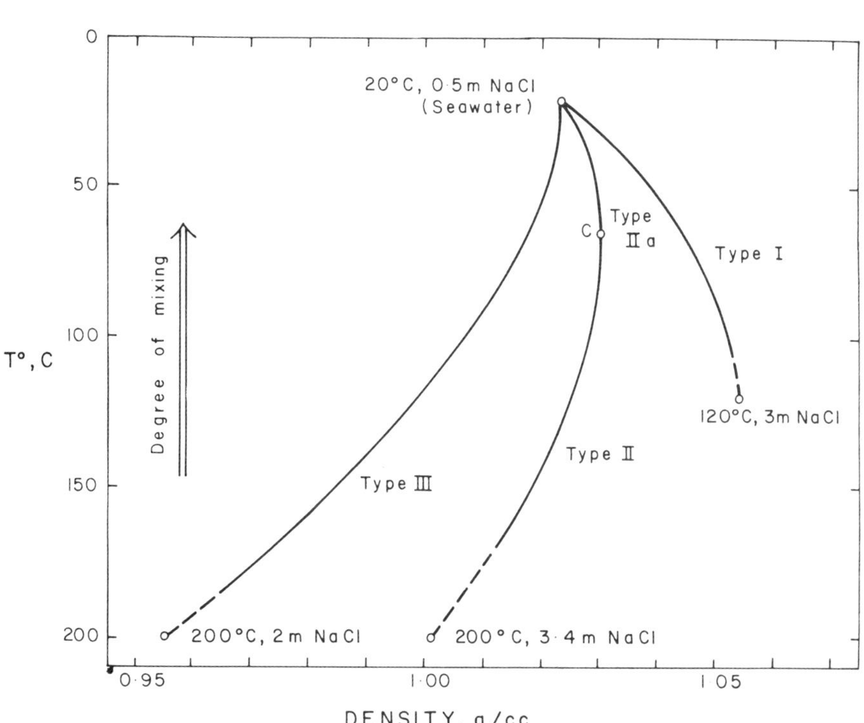

Figure 15 *The evolution of brines (NaCl solutions) as they mix with seawater, showing temperature and density. Note Type II changes at point C forming brine Type IIa. (After Sato, 1972).*

Metallogenic Epochs and Provinces

Metallogenic provinces and epochs are of prime importance to economic geologists for conceptual regional exploration. Major metallogenic epochs for intractonic ore deposits are the Proterozoic and the Permian periods. The age of these deposits coincides with extensive deposition of red beds, which are in turn a function of tectonic and paleogeographic setting. Metallogenic provinces such as the Kupferschiefer and the Zambian Copperbelt occur in major intracratonic basins containing many deposits. In the case of flysch basin deposits, the major metallogenic epoch is the Devonian-Mississippian with two major provinces evident. These are the Antler equivalents in the Cordillera (Boucot *et al.*, 1974) and the Variscan geosyncline in Europe (Krebs, 1976). Flysch sequences show a close association with major tectonic events. For example, the Proterozoic Sullivan deposit shows an association with rifting (Stewart, 1972). The metallogenic epoch for the platform-marginal deposits is the Ordovician-Silurian, and corresponds to a period of world-wide black shale deposition (Berry and Wilde, 1978). Metallogenic provinces appear related to major starved basins, such as the Selwyn Basin, where sub-basins occurred down slope of exhalative sites.

Conclusions

Sedimentary-type stratiform ore deposits are those stratiform deposits occurring in clastic sequences with no strong volcanic association. Three different tectono-sedimentary environments have been considered — intracratonic, flysch and platform-marginal (Figure 16). Brines generated within these environments may precipitate sulphides during diagenesis or be exhaled onto the sea floor and precipitate sulphides and sulphates in sub-basins. Overprinting is common because of the evolutionary nature of the deposits before lithification of the surrounding sediments. Furthermore, subsequent post-lithification metamorphism and structural events make elucidation of the original textures of many deposits more difficult. More detailed stable isotope, fluid inclusion and organic geochemical studies are required if we are to have the required data for detailed unified models.

Acknowledgements

D.F. Sangster, V. Hollister and D.N. Hillhouse reviewed an early draft and made comments, some of which have been incorporated in the final manuscript.

References

Anderson, G.M., 1973, The hydrothermal transport and deposition of galena and sphalerite near 100°C: Economic Geology, v. 68, p. 480-492.

Anderson, G.M., 1978, Basinal brines and Mississippi Valley-type ore deposits: Episodes, v. 1978, no. 2, p. 15-19.

Anger, G., Nielsen, H., Puchelt, H. and Ricke, W., 1966, Sulphur isotopes in the Rammelsberg ore deposit (Germany): Economic Geology, v. 61, p. 511-536.

Annels, A.E., 1979, The genetic relevance of recent studies at Mufulira mine, Zambia: Societe Geologique de Belgique, Annales, v. 102, p. 431-449.

Argall, G.O., 1975, Eagle-Picher strips 45 feet to mine 11 inches of 1.92 percent copper: World Mining, v. 28, no. 5, p. 40-42.

Berry, W.B.N. and Wilde, P., 1978, Progressive ventilation of the oceans — an explanation for the distribution of the lower Paleozoic black shales: American Journal of Science, v. 278, p. 157-175.

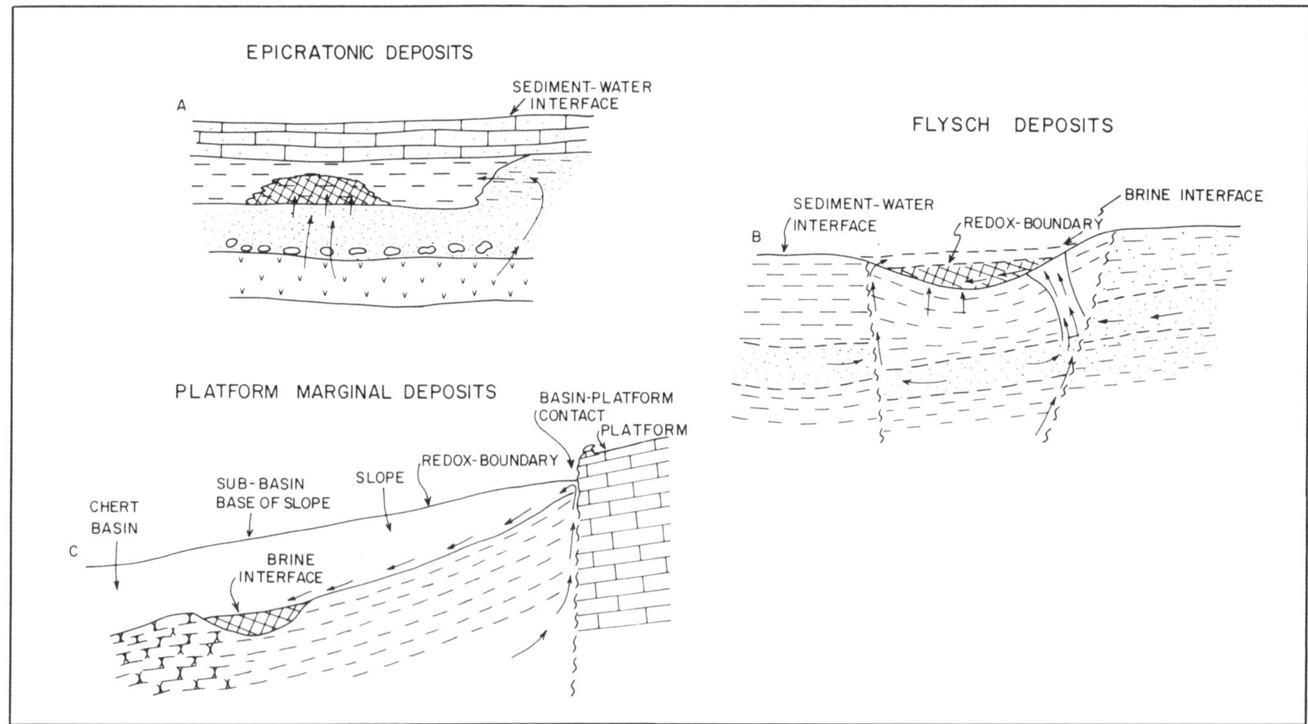

Figure 16 *Generalized models of formation for sedimentary-type stratiform sulphide deposits. Unlabelled arrows show direction of brine migration. Deposits formed in such a manner are synsedimentary and the models show what may be best termed the sedimentary-exhalative model of ore formation (Morganti, 1977).*

Boucot, A.J., Dunkle, D.H., Potter, A., Savage, N.M. and Rohr, D., 1974, Middle Devonian orogeny in western North America?: a fish and other fossils: Journal of Geology, v. 82, p. 691-708.

Brown, A.C., 1971, Zoning in the White Pine copper deposit, Ontonagon County, Michigan: Economic Geology. v. 66, p. 543-573.

Brown, A.C., 1974, The copper province of northern Michigan, U.S.A., *in* Bartholome, P., ed., Gisements stratiformes et provinces cuprifères: Societie Geologique de Beligique, Liege.

Carpenter, A.B., Trout, M.L. and Pickett, E.E., 1974, Preliminary report on the origin and chemical evolution of lead- and zinc-rich oil field brines in central Mississippi: Economic Geology, v. 69, p. 1191-1206.

Carne, R.C., 1979, Geological setting and stratiform mineralization, Tom claims, Yukon Territory: Department Indian and Northern Affairs, Paper 1979-4, 30 p.

Chapman, R.E., 1972, Primary migration of petroleum from clay source rocks: American Association of Petroleum Geologists, Bulletin, v. 56, p. 2185-2191.

Conybeare, C., 1979, Lithostratigraphic analysis of sedimentary basins: Academic Press, New York, 555 p.

Croxford, N.J.W., and Jephcott, S., 1972, The McArthur lead-zinc-silver deposit, N.T.: Australasian Institute of Mining and Metallurgy, Proceedings, no. 243, p. 1-26.

Deans, T., 1948, The Kupferschiefer and the associated mineralization in the Permian of Silesia Germany, and England: 18th International Geological Congress, Great Britain, pt. 7, p. 340-352.

Ensign, C.O., White, U.S., Write, J.C., Patrick, J.L., Leone, R.J., Hathaway, D.J., Trammell, J.W., Fritts, J.J. and Write, T.L., 1968, Copper deposits in the Nonesuch shale, White Pine, Michigan, *in* Ridge, J.D., ed., Ore deposits of the United States 1933-1967. (Graton-Sales vol.): American Institute of Mining and Metallurgical Engineers, New York, p. 460-488.

Fleischer, V.D., Garlick, W.G. and Haldlane, R., 1976, Geology of the Zambian Copper belt, *in* Wolf, K.H., *ed.,* Handbook of Strata-bound and Stratiform Ore Deposits: Elsevier, Amsterdam, v. 6, p. 223-352.

Finlow-Bates, T. and Stumpfl, E.F., 1979, The copper and lead-zinc-silver orebodies of Mount Isa Mine, Queensland: Products of one hydrothermal system: Societé Géologique de Belgique, Annales, v. 102, p. 497-517.

Freeze, A.C., 1966, On the origin of the Sullivan ore body, Kimberly, B.C., *in* Tectonic History and Mineral Deposits of the Western Cordillera: Canadian Institute of Mining and Metallurgy, Special Volume 8, p. 263-294.

Garlick, W.G., 1961, The syngenetic theory, *in* Mendelsohn, F., ed., The Geology of the Northern Rhodesian Copper-belt: McDonald, London, p. 146-165.

Gunzert, G., 1969, Altes Und Neues Lager am Rammelsberg de Goslar: Erzmetall, v. 22, p. 1-10.

Helmstaedt, H., Eisbacher, G.H. and McGregor, J.A., 1979, Copper mineralization near an intra-Rapitan unconformity, Nite copper prospect, Mackenzie Mountains, Northwest Territories, Canada: Canadian Journal of Earth Sciences, v. 16, p. 50-59.

Jambor, J.L., 1979, Mineralogical evaluation of proximal-distal features in New Brunswick massive-sulphide deposits: Canadian Mineralogist, v. 177, p. 649-664.

Johnson, K.S., 1976, Permian copper shales of southwestern Oklahoma, *in* Johnson, K.S. and Croy, R.L., eds., Stratiform copper deposits of the midcontinent region: a symposium: Oklahoma Geological Survey, Circular 77, p. 3-14.

Jung, W. and Knitzschke, G., 1976, Kupferschiefer in the German Democratic Republic (GDR) with special reference to the Kupferschiefer deposit in the southeastern Harz Foreland: *in* Wolf, K.H., ed., Handbook of Strata-bound and Stratiform Ore Deposits: Elsevier, Amsterdam, v. 6, p. 353-406.

Krebs, W., 1976, Geology of European strata-bound lead-zinc-copper deposits: Canadian Society of Petroleum Geologists, Seminar, April 7-9, 1976, Calgary, Alta.

Lambert, I.B., 1976, The McArthur zinc-lead-silver deposit: features, metallogenesis and comparisons with other stratiform ores, *in* Wolf, K.H., ed., Handbook of Strata-bound and Stratiform Ore Deposits: Elsevier, Amsterdam, v. 6, p. 535-585.

Large, D., 1979, Proximal and distal stratabound ore deposits. A discussion of the paper by I.R. Plimer — Mineralium Deposita v. 13, p. 345-353 (1978): Mineralium Deposita, v. 14, p. 123-124.

Lydon, J.W., Lancaster, R.D. and Karkkainen, P., 1979, Genetic controls of Selwyn Basin stratiform barite/sphalerite/galena deposits: an investigation of the dominant barium mineralogy of the Tea deposit, Yukon: Geological Survey of Canada, Paper 79-1B, p. 223-229.

Mathias, B.V. and Clark, G.J., 1975, Mount Isa copper and silver-lead-zinc ore bodies — Isa and Hilton mines, *in* Knight, C.L., ed., Economic Geology of Australia and Papua New Guinea: Australasian Institute of Mining and Metallurgy, Monograph Series 5, p. 351-372.

Morganti, J.M., 1977, Howards Pass: an example of a sedimentary-exhalative base metal deposit: Geological Association of Canada—Mineralogical Association of Canada, Program with Abstracts, v. 2, p. 38.

Morganti, J.M., 1979, The geology and ore deposits of the Howards Pass area, Yukon and Northwest Territories: The origin of basinal sedimentary stratiform sulphide deposits, Unpublished Ph.D. Thesis, University of British Columbia, Vancouver, B.C., 327 p.

Murray, W.J., 1975, McArthur River H.Y.C. lead zinc and related deposits N.T., *in* Knight, C.L., ed., Economic geology of Australia and Papua New Guinea: Australasian Institute of Mining and Metallurgy, Monograph Series 5, p. 351-372.

Orr, W.L., 1977, Geologic and geochemical controls on the distribution of hydrogen sulphide in natural gases: Advances in Organic Geochemistry, p. 572-597.

Plimer, I.R., 1978, Proximal and distal stratabound ore deposits: Mineralium Deposita, v. 13, p. 345-353.

Renfro, A.R., 1974, Genesis of evaporite associated stratiform metalliferous deposits — a sabkha process: Economic Geology, v. 69, p. 33-45.

Rentzsch, J., 1974, The Kupferschiefer in comparison with the deposits of the Zambian Copperbelt, *in* Bartholome, P., ed., Gisements stratiformes et provinces cuprifères: Societé Géologique de Belgique, Liege.

Roedder, E., 1976, Fluid-inclusion evidence on the genesis of ores in sedimentary and volcanic rocks, *in* Wolf, K.H., ed., Handbook of Strata-bound and Stratiform Ore Deposits: Elsevier, Amsterdam, v. 2, p. 67-110.

Roedder, E., 1979, Fluid inclusion evidence on the environments of sedimentary diagenesis, a review, *in* Scholle, P.A. and Schluger, P.R., eds., Aspects of diagenesis: Society of Economic Paleontologists and Mineralogists, Special Publication No. 26, p. 89-107.

Rose, A.W., 1976, The effect of cuprous chloride complexes in the origin of red bed copper and related deposits: Economic Geology, v. 71, p. 1036-1048.

Sato, T., 1972, Behaviours of ore-forming solutions in seawater: Mining Geology, v. 22, p. 31-42.

Smith, C., 1978, Geological setting of Jason and Tom deposits, MacMillan Pass area, eastern Yukon: presented at Northern Geoscience Conference, Dec. 1978, Whitehorse, Y.T.

Stanton, R.L., 1972, Ore Petrology: McGraw-Hill, New York, 713 p.

Stewart, J.H., 1972, Initial deposits of the Cordilleran Geosyncline: evidence of Late Precambrian (>850 m.y.) continental separation: Geological Society of America, Bulletin, v. 83, p. 1345-1360.

Tissot, B.P. and Welte, D.H., 1978, Petroleum Formation and Occurrence: Springer-Verlag, New York, 538 p.

Turner, J.S. and Gustafson, L.B., 1978, The flow of hot saline solutions from vents in the sea floor — some implications for exhalative massive sulphide and other ore deposits: Economic Geology, v. 73, p. 1082-1100.

Wakefield, J., 1978, Samba: a deformed porphyry-type copper deposit in the basement of the Zambian Copperbelt: Institution of Mining and Metallurgy, Transactions, v. 87, p. B43-B52.

Walker, R.G., 1976, Turbidites and associated coarse clastic deposits: Geoscience Canada, v. 3, p. 25-36.

Walker, R.G., 1979, ed., Facies Models: Geological Association of Canada, Geoscience Canada Reprint Series 1, 211 p.

White, W.S. and Wright, J.C., 1966, Sulphide-mineral zoning in the basal Nonesuch shale, northern Michigan: Economic Geology, v. 61, p. 1171-1190.

Williams, N., 1976, The formation of sedimentary-type stratiform sulphide deposits, Unpublished Ph.D. dissertation, Yale University, 330 p.

Williams, N., 1978, Studies of the base metal sulphide deposits at McArthur River, Northern Territory, Australia: 1. The Cooley and Ridge deposits: Economic Geology, v. 73, p. 1005-1035.

Originally published in *Geoscience Canada* v. 8 Number 2 (June 1981)

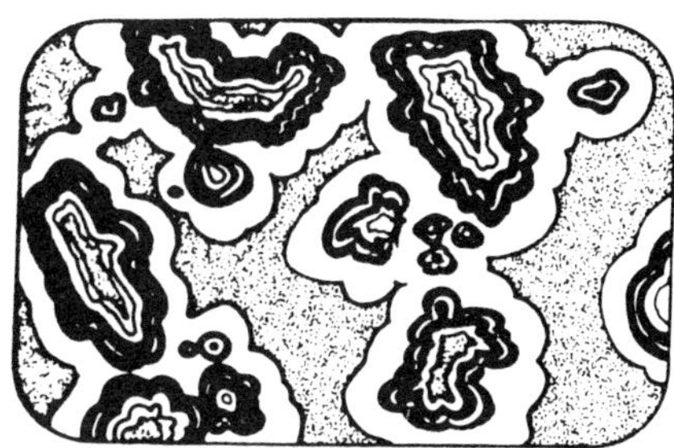

Mississippi Valley-Type Lead-Zinc Deposits

G.M. Anderson
Department of Geology
University of Toronto
Toronto, Ontario M5S 1A1

R.W. Macqueen
Geological Survey of Canada
3303 - 33rd St. N.W.
Calgary, Alberta T2L 2A7

Introduction

Carbonate-hosted lead-zinc deposits of the Mississippi Valley-type (MVT) appear to be at one end of a spectrum of base-metal ore deposit types which form in sediments some time during the lifetime of a sedimentary basin. The spectrum of deposits includes red-bed copper, shale-hosted lead-zinc and sandstone-hosted lead, as well as MVT deposits. As in the case of petroleum and natural gas, these ore deposits are now generally viewed as being a normal part of the evolution of a sedimentary basin. Fluids move in sedimentary basins, both during basin formation and long afterward, in response to hydrostatic gradients established by compaction, thermal gradients, topographic relief, deformation and other factors, and they are capable of transporting and depositing metals. This much is easy to say; elucidating the details is a little more difficult.

Recent review articles discuss the broader relations between the various types of sediment-hosted deposits mentioned above (Gustafson and Williams, 1981, Bjørlykke and Sangster, 1981; Sangster, 1983). They demonstrate that hypotheses involving synsedimentary, diagenetic and epigenetic processes are all alive and well for one or another of these deposit types and that although certain genetic factors are reasonably clear for each type, the place of each in relation to the others and to the overall process of basin evolution is still not clear in many respects, as considered below.

MVT deposits form the principal source of lead and zinc in the United States, where there are four main districts (Figure 1): central and east Tennessee, southeast Missouri, the Tri-state area (Missouri, Oklahoma and Kansas) and the Wisconsin area of the Upper Mississippi River valley. All of these districts, except east Tennessee, are located in essentially undeformed platformal carbonates. Important Canadian deposits (Figure 2) include Pine Point of the Interior Platform, NWT (Skall, 1975; Macqueen and Powell, 1983; Rhodes *et al.*, 1984; Krebs and Macqueen, 1984; Powell and Macqueen, 1984); Polaris (Kerr, 1977) and Nanisivik of the northern platform, NWT (Olson, 1984); Gays River of the Appalachian platform (Akande and Zentilli, 1984; Ravenhurst *et al.*, 1987); and Daniel's Harbour of the western Newfoundland platform (Collins and Smith, 1975; Coron, 1982). Promising Cordilleran Orogen deposits in the Rocky Mountain Belt and possibly of future ore grade include Robb Lake in north eastern British Columbia (Macqueen and Thompson, 1978; Manns, 1981) and Gayna River in the Mackenzie Mountains, NWT (Hardy, 1979). Broadly, these deposits are stratabound — confined to individual stratigraphic units. Not included are those carbonate-hosted deposits which are stratiform or bed-like and appear to be of syngenetic or early diagenetic origin (*e.g.*, Alpine Triassic deposits of Austria and Poland).

Recent summary articles on aspects of MVT deposits in general, besides those mentioned above, include Anderson (1978, 1983), Macqueen (1979), Heyl *et al.* (1974), Ohle (1959, 1980), Sangster and Lancaster (1976), Sverjensky (1986) and Wolf (1976, 1981). A recent symposium volume contains abundant new data on MVT deposits (Kisvarsanyi *et al.*, 1983). Mention must also be made of Economic Geology Monograph No. 3 (Brown, 1967), which still contains a wealth of useful information on MVT deposits.

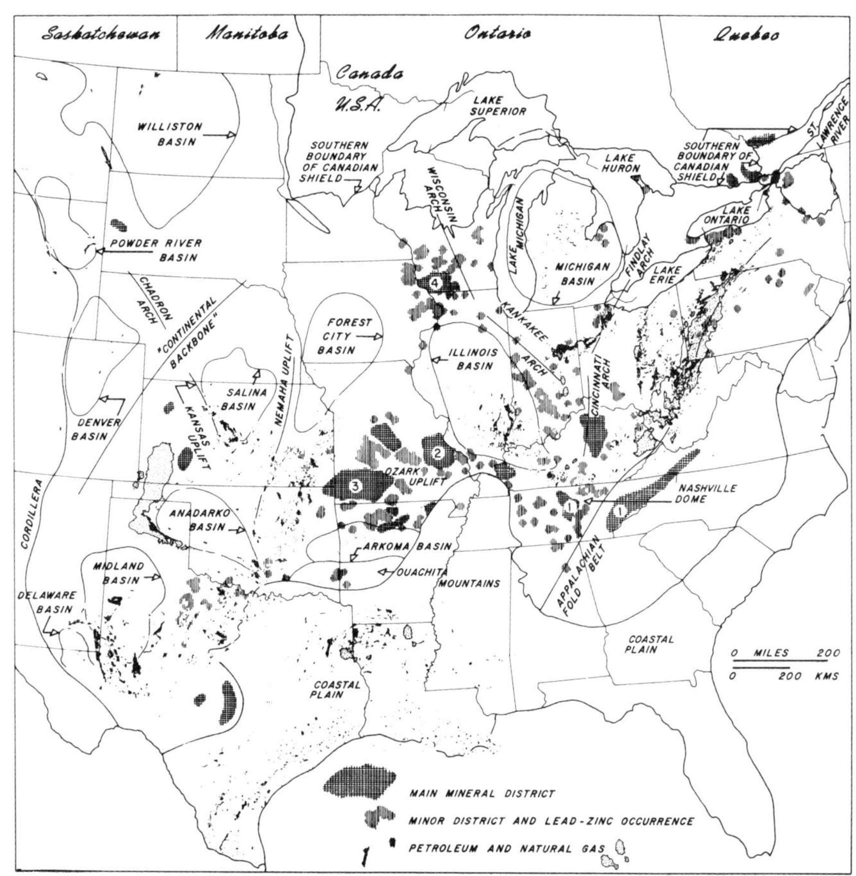

Figure 1 *Map of US showing major lead-zinc districts, petroleum and natural gas areas, and major basins. Lead-zinc districts:* **1** *- Tennessee,* **2** *- Southeast Missouri,* **3** *- Tri-state,* **4** *- Upper Mississippi or Wisconsin. Note concentration of metal areas between basins, and almost complete lack of overlap between metal and petroleum areas. (From Anderson, 1978).*

Tonnage and grade of a number of MVT districts and deposits are given by Gustafson and Williams (1981) and Barnes *et al.* (1981). It is important to realize that MVT ore deposits are simply unusually large representatives of an ubiquitous phenomenon: occurrences of sphalerite and galena are commonplace in carbonate successions.

Characteristics

Recognizing general characteristics minimizes differences, some of which may be highly important. MVT deposits are diverse! The following generalizations (some of them tentative) are based on Ohle's (1959) pioneer work, updated over the past two decades. In sum, these generalizations describe the MVT ore deposit class.

Setting. (1) Most deposits occur in carbonate rocks, with a strong bias toward dolomites. (2) Deposits occur in most sedimentary basins in all parts of the world. Host rocks range in age from Proterozoic to Cretaceous, although many fewer deposits are known in the Proterozoic, Jurassic and Cretaceous than in the Cambro-Ordovician and Carboniferous, which to 1962 contained about 80% of MVT deposits in the coterminous United States (Beales and Onasick, 1970). Deposits in Cenozoic rocks are curiously lacking. (3) Deposits tend to be found at or near the edges of basins as presently preserved or on arches between basins (Figure 1). (4) They may occur in relatively undisturbed platformal carbonates or within foreland fold and thrust belts. (5) Individual deposits, similar to one another in their occurrence, tend to occur in districts which may be distributed over hundreds of square kilometres (Figure 1). This argues against strictly local sources for metals and sulphur. (6) In most districts, local igneous rocks are unknown. (7) Host carbonate rocks are unmetamorphosed. (8) For settings located in relatively undisturbed platformal settings, stratigraphic evidence suggests that mineralization took place at relatively shallow depths, perhaps a few hundred to ~1000 m, thus involving pressures not exceeding several hundred atmospheres. Geothermal gradients of 25-30°C·km^{-1} are typical in these environments, yielding average host rock temperatures that are significantly less than 100-150°C; thus, either a heat source at shallow depths is a requisite or heated fluids must have migrated from deeper sources (*see* White, 1974; Macqueen and Powell, 1983). (9) Some deposits appear to be strongly controlled by unconformity-associated features such as paleokarst terrane (*e.g.*, East Tennessee; Daniel's Harbour; Nanisivik; Pine Point, Figures 3 and 4). (10) Some deposits tend to be localized along facies fronts between platformal carbonates and basinal shales (*e.g.*, Robb Lake). (11) Although early workers commonly suggested that deposits were located within carbonate reef masses, subsequent work on many deposits

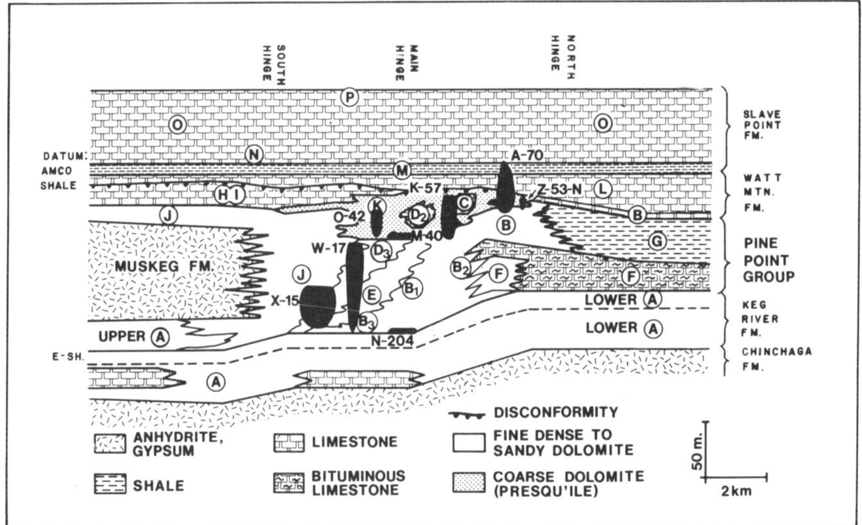

Figure 3 *Schematic south-north cross-section through Middle Devonian Pine Point property, NWT, Canada, showing: division into lithologic facies:* **A - P** *main rock types;* **E** *- shale markers (Keg River Fm.) and Amco Shale marker (Watt Mountain Fm.) which indicate influence of hinge zones identified across top of diagram; Muskeg evaporites of Elk Point Basin to south, and* **G** *facies shales of shale basin to north; barrier complex of facies* **A-L***; unconformity between facies* **H/I** *and* **L***, suggested to control karsting in underlying units; and projected stratigraphic locations of some of the major orebodies (A-70, K-57, X-15, etc.) Two facies* **(B, F)** *are rich in indigneous organic matter. Organic geochemical studies of indigenous organic matter, heavy oil and bitumen demonstrate that the carbonate barrier on the property has a low temperature history (~60°C maximum), and that the orebodies, with fluid inclusion filling temperatures to ~100°C, represent thermal anomalies (Macqueen and Powell, 1983). An attractive explanation for the origin of the orebodies and the pyrobitumen which is associated with them is abiogenic reduction of sulphate, a mechanism with which the sulphur isotope data are compatible (Powell and Macqueen, 1984). This is the "mixing" model described herein, with a local source for H_2S. Considering carbonate-hosted lead-zinc deposits as a whole, Pine Point is perhaps atypical in that it is situated between an evaporite basin to the south and a shale basin to the north, and the barrier may have acted as a conduit localizing fluid flow over a large area (Hitchon, 1969). Diagram from Macqueen and Powell (1983). Original diagram and lithologic facies after Skall, 1975; and Kyle, 1981. More detailed cross-sections reflecting increased knowledge of the stratigraphy of Pine Point can be found in Rhodes* et al. *(1984).*

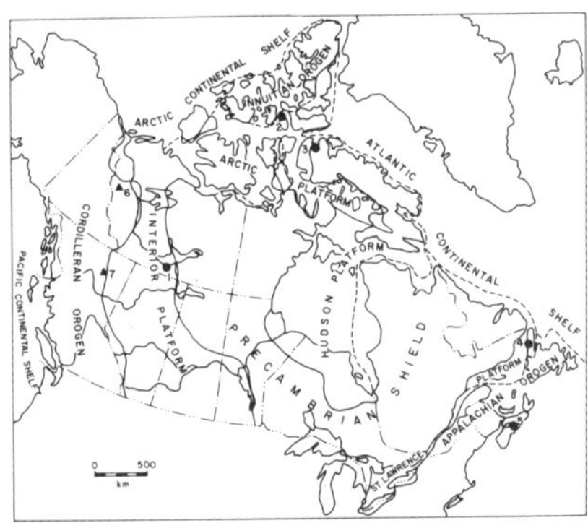

Figure 2 *Location of some major Canadian MVT lead-zinc deposits. Solid circles are existing mines, although only Pine Point, Nanisivik and Daniel's Harbour are currently producing; solid triangles represent possible future mines.* **1** *- Pine Point,* **2** *- Polaris,* **3** *- Nanisivik,* **4** *- Daniel's Harbour,* **5** *- Gays River,* **6** *- Gayna River,* **7** *- Robb Lake. Also see Macqueen (1976), and Sangster and Lancaster (1976) for the location of many new carbonate-hosted lead-zinc showings in the Rocky Mountain Belt.*

demonstrates that actual reef-hosted deposits are minor. Deposits are closely controlled by the prior development of porosity, and thus may be located in platformal carbonates of biostromal character or back-reef or fore-reef settings (*e.g.*, Pine Point, Figure 3; Skall, 1975; Rhodes *et al.*, 1984).

Deposits. (1) Most deposits have relatively simple mineralogy, with galena and/or sphalerite as the main ore minerals (Figure 4), nearly always accompanied by pyrite and/or marcasite. Barite and fluorite are common in some districts. Chalcopyrite is a minor to very minor associate of some deposits. Commonly, galena is low in silver and sphalerite is low in iron. (2) Deposits clearly are epigenetic: sulphides were emplaced in pre-existing pore spaces commonly developed within breccias or paleokarst topography which, in turn, developed within lithified carbonates (Figures 4 and 5). (3) Studies of fluid inclusions within coarsely crystalline sphalerite, barite and carbonates have established two important facts: (a) mineralization temperatures on average ranged from 80°C to 200°C and (b) ore-bearing fluids were highly saline Na-Ca-Cl brines, 5-10x the salinity of seawater (Roedder, 1976, 1979). (4) Organic material in the form of kerogen or bitumen in the host rocks and/or petroleum in fluid inclusions is very commonly observed in MVT districts (Figures 4E and 5B). (5) Isotope studies show that sulphur is generally heavy and with fairly wide-ranging values. This establishes a crustal, ultimately seawater, origin involving sulphate reduction at some time in the genetic story. Bacterial action is no longer believed to be the only viable mechanism of sulphate reduction below 100°C but is still a distinct possibility in many districts, as long as the site of reduction is at some distance from the site of sulphide deposition, since deposition temperatures are usually too high for the bacteria (Ohmoto and Rye, 1979; Trudinger *et al.*, 1985). Lead isotopes can also show a considerable range in any one district and are commonly highly radiogenic, yielding future ages (negative model ages). Data are generally consistent with a basement or reworked sediment (crustal) source for the lead, but do not point to specific sources (Doe and Zartman, 1979). Both the sulphur and lead isotope data

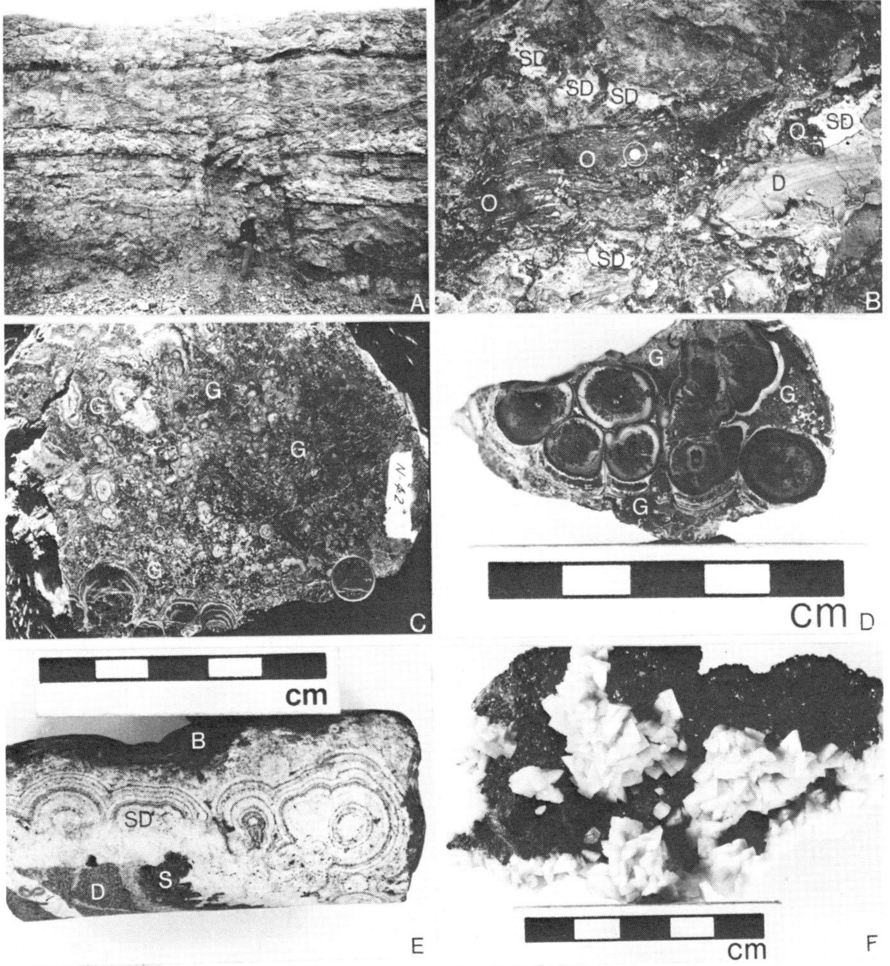

Figure 4 *All photos from Pine Point property, NWT.*
(A) *sagged solution breccia in Presqu'ile dolomite (facies K; Skall, 1975), orebody N-38A; note abundant white sparry dolomite.*
(B) *close view of breccia in main ore zone, orebody N-38A;* ***D****, host dolomite;* ***SD****, white sparry dolomite;* ***O****, high-grade banded and colloform lead-zinc ore; 25¢ piece (circled) gives scale.*
(C) *high-grade ore showing colloform sphalerite (light to medium grey), infilled with massive galena (**G**, dark grey), orebody N-42; 10¢ piece gives scale.*
(D) *close view of colloform sphalerite infilled with massive galena,* ***G****, orebody N-42.*
(E) *colloform white sparry dolomite,* ***SD****, alternating with black pyrobitumen, overlain by soft bitumen,* ***B****; also shows host dolomite,* ***D****; and sphalerite mineralization,* ***S****; drillcore.*
(F) *euhedral black sphalerite crystals intergrown with and overgrown by white sparry dolomite ("saddle dolomite" of Radke and Mathis, 1980); orebody M-40.*

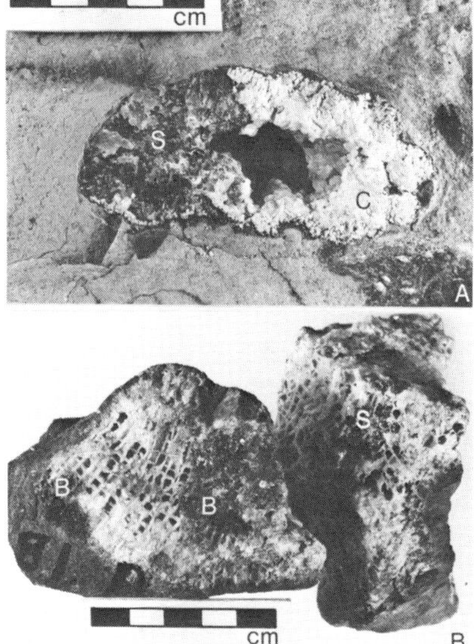

Figure 5 **(A)** *vug in dolomite of Middle Silurian Eramosa Member, Lockport-Amabel Formation, Guelph, Ontario, partly filled with sphalerite,* ***S****, and calcite,* ***C****.*
(B) *Location as (A); favositid corals showing local infill with sphalerite,* ***S****, right specimen; and bitumen,* ***B****, left specimen Eramosa rocks are rich in organic matter, which appears to have provided the H_2S for sphalerite and local galena precipitation (Tworo, 1985).*

furnish important constraints on models for individual areas, but cannot be expected to supply unique answers independent of other observations, anymore than can fluid inclusion or alteration studies. (6) Open space, developed by a variety of mechanisms in carbonate (Choquette and Pray, 1970), seems to be a prime requisite for the development of an economic deposit (Callahan, 1967) (also see Figures 3, 4 and 5). How much carbonate dissolution accompanies ore deposition is a matter of controversy (Heyl *et al.*, 1959; Beales, 1975, Anderson, 1983).

Model of Origin

Collectively, the above characteristics support the sedimentary-diagenetic model of origin. The now-classic statement of this model is that of Jackson and Beales (1967; also see Beales and Jackson, 1966), who used the Pine Point deposits as an example. In this model, basin-derived fluids which acquire heat, metals and other solutes during their travels, deposit sulphides in the carbonates they encounter as they emerge from deeper parts of the basin. Fluids driven by sediment compaction derive their metals through brine leaching, carry them as chloride or organic complexes and precipitate them as sulphides where H_2S is encountered. Jackson and Beales (1967) viewed H_2S as filtering into the migration route from nearby evaporites, where it was produced by sulphate-reducing bacteria possibly in the presence of petroleum. This model has survived remarkably well for the Pine Point deposits and has been applied in many other areas with various modifications. In fact, most subsequent discussions of MVT deposit origins accept the sedimentary-diagenetic or "basin evolution" approach and can be seen as variations of the Jackson-Beales model, but with differences as to source of metals, fluid drive mechanism, solution chemistry and mechanism of precipitation.

Brines found in fluid inclusions of the minerals of MVT deposits are so similar to those encountered in petroleum exploration boreholes in sedimentary basins, and deposits and basinal fluids are commonly so far removed from known igneous rocks, that the formerly popular magmatic-hydrothermal hypothesis of origin is no longer widely held.

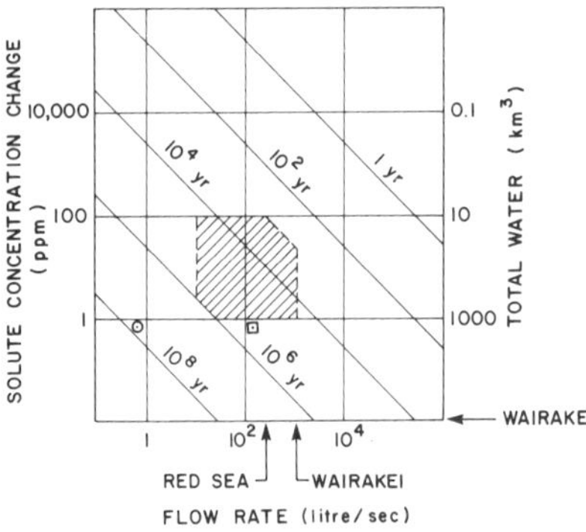

Figure 6 *Relationship between flow-rate, fluid volumes, concentration change and time for deposition of 20 million tons of 5% ore. Also shown are flow rates of the Red Sea, and Wairakei geothermal system (New Zealand). Geologically more reasonable conditions are shown in the shaded area, and are based on flow rates within an open cross-sectional area of 12,250* m² *(Roedder, 1960). As noted in the text, the specific discharge of 1.8* $m^3 \cdot yr^{-1}$ *per* m² *and 0.72 ppm metal modelled by Garven and Freeze (1982) results in conditions plotted at the small square using their cross-sectional area of 2 x* 10^6 m², *or at the small circled point using Roedder's (1960) cross-sectional area value of 12,150* m². *Original concept and diagram from Roedder (1960); used by Anderson (1978).*

Fluid Origins and Movement

Since sediment porosity decreases from 70-80% at the sediment-water interface to 0-20% at a depth of 3000 m or so, there is clearly an upward movement of large volumes of fluid caused by compaction. Basinal fluids invariably become more saline and warmer with depth either through shale filtration or evaporite dissolution (another area of controversy) and the upward movement of warm brines is generally believed to be responsible for the migration of petroleum from source beds to physical traps (Burst, 1976; Roberts and Cordell, 1980). It is natural to think along similar lines for the analogous movement of metals from source beds to chemical traps (Garrard, 1977). This analogy is made even more appealing for those cases (*e.g.*, Pine Point) where abundant bitumen is present and is suggested to be involved in the reduction of sulphate (Powell and Macqueen, 1984). No commercial metallic deposits are associated with petroleum reservoirs, however, and the geographic distributions of each are antipathetic, at least in the US (Figure 1).

Seawater has a normal salinity of 35 parts per thousand (‰), whereas brines from the deeper parts of sedimentary basins may reach salinities of 150-300‰, a dramatic increase. Brines with the highest salinities involve such a large increase as compared with seawater, that derivation of salts from evaporites seems essential. This could be either dissolution of salt beds, perhaps by surface-derived ground waters or by expulsion of interstitial fluids from evaporite beds.

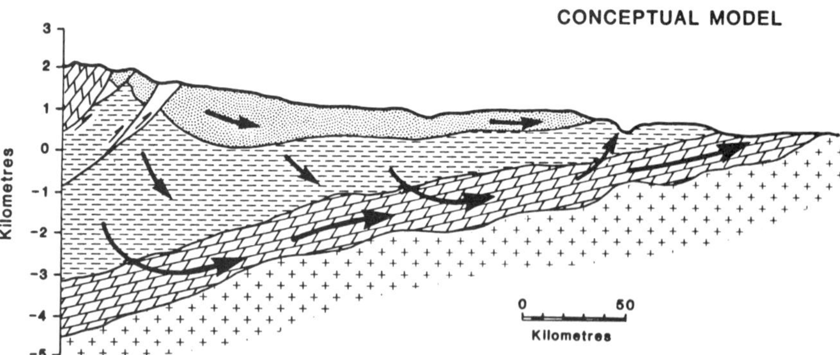

Figure 7 *Conceptual model of fluid in a carbonate unit which could be the locus of a stratabound orebody, as given by Garven and Freeze (1982, 1984a). Precambrian basement (+ pattern) is overlain by a carbonate unit which localized fluid flow from overlying shale units; compactional fluid flow through carbonate unit is enhanced by topographic cross-formational flow (Toth, 1980) from elevated thrust-faulted landmass to the left. Garven's and Freeze's (1982, 1984a,b) approach to quantify this conceptual model is to use finite element computer programs to solve the coupled equations of fluid flow, heat transport and mass transport for a series of two-dimensional cross-sections (normal to the figure) representative of sedimentary basins. Metals, leached from a source unit or units in the shales, spread throughout the basin as a function of time and flow parameters, and concentrate at the discharge end.*

Isotopic studies of oxygen and hydrogen show that the water in basinal brines is neither concentrated seawater nor meteoric water with added salts (Taylor, 1974; White, 1974). In each of five major sedimentary basins in North America, brine compositions appear to be best explained by the brines being complex mixtures of seawater and meteoric water, modified by rock-water interactions within the basin and possibly by other, unknown factors (White, 1974). Clearly, brines do not have a simple origin.

The mechanism and timing of migration of petroleum remains one of the most enigmatic features in basin evolution and much the same can be said of metal migration. In fact, the metal migration puzzle is probably worse; although petroleum is known to come from organic-rich rocks, practically every rock in the stratigraphic column has been proposed at one time or another as a source rock for metals! In addition, although compaction probably does produce large volumes of the right kind of fluids, it restricts deposit origins to an early stage of the history of the basin, and it is not at all clear how compaction, at the depths of 2 and 3 km required to give the requisite temperatures and salinities, would result in the flow of that fluid through the shallower carbonate units at the basin edge.

As shown in Figure 6, large deposits probably require on the order of thousands km^3 of fluid. For example, the Viburnum trend in SE Missouri contains 3×10^{12} g Pb (Gustafson and Williams, 1981, fig. 3). Precipitated at the rate of 10 ppm, this requires 3×10^{18} g or 3000 km^3 of water (saline fluid). Rickard *et al.* (1979) calculated that 2.4×10^{18} g water was needed to produce the Laisvall deposit, using 1 ppm Pb. Hitchon (1968) estimated that pore water in the Western Canada sedimentary basin now occupies 265,000 km^3, a truly impressive figure, especially since at least some of it contains considerable zinc (Table 1; Billings *et al.*, 1969).

If we assume that compaction at depths of 2-3 km results in an average of 0.1 litres of fluid per metre of burial (Hanor, 1979, based on Johns and Shimoyama, 1972), the 2.242×10^6 km^3 of sediments in the Western Canada basin (Hitchon, 1968) would produce approximately 200,000 km^3 of fluid, probably with a minimum temperature between 50°C and 100°C and a salinity of at least 15 wt.% (Hanor, 1979). Even allowing for the fact that the Western Canada basin is bigger than most, and that not all the fluids produced will be funnelled through an ore-depositing environment, it seems clear that we are dealing with a viable source of fluids. (For comparison, the oceans of the world contain 1.4×10^9 km^3 water).

As for the hydraulic drive required to flush these brines to higher levels, Toth (1980) suggested that topography-driven cross-formational flow can penetrate to deep levels, and over long periods of time can be involved in both petroleum and metal migration. This would free the timing of ore formation from the early basin formation stage apparently required if compaction alone is called upon to drive fluids. Another suggestion is that of Hanor (1979) who calculated that, at deeper levels in sedimentary basins, the observed increase in the density of brines due to increasing salt content is almost exactly counterbalanced by the decrease in density due to rising temperatures. Any slight increase in temperature at depth could produce gravitationally buoyant brines having a tendency to move upward.

Clearly, modelling of time-temperature-salinity-fluid flow history of sedimentary basins is a vital area for research (Jarvis and McKenzie, 1980). With respect to MVT deposits, Sharp (1978) pioneered numerical simulation of a compacting sedimentary basin, obtaining predicted one-dimensional fluid flow rates along a fault zone within the basin modelled. As noted by Cathles (1981), in a general review of fluid flow and the genesis of hydrothermal ore deposits, Sharp's (1978) data support the stratafugic, Jackson-Beales hypothesis for the genesis of MVT deposits. Garven and Freeze (1982, 1984a, b) applied two-dimensional numerical modelling to the problem by simultaneous solution of equations of fluid flow, heat transport, mass transport and chemical reactions, based on modern ground water and contaminant migration concepts. Figure 7 (from Garven and Freeze, 1982, 1984a) shows the general configuration of the basin and fluid flow regime they considered. The modelling allows one to follow the spreading of aqueous metal from a source bed in the shale unit throughout the rest of the basin. Their gravity-driven circulation model is capable of enormous variations, but the particular results cited in their 1982 paper give a specific discharge rate of 1.8 $m^3 \cdot yr^{-1}$ per m^2 at the deposition site. Using a sub-1 ppm level of metal precipitation (0.72 ppm), the volumes and times required for precipitation of 10^6 tons of metal (or 20×10^6 tons of 5% ore) are shown in Figure 6 for two different cross-sectional areas. These lie outside Roedder's (1960) chosen "reasonable" conditions primarily due to the low metal concentration used (shaded in Figure 6), but obviously the method shows great promise in constraining models of ore genesis. Garven (1985) has attempted this for Pine Point.

Table 1 Naturally occurring metal-bearing brines.

Location	Depth	Temperature	Metal Content (ppm) Zn	Metal Content (ppm) Pb	Reference
Gulf Coast	~ 2400-4000 m	100-500°C	360	100	Carpenter *et al.* (1974)
Northern Alberta	to ~ 1000 m	~ 76°C	19	—	Billings *et al.* (1969)
Cheleken Peninsula, USSR	to 1000 m	74-97.5°C	2.7	5.4	Lebedev (1973); White (1981)
Red Sea	sea floor	~ 56°C	5.4	0.63	White (1981)
Salton Sea, California	~ 1100 m	~ 340°C	540	102	Helgeson (1967); White (1981)
East Pacific Rise (21°N)	sea floor	350-410°C	unknown	unknown	Hekinian *et al.* (1980)

Ore Fluids

Fluid inclusion evidence from MVT deposits provides some of the best constraints on the origin of these deposits. There are more data on fluid inclusions from MVT deposits than any other ore deposits (Roedder, 1979). Inclusions studied reside in sphalerite, carbonates, fluorite and barite (this latter mineral commonly gives rather different results). The similarity of the fluid inclusion data from deposit to deposit is remarkable, regardless of whether the deposit consists largely of galena or sphalerite or fluorite (Roedder, 1979). The fluid inclusion data reveal the following characteristics (Roedder, 1976, 1979): (a) density is always greater than 1, and commonly greater than 1.1; (b) salinity is usually greater than 15 wt.% salts (150 ppt, implying a four-fold or greater increase over seawater), but NaCl "daughter" crystals are almost unknown, indicating the presence of appreciable amounts of ions in the inclusions other than Na and Cl; (c) the inclusions contain concentrated solutions of Na and Ca chlorides, with minor amounts of K, Mg, and Br, and (locally), heavy metals such as Cu, Zn; (d) organic matter is common, as methane or similar gases, or immiscible oil-like droplets, or in solution within the brines; (e) fluid inclusion filling temperatures cover the range of about 80°C to 200°C, but are most commonly in the range of 100°C to 150°C.

K/Na ratios have been determined on a number of fluids from MVT deposit inclusions, and they are all higher than the highest values in oilfield waters (Roedder, 1979). This could mean that the fluids are at least partly composed of interstitial fluids from evaporite beds, enriched in residual K. An alternative explanation, that the temperatures were generally higher for MVT fluids than for oilfield brines (resulting in a different exchange ratio with the sediments), does not seem to be borne out by comparing the temperature ranges involved: they are very similar.

There is scope for further work on the organic components of these inclusions, given the great progress made in our understanding of organic geochemistry (*e.g.*, Tissot and Welte, 1984; Macqueen and Powell, 1983). Meanwhile the fluid inclusion data are important components of any theory of origin of these deposits. For example, the high temperatures and salinities that characterize MVT deposits rule out postulated origins which feature cold surface waters alone as ore-forming fluids, including seawater or ground water.

Metals. Experimental work and direct observation of natural brines demonstrate that chloride-rich brines have the ability to leach trace quantities of metals from rocks through which they flow (*e.g.*, Ellis, 1968; Carpenter *et al.*, 1974). This may involve either desorption of loosely bound metals, or release of metals on recrystallization of certain mineral species (Helgeson, 1967). A third possibility is the release of metals from metal-organic complexes through thermal alteration or destruction of such complexes.

There is no agreement on which kinds of rocks may act as metal sources. Many workers have postulated shale sources because shales normally are (relatively) enriched in trace metals, and because changes in clay mineralogy and structure with increasing burial and temperature could release such metals (*e.g.*, Macqueen, 1976). Some have advocated a carbonate source for metals, with metals being released to aqueous solutions during the replacement of initial metastable forms aragonite and high-magnesium calcite by the stable forms, low Mg-calcite and dolomite (Dunsmore and Shearman, 1977). These carbonate mineralogical changes appear to occur too early in the diagenesis of sediments to be of much use as major metal sources for MVT deposits. A third possibility as a metal source is evaporite beds, as advocated by Davidson (1966) and recently revived by Thiede and Cameron (1978) from their study of metal concentrations in the Middle Devonian Elk Point evaporite sequences of the Western Canada basin. Their work suggests a possible explanation for one of the intriguing problems associated with MVT deposits — why they have so little in copper and iron compared with zinc and lead. The Elk Point data suggest that copper remains in the brine during the evaporation process, whereas lead and zinc enter solid phases such as gypsum and anhydrite. Copper-rich residual brines could lead to the formation of red-bed copper deposits on a local scale, whereas lead and zinc would be mobilized much later by circulating connate fluids which dissolved evaporites, becoming highly saline in the process. An alternative idea, mentioned by Anderson and Garven (1987) is that transport takes place in brines in which sulphate is slowly reduced to H_2S, or where H_2S is gradually acquired along the flow path, but that this H_2S is buffered at a very low level by the Cu content of the solution. This buffering action gradually depletes the solution of Cu, leaving the more soluble Pb and/or Zn untouched.

Chloride brines with geologically significant metal contents are known in a number of areas (Table 1). Gulf Coast, northern Alberta and Cheleken Peninsula brines occur in normal sedimentary basins; others shown in Table 1 are related to near-surface volcanic rocks or abnormal heat sources unknown in Mississippi Valley-type deposit settings. Nevertheless, these occurrences verify that metal transport in chloride-rich brines occurs widely in nature, something hardly suspected a couple of decades ago (White, 1981). Hitchon (1977, 1980) has shown that the content of effectively all components of geothermal brines of the type shown in Table 1 falls within the concentration limits of sedimentary basin formation waters. The high salinity and high metal content of the geothermal brines are matched by the deeper, hotter more saline formation waters (Hitchon, 1977, 1980). Hitchon (1977) could find no systematic differences in the composition of these brines that could not be reasonably attributed to differences in temperature and host rocks. Nevertheless, lead and zinc are not common in oilfield waters; minor element compositions of 832 oilfield waters showed none with Pb or Zn in excess of 1 ppm (Rittenhouse *et al.*, 1969).

Sulphur. Most sub-surface brines contain low amounts of sulphur, in the range of a few tens to a few thousands of parts per million. This is nearly always present as sulphate, rather than the reduced sulphide which is required to precipitate a sulphide ore deposit. Brines which contain even a small amount of H_2S are invariably very low in metal content, at least over the temperature ranges which appear to characterize MVT deposits ($< \sim 200°C$).

Transport

Brine Chemistry. A major geochemical problem in MVT deposits is whether the metals and reduced sulphur (H_2S and perhaps HS^-) were transported together to the site of deposition, or whether the metals were carried in an essentially H_2S-free solution and were precipitated by the addition of H_2S as the solution passed through the carbonate host rocks. Both experimental and geological observations bear on this point.

The parameters controlling the solubility of ZnS and PbS in NaCl-rich brines at 100-150°C are reasonably well understood (Barrett and Anderson, 1982), because there is a satisfying level of agreement between experimentally measured and theoretically calculated metal values under controlled conditions. This allows calculation of metal values of brines under various proposed geological conditions, and the general conclusion is that ZnS and especially PbS solubilities in the presence of significant H_2S concentrations are too low to allow transport of ore-forming quantities of both metals and H_2S in the same solution (Anderson, 1975, 1983). A possible exception to this generalization has been provided by Sverjensky *et al.* (1979) who, on the basis of lead and sulphur isotope data on galena from the Buick Mine, southeast Missouri, suggested that lead and sulphur were transported together in the same solution. Either the sulphur was present as sulphate, or if as sulphide, unrealistically low pH levels are required as noted.

Experimental work to date has only considered inorganic complexing, principally chloride. It is possible that as yet unknown complexing effects are important, and the most likely candidate at the moment is organic complexing. The concentration of organic ligands is small, but their effects could be large (Barnes *et al.*, 1981; Manning, 1986). Other aspects of brine chemistry are considered by Sverjensky (1984).

Deposition of Sulphides

Depositional mechanisms depend fundamentally on the nature of the metal-transporting solutions, and particularly the source of sulphide as noted above. Generally we can recognize two cases: one in which the metals and reduced sulphur (H_2S, HS^-) travel together in the ore-forming solution ("non-mixing" model), and a second case in which the sulphide is derived at the site of ore deposition, possibly by reduction of sulphate carried by the ore-forming solution ("mixing" model). Table 2 provides a summary of possible depositional mechanisms for each case.

"Non-Mixing" Model. Metals and reduced sulphur travel together. In this case, we appear to have a transport problem because metal solubilities in solutions containing appreciable amounts of reduced sulphur are so low. If this model is to operate under realistic geological conditions (*i.e.*, at pH values of 4-5 or greater, Figure 8), some form of metal transport other than chloride complexing is required, such as organic complexing. Bisulphide complexes (HS^-) for Pb have been ruled out (Hamann and Anderson, 1978; Giordano and Barnes, 1981). If this model should apply, Anderson (1975) has shown that precipitation of sulphides could occur through cooling, dilution by ground water, or by pH change. Geological evidence tending to support this model includes such features as the district-wide sphalerite banding recognized by McLimans *et al.* (1980) in the southwestern Wisconsin district (Figure 1). They argue that individual local sources of sulphur could not produce such a consistent feature. In some areas, *e.g.*, SE Missouri (Figure 1), extensive dissolution of previously deposited sulphides has taken place. This is difficult to explain if the sulphides are as insoluble as advocated by the "mixing" model school of thought. Large, well-formed sulphide crystals also have been considered by some workers to indicate the absence of large degrees of supersaturation or large concentration gradients, both more likely features of the "mixing" models.

"Mixing" Model. Here, sulphide is supplied at the site of deposition by a number of possible mechanisms, as seen in Table 2. Because low pH conditions are required to transport metals and sulphides together, Anderson (1975), Beales (1975) and others have favoured this model. "Mixing" models which involve adding H_2S to metal-bearing brines at the site of ore deposition (*e.g.*, by brine-derived or locally derived sulphate being reduced by reactions involving organic matter) are much more attractive hydrodynamically than the mixing of two transported solutions, one carrying metals and the other carrying reduced sulphur.

The only reasonable source of reduced sulphur is sulphate, either from locally available sulphate minerals or dissolved brine-transported sulphate. Organic matter, common in many MVT settings, may be involved in sulphate reduction at the temperatures postulated (Table 2). Bacterial sulphate reduction is unlikely at temperatures above ~80°C. Thermal degradation of petroleum to yield sulphide has yet to be evaluated for most settings, but is not a likely factor at Pine Point (Macqueen and Powell, 1983).

Bringing metals and H_2S together at the site of ore deposition should be capable of generating large concentration gradients, high degrees of supersaturation, and relatively rapid precipitation of sulphides. These factors in turn should result in very fine-grained crystals; dendritic, hopper and other high-energy crystal forms; and isotopic and compositional disequilibrium. The sulphides at Pine Point show evidence of relatively rapid deposition and crystal growth in the form of dendritic and hopper galena, which is consistent with the addition of H_2S to metal-bearing brines. Unlike many settings, Pine Point has an abundance of locally available sulphate. All of these features have been found in places in some MVT deposits, but they are by no means typical.

Recent work at Pine Point indicates that H_2S was derived locally through

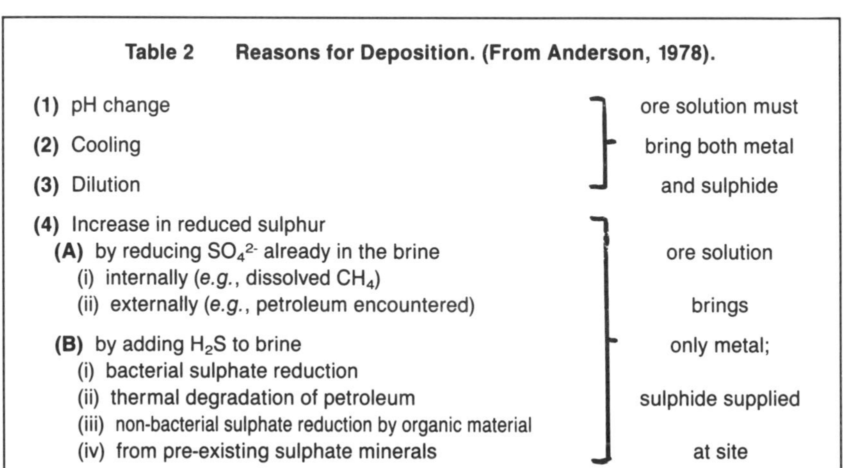

Table 2 Reasons for Deposition. (From Anderson, 1978).

Reason	
(1) pH change **(2)** Cooling **(3)** Dilution	ore solution must bring both metal and sulphide
(4) Increase in reduced sulphur **(A)** by reducing SO_4^{2-} already in the brine (i) internally (*e.g.*, dissolved CH_4) (ii) externally (*e.g.*, petroleum encountered) **(B)** by adding H_2S to brine (i) bacterial sulphate reduction (ii) thermal degradation of petroleum (iii) non-bacterial sulphate reduction by organic material (iv) from pre-existing sulphate minerals	ore solution brings only metal; sulphide supplied at site

thermochemical (abiological) reactions involving sulphate, bitumen and H_2S (Macqueen and Powell, 1983; Powell and Macqueen, 1984). Sulphur isotope data from metallic sulphides, sulphate and bitumens at Pine Point are consistent with such a process having operated, as are mass balance calculations. To date, however, a biological reduction of sulphate has not been demonstrated in the laboratory at temperatures below about 250°C (Trudinger *et al.*, 1985). Possibly as yet unknown catalysts are involved: further work is in progress. For Pine Point, fluid inclusions contain significant amounts of Ca and S but lack anhydrite, indicating that the S is reduced and thus lending support to a mixing model of sulphide precipitation (Haynes and Kesler, 1987).

On the other hand, the "mixing" process may take place extremely slowly, say by diffusion of H_2S through wall rocks into the hydrothermal system, resulting in typical well-formed and large sulphide crystals considered by some workers to be indicative of the operation of the "non-mixing" model! The problem is not solved! If large sulphide crystals are involved, however, slow addition of reduced sulphur is required to prevent widespread nucleation and the deposition of a sludge of extremely fine-grained sulphide crystals ("dumping" of sulphide). This also implies only very slight degrees of supersaturation. In connection with these kinds of statements, it has to be pointed out that due to a complete lack of data on the effect of growth rate on the morphology of sulphide crystals, we really do not know what "slow" or "fast" means in terms of months or years. Maybe "slow" is geologically "fast".

If the H_2S is produced locally, one should be able to see some evidence of this, such as the presence of gypsum or anhydrite as well as a reducing agent such as organic material. The association of bitumen with MVT deposits is quite characteristic, but association with sulphates is less so, except for Pine Point. Beales and Hardy (1980) have argued that for MVT deposit sites where sulphates are missing, their former presence ("occult gypsum") can be demonstrated by gypsum inclusions in dolomite. Their interpretation is supported by the work of Radke and Mathis (1980), who suggested that white sparry dolomite (their "saddle dolomite"), so common in many MVT settings, is related to sulphate reduction.

Depositional Setting. One of the most striking features of MVT deposits is their association with carbonates, and further, with zones of extensive localized solution and collapse of carbonates (Ohle, 1985). This is often karst-related, but not always so, and considerable variation seems to exist in the relative timing of carbonate solution and sulphide precipitation. Where sulphide brecciation is observed, post-sulphide carbonate solution is inferred. Such solution could be simultaneous with sulphide deposition, however, and the common association of ore and solution features leads to the hypothesis that precipitation is caused by the pH change occurring where the metal-bearing solution encounters the carbonate rocks. In this case, however, since reaction with carbonates is very rapid, we should find the deposits strongly associated with facies changes, whereas they are commonly surrounded by carbonate rocks.

Although the ore solution may be in equilibrium with carbonate during transport, deposition of sulphides from chloride complexes will inevitably release acid, for example,

$ZnCl_2\,(aq) + H_2S = ZnS + 2HCl\,(aq)$.

This acid could not escape from the system without dissolving carbonate. Thus, sulphide deposition itself should help in creating solution collapse and open space, and the association between ores and collapse zones including brecciated ore becomes easier to understand. Slight cooling in the ore zone would also result in carbonate dissolution, but leaves open the question of why solution and collapse are often so closely related to the sulphide zones.

Either, or both, fluorite and barite are commonly associated with MVT deposits. Holland and Malinin (1979) discussed the solubility of these minerals and mentioned that decrease in temperature is probably a common reason for precipitation of both. Very small

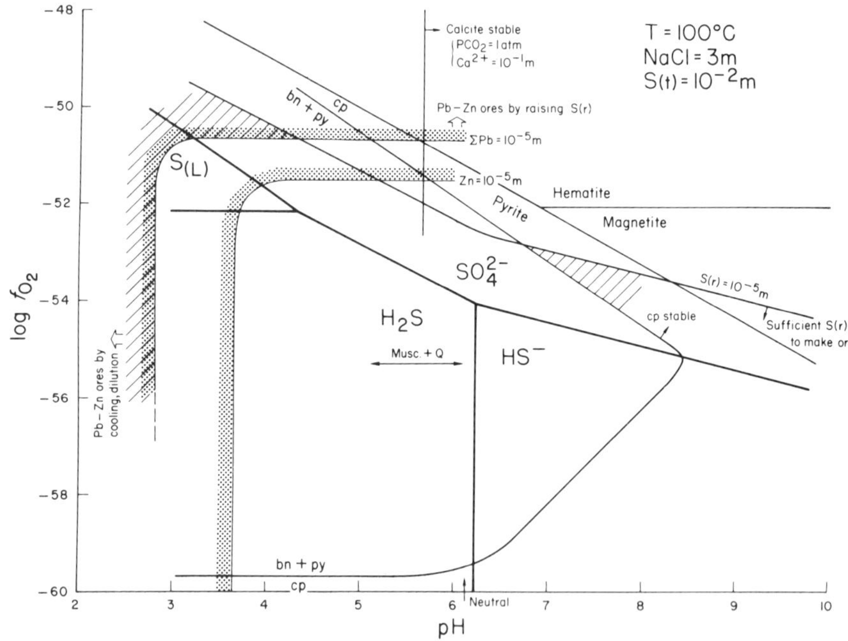

Figure 8 *Sphalerite and galena solubilities and mineral stability fields as a function of oxidation state (log fO_2) and acidity (pH). The diagram indicates how experimentally-derived solubility data can be related to mineral stability fields in attempting to understand conditions of transport and deposition. S(r), S(t), S(L): reduced, total and liquid sulphur, bn = bornite, py = pyrite, cp = chalcopyrite. Musc + Q: pH stability range of muscovite + quartz. The left-hand hachured area is the range of conditions where more than 2 ppm of Pb can be transported with sufficient H_2S to provide ore. This is generally below a pH of 3 except for a little wedge up to pH 4. Alternatively, hundreds of ppm Pb can be transported above fO_2 of 10^{-51} with sulphur as sulphate. Sphalerite is more soluble than galena, requiring a less acid pH to achieve the same concentration as shown by the stippled 10^{-5} Zn contour. The assumed conditions of course may be oversimplified, as in most models. (Modified from Anderson, 1975).*

temperature changes are likely in the MVT situation, but if sulphide precipitation was accompanied by carbonate dissolution, the increase in calcium would precipitate fluorite if the solution was near saturation with fluorite. Nordstrom and Jenne (1977) have shown that this is the case for geothermal waters in the western US. Thus Ca^{2+} is released by

$CaCO_3 + 2H^+ = Ca^{2+} + CO_2 + H_2O$

and then reacts with F ions:

$Ca^2 + 2F^- = CaF_2$.

The presence of barite in MVT ores has been used as an argument against an oxidized ore solution, since the solubility of barite is so small in SO_4^{2-} bearing solutions. Strictly speaking, it is an argument against an SO_4^{2-}-bearing brine, not against an oxidized brine. A low-SO_4^{2-}, Ba^{2+}-bearing brine, on encountering a region in its flow path rich in H_2S, would be expected to precipitate barite either on the periphery of the H_2S zone or above it, since an H_2S-rich zone would be surrounded by an oxidized SO_4^{2-}-rich zone. This is the case for example in the Pennines (Sawkins, 1966). The dissolution and precipitation of carbonates, fluorite and barite is discussed more fully by Anderson (1983).

Exploration Philosophy

The discovery of many carbonate-hosted lead-zinc showings in the Rocky Mountain Belt in the 1970s (Macqueen, 1976; Sangster and Lancaster, 1976) demonstrates that MVT deposit characteristics discussed above are instrumental in exploration. What are the general exploration concepts involved? Favourable ground includes carbonate successions developed on the flanks of large, deep basins (*e.g.*, Pine Point, Polaris, Daniel's Harbour, *etc.*). In these basin margin settings, unconformities and possible karsts are "good signs". Basement control in the form of arching, flexing or faulting may provide fluid escape routes or heat sources or both, also "good signs". Dolomitized sequences, and the presence of evaporites and organic matter are all favourable indicators. Carbonate fronts between basinal shales and platformal carbonates are also promising. The major control on the MVT ore deposits is not merely a structural or stratigraphic trap, but appears to be the happy coincidence of large volumes of pore space, base metals, and a source of sulphide or a precipitating agent. Promising platformal successions where these requisites may be met include the lower Paleozoic of the Mackenzie River area, NWT, the Silurian of Ontario (Tworo, 1985), and the Ordovician of Manitoba. As we gain increased understanding of existing deposits such as Pine Point, we will undoubtedly be in a better position to prospect such vast areas as the Lower Mackenzie River, and southern Manitoba and Ontario.

Problems

Of all the problems noted, perhaps *timing* is the most critical. For most MVT deposit settings, we have almost no idea when mineralization occurred. Beales *et al.* (1980), using paleomagnetism, obtained a Late Pennsylvanian age (paleomagnetic pole) for a deposit hosted by Cambrian carbonate rocks in the Viburnum trend, SE Missouri. Isotopic dating of microgram quantities of Rb, Sr and U in galenas are exceptionally painstaking, but potentially valuable, dating methods (R.L. Armstrong, pers. comm., 1980). K/Ar dating is another promising approach (York *et al.*, 1980). Better knowledge of timing will permit us to make much more conclusive statements about where mineralization fits in the overall scheme of basin evolution. Chemistry remains a problem: why do these deposits contain only lead and zinc in economic quantities? Is it the gathering mechanisms or the precipitating mechanism that chooses the lead and zinc (Meyer, 1981)? All evidence at the moment points to source rock control.

What is the local and regional *hydrology* of particular MVT districts and deposits? In particular, what flow paths are involved and over what dimensions and time durations did they operate? Hydrologists are most comfortable with the dynamic systems encountered in modern ground-water studies, but it is clear that their approach has much to offer. Might the fluids responsible for one type of base metal deposit have also formed another type earlier/deeper in their evolution, due to differing physical and chemical conditions? Can the fluids which move petroleum also be ore-forming fluids (*e.g.*, Manning, 1986)?

Are *evaporites* critical to the origin of MVT deposits, either in the fluid migration routes or at the deposition site, as suggested by the high salinities in the fluid inclusions? Why are MVT deposits almost invariably associated with dolomites (dolostones of some authors) rather than limestones? How much *solution of carbonate* is there at any particular property or in any district and what has caused it? What is the contribution of *organic matter* to the story: do hydrocarbons or hydrocarbon/sulphate reactions govern the location of many or most MVT deposits by supplying H_2S locally? What causes *metal zoning* in districts or at individual properties: Pine Point, for example, tends to have lead-cored orebodies with zinc and iron envelopes (Kyle, 1981) — why? What *ground preparation mechanisms* govern the location of particular deposits? What role do *broad-scale tectonic processes*, for example, plate margin interactions or spreading rate sea-level changes, play in the story? Some of these questions have implications for prospecting, whereas others are more academic.

Three areas seem to us to offer outstanding promise in MVT deposit research. These are dating of deposits, either by paleomagnetic means or by direct isotopic dating; further observational and experimental assessment of the role of organic matter in metal-organic complexing reactions or in generating or providing H_2S; and, hydrogeological modelling of MVT deposit basinal systems. There is no shortage of intriguing problems to study!

Acknowledgements

It is a pleasure to acknowledge the support and interest of colleagues, including some MVT workers who are more knowledgeable than we are about specific settings. We are appreciative of Natural Sciences Engineering and Research Council of Canada and Department of Energy, Mines and Resources support of our work. The manuscript has benefited from suggestions made by F.W. Beales, R.H. McNutt and R.G. Roberts, but we are responsible for any errors which remain.

References

Akande, S.O. and Zentillli, M., 1984, Geologic, Fluid Inclusions and Stable Isotope Studies of the Gays River Lead-Zinc Deposit, Nova Scotia, Canada: Economic Geology, v. 79, p. 1187-1211.

Anderson, G.M., 1975, Precipitation of Mississippi Valley-type ores: Economic Geology, v. 70, p. 937-942.

Anderson, G.M., 1978, Basinal brines and Mississippi Valley-type ore deposits: Episodes, no. 2, p. 15-10.

Anderson, G.M., 1983, Some geochemical aspects of sulphide precipitation in carbonate rocks, *in* Kisvarsanyi, G., Grant, S.K., Pratt, W.P. and Koenig, J.W., eds., International Conference on Mississippi Valley-Type Lead-Zinc Deposits, Proceedings Volume: University of Missouri-Rolla, Rolla, Missouri, p. 61-76.

Anderson, G.M. and Garven, G., 1987, Sulfate-Sulfide-Carbonate Associations in Mississippi Valley-Type Lead-Zinc Deposits: Economic Geology, v. 82, p. 482-488.

Barnes, H.L., Adams, S.S. and Rose, A.W., 1981, Ores formed by Diagenetic and Metamorphic Processes, *in* Mineral Resources: Genetic Understanding for Practical Applications: Studies in Geophysics, National Academy Press, Washington, D.C., p. 73-81.

Barrett, T.J. and Anderson, G.M., 1982, The solubility of sphalerite and galena in NaCl brines: Economic Geology, v. 79, p. 1923-1933.

Beales, F.W., 1975, Precipitation mechanisms for Mississippi Valley-type deposits: Economic Geology, v. 70, p. 943-948.

Beales, F.W. and Hardy, J.W., 1980, Criteria for the recognition of diverse dolomite types with an emphasis on studies of host rocks for Mississippi Valley-type ore deposits, *in* Zenger, D.J., Dunham, J.B. and Ethington, R.L., eds., Concepts and Models of Dolomitization: Society of Economic Paleontologists and Mineralogists, Special Publication No. 28, p. 197-213.

Beales, F.W., Jackson, K.C., Jowett, E.C., Pearce, F.W. and Wu, Y., 1980, Paleomagnetism applied to the study of timing in stratigraphy with special reference to ore and petroleum problems, *in* Strangway, D.W., ed., The Continental Crust and Its Mineral Deposits: Geological Association of Canada, Special Paper 20, p. 789-804.

Beales, F.W. and Jackson, S.A., 1966, Precipitation of lead-zinc ores in carbonate rocks as illustrated by Pine Point ore field: Canadian Institute of Mining and Metallurgy, Transactions, V. 75, p. B278-B285.

Beales, F.W. and Onasick, E.P., 1970, The stratigraphic habitat of Mississippi Valley-type ore bodies: Institution of Mining and Metallurgy, Transactions, v. 79, p. B145-B154.

Billings, G.K., Kessler, S.E. and Jackson, S.E., 1969, Relation of zinc-rich formation waters, northern Alberta, to the Pine Point ore deposit: Economic Geology, v. 64, p. 385-391.

Bjørlykke, A. and Sangster, D.F., 1981, An overview of sandstone lead deposits and their relationship to red-bed copper and carbonate-hosted lead-zinc deposits: Economic Geology, 75th Anniversary Volume, p. 179-213.

Brown, J.S., 1967, ed., Genesis of Stratiform Lead-Zinc-Barite-Fluorite Deposits: Economic Geology, Monograph 3, 443 p.

Burst, J.F., 1976, Argillaceous sediment dewatering: Annual Review of Earth and Planetary Sciences, v. 4, p. 293-318.

Callahan, W.H., 1967, Some spatial and temporal aspects of the localization of Mississippi Valley-Appalachian type ore deposits, *in* Brown, J.S., ed., Genesis of Stratiform Lead-Zinc-Barite-Fluorite Deposits: Economic Geology, Monograph 3, p. 14-19.

Carpenter, A.B., Trout, M.L. and Pickett, E.E., 1974, Preliminary report on the origin and evolution of lead and zinc-rich oil field brines in Central Mississippi: Economic Geology, v. 69, p. 1191-1206.

Cathles, L.M., 1981, Fluid flow and genesis of hydrothermal ore deposits: Economic Geology, 75th Anniversary Volume, p. 424-457.

Choquette, P. and Pray, L.C., 1970, Geologic nomenclature and classification of porosity in sedimentary carbonates: American Association of Petroleum Geologists, Bulletin, v. 54, p. 207-250.

Collins, J.A. and Smith, L., 1975, Zinc deposits related to diagenesis and intrakarstic sedimentation in the Lower Ordovician St. George Formation, western Newfoundland: Canadian Society of Petroleum Geologists, Bulletin, v. 23, p. 393-427.

Coron, C.R., 1982, Facies relations and ore genesis of the Newfoundland zinc mines deposit, Daniel's Harbour, western Newfoundland, unpublished Ph.D. thesis, University of Toronto.

Davidson, C.F., 1966, Some genetic relationships between ore deposits and evaporites: Institution of Mining and Metallurgy, Transactions, v. 75, p. B216-B225.

Doe, B.R. and Zartman, R.E., 1979, Plumbotectonics, the Phanerozoic, *in* Barnes, H.L., ed., Geochemistry of Hydrothermal Ore Deposits, Second Edition: Wiley, New York, p. 22-70.

Dunsmore, H. and Shearman, D.J., 1977, Mississippi Valley-type lead-zinc orebodies: a sedimentary and diagenetic origin, *in* Garrard, P., ed., Proceedings of the forum on oil and ore in sediments: Imperial College, London, England, p. 189-201.

Ellis, A.J., 1968, Natural hydrothermal systems and experimental hot water/rock interaction: reactions with NaCl solutions and trace metal extraction: Geochimica et Cosmochimica Acta, v. 32, p. 1313-1363.

Garrard, P., 1977, ed., Proceedings of the forum on oil and ore in sediments: Imperial College, London, England, 205 p.

Garven, G., 1985, The role of regional fluid flow in the genesis of the Pine Point deposit, Western Canada Sedimentary basin: Economic Geology, v. 80, p. 307-324.

Garven, G. and Freeze, R.A., 1982, The role of regional groundwater flow in the formation of ore deposits in sedimentary basins: a quantitative analysis, *in* Ozoray, G., ed., Proceedings of the Second National Hydrogeological Conference: International Association of Hydrogeologists, Canadian Chapter, p. 59-68.

Garven, G. and Freeze, R.A., 1984a, Theoretical analysis of the role of ground water flow in the genesis of stratabound ore deposits. 1. Mathematical and Numerical Model: American Journal of Science, v. 284, p. 1085-1124.

Garven, G. and Freeze, R.A., 1984b, Theoretical analysis of the role of ground water flow in the genesis of stratabound ore deposits. 2. Quantitative Results: American Journal of Science, v. 284, p. 1125-1174.

Giordano, T.H. and Barnes, H.L., 1981, Lead transport in Mississippi Valley-type ore solutions: Economic Geology, v. 76, p. 2200-2211.

Gustafson, L.B. and Williams, N., 1981, Sediment-hosted stratiform deposits of copper, lead, and zinc: Economic Geology, 75th Anniversary Volume, p. 139-178.

Hamann, R.J. and Anderson, G.M., 1978, Solubility of galena in sulphur-rich solutions: Economic Geology, v. 73, p. 96-100.

Hanor, J.S., 1979, The sedimentary genesis of hydrothermal fluids, *in* Barnes, H.L., ed., Geochemistry of Hydrothermal Ore Deposits, Second Edition: Wiley, New York, p. 157-172.

Hardy, J.L., 1979, Stratigraphy, brecciation, and mineralization: Gayna River, Northwest Territories, unpublished M.Sc. thesis, University of Toronto, 461 p.

Haynes, F.M. and Kesler, S.E., 1987, Chemical evolution of brines during Mississippi Valley-type mineralization: Evidence from East Tennessee and Pine Point: Economic Geology, v. 82, p. 53-71.

Hekinian, R., Fevrier, M., Bischoff, J.L., Picot, P. and Shanks, W.C., 1980, Sulphide deposits from the East Pacific Rise near 21°N: Science, v. 207, p. 1433-1444.

Helgeson, H.C., 1967, Silicate metamorphism in sediments and the genesis of hydrothermal ore solutions, *in* Brown, J.S., ed., Genesis of Stratiform Lead-Zinc-Barite-Fluorite Deposits: Economic Geology, Monograph 3, p. 333-342.

Heyl, A.V., Agnew, A.F., Lyons, E.J. and Behre, C.H., Jr., 1959, The geology of the upper Mississippi Valley lead-zinc district (Illinois-Iowa-Wisconsin): United States Geological Survey, Professional Paper 309, 310 p.

Heyl, A.V., Landis, G.P. and Zartman, R.E., 1974, Isotopic evidence for the origin of Mississippi Valley-type mineral deposits: a review: Economic Geology, v. 67, p. 992-1006.

Hitchon, B., 1968, Rock volume and pore volume data for plains region of Western Canada sedimentary basin between latitudes 49° and 60°N: American Association of Petroleum Geologists, Bulletin, v. 52, p. 2318-2323.

Hitchon, B., 1969, Fluid flow in western Canada sedimentary basin, I. Effect on topography: Water Resources Research, v. 5, p. 186-195.

Hitchon B., 1977, Geochemical links between oil fields and ore deposits in sedimentary rocks, *in* Garrard, P., ed., Proceedings of the forum on oil and ore in sediments: Imperial College, London, England, p. 1037.

Hitchon, B., 1980, Some economic aspects of water-rock interaction, *in* Roberts, W.H., III and Cordell, R.J., eds., Problems of Petroleum Migration: American Association of Petroleum Geologists, Studies in Geology No. 10, p. 461-508.

Holland, H.D. and Malinin, S.G., 1979, The solubility and occurrence of non-ore minerals, *in* Barnes, H.L., ed., Geochemistry of Hydrothermal Ore Deposits, Second Edition: Wiley, New York, p. 461-508.

Jackson, S.A. and Beales, F.W., 1967, An aspect of sedimentary basin evolution: the concentration of Mississippi Valley-type ores during the late stages of diagenesis: Canadian Society of Petroleum Geologists, Bulletin, v. 15, p. 393-433.

Jarvis, G.T. and McKenzie, D.P., 1980, Sedimentary basin formation with finite extension rates: Earth and Planetary Science Letters, v. 48, p. 42-52.

Johns, W.D. and Shimoyama, A., 1972, Clay minerals and petroleum forming reactions during burial and diagenesis: American Association of Petroleum Geologists, Bulletin, v. 56, p. 2160-2167.

Kerr, J.W., 1977, Cornwallis lead-zinc district: Mississippi Valley-type deposits controlled by stratigraphy and tectonics: Canadian Journal of Earth Sciences, v. 14, p. 1402-1426.

Kisvarsanyi, G., Grant, S.K., Pratt, W.P. and Koenig, J.W., 1983, eds., International Conference on Mississippi Valley-Type Lead-Zinc Deposits, Proceedings Volume: University of Missouri-Rolla, Rolla, Missouri, 603 p.

Krebs, W. and Macqueen, R.W., 1984, Sequence of Diagenetic and Mineralization Events, Pine Point Lead-Zinc Property, Northwest Territories, Canada: Canadian Society of Petroleum Geologists, Bulletin, v. 32, p. 434-464.

Kyle, J.R., 1981, Geology of the Pine Point lead-zinc district, *in* Wolf, K.H., ed., Handbook of Strata-Bound and Stratiform Ore Deposits: Elsevier, Amsterdam, v. 9, p. 643-741.

Lebedev, L.M., 1973, Minerals of contemporary hydrotherms of Cheleken: Geochemistry International, v. 9, p. 485-504.

Macqueen, R.W., 1976, Sediments, zinc and lead, Rocky Mountain Belt, Canadian Cordillera: Geoscience Canada, v. 3, p. 71-81.

Macqueen, R.W., 1979, Base metal deposits in sedimentary rocks: some approaches: Geoscience Canada, v. 6, p. 3-9.

Macqueen, R.W. and Powell, T.G., 1983, Organic Geochemistry of the Pine Point Lead-Zinc Orefield and Region, Northwest Territories, Canada: Economic Geology, v. 78, p. 1-25.

Macqueen, R.W. and Thompson, R.I., 1978, Carbonate-hosted lead-zinc occurrences in northeastern British Columbia with emphasis on the Robb Lake deposit: Canadian Journal of Earth Sciences, v. 15, p. 1737-1762.

Manning, D.A.C., 1986, Assessment of the role of organic matter in ore transport processes in low-temperature base-metal systems: Institution of Mining and Metallurgy, Transactions, v. 95, p. B195-B200.

Manns, F.T., 1981, Stratigraphic aspects of the Silurian-Devonian sequence hosting zinc and lead mineralization near Robb Lake, northeastern British Columbia, unpublished Ph.D. thesis, University of Toronto, 334 p.

McLimans, R.K., Barnes, H.L. and Ohmoto, H., 1980, Sphalerite stratigraphy of the upper Mississippi Valley zinc-lead district, southwestern Wisconsin: Economic Geology, v. 75, p. 351-361.

Meyer, C., 1981, Ore-forming processes in geologic history: Economic Geology, 75th Anniversary Volume, p. 6-41.

Nordstrom, D.K. and Jenne, E.A., 1977, Fluorite solubility equilibria in selected geothermal waters: Geochimica et Cosmochimica Acta, v. 41, p. 175-188.

Ohle, E.L., 1959, Some considerations in determining the origin of ore deposits of Mississippi Valley-type: Economic Geology, v. 54, p. 769-789.

Ohle, E.L., 1980, Some considerations in determining the origin of ore deposits of the Mississippi Valley-type-Part II: Economic Geology, v. 75, p. 161-172.

Ohle, E.L., 1985, Breccias in Mississippi Valley-Type Deposits: Economic Geology, v. 80, p. 1736-1752.

Ohmoto, H. and Rye, R.O., 1979, Isotopes of sulphur and carbon, *in* Barnes, H.L., ed., Geochemistry of Hydrothermal Ore Deposits, Second Edition: Wiley, New York, p. 509-567.

Olson, R.A., 1984, Genesis of paleokarst and strata-bound zinc-lead sulphide deposits in a Proterozoic dolostone, Northern Baffin Island, Canada: Economic Geology, v. 79, p. 1056-1103.

Powell, T.G. and Macqueen, 1984, Precipitation of sulphide ores and organic matter: Sulphate reactions at Pine Point, Canada: Science, v. 224, p. 63-66.

Radke, B.M. and Mathis, R.L., 1980, On the formation and occurrence of saddle dolomite: Journal of Sedimentary Petrology, v. 50, p. 1149-1168.

Ravenhurst, S.E., Reynolds, P.H., Zentilli, M. and Akande, S.O., 1987, Isotope constraints on the genesis of Zn-Pb mineralization at Gays River, Nova Scotia, Canada: Economic Geology, v. 82, p. 1294-1308.

Rhodes, D., Lantos, E.A., Lantos, J.A., Webb, R.J. and Owens, D.C., 1984, Pine Point orebodies and their relationship to the stratigraphy, structure, dolomitization and karstification of the Middle Devonian Barrier Complex: Economic Geoogy, v. 79, p. 991-1055.

Rickard, D.T., Wilden, M.Y., Marinder, N.E. and Donnelly, T.H., 1979, Studies on the genesis of the Laisvall sandstone lead-zinc deposits, Sweden: Economic Geology, v. 74, p. 1255-1285.

Rittenhouse, G., Fulton, R.B., III, Grabowski, R.J. and Bernard, J.L., 1969, Minor elements in oil-field waters: Chemical Geology, v. 4, p. 189-209.

Roberts, W.H., III and Cordell, R.J., 1980, eds., Problems of petroleum migration: American Association of Petroleum Geologists, Studies in Geology No. 10, 273 p.

Roedder, E., 1960, Fluid inclusions as samples of the ore-forming fluids: International Geological Congress, Copenhagen, Report Part 16, p. 218-229.

Roedder, E., 1976, Fluid inclusion evidence in the genesis of ores in sedimentary and volcanic rocks, *in* Wolf, K.H., ed., Handbook of Strata-Bound and Stratiform Ore Deposits: Elsevier, Amsterdam, v. 4, p. 67-110.

Roedder, E., 1979, Fluid inclusion evidence on the environments of sedimentary diagenesis, a review, *in* Scholle, P.A. and Schluger, P.R., eds., Aspects of Diagenesis: Society of Economic Paleontologists and Mineralogists, Special Publication 26, p. 89-107.

Sangster, D.F., 1983, Mississippi Valley-Type Deposits: A Geological Mélange, *in* Kisvarsanyi, G., Grant, S.K., Pratt, W.P. and Koenig, J.W., eds., International Conference on Mississippi Valley-Type Lead-Zinc Deposits, Proceedings Volume: University of Missouri-Rolla, Rolla, Missouri, p. 7-19.

Sangster, D.F. and Lancaster, R.D., 1976, Geology of Canadian lead and zinc deposits: Geological Survey of Canada, Paper 76-1A, p. 301-310.

Sawkins, F.J., 1966, Ore deposition in the north Pennine ore field, in the light of fluid inclusion studies: Economic Geology, v. 61, p. 385-401.

Sharp, J.J., Jr., 1978, Energy and momentum transport model of the Ouachita Basin and its possible impact on the formation of economic mineral deposits: Economic Geology, v. 73, p. 1057-1068.

Skall, H., 1975, The paleoenvironment of the Pine Point lead-zinc district: Economic Geology, v. 73, p. 22-45.

Sverjensky, D.A., 1984, Oil field brines as ore-forming solutions: Economic Geology, v. 79, p. 23-37.

Sverjensky, D.A., 1986, Genesis of Mississippi Valley-type lead-zinc deposits: Annual Reviews of Earth and Planetary Sciences, v. 14, p. 177-199.

Sverjensky, D.A., Rye, R.O., and Doe, B.R., 1979, The lead and sulphur isotope composition of galena from a Mississippi Valley-type deposit in the New Lead Belt, Southeast Missouri: Economic Geology, v. 74, p. 149-153.

Taylor, H.P., 1974, The application of oxygen and hydrogen isotope studies to problems of hydrothermal and alteration and ore deposition: Economic Geology, v. 69, p. 843-883.

Thiede, D.S. and Cameron, E.N., 1978, Concentration of heavy metals in the Elk Point evaporite sequence, Saskatchewan: Economic Geology, v. 73, p. 405-415.

Tissot, B.P. and Welte, D.H., 1984, Petroleum Formation and Occurrence, Second Edition: Springer-Verlag, Berlin, 699 p.

Toth, J., 1980, Cross-formational gravity-flow of ground water: a mechanism of transport and accumulation of petroleum (the generalized hydraulic theory of petroleum migration), *in* Roberts, W.H., III and Cordell, R.J., eds., Problems of Petroleum Migration: American Association of Petroleum Geologists, Studies in Geology No. 10, p. 121-167.

Trudinger, P.A., Chambers, L.A. and Smith, J.W., 1985, Low temperature sulphate reduction: biological *versus* abiological: Canadian Journal of Earth Sciences, v. 22, p. 1910-1918.

Tworo, A.G., 1985, The nature and origin of lead-zinc mineralization, Middle Silurian dolomites, southern Ontario, unpublished M.Sc. thesis, University of Waterloo, Waterloo, Ontario, 276 p.

White, D.E., 1974, Diverse origins of hydrothermal ore fluids: Economic Geology, v. 69, p. 954-973.

White, D.E., 1981, Active geothermal systems and hydrothermal ore deposits: Economic Geology, 75th Anniversary Volume, p. 392-423.

Wolf, K.H., 1976, ed., Handbook of Strata-Bound and Stratiform Ore Deposits, Part I, Principles and General Studies (vols. 1-4); Part II, Regional Studies and Specific Deposits (vols. 5-7): Elsevier, Amsterdam.

Wolf, K.H., 1981, ed., Handbook of Strata-Bound and Stratiform Ore Deposits, Part III (vols. 8-10): Elsevier, Amsterdam.

York, D., Hanes, J.A., Kuybida, P., Hall, C.M., Kenyon, W.J., Masliwec, A., Scott, S.D. and Spooner, E.T.C., 1980, The direct dating of ore minerals (Abstract): EOS, v. 61, p. 399.

Accepted 27 January 1982
Originally published in
Geoscience Canada v. 9 Number 2
(June 1982)
Revised December 1987

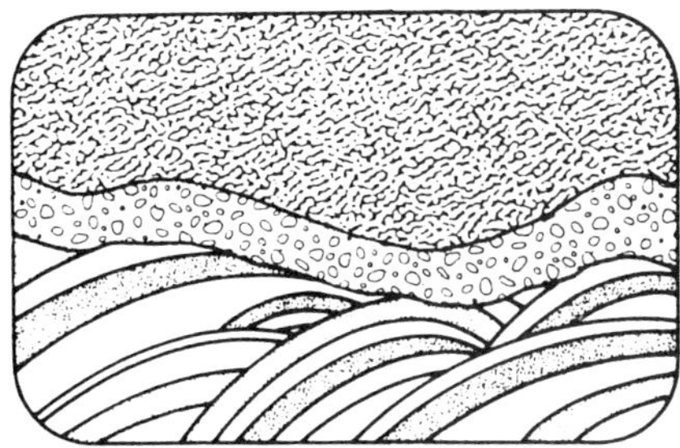

Genetic Considerations Relating to Some Uranium Ore Deposits

J.E. Tilsley
Consulting Geologist
5 Steeplechase Avenue
Group Box 115
R.R. #2
Aurora, Ontario L4G 3G8

Introduction

This paper discusses briefly some general aspects of the behaviour of uranium minerals through geologic time and the genetic significance of these aspects in a selection of environments that have been shown by exploration results to host economic uranium concentrations.

First, we review the changes in chemical behaviour of uranium and uranium minerals as the earth evolved and relate these changes to the kinds of deposits that may be expected in various geologic environments through time. Second, we then consider some aspects of a model for the formation of uranium ores in Proterozoic conglomerates and the limiting factors that confine conglomerate ores to sediments older than about 2200 Ma. Next, a genetic model for sandstone-hosted uranium ores is reviewed and several areas where deviation from conventional wisdom may be useful are suggested. Finally, a model for concentration of uranium in carbonaceous pelites younger than 2200 Ma is proposed, including the mechanism that may contribute to the development of protores and ores in these rocks.

Evolution of the Earth and Uranium Concentration

Summary of Geologic Evolution of the Earth. The subject of geologic evolution of the earth has been dealt with by many authors (*e.g.*, Veizer, 1973, 1976; Ronov, 1964; Roscoe, 1973; Cloud, 1969, 1972, 1972, 1976; Ronov and Migdisov, 1971; Rubey, 1951; Ronov and Yaroskewski, 1969; Holland, 1962, 1974). As may be expected, interpretations vary, particularly with regard to the details of the early part of this evolution, but we can be reasonably confident that significant changes have taken place both in the chemical composition of the crust and in crustal evolutionary processes. The following summary attempts to synthesize the work of these authors.

The composition of the upper mantle prior to pre-Azoic time (see Figure 1) is thought to have approached that of basalt. This basalt is interpreted to be the result of differentiation of chondritic material during the initial stages of formation of the Earth. The atmosphere probably consisted of water vapour and nitrogen (~95%) and the remainder of inert gases, and HCl, HF, H_2S, CO, CO_2, CH_4, *etc.*, all derived from the differentiation process.

The decrease in the surface temperature to below 100°C at about 4500 Ma resulted in condensation of water vapour and formation of oceans, initially of low pH but probably rose quickly to pH 7-9.

The consequent change in composition of the atmosphere during the Azoic resulted in methane, carbon monoxide and inert gases becoming predominant. While oxygen may have been produced photochemically in small amounts in the upper part of the atmosphere from CO and H_2O, it was used up in the oxidation of NH_4, CH_4 and sulphur. Volcanic activity continued to contribute water vapour and acids which precipitated as an original "acid rain" that leached alumina, ferrous iron, alkaline earths and alkalis from the exposed rocks. As volcanic activity decreased with time, the pH of atmospheric and surface waters rose and the rate of silicate-water reactions and the solubilities of silicates in surface waters decreased. Sediments from this time are neither widely distributed nor well preserved. It is probable that there was little topographic relief and few coarse clastic accumulations, which, because the crust was essentially mafic, would have been greywackes rather than arenites. Nevertheless, quartzites formed *in situ* as a result of weathering of mafic volcanic rocks may have been widespread.

Lithologically, Archean time extends from about 3800 to 2600 Ma. During the early part of this period, continental nuclei were still made up of essentially mafic volcanic rocks, although intermediate and acid flows are recognized. Weathering of these complexes produced aluminous greywackes. Ferruginous sandstones and, locally, quartzites were also deposited, but neither carbonate rocks nor arkoses were common. Metamorphism of these derived sediments produced the granites which are now seen to separate the ancient

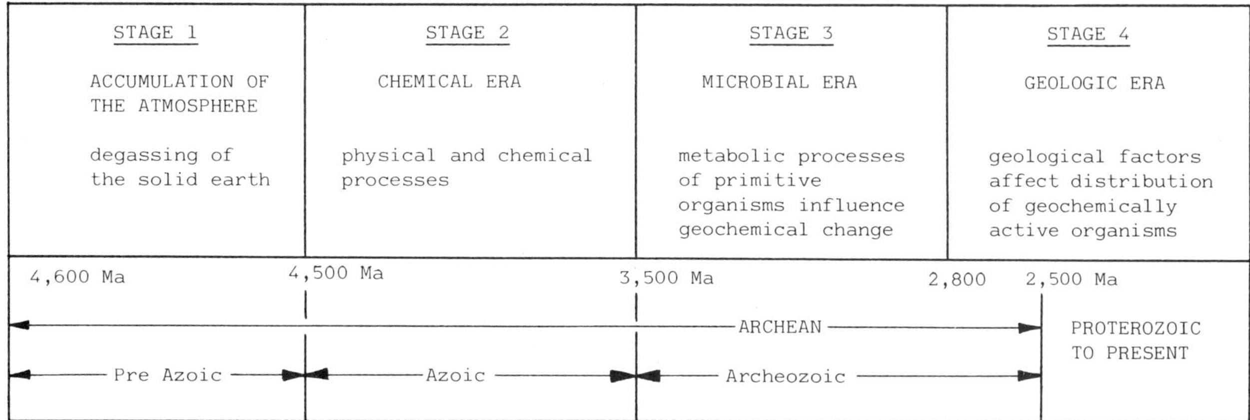

Figure 1 *Geologic time and evolution of the atmosphere. (After data by Walker, 1977).*

continental nuclei. It is estimated by Ronov (1964) that this granitization took place toward the end of the Archean and that 50-60% of all granites were developed at that time. The energy required to cause the extensive granitization of the late Archean (and early Proterozoic) is enormous. Its source is a matter for considerable debate. Such extensive and intensive activity has never been repeated.

The tectonic activity of late Archean and early Proterozoic produced substantial topographic relief, and conglomerates, commonly widespread, were deposited adjacent to mountain ranges and highlands, sometimes upon weathered Archean rocks. Granitic terrain eroded to produce arkoses, arenites and aluminous shales. Mafic volcanics, or their metamorphic equivalents, contributed greywackes and high alumina shales. Chemical sediments, in addition to carbonate rocks, included siliceous and ferruginous oozes. Organic matter began to be preserved in pelitic sediments.

During the late Archean, the atmosphere evolved toward the nitrogen-oxygen combination of today, beginning perhaps as early as 3000 Ma with the development of the first blue-green algae and photosynthesis. At this time, biogenic oxygen was probably consumed immediately in oxidation of carbon monoxide, ammonia, methane and sulphur. Only when the atmosphere was essentially purged of these components could an increase in free oxygen develop, resulting in an increase in the carbon dioxide content of both the oceans and the atmosphere. This change is reflected in the initiation of widespread magnesium and calcium carbonate deposition.

At about 2200 Ma, the oxygen content of the atmosphere passed through a critical level. There is some question what this was, but it was certainly small, perhaps on the order of 1% of the present concentration. Exposure of ferrous iron to the atmosphere resulted in the formation of red iron oxides (Walker, 1977; Cloud, 1976) in continental sediments. This is perhaps the most significant indicator of oxygen level in the evolving atmosphere. It will be noted that red iron oxides exist in Archean marine rocks, but in this environment they are considered indicative of locally available biogenic oxygen rather than general atmospheric oxygenation (Mason and Von Brunn, 1977).

Widespread oxidation of iron would have provided an oxygen demand of enormous size (one tonne of Fe_2O_3 contains about 300 kg O_2). Carbon monoxide and methane in the atmosphere, and ammonia in seawater, also presented an oxygen demand and these factors may have delayed the oxidation of sulphur and hydrogen sulphide in ocean waters. Sulphate deposits (anhydrite and gypsum) in sediments are rare prior to 1500 Ma and become common during the Paleozoic. When these demands had been satisfied, the oxygen content of the atmosphere began to increase through time to the present 20.9%.

An approximate equilibrium between carbon dioxide in the atmosphere and in the ocean was probably attained shortly after oxygenation of the atmosphere, although a gradual decrease in CO_2 content through the Paleozoic, with a sharp reduction at the beginning of the Mesozoic, is indicated by an increasing Ca/Mg ratio in carbonate rocks.

The Behaviour of Uranium Through Time

It is very likely that the availability and mobility of uranium and the probability of uranium concentration have changed through time with the changing physicochemical conditions outlined above (Figure 2).

In the early Archean, the availability of uranium in the crust would have reflected the average abundance of the metal in basaltic rocks. Basalts of all ages today contain, on average, less than 0.6 ppm uranium. Extrapolation based on the half-life of ^{238}U and ^{235}U to about 4500 Ma, and assuming no "enrichment" in the basalts sampled, indicates that the original uranium content would have been about 1.5 ppm. As long as the crust was essentially basaltic, the probability of developing uranium concentrations that might be of economic interest (500 ppm or more)

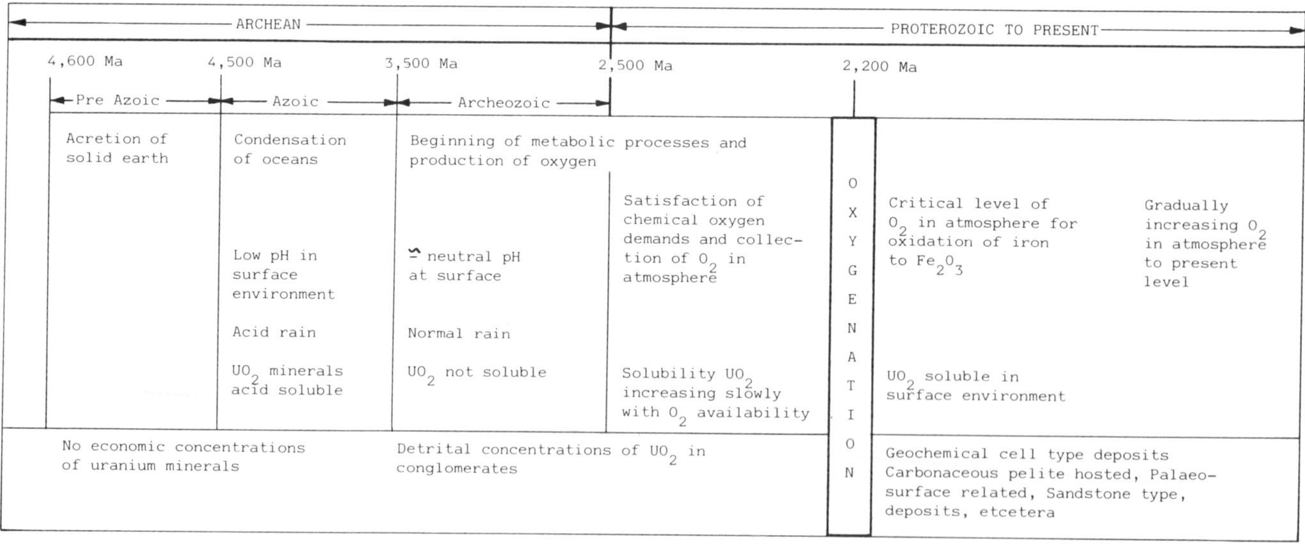

Figure 2 *Behaviour of uranium through geologic time.*

appears to be very low. Weathering of basalts in the early Archean, probably resulted in solution of uranium by "acid rain" and re-precipitation on neutralization of runoff on entry into the primordial seas, so some minor enrichment of uranium in early sediments seems possible.

During the middle and late Archean, hydrolysis of silicates became the predominant weathering reaction. Uranium minerals, *e.g.* juvenile UO_2, would be essentially insoluble in surface waters and, although not notably enriched in the source rock, could be expected to occur as a heavy mineral in clastic sediments.

Granitization of sedimentary piles with introduction of volatiles, alkalis and metals, appears to have been necessary to provide the enrichments in source rocks required for development of paleoplacers of the Dominion Reef and Elliot Lake type. Suitable source granites may have developed as early as 3100 Ma (Allsopp, 1964), and more were added in the late Archean and early Proterozoic. Evidence for primary hydrothermal or volcanogenic uranium concentrations of economic interest in either the Archean or early Proterozoic has yet to be found. However, sufficient uranium was available in the more evolved magmas (syenites, *etc.*), beginning in the middle Proterozoic, to produce volcanogenic deposits, such as near Baker Lake in the District of Keewatin.

It has been noted earlier that oxygen concentration in the atmosphere in early Proterozoic time rose through a critical level at about 2200 Ma. This change is of major significance in the genesis of uranium ore deposits, since from that time onward simple uranium oxides were no longer stable in near-surface environments. Non-thorian uraninites could no longer occur as detrital minerals in alluvial deposits, and formation of conglomerate ores of the Dominion Reef and Elliot Lake type became impossible. With increasing atmospheric oxygen content, the mobility of uranium became similar to that known today. The metal entered surface and ground waters in the zone of weathering, some to be carried by surface drainage to the sea where it was concentrated in shales or carbonaceous pelites, occasionally to economically interesting levels, or dispersed into the ocean waters. Ground waters carried uranium into anoxic conditions at depth where concentration through reduction was possible.

In the near-surface environment, other uranium concentrations have developed in coals, in calcretes and gypcretes, and syngenetically in paleostream channels due to reducing conditions associated with decomposing organic material.

The mobility that uranium attained with oxygenation of the atmosphere permits a variety of "geochemical cell" concentrations including classic "roll-front" and "tabular" sandstone-type deposits of the western United States and the paleosurface-related deposits of the Athabasca Basin. Additionally, deposits near Limoges, France, and several of the Portuguese deposits show evidence of concentration, or at least enrichment, due to downward-flowing uranium-bearing waters.

Exploration results to date suggest that there are two such periods of maximum primary chemical concentration of uranium in sedimentary rocks. The earlier concentration took place in the post-oxygenation early Proterozoic, probably between 2000 and 1800 Ma (*e.g.*, Saskatchewan, Northern Australia). The second period is in the early Paleozoic, typified by the Swedish Alum shales of the Middle and Upper Cambrian.

Metamorphic recycling of uraniferous sediments produced further enrichment in portions of the resulting granites. The "shale-granite-shale" cycle could be repeated several times. Metamorphism of uranium-enriched rocks of Proterozoic age (post-oxygenation or younger) has been noted to produce hydrothermal veins (metamorphic hydrothermal) that show economic promise. Metamorphic hydrothermal uranium-bearing polymetallic veins and simple uranium (uraninite) systems that date from Proterozoic to Cenozoic ages are also known.

<u>Another Classification of Uranium Deposits</u>

Major Uranium Deposit Types	
Surface Related	**Deep Seated**
1. Pre-oxygenation conglomerate	
2. Sandstone-hosted geochemical cell a) Roll front b) Tabular c) Channel - epigenetic d) Channel - syngenetic	
3. Carbonaceous pelite-hosted a) Syngenetic ores b) Protores c) Subsequently enriched protores 1. Weathering	2. Metamorphism 3. Autogenous Processes
4. Paleosurface-related	
Minor Uranium Deposits Types	
Surface Related	**Deep Seated**
	5. Magmatic a) Plutonic 1. Internal Enrichment 2. Peripheral concentration 3. Classical hydrothermal b) Shallow intrusive 1. Degassing channels 2. Quenched gas hydrothermal 3. Enriched extrusives (protore) 4. Enriched dykes, sills, etc. 5. Carbonatite 6. Alkaline complex exotic differentiates c) Post-intrusive convection cell 1. Cooling magma-related - hydrothermal 2. Radiogenic heat-related - hydrothermal
	6. Metamorphic a) Anatectic b) Contact c) Hydrothermal
7. Uraniferous coals	
8. Calcrete and gypcrete-hosted	
9. Sedimentary phosphates	
10. Surface enrichments	
11. Other	
	12. Other

Classification of Uranium Deposits

General

In this paper, uranium deposits will be classified in two general categories: "Major Deposits" and "Minor Deposits" (Table 1). The Major Deposits include Precambrian conglomerates and sandstone-hosted ores of the western United States. Vein-type deposits, including the Athabasca and Northern Territory, Australia, deposits are also listed in this group. All other deposits are classified herein as "Minor Deposits".

Other classifications of uranium deposits have been proposed (*e.g.*, McMillan, 1976, 1977; Barnes and Ruzicka, 1972; Derry, 1980). There is by no means complete agreement on what classification scheme or genetic model may be appropriate for individual ore deposits or types of ore deposits. My experience suggests that classification of uranium orebody types with reference to their relationships to surface or deep-seated processes provides useful guides for exploration.

Modelling Criteria

The approach taken in this paper is based on the fundamental requirements that must be met if unusual concentrations of metals are to develop. These requirements are: an adequate source of readily available metal(s) within the environment under consideration; an effective collection and transportation system while the metal(s) are readily available; exposure of the metal(s) to an environment where conditions permit their deposition and concentration; preservation of that concentration through time (Figure 3). If any one of these requirements is not fulfilled, there can be no orebody.

A variety of models may be proposed for a given orebody or type of orebody, and each may have its merits. In this discussion, however, only one model will be presented for each of the deposit types considered; each model best fits my experience and particular biases.

These models are not descriptions of particular orebodies, but general frames of reference which may be applied to the understanding of, and exploration for, the deposit type. However, reference to particular deposits by way of illustration is unavoidable. It is important to emphasize that use of formalized models of ore deposits should be undertaken with some care since many of them, uranium deposits in particular, are likely to be polygenetic (McMillan, 1980) and remarkably sensitive to physical and chemical modification due to metamorphism, weathering, sub-surface fluid movement and, under appropriate conditions, to radiation and thermal effects originating in the radioactive ores themselves.

Major Uranium Deposits

A review of the major uranium deposits of the world shows that the majority of them are in geologic environments that indicate a definite genetic relationship to near-surface processes. Deep-seated geologic processes are an important part of the geochemical recycling of uranium. Although, there are notable exceptions, generally these processes have not produced deposits and reserves comparable in size to those contained in the surface-related deposits.

Uranium production of approximately 150,000 tonnes has been realized from pre-oxygenation Precambrian *conglomerates* of the Elliot Lake area, Canada, and the Witwatersrand in the Republic of South Africa.

Sandstone-hosted ores are the most important source of uranium in the United States. Production to date totals about 300,000 tonnes. Similar types of deposits, some of major proportions, have been discovered in other areas (for example, Argentina in South America, Niger in Africa).

Some post-oxygenation *Proterozoic sediments* are known to be particularly enriched in uranium. Sandstones, shales, carbonaceous pelites, and detrital (?) carbonate horizons, that contain apparently syngenetic concentrations of uranium, have been identified in many parts of the world. With a few notable exceptions, carbonaceous pelites seem to be the most favourable host rocks. Reasonably assured reserves in this environment presently are roughly 350,000 tonnes U.

Exploration in Saskatchewan along the unconformity between middle Proterozoic (~1450 Ma) continental sediments and older Precambrian basement rocks, has proven to be particularly successful in locating high-grade concentrations of uranium in either the basement, or cover rocks, or both, adjacent to the paleosurface.

On the basis of production realized and assured reserves, deposits in these four environments are the most important source of uranium.

Minor Uranium Deposits

Notwithstanding economically significant uranium production or reserves in some individual deposits of the kinds mentioned in the following paragraphs, none of these types has aggregate production or reserves recoverable under current economic conditions comparable to any one of the four major types listed earlier.

These deposit types can be divided into two general classes: those formed directly or indirectly as the result of surface-related (weathering) processes and those related to "deep-seated" magmatic or metamorphic processes.

Deposits related to magmatic activity include concentrations within magmatic material, concentrations peripheral

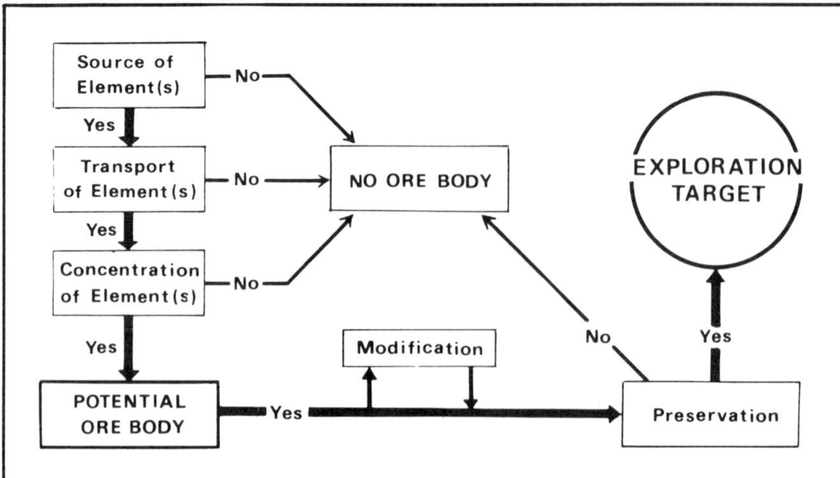

Figure 3 *Conceptual model for the formation of an ore deposit.*

intrusions as a result of pneumatogenic activity, and classically defined hydrothermal veins. Uranium mineralization can also be associated with extrusion of lavas and energetic degassing related to volcanism and shallow intrusion. Circulation of meteoric waters driven by thermal energy related to intrusives can also result in veins that may be classified as hydrothermal. Deposition in both ascending and descending segments of such a convection cell is possible (Fehn *et al.*, 1978).

In addition to contact metamorphic mineralization, metamorphic processes can result in derived fluids collecting, transporting and depositing uranium in fault or fracture systems in concentrations of potential economic interest. Metamomorphic-hydrothermal deposits of this kind are found, for example, in northwestern Europe and northeastern Canada. *In situ* granitization of uranium-enriched shales may produce anatectic deposits such as at Charlebois Lake, Saskatchewan, or near Rössing in Namibia.

In the surface-related minor deposit category, a number of types of mineral concentration have resulted in commercial production, or potential producers. Among these are uranium-enriched coals of various Phanerozoic ages, Late Tertiary and Recent calcretes, gypcretes and secondary enrichments.

Sedimentary phosphates are known to contain 10-100 ppm uranium. None have been mined strictly for their uranium content, and projections are that sedimentary phosphates can produce uranium only as a by-product at relatively high cost.

A variety of other uranium concentrations exist as a result of surface-related processes. The possibility of production from uraniferous organic lake-bottom sediments is being studied in some taiga environments. Certain guano deposits carry more than 1000 ppm uranium. Residual phosphates in karst structures may be sources of limited, but significant, quantities of the metal.

Review of Major Uranium Deposit Types

Proterozoic Conglomerates

Pre-oxygenation Proterozoic conglomerates have been identified in a total of eight localities in North and South America, Southern Africa and Australia. Additional ores may occur in India, but data are scarce.

Uranium production has come from the Elliot Lake conglomerates where it is the main commercial product (along with minor yttrium, but no gold), and from the Witwatersrand gold mines where the metal is recovered chiefly as a by-product or co-product. None of the other pre-oxygenation conglomerates have been the source of commercial production, although certain of the Brazilian conglomerates have uranium concentrations that approach economic grades. The Australian conglomerates carry, on average, less than 50 ppm U. Although the distribution of pre-oxygenation Proterozoic conglomerates is relatively restricted, over 25% of the uranium mined up to 1979 has come from these rocks in the Canadian Lower Huronian Supergroup and the South African Witwatersrand Basin.

Uranium has been produced from four paleo-drainage systems in the Huronian rocks of the Elliot Lake area. In the Witwatersrand Basin, uranium has been recovered from mines in five of the six fans that have been exploited for gold.

Source of Metal. The uranium minerals concentrated in both the Canadian and South African conglomerates were derived from uraniferous pegmatites in the sediment source areas (Robertson, 1978).

Release of Metal. Uraninite grains were probably released from the pegmatites as a result of chemical decomposition of the feldspars (hydrolysis) and subsequent mechanical disintegration of the rocks. As mentioned earlier, prior to about 2200 Ma, the atmosphere was essentially anoxic. The surface environment was reducing to the extent that uraninites, particularly those containing more than 1-2% ThO_2, were not soluble in runoff or ground water. It was possible, therefore, for uraninite to accumulate in eluvial and alluvial environments in the same fashion as other heavy mineral grains.

Transport of Metal. The mineral grains freed by weathering moved with other sediment from the provenance area in the drainage system as part of the bed load.

Concentration of Metal. Uraninite collected preferentially in locations within the drainage system where energy conditions were favourable. The subject of the energy requirements for deposition is well covered in papers by Pretorius (1976) and Theis (1979) among others.

Mineral distribution within the conglomerate units is a reflection of these energy conditions and is often indicated on a broad scale by changing uranium-thorium ratios both along and across the axis of deposition of the units and also, on a smaller scale, vertically through individual conglomerates (Robertson, 1962). The oldest conglomerate ores, such as those of the Dominion Reef (Lower Witwatersrand Supergroup) and the Elliot Lake camp (basal Huronian Supergroup) appear to be the result of high energy concentrations. These lie within coarse clastic sediments of the main drainage axis of the fluvial system or important channels of the alluvial fan.

Reworking of older conglomerates exposed along the margins of the Witwatersrand Basin during structural down-warping and regression is discussed by Pretorius (1976). A similar down-warping of the Elliot Lake Basin as sedimentation progressed is reported by Robertson and Douglas (1970). Erosion of some parts of the ore beds is observed in both the Nordic and Quirke fans but presently economic reconcentration in younger (Middle Mississagi) sediments is not known.

In the Witwatersrand Basin stratigraphic section above the Dominion Reef, evidence of primitive plant growth becomes progressively more common in these pre-oxygenation Proterozoic sediments. Both uraninite and native gold were collected in the lower energy environments of the mid-fan and fan base by the "golden fleece" effect of these algal mats. Similar, low-energy, detrital mineralization is not common in the Huronian of Canada, although Theis (1979) reports carbon bands in mid-fan conglomerates from Quirke Lake that contain notable enrichment of detrital grains of uraninite. There is also evidence of chemical replacement of organic structures in the Witwatersrand Basin algal mats by gold and uranium. This has not been noted in the Canadian ores, perhaps because exploration seems to have been confined largely to the high-energy fluvial and fan head environments.

The conglomerates of Belo Horizonte, Brazil, (Ramos *et al.*, 1974;

Robertson, 1974) are stratigraphically high in the pre-oxygenation Proterozoic section. They lie within less than 200 m of the base of the red hematite ores of the Itabira Formation. The association of organic matter with uranium values has been observed, but due to the relatively high-energy fluvial environment of deposition, mineral concentration appears to be chiefly by mechanical segregation in the bed load.

The pre-oxygenation Proterozoic conglomerates at Jacobina, Brazil, are less well developed than those mentioned above. The environment is relatively high-energy braided fluvial. The conglomerates are notably auriferous (White, 1961), but are low in uranium. The source area was apparently uranium deficient.

Modification. The conglomerates from which the bulk of production has been realized are very mildly metamorphosed. There is little redistribution of metals although some mobility is evident. Theis (1979) shows that the Elliot Lake "brannerites" are the result of replacement of TiO_2 by UO_2. This may have been a diagenetic process.

Dynamic metamorphism of Huronian conglomerates at Agnew Lake, Canada, has resulted in stretching of pebbles accompanied by greenschist-grade thermal metamorphism. There has been only slight redistribution of uranium.

Preservation. Metamorphism sufficiently strong to destroy a conglomerate as a definable rock unit could be expected to result in dispersion of the uranium. Such a situation has not been identified. Erosion and severe metamorphism are the mechanisms most likely to destroy deposits of this nature. The host rocks therefore must lie in regions of the crust that have been tectonically stable since deposition.

Discussion

The distribution of pre-oxygenation Proterozoic conglomerates is restricted on a world-wide basis. Most of the potentially productive basins appear to have been identified and explored rather intensively. Thus, while this environment has proven to be a notable source of uranium and the presently known ore fields will continue to produce into the twenty-first century, the prospect for additional discoveries of the scale of the Elliot Lake or Witwatersrand types is not particularly good. Nevertheless, exploration for "distal-type" (Pretorius, 1976) concentrations may be warranted in the upper part of the pre-oxygenation Proterozoic section.

Sandstone-Hosted Ores

This class of uranium ore is of major importance on a world-wide basis. Over 300,000 tonnes U_3O_8 has been produced from sandstone-hosted ores in the United States of America alone. Significant deposits are also known in Argentina, South America, Japan, Australia (Westmorland) and Africa (Niger and Gabon). The subclasses include classical "roll-fronts", "tabular bodies" and channel-type ore deposits, most commonly developed in permeable sandstones.

The geometry of these bodies is variable depending on a number of sedimentary and hydrological controls. The concentration mechanisms is the same in all cases. Oxidizing solutions carrying dissolved uranium as the uranyl (U^{6+}) carbonate complex enter a reducing environment where U^{4+} ion is dominant and uraninite (UO_2) is precipitated. Deposition of uraninite occurs at the so-called oxidation/reduction front. The most visibly obvious oxidation/reduction front is that of iron, since the reduced minerals are grey-green and the oxidized minerals are reddish-brown. The uranyl complex is reduced and precipitated from solution at an Eh close to that of the Fe^{2+}/Fe^{3+} boundary, so it is visibly associated with the iron oxidation/reduction front. Other elements (such as V, Se, and Mb), that may be associated with uranium in this sort of environment, form minerals that have oxidation/reduction boundaries slightly displaced from those of uranium or iron. This is reflected by the position of their maximum concentrations at some distance ahead or behind the Fe^{2+}/Fe^{3+} boundary.

The oxidation/reduction boundaries migrate within the sandstone at a velocity which is a function of the rate at which free oxygen is supplied to the system by ground water and the oxygen-demand of the sediments through which the ground water flows. Organic carbon and ferrous iron are the chief oxygen demanding agents in the system. The low capacity of water to transport free oxygen (8-10 $g \cdot tonne^{-1}$) results in the rate of advance of an oxidation/reduction front being orders of magnitude less than the rate of groundwater flow. For example, the oxidation of 1% FeO to Fe_2O_3 would require about 10,000 g O_2 per m^3 of sediment, or ~ 1000 tonnes of water passing through that volume of sandstone. Given a groundwater flow velocity of 0.5 $m \cdot yr^{-1}$ and a porosity of 20%, this would require 10,000 years for the front to advance one metre (rate ~ 0.1 $mm \cdot yr^{-1}$).

The rate at which oxidation/reduction fronts move in sediments is relatively slow, but in terms of geologic time, these fronts are transient. Economic deposits tend to be confined to rocks younger than the Carboniferous, and although there is some evidence of possible "fossil roll-fronts" in rocks up to about 2000 Ma, the Mesozoic and Cenozoic sediments contain the bulk of the mineable reserves.

The literature contains extensive information on the American sandstone-hosted ores and some data are available on similar deposits in other parts of the world. Papers by Adler (1974) and Rackley (1976) are particularly informative.

Source of Metal. Most of the uranium now concentrated in sandstones probably originated in source areas topographically higher (*i.e.*, up groundwater flow) than the sediments at the time of ore formation. In these highlands, uranium could be derived by weathering from acid igneous rocks, pyroclastic rocks, lavas and, occasionally, from uranium-enriched metamorphic rocks or older sediments. In many situations, tuffaceous components of the host rocks are believed to have provided much of the metal.

Release of Metal. Chemical weathering processes, at the surface and under special topographic conditions at some substantial depth, are responsible for release of metal from its source rocks. Devitrification of volcanic material is also considered to be an important factor where extrusive rocks appear to be a likely source.

Transportation. Metal is carried in solution in surface and ground waters into and within the sediments along aquifers that may extend to great depth below the erosion surface.

Concentration of Metal. Uranium is removed from oxidizing solutions by reduction at the oxidation/reduction interface mentioned earlier. The interface develops in permeable rocks containing ferrous iron, *e.g.*, as pyrite, and perhaps some carbonaceous material. Although this type of deposit may form at considerable depth, the oxygen is derived from surface waters that act as the transportation medium for both the oxygen and metals.

Modification. As stated previously, oxidation/reduction boundaries are mobile within the host sediment during the time that significant volumes of ground water are available. Through time, such concentrations may be completely flushed from the section.

Changes in the recharge-discharge system may curtail ground-water flow through the host rocks. Mineralization on the oxidation/reduction front remains stable although the oxidized portion of the sediment through which the front has moved may have been reduced again, leaving little visible indication of its previously oxidized condition.

Preservation. Survival through time of mineralization concentrated on an oxidation/reduction boundary requires that the hydrological system that was active during formation is changed, so that free oxygen is not brought into contact with the mineral concentration at depth. The interface must be protected from weathering and erosion. Metamorphism of the section is likely to cause dispersion of the metal values. There are no commercial uranium deposits of obvious oxidation/reduction front origin in metamorphosed rocks.

Discussion

While in the past, there has been rather strong emphasis on the geometry of this type of deposit, *e.g.*, "roll-front"; "tabular deposits", *etc.*, consideration of the genetic model should not be limited by geometric constraints. The shape of the oxidation/reduction front and the form of associated mineralization are the result of variations in attitude, permeability and chemical composition of the host rocks, and with the volume, pressure and chemistry of the ground water flowing through the system.

The term "sandstone-type" deposit has been applied to any uranium deposit developed in sandstone-host rock. Not all the deposits included in this classification have formed in the epigenetic manner discussed. Syngenetic concentration in stream channels during the time of active sedimentation, or at least prior to the onset of diagenesis, is not uncommon. Recognition of this variant in the early stages of exploration is important for effective prospecting. The concentration process is by reduction, usually initiated by decaying organic material in the stream bed sediment and commercial grade mineralization is usually confined to the active channel sediments, as opposed to the finer argillaceous overbank or laterally deposited material. Commonly, however, when associated with mineralized channels, these finer sediments are enriched in uranium as compared with argillaceous material at similar stratigraphic positions but remote from the main drainage axis. This latter factor can be used in exploration.

The Blizzard deposit in British Columbia is an example of this type of deposit. Elsewhere in Canada, successful exploration for this type of deposit has not been successful. Despite this, sandstone-hosted epigenetic "geochemical cell" deposits are an excellent exploration target in much of the world.

Carbonaceous Pelite-Hosted Uranium Deposits

Uranium deposits hosted by post-oxygenation early Proterozoic pelitic sediments have been discovered in North America and Australia during the past ten years. Some of these have proven to contain particularly important uranium reserves. In many other localities, pelitic sediments of comparable age laid down in similar depositional environments are locally uranium-enriched.

In some classification schemes, these deposits have been called vein-type. In the Northern Territory, Australia, discoveries have been made close to the exposure of an unconformable contact between metamorphosed pelitic sediments and younger, undisturbed continental clastic rocks. This led to the conclusion that the deposits were genetically related to the unconformity, and resulted in the term "uncomformity-related veins". However, some pelite-hosted deposits, such as the Kitts deposit in Labrador, show no relationship to a younger unconformity.

Various genetic models have been proposed. Some involve hydrothermal processes (*e.g.*, Morton and Beck, 1978; Little, 1974; Munday, 1978); others suggest that uranium has been released from the pelites into solutions derived during metamorphism and that these solutions have migrated toward and along the unconformity to locations where they were trapped and the contained metal precipitated (Eupene *et al.*, 1976; Ryan, 1977).

U-Pb and Pb-Pb dating of these deposits give ages that, as expected, are generally significantly younger than the host and cover rocks. This has led to the belief that mineralization postdates deposition of both (*e.g.*, Hoeve, 1978; Hoeve and Sibbald, 1978; Kirchner and Tan, 1978), however, the dates could also reflect recrystallization of the uranium minerals during a metamorphic event.

There is mounting evidence that many of these deposits have only an accidental spatial relationship to the unconformity, and that they have a polygenetic origin (McMillan and Heenan, 1980) as reflected by the model presented herein. This model assumes a primary syngenetic concentration of the metal in carbonaceous pelitic sediments to produce protores and, sometimes, ores. Subsequent modification due to metamorphism is probable in many cases and protores may become sufficiently enriched to be recovered economically. Autogenic processes may produce thermal anomalies that, under certain conditions, can be responsible for significant enrichment (Figure 4).

Environmental Constraints. This sort of uranium concentration can take place only subsequent to oxygenation of the atmosphere at about 2200 Ma. The mechanism envisaged could operate at any subsequent time. Exploration experience, however, indicates that early Proterozoic sediments are the major collectors, and to date, the only ore hosts, although sections of the Cambrian Kolm shales near Ranstad, Sweden, show some economic potential.

Source of Metal. The source of uranium for syngenetic concentration in carbonaceous pelites is generally accepted to be moderately to strongly enriched granites or metamorphic rocks lying within the drainage basin that produced the sediments (Hegge, 1977; Hegge and Rowntree, 1978). Alternatively, in some areas, uranium-bearing volcanic rocks, which are penecontemporaneous with the graphitic sediments that host ores or protores, appear to be likely source rocks (McMillan, 1976; Gandhi, 1978). The preferred provenance area would have, ideally, gentle topography and a moist, mild to tropical climate, without extreme variation in rainfall and runoff.

Release of Metal. Uranium was released to percolating rain water by destruction of uranium-bearing minerals unstable in the prevailing oxidizing environment. Uranium contained in uraninite would be almost totally available, as would most of the metal bound in silicate minerals. Uranium present in certain refractory minerals, such as zircon and monazite, would not have been released in any appreciable quantity.

Uranium released by weathering probably appeared in solution as positively charged complexes in environments having a pH of less than 5 and as uncharged or negatively charged uranyl carbonate complexes at pH greater than 5 (Langmuir, 1978).

Transport. Transportation of the metal from point of release to point of concentration was via surface and groundwater flow to the surface drainage systems that also served to transport the clastic sediments which act as host rocks.

Concentration. Uranium in solution in surface waters was carried into a shallow marine, lagoonal environment, where mixing of fresh and salt water would be relatively efficient. In order that precipitation and collection of metal take place, the dissolved metal species must come into contact with organic complexes and clays or migrate into an environment where the Eh is more negative than the upper stability limit of the compatible solid phase, usually uraninite (Langmuir, 1978; Hostetler and Garrels, 1962).

Muds containing decaying organic material characteristically produce anoxic conditions due to a very high oxygen demand, accompanied by low Eh and low pH (Bass-Becking *et al.*, 1960). These conditions are ideal for the fixation of uranium as UO_2 and in organic complexes. The low Eh of the muds contrast with relatively more oxidizing Eh of fresh surface waters. A relatively steep potential gradient will exist between the carbonaceous muds and the waters passing above, and the mixing of surface runoff and salt or brackish water will produce a weak electrolytes. Migration of positively charged metal-bearing complexes from the water into the sediment may account for a significant portion (50-60%) of the uranium collected in this environment. The ideal concentration locality would be a back-reef, sabkha or barrier island lagoonal environment in which large volumes of surface water could flow slowly at shallow depth across an extensive zone at intertidal and subtidal mud flats.

The degree of concentration of uranium in these sediments from solution is a function of the volume of water exposed to the host material through time, the amount of metal in solution, and the efficiency of the collecting process. Since these factors are likely to vary from one part of a sedimentary basin to another, local variation in the degree of uranium enrichment is to be expected. It is unlikely that significant additional uranium is introduced into the clays immediately after deposition or during early diagenesis because of low permeability in argillaceous sediments at these stages of development.

Enrichment of metals in clay-rich carbonaceous sediments probably indicate rather special conditions in both the local drainage area and in the area of metal accumulation. Concentrations of economic interest will not necessarily be found in pelitic horizons that have an anomalously high abundance of uranium. Conversely, similar rocks shown to contain only "average" uranium abundances may have local areas of enrichment worthy of exploration attention. If the constraints and requirements discussed above are satisfied, syngenetic ores or protores may develop.

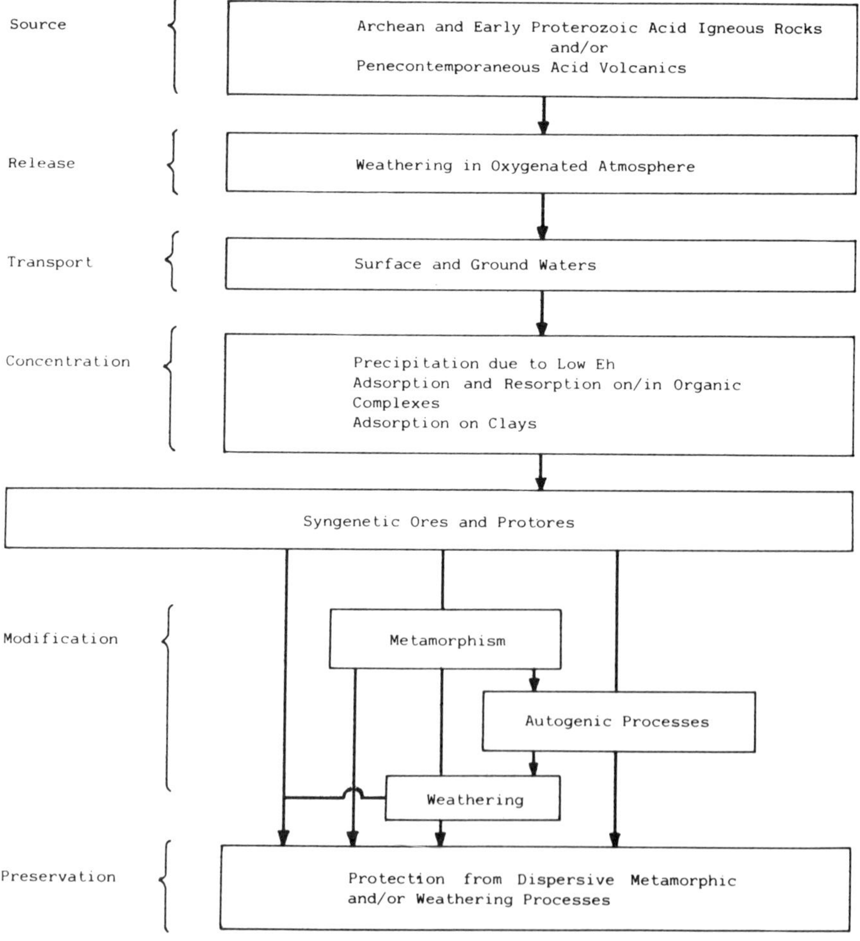

Figure 4 *A genetic model for carbonaceous pelite-hosted uranium deposits.*

Modification. Sediments, ores and protores peculiar to the environments under consideration, may be preserved through time relatively unchanged. Others may be modified by a variety of processes and some may be destroyed.

Weathering. Carbonaceous pelite-hosted uranium concentrations may be exposed at surface and undergo weathering at any time. Enrichment is possible below the limit of free oxygen in most environments.

In wet-dry or arid environments, stable phosphate minerals may develop at surface, but also at considerable depth. Although economic concentrations due to weathering of carbonaceous pelites are not known, these rocks would appear to be particularly favourable sources for weathering-related secondary enrichment.

Metamorphism. Metamorphism to greenschist facies will produce graphite from organic material in the sediment (Stapleton, 1978). Although organic complexes may contain appreciable quantities of uranium, the crystal lattice of graphite will not accommodate any foreign elements other than boron, and uranium will be expelled. Dehydration of clays during metamorphism will also expel metals, and in addition, produce aqueous fluids suitable as transportation media. These fluids may carry uranium (and other metals) into zones of lower confining pressure where precipitation, and possibly concentration, can take place.

Metamorphism intense enough to cause local development of pegmatoids will result in remobilization and possibly concentration of metals in the mobile phases (*e.g.*, Charlebois Lake, Saskatchewan; Rössing, Namibia). However, dispersion of major syngenetic concentrations of uranium by metamorphism, short of complete re-melting of the sedimentary pile, is unlikely.

Low-temperature thermal metamorphism resulting from heat flow anomalies also holds intriguing possibilities. Such thermal anomalies may be related to steeply inclined graphitic zones and these may play some role in redistribution or enrichment of ores and protores.

Authigenic Process. Remobilization of minerals that have been stable in any geologic environment requires energy which may be supplied by the processes noted earlier. Another source of energy available for modification of radioactive ores is the heat of radioactive decay. Historically, consideration of this energy has been confined largely to relatively small deposits of the sandstone type (*e.g.*, Okla, Gabon). The volume of ore available in these deposits is restricted and the geometry of the ore zones is not conducive to temperature rise.

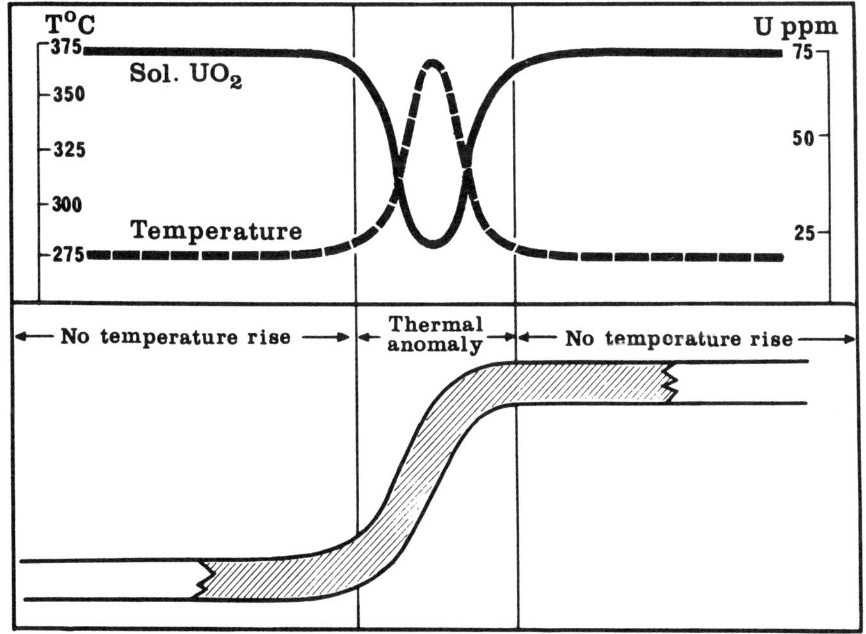

Figure 5 *Relationship between attitude of mineralized body temperature rise and uranium solubility.*

In other situations, however, such as the high-grade deposits of the Athabasca Basin type, or in large, lower grade bodies that are steeply inclined, radiogenic heat becomes important. Convection cells driven by the heat of radioactive decay may be responsible for redistribution of uranium into overlying rocks, *e.g.*, Midwest Lake, Saskatchewan (Tilsley, 1980).

If one considers a situation where a tabular syngenetic uranium concentration has been folded into a nearly vertical position, the cross-section through which the thermal energy of radioactive decay can be transmitted toward surface would be greatly reduced resulting in a temperature rise. The magnitude of this temperature rise is a function of the volume of uranium present, the horizontal cross-section of the mineralized zone and the depth of cover. For example, the Jabiluka II orebody contains about 200,000 tonnes U and is steeply inclined in cross-section. Assuming 1,000 metres of cover, a temperature in the central part of the ore zone of 60-90°C above ambient is possible. The local temperature anomaly could be responsible for initiation of a pore-water convection cell (Figure 5).

If we assume an ambient temperature of about 275°C and an internal temperature rise of 60°C, the centre of the zone might attain a temperature of 335°C. In higher temperature environments, *e.g.*, greenschist facies grade, the solubility of UO_2 in aqueous solutions reaches a maximum. A temperature rise above about 270°C at 750 bars results in a rather rapid drop in solubility. Reference to Lemoine's data (1975) suggests a drop in solubility of UO_2 at 335°C of about 40% from the maximum solubility at 265°C (Figure 6). Fluids circulating from the fringes of the orebody would therefore tend to collect uranium in the low temperature zone and deposit UO_2 toward the centre of the orebody with rising temperature. Depletion of UO_2 might also occur in the upper part of the mineralized zone as fluid temperature fell toward 270°C. If additional temperature drops, say to 200°C, were experienced outside the ore zone, a re-precipitation of UO_2 would be expected.

There are a multitude of possibilities with regard to redistribution/concentration by this mechanism. Leaching or precipitation of UO_2 in this model is strictly temperature dependent. Location of the minerals redistributed in this manner is simply a function of fluid flow path and paleotemperature patterns.

The temperature of the orebody is the critical factor in this concentration/dispersion model. In any given orebody, variation in temperature due to the process described is a function of depth of cover. The periodic redistribution of uranium within these ore zones, as suggested by the spread of isotopic age determinations, may reflect changing cover thickness due to sedimentation and erosion rather than assumed regional metamorphic or intrusive events for which there is often little evidence (Table 2).

Preservation. Preservation of this type of deposit and any of its sub-types depends upon protection from the extremes of the modification processes discussed. Weathering and metamorphism to the degree of anatexis are the most efficient destructive mechanisms. Fluid convection driven by geothermal anomalies of one type or another could conceivably cause total dispersion of a mineral concentration. However, we know of no example of either total or partial depletion of mineralization attributable to such activity. Internal temperature rise that might be responsible for driving a convection cell at temperatures compatible with redistribution of UO_2 could not be maintained beyond some minimum residual grade. Total self-destruction is not possible.

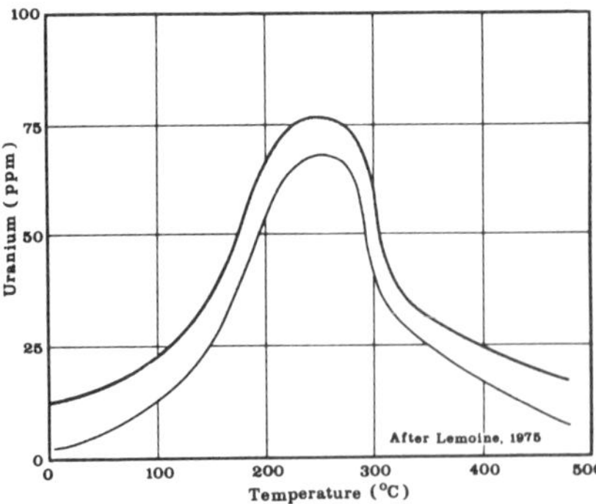

Figure 6 *Solubility of UO_2 at 750 bars.*

Discussion

This proposed class of uranium deposit includes such major ore bodies as Jabiluka II and Koongarra in the Northern Territory, Australia, near-ores at Ranstad, Sweden, and Kitts, Labrador, as well as numerous enrichments in carbonaceous pelites of post-oxygenation Proterozoic age and younger.

Syngenetic concentration of uranium in carbonaceous pelites deposited in relatively restricted shallow-water environments appears to be a significant ore- and protore-forming process. Protores may be subsequently enriched more or less *in situ* due to regional or local thermal metamorphism or by secondary concentration related to weathering or to authigenic processes.

Exploration experience has shown that post-oxygenation Proterozoic carbonaceous pelites are the most favourable rocks for this type of uranium concentration. Younger carbonaceous pelites, which have been subjected to conditions similar to those that produced ores in Proterozoic rocks, would appear to be attractive exploration environments, but, to date, there are no discoveries in younger rocks of economic mineralization similar in scale to those known in the Northern Territory, Australia.

Table 2 **Authigenic enrichment of syngenetic uranium ores**

Genetic Sequence	
Stage 1	Syngenetic concentration of U in carbonaceous pelites
Stage 2	Metamorphism and folding; ambient temperature ~ 250°C
Stage 3	Development of thermal anomaly due to radiogenic heat in steeply dipping mineralized body; internal temperature ~ 335°C
Stage 4	Circulation of pore fluids; collection of UO_2 from fringes of temperature anomaly; precipitation of UO_2 toward interior of mineralized zone
Stage 5	Reduction of ambient temperature due to cessation of orogeny responsible for metamorphism or reduction of cover rock thickness by erosion
Subsequent Activity	Possible re-initiation of activity due to increasing depth of cover; cessation due to erosion of cover; may be repeated several times

References

Adler, H.H., 1974, Concepts of uranium ore formation in reducing environments in sandstones and other sediments, *in* Formation of uranium ore deposits: International Atomic Energy Agency, Vienna, p. 531-549.

Allsopp, H.L., 1964, Rubidium/strontium ages from the Western Transvaal: Nature, v. 204, p. 361-363.

Barnes, F.Q. and Ruzicka, V., 1972, A genetic classification of uranium deposits: 24th International Geological Congress, Sect. 4, Montreal, p. 159-166.

Bass-Becking, F.L.M., Kaplan, J.R. and Moore, D., 1960, Limits of the natural environment in terms of pH and oxidation/reduction potentials: Journal of Geology, v. 68, p. 243-284.

Cloud, P.E., 1969, Pre-Paleozoic sediments and their significance for organic chemistry, *in* Eglinton, G., Murphy, M.T.J., eds., Organic Geochemistry: Springer-Verlag, New York, p. 772-786.

Cloud, P.E., 1971, The Precambrian: Science, v. 173, p. 851-854.

Cloud, P.E., 1972, A working model of the primitive earth: American Journal of Science, v. 272, p. 537-548.

Cloud, P.E., 1976, Major features of crustal evolution: Geological Society of South Africa, Annexo 59, p. 33.

Derry, D.R., 1980, Uranium deposits through time, *in* Strangway, D.W., ed., Continental Crust and Its Mineral Deposits: Geological Association of Canada, Special Paper 20, p. 625-632.

Eupene, G.S., Fee, P.H. and Colvill, R.G., 1976, Ranger one uranium deposits, *in* Knight, C.L., ed., Economic Geology of Australia and Papua, New Guinea: Australasian Institute of Mining and Metallurgy, p. 300-317.

Fehn, U., Cathles, L.M., Holland, H.D., 1978, Hydrothermal convection and uranium deposits in abnormally radioactive plutons: Economic Geology, v. 73, p. 1556-1566.

Gandhi, S.S., 1978, Geological setting and genetic aspects of uranium occurrences in the Kaipokok Bay-Big River area, Labrador; Economic Geology, v. 73, p. 1492-1522.

Hegge, M.R., 1977, Geologic setting and relevant exploration features of the Jabiluka uranium deposits: Canadian Institute of Mining and Metallurgy, Bulletin, v. 70, No. 788, p. 50-61.

Hegge, M.R. and Rowntree, J.C., 1978, Geologic setting and concepts on origin of uranium deposits in the East Alligator River region, N.T., Australia: Economic Geology, v. 73, p. 1420-1429.

Hostetler, P.B. and Garrels, R.M., 1962, Transportation and precipitation of uranium and vanadium at low temperatures, with special reference to sandstone-type uranium deposits: Economic Geology, v. 57, p. 137-167.

Hoeve, J., 1978, Uranium metallogenetic studies, Rabbit Lake: Mineralogy and geochemistry, *in* Christopher, J.E. and Macdonald, R., eds., Summary of investigations, 1978: Saskatchewan Geological Survey, p. 61-65.

Hoeve, J. and Sibbald, T.I.I., 1978, On the genesis of Rabbit Lake and other unconformity-type uranium deposits in northern Saskatchewan, Canada: Economic Geology, v. 73, p. 1450-1473.

Holland, W.D., 1964, On the chemical composition of the terrestrial and cytherean atmospheres, *in* Brancazio, P.J. and Cameron, A.G.W, eds., The Origin of and Evolution of Atmospheres and Oceans: John Wiley, New York, p. 86-101.

Kirchner, G. and Tan, B., 1979, Prospektion, Exploration and Entwicklung der Uranlagerstatte Key Lake, Kanada: Erzmetal, Bd. 30 (1977) (H-12), p. 584-489.

Langmuir, D., 1978, Uranium solution-mineral equilibria at low temperatures with application to sedimentary ore deposits: Geochimica et Cosmochimica Acta, v. 42, p. 547-569.

Lemoine, A., 1975, Contribution l'étude du comportement de UO_2 en milieu aqueux à haute température et haute pression, unpublished Ph.D. Thesis, Nancy, France, 111 p.

Little, H.W., 1974, Uranium in Canada: Geological Survey of Canada, Paper 74-1, p. 137-139.

McMillan, R.H., 1976, Metallogenesis of Canadian Uranium Deposits: A Review, *in* Geology, Mining and Extractive Processing of Uranium: Institution of Mining and Metallurgy, Special Publication, p. 43-55.

McMillan, R.H., 1977, Uranium in Canada: Canadian Society of Petroleum Geologists, Bulletin, v. 26, p. 1222-1249.

McMillan, R.H. and Heenan, P.R., 1980, Uranium Explorations in Canada, 1979, *in* Markets for Canadian Uranium. Proceedings of a Seminar held 25 October 1979, Calgary, Alberta: Canadian Energy Research Institute, Calgary, Alberta, p. 63-76.

Mason, T.R. and Von Brunn, V., 1977, 3 G yr-old stromatolite from South Africa: Nature, v. 266, p. 47-49.

Morton, R.D. and Beck, L.S., 1978, The origin of the uranium deposits of the Athabasca Region, Saskatchewan, Canada (Abstract): Economic Geology, v. 73, p. 1408.

Munday, R.J., 1978, Uranium mineralization in northern Saskatchewan: Canadian Institute of Mining and Metallurgy, Bulletin, v. 71, no. 791, p. 76.

Pretorius, D.A., 1976, Gold in the Proterozoic sediments of South Africa: systems, paradigms and models, *in* Wolf, K.H., ed., Handbook of Strata-Bound and Strataform Ore Deposits: Elsevier, Amsterdam, v. 7, p. 29-87.

Rackley, R.I., 1976, Origin of Western-States type uranium mineralization, *in* Wolf, K. H., ed., Handbook of Strata-Bound and Strataform Ore Deposits, Elsevier, Amsterdam, v. 7, p. 89-156.

Ramos, J., de Andrade, R. and Fraenkel, M.O., 1974, Uranium occurrences in Brazil, *in* Formation of Uranium Ore Deposits: International Atomic Energy Agency, Vienna, p. 637-658.

Robertson, D.S., 1962, Thorium and uranium variations in Blind River ores: Economic Geology, v. 57, p. 1175-1184.

Robertson, D.S., 1974, Proterozoic units as fossil time markers and their use in uranium prospecting, *in* Formation of Uranium Ore Deposits: International Atomic Energy Agency, Vienna, p. 495-512.

Robertson, D.S. and Douglas, R.F., 1970, Sedimentary uranium deposits: Canadian Institute of Mining and Metallurgy, Transactions, v. 73, p. 109-118.

Robertson, J.A., 1978, Uranium deposits in Ontario, *in* Kimberley, M.M., ed., Uranium Deposits: Their Mineralogy and Origin: Mineralogical Association of Canada, Short Course Handbook #3, p. 229-280.

Ronov, A.B., 1964, Common tendencies in the chemical evolution of the earth's crust, oceans and atmosphere: Geokhimiya, Transactions, v. 8, p. 715-743.

Ronov, A.B. and Migdisov, A.A., 1971, Evolution of the chemical composition of the rocks in the shields and sediment cover of the Russian and North American platforms: Sedimentology, v. 16, p. 137-158.

Ronov, A.B. and Yaroshewski, A.A., 1969, Chemical composition of the earth's crust, *in* Hart, P.J., ed., The Earth's Crust and Upper Mantle, American Geophysical Union, p. 37-57.

Roscoe, S.M., 1973, The Huronian Supergroup, a paleo-Aphebian succession showing evidence of atmospheric evolution, *in* Young, G.M., ed., Huronian Stratigraphy and Sedimentation: Geological Association of Canada, Special Paper 12, p. 31-47.

Rubey, W.W., 1951, Geologic history of sea water: An attempt to state the problem: Geological Society of America, Bulletin, v. 62, p. 1111-1148.

Ryan, B.R., 1977, Uranium in Australia, *in* Jones, M.J., ed., Geology, Mining and Extractive Processing of Uranium: Institute of Mining and Metallurgy, v. 86, p. 24-42.

Stapleton, R.P., 1978, Organic metamorphism and uranium occurrences in the Beaufort Group of South Africa: Economic Geology, v. 73, p. 283-285.

Theis, N.J, 1979, Uranium-bearing and associated minerals in their geochemical and sedimentological context, Elliot Lake, Ontario: Geological Survey of Canada, Bulletin 304, 50 p.

Tilsley, J.E., 1980a, Genetic considerations relating to some uranium ore deposits: Geoscience Canada, v. 7, p. 143-148.

Tilsley, J.E., 1980b, Continental weathering and development of paleo-surface-related uranium deposits: Some genetic considerations, *in* Ferguson, J. and Goleby, A.B., eds., Proceedings of International Uranium Symposium on the Pine Creek Geosyncline: International Atomic Energy Agency, Vienna, p. 721-732.

Veizer, J., 1973, Sedimentation in geologic history: Recycling vs. evolution or recycling with evolution: Contributions to Mineralogy and Petrology, v. 38, p. 261-278.

Veizer, J., 1976, The nature of $^{18}O/^{16}O$ and $^{13}C/^{12}C$ secular trends in sedimentary carbonate rocks: Geochimica et Cosmochimica Acta, v. 40, p. 1387-1395.

Walker, J.C.C., 1977, Evolution of the Atmosphere: MacMillan and Co., New York, p. 264-278.

White, M.G., 1961, Origin of uranium and gold in the quartzite-conglomerate of the Serra de Jacobina, Brazil: United States Geological Survey, Professional Paper 424-B, p. B-9.

Accepted, as revised, 4 September 1980 and 11 December 1980.
Originally published (in two parts) in *Geoscience Canada* v. 7 Number 4 and v. 8 Number 1
(December 1980 and March 1981)
Revised December 1987

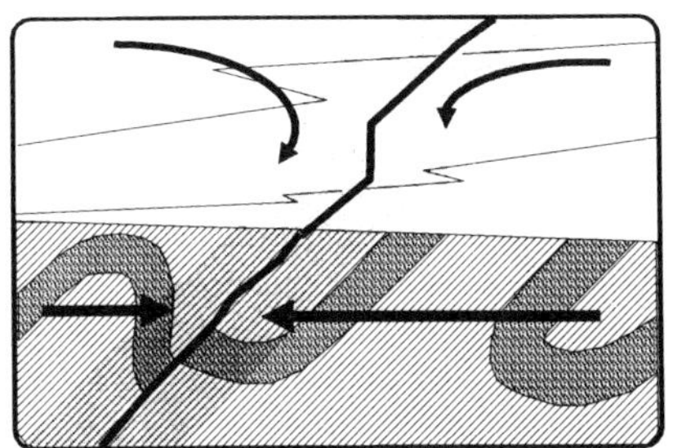

Unconformity-type Uranium Deposits

Soussan Marmont
Precambrian Geology Section
Ontario Geological Survey
1033 - 77 Grenville Street
Toronto, Ontario M7A 1W4

Introduction

Uranium is a relatively mobile, lithophile element occurring in nearly all major lithologies, and has an average crustal abundance of two to four ppm. As with many other metals, economic concentrations of uranium show a distinct time-bound nature (Robertson *et al.*, 1978) (Figure 1). In the Precambrian, anomalous concentrations of uranium are found in two specific geological settings, separated in time by the oxygenation event of the earth's atmosphere at about 2600-2200 Ma. Prior to this event, paleoplacer uranium deposits hosted by quartz-pebble conglomerates, in the Witwatersrand Supergroup, South Africa, and in the Huronian Supergroup at Elliot Lake, Canada, were formed as a result of mechanical transportion of detrital uraninite grains. Subsequent to atmospheric oxygenation, however, hexavalent uranium was dissolved and transported as uranyl complexes in aqueous solutions. In the period between 1800 Ma to 1200 Ma, extensive concentrations of uranium, which are spatially, and most probably genetically, related to paleo-weathering surfaces were formed and gave rise to a new type of deposits, generally referred to as the "Unconformity-type" uranium deposits. These Middle to Upper Proterozoic rocks host a significant component of the western world's uranium resources; for example, the Athabasca deposits, of northern Saskatchewan, Canada, constitute about 10% of the western world's low cost reserves (Sibbald and Quirt, 1987). To date, major deposits of this type are known in northern Saskatchewan and the Northwest Territories of Canada, the Northern Territory of Australia and parts of West Africa. Unlike the Early Proterozoic paleoplacers and the Phanerozoic sandstone-hosted roll-front deposits, the typical unconformity-type orebodies contain extremely high concentrations of uranium, which make them a more attractive mining proposition.

The discovery of a rich orebody at Rabbit Lake, Saskatchewan in 1968 and the increased world demand for uranium of the early to mid-1970s, triggered an avid search for this type of deposits. As a result, in Saskatchewan, numerous deposits such as Cluff Lake D Zone (Lainé, 1986), Key Lake (de Carle, 1986), Midwest (Ayres *et al.*, 1983), McClean (Saracoglu *et al.*, 1983), Dawn Lake (Clarke and Fogwill, 1986), and Cigar Lake (Fouques *et al.*, 1986) were discovered between 1969 and 1981 (Figure 2). In Australia, world-class deposits such as Koongarra, Ranger, Nabarlek and Jabiluka were discovered in the 1970s (Nash, 1978)(Figure 3). A decline in uranium prices since the late 1970s has substantially slowed the exploration and exploitation rate of these deposits throughout the world.

The following is a summary of the main features of this group of deposits, based on descriptions of those in northern Saskatchewan, Canada and the Northern Territory, Australia.

Regional Setting

Northern Saskatchewan, Canada. Part of the Churchill Province of the Canadian Shield, the Archean basement in this area consists of granitoid and gneissic terranes which are overlain by a sequence of Lower Proterozoic (Aphebian) metasediments. The crystalline basement has been divided into several lithostructural domains with different compositions, structures and metamorphic grades (Langford, 1986; Hoeve *et al.*, 1980; Lewry and Sibbald, 1979)(Figures 2 and 4). The most important uranium deposits are located within

Figure 1 *Time-bound character of uranium deposits. (After Robertson* et al.*, 1978).*

the Cree Lake Mobile Zone (bounded to the east by the Needle Falls Shear Zone) and the Western Craton, specifically where these terranes are covered by the Upper Proterozoic (Helikian) sediments of the Athabasca Group.

The Cree Lake Mobile Zone, which is divided into Wollaston, Mudjatik and Virgin River domains, consists of granitoid and gneissic domes surrounded by Aphebian supracrustal rocks, all strongly deformed by Hudsonian tectonism and metamorphosed to upper amphibolite grade. Sibbald and Quirt (1987) and Lewry and Sibbald (1980) define four lithostratigraphic units within the supracrustal package (Wollaston Group) which is best developed in the Wollaston domain. The succession comprises a lower package of arkose, quartzite and pelite (only present at the eastern margin of the Cree Lake Mobile Zone) overlain by graphitic pelite (which serves an important role in EM surveys) interlayered with calc-silicate and minor marble. These rocks are in turn overlain by an upper meta-arkosic unit which includes calc-silicate and pelite and a final assemblage of amphibolite-quartzite. The arkosic and pelitic units have been metamorphosed to feldspathic gneisses and biotite-schists, respectively. In the Rabbit Lake area, the thickness of the Wollaston Group has been estimated to be at least 3-4 km (Hoeve and Sibbald, 1978). The depositional environment of the metasediments has been described as shallow water, marginal marine conditions (Nash *et al.*, 1981). However, Hoeve and Sibbald (1978) note that the Hudsonian Orogeny has obscured many of the primary sedimentary features and suggest the presence of two distinct sedimentary environments:

(1) deposition of pelite, quartzite and meta-arkose under stable tectonic conditions, as a westwardly transgressional, massive sequence over the Archean craton, and (2) meta-arkose, quartzite, amphibolite and coarse clastic sediments which were deposited under less stable conditions, possibly during uplift, to the east.

Many stages of deformation during the Hudsonian Orogeny are recorded in the Wollaston Group. An early foliation, commonly parallel to lithological contacts and compositional layering, is developed in both the Archean and Aphebian rocks. The penetrative fabric was subsequently folded about northeast-trending axial planes. Late-Hudsonian, left-lateral, strike-slip faults and post-Athabasca Group thrust faults terminated the structural history of the supracrustal rocks (Sibbald, 1983).

The metamorphosed Archean-Aphebian basement is unconformably overlain by the Athabasca Group (Helikian) which comprises four marine transgressive sequences and one fluvial regressive sequence (Ramaekers, 1983). The lithologies vary from "poorly sorted" to "well sorted", clay-rich sandstone, siltstone and mudstones, intercalated with conglomerate. The conglomerate is more prevalent at the base of the succession (Hoeve *et al.*, 1980). Ramaekers (1976, 1983) demonstrates that, within the basal conglomeratic members, paleocurrent directions indicate two prominent radiating fluvial fans, one deriving from northeast and the other from east; however, local exceptions are reported, such as at the Collins Bay deposits where the current was to the north-northwest (Jones, 1980). Within the sandy sediments, Ramaekers (1983) points out that the paleocurrent indications direct to the west and suggests that the environment of deposition was an alluvial plain of a system of braided streams with intermittent lakes.

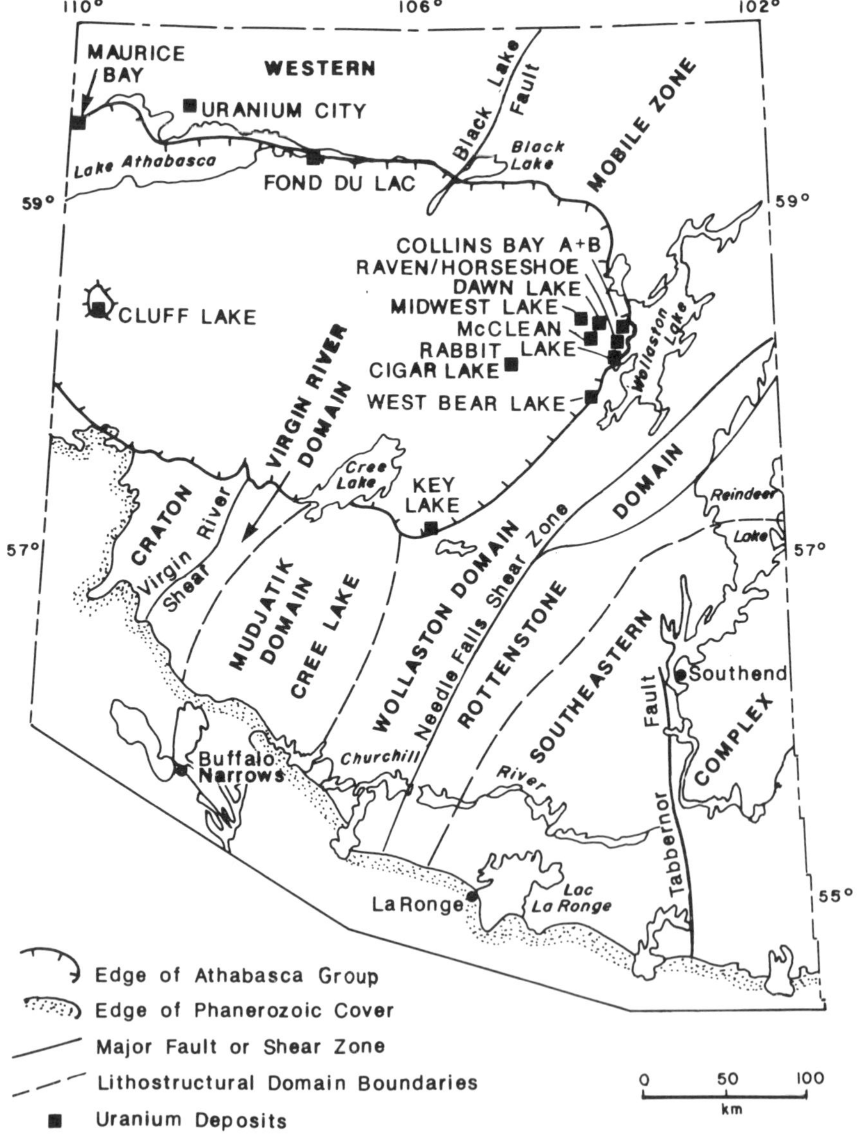

Figure 2 *Location of major uranium deposits in northern Saskatchewan. (Modified after Hoeve et al., 1980).*

The Athabasca Group sediments are highly hematitic, with the hematite content normally between 1 to 2%, but in places as high as 30%. Hoeve *et al.* (1980) suggest a complex history of diagenesis for the sediments whereby oxidation and alteration, similar to processes active in recent red beds, started soon after deposition, and continued for a long time thereafter. This is witnessed by hematitization, kaolinitization and fracture-fillings by quartz and illite of a swarm of diabase dykes which cross-cut the Athabasca Group. The Athabasca Group sediments have been dated at 1484 ± 55 and 1459 ± 4 Ma by $^{40}Ar/^{39}Ar$ method (Bray *et al.*, (1987). Armstrong and Ramaekers (1985), using Rb/Sr method, have obtained similar ages for the sediments. They also report Rb/Sr ages of 1.31 ± 0.07 Ga and 1.16 ± 0.04 Ga for the diabase dykes.

The unconformity at the base of the Athabasca Group is marked by a weathered paleosoil profile averaging 20-40 m and in places reaching up to 100 m in thickness (Langford, 1986). Where complete, the regolith is zoned with a green chloritic zone in the lower parts of the soil profile grading to an overlying red and white zone consisting primarily of kaolinite and illite. Hematite staining is pervasive throughout the upper portion of the profile (Hoeve *et al.*, 1980; Langford, 1986). In most places, the regolith horizon has been well preserved by the Athabasca Group sediments, but in rare cases, such as the Rabbit Lake deposit, the paleo-weathered surface is absent from the upthrown block of the Rabbit Lake Fault, but present on the downthrown block (Hoeve and Sibbald, 1978). Macdonald (1981), in documenting the characteristics of the regolith, shows many similarities between this unit and present-day laterites, and attributes their differences to the lack of land vegetation in the pre-Helikian period. The regolith has been dated at 1482 ± 49 and 1453 ± 49 Ma ($^{40}Ar/^{39}Ar$) by Bray *et al.* (1987).

Northern Territory, Australia. The major uranium deposits of northern Australia are located in the East Alligator Rivers area within the Pine Creek Geosyncline (Figure 3). The Archean to Lower Proterozoic Nanambu Complex, which forms the basement, consists of a wide range of granitic and metamorphic lithologies. Needham and Stuart-Smith (1980), using radiometric ages (U-Pb of zircon and monazite, and whole rock Rb/Sr and K/Ar) published by Page *et al.* (1980), divided the Nanambu Complex into massive to foliated granites (2500-2400 Ma), grading to foliated gneisses (1980-1800 Ma) surrounded

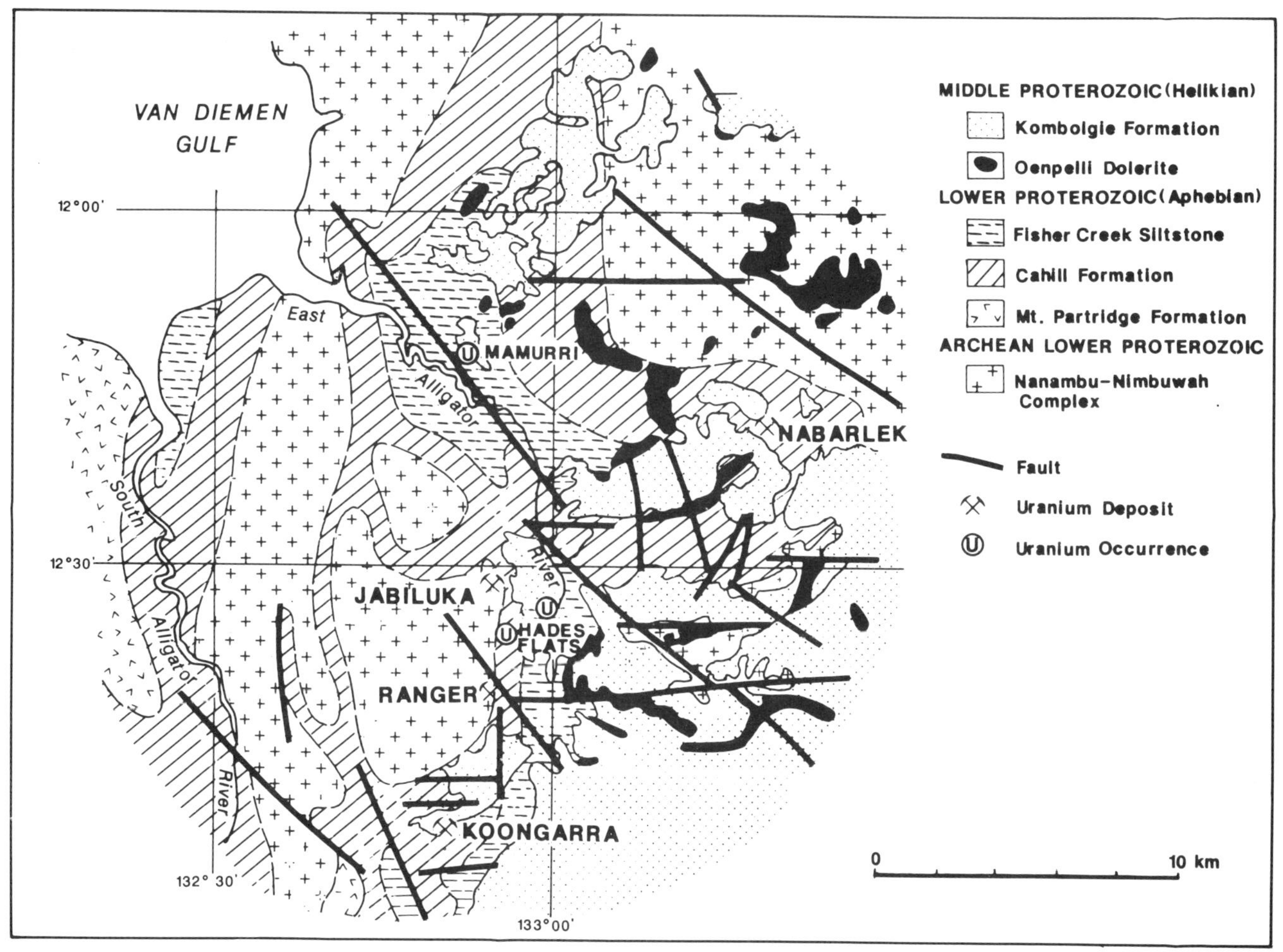

Figure 3 *General geology and major uranium deposits of the East Alligator Rivers area, N.T. (After Hegge and Rowntree, 1978).*

by a sequence of younger (1800 Ma) gneisses, migmatites and metasediments which were derived from the Lower Proterozoic Kakadu Group. The basement rocks were metamorphosed to amphibolite grade, isoclinally folded and uplifted during an orogenic event at 1800 Ma (Needham and Stewart-Smith, 1980). The Cahill Formation, which is host to the majority of the uranium deposits, overlies the Nanambu Complex and the Kakadu Group. Nash and Frishman (1981) and Needham and Stewart-Smith (1980) divide the 3,000 m thick Cahill Formation into a lower member (200-500 m thick), consisting of Mg-rich marble, schist and gneisses and an upper member of carbonaceous pelite and impure sandstone, metamorphosed to quartz-biotite schist and gneiss.

The Kombolgie Formation (one of the major units of the Middle Proterozoic Carpentarian sediments) unconformably overlies the Archean and Lower Proterozoic rocks. It is similar to the Athabasca Group sediments, in that it consists predominantly of a sequence of well-sorted fluviatile sandstones. Page *et al.* (1980) have dated volcanic units intercalated with the Kombolgie sandstone at 1648 ± 29 Ma (Rb/Sr).

Other districts. In addition to the two major uranium fields in northern Saskatchewan and Northern Territory of Australia, a number of small discoveries and high potential occurrences have been reported in the Keewatin District of Canada (northeastern sub-arctic region). Curtis and Miller (1980) describe the regional setting of uranium showings associated with the Thelon, Baker Lake and Dubawnt basins. Similar to the larger uranium fields, these occurrences are hosted by Lower Proterozoic (Aphebian) shelf facies and Middle Proterozoic continental metasediments (Wright, 1967). The latter are intercalated with and overlain by alkalic and calc-alkalic volcanics and fluviatile metasediments. The metamorphosed and folded basement and the Aphebian rocks are unconformably overlain by mature sandstones (including a basal conglomeratic unit) of the Thelon, Dubawnt and Baker Lake basins.

Deposit Characteristics

Stratigraphic location and host lithologies. Unconformity-type uranium deposits occupy a very specific stratigraphic location. Exploration leading to the early discoveries of this group of deposits, such as a number of large orebodies in the East Alligator Rivers of Northern Territory, was based on the model for the deposits of the Rum Jungle area, also in the Northern Territory, which were discovered in 1949 (Fraser, 1980). As a result, the initial search for the deposits was largely focussed on the Lower Proterozoic metasediments. In the Rabbit Lake deposit (Figure 2), host lithologies consist of interlayers of calc-silicates and meta-arkoses accompanied by massive meta-arkose, segregation pegmatites, plagioclasite and biotite microgranite (Hoeve and Sibbald, 1978; Sibbald, 1983). Similarly, the A zone of the Collins Bay deposits (Jones, 1980), a portion of the Cigar Lake (Fouques *et al.*, 1986), the N Zone, Claude and Dominique deposits of Cluff Lake (Hoeve *et al.*, 1980), and the Midwest deposit (Wray *et al.*, 1981) are hosted by carbonaceous metasediments, calc-silicates, anatectic gneisses and granite-segregation pegmatites. Nash *et al.* (1981) indicate that most East Alligator Rivers deposits are hosted by, or are in close proximity to, the carbonaceous metasediments of the Cahill Formation.

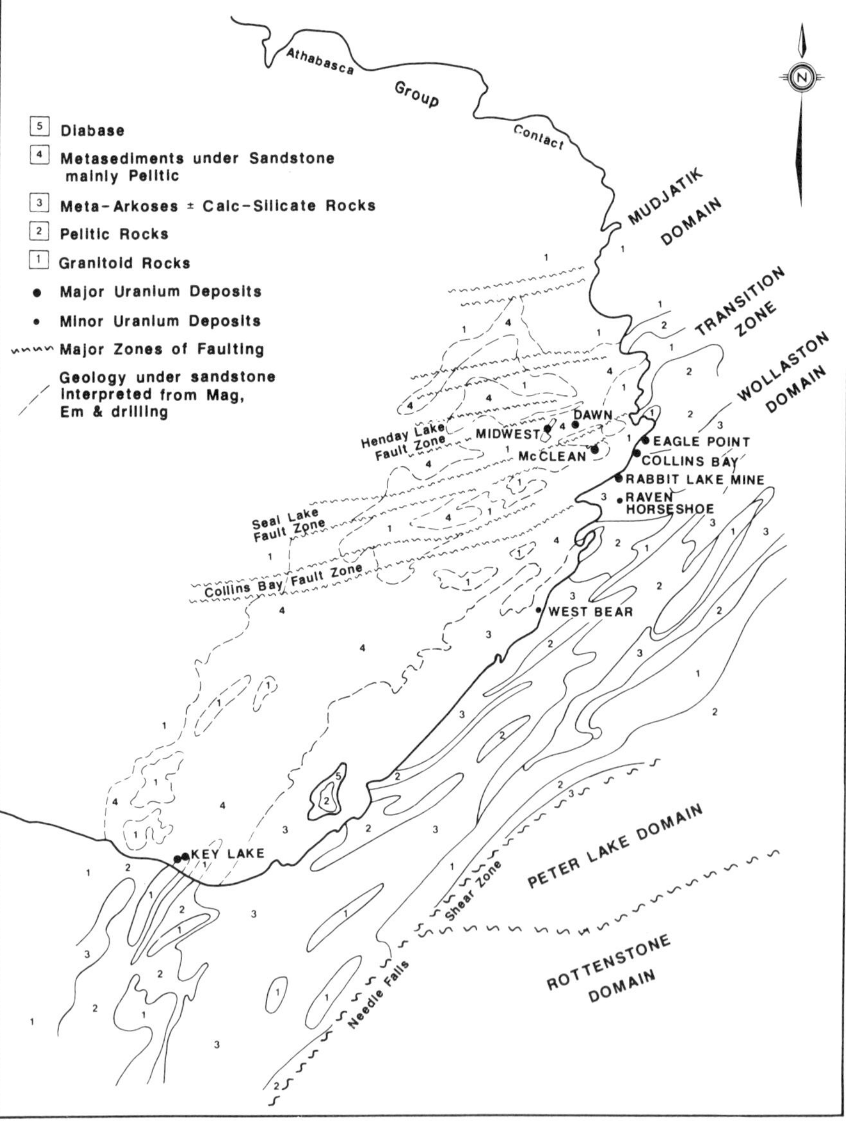

Figure 4 *Generalized geology of the eastern Athabasca uranium field. (After Fogwill, 1981).*

Because of the close spatial association with these rocks, Needham and Stewart-Smith (1980) concluded that these lithologies played an important role in inducing precipitation of uranium from fluids. However, later discoveries of most of the Saskatchewan deposits showed that, in addition to the crystalline basement, the unconformably overlying fluvial sediments are prospective ground for substantial mineralization. Whereas deposits, such as Rabbit Lake, Collins Bay A Zone (Figure 2), Jabiluka I and II, Koongarra and Nabarlek (Figure 3), are entirely restricted to the Lower Proterozoic metasediments, deposits such as Midwest, Dawn Lake, McClean, Collins Bay B zone and Key Lake straddle the pre-Helikian unconformity (Figure 4). It should be pointed out, however, that in deposits such as Jabiluka I and Ranger the present-day exposure level coincides with the paleo-erosional surface and hence, if there was any mineralization in the overlying sub-aerial sediments it has been eroded. In fewer cases such as the D zone of the Cluff Lake deposits, with the exception of minor mineralization in basement fractures, all of the mineralization occurs above the sub-Athabasca unconformity. It is apparent that this group of uranium deposits is spatially restricted to areas immediately above and below the unconformity. The extent of mineralization with respect to the paleo-erosional surface is variable. In cases such as the Rabbit Lake deposit, mineralization continues to a depth of 200 m below and, at Midwest, up to 200 m above the unconformity (Hoeve *et al.*, 1980, Sibbald, 1983). An average range of 70-100 m of mineralization on either side of the unconformity is fairly consistent in most deposits.

Hoeve *et al.* (1980) note that some of the NE-trending diabase dykes and sills cutting the Athabasca Basin may also be host to some mineralization.

Structural features and orebody form. In most unconformity-type deposits many episodes of pre-, syn- and post-mineralization deformation have been documented. Often, the siting of the deposits is either related to reverse faults or to normal faults (Sibbald, 1987; Hoeve *et al.*, 1980; Jones, 1980; Ayres *et al.*, 1983). For instance, at the Rabbit Lake deposit, NNE-trending faults have been cut off and reactivated by a set of low-angle, ENE-trending, reverse faults (Hoeve *et al.*, 1980). One of these faults, the Rabbit Lake Fault dips 30°SE and shows a vertical displacement of at least 75 m (Hoeve and Sibbald, 1978). The Rabbit Lake deposit is located in the upthrust block of this low-angle reverse fault (Hoeve and Sibbald, 1978; Nash *et al.*, 1981). Similarly, at the Koongarra deposit, reverse faulting has placed the Lower Proterozoic Cahill Formation above the Middle Proterozoic Kombolgie Formation. This reverse fault dips 60°SE and is filled with brecciated mineralized rocks (De Voto, 1978). At the B zone of the Collins Bay deposits, highly altered segments of sub-Athabasca rocks, having already been altered to sericite and kaolinite, were thrust upon or squeezed into the Athabasca Group sandstone as irregularly shaped dykes (Jones, 1980).

The best example of mineralization associated with normal faulting is at the Midwest deposit (Sibbald and Quirt, 1987).

In some deposits, both normal and reverse faulting have been reported; Nash and Frishman (1981) describe three distinct host structures from the Ranger orebody: (a) low-angle reverse or thrust faults; (b) high-angle normal faults; and (c) carbonate thinning and collapse breccias.

Both reverse and normal faults are often best developed along planes of weakness such as contacts of lithological units with differing competencies. Sibbald and Quirt (1987) give examples of this phenomenon from Key Lake and Collins Bay A and B zones where contacts between less competent graphitic pelites and more brittle granitoid gneisses have been the focus of high strain. Evidence of cataclasis, and intermittent brittle and ductile deformation, associated with mineralization are reported from many deposits. Hoeve *et al.* (1980) suggest that in many cases, structures hosting mineralization consist of veins, veinlets, open space-fillings and breccias. They emphasize that, as in the case of Rabbit Lake, although the breccia was initially tectonically induced, subsequent dissolution of carbonates generated a collapse breccia. Similarly, removal of carbonates by silicification, resulting in collapse breccias which host mineralization at the Ranger deposit, have been described by Hegge and Rowntree (1978). Ewers and Ferguson (1980) further emphasize the significance of continued re-brecciation at this deposit.

At the McClean deposits, Wallis *et al.* (1984) and Jagodits *et al.* (1986) report extensive fracturing, faulting and brecciation associated with orebodies. They note, however, that the zones of intense fracturing extend far beyond the mineralized zones, specifically within the rocks of the Athabasca Group, and draw attention to the fact that although intense brittle failure is significant and necessary to provide channelways, it is not a unique exploration guide, as in some cases the barren rocks may have a denser pattern of fractures than the deposit itself.

In addition to the structures related to brittle failure, features representing advanced stages of ductile deformation have been documented in many deposits. The most detailed description of such features is given by Dahlkamp (1978) for the Key Lake deposit (Figure 5) where the host lithologies are divided on the basis of their state of deformation. The Aphebian metasediments, where undeformed, consist of carbonaceous metapelite, biotite-plagioclase-cordierite gneiss and a coarse-grained anatectic gneiss/pegmatite. The deformed host rocks are described as various, highly altered mylonites. Dahlkamp (1978) points out that because of the intensity of deformation, with the exception of the biotite-plagioclase-cordierite gneiss which is the parent rock to a sericitic-chloritic mylonite, precursors to the other deformed host lithologies are not known. Von Pechmann (1981) also emphasizes the fine-grained nature of the rocks at Key Lake, as a result of tectonism. Wallis *et al.* (1984) make reference to some mylonitic textures, kink folds and crenulation that existed prior to the formation of the regolith, at the McClean deposit. From the Cigar Lake deposit, Fouques *et al.* (1986) describe blastomylonitic textures in a "quartz-eye" metapelite underlying the ore zone and Ayres *et al.* (1983) report ribbon-like textures in the porphyroblastic metapelites at the Midwest deposit.

At the McClean deposit, Wallis *et al.* (1986), show that many generations of channelways, which played an important role in focussing mineralizing fluids, were developed during the deformational history of the deposit. They propose a sequence as follows: (a) in the crystalline basement, numerous brittle structures formed after the Hudsonian Orogeny (1850-1750 Ma) and prior to the formation of the regolith (± 1600 Ma) and the Athabasca Group sediments. Some of these structures, however, have subsequently been reactivated; (b) in the Athabasca Group, structures consisted of joints, fractures, faults and the inherent porosity/permeability of bedding planes and conglomeratic strata; (c) mineralization-associated structures, superimposed on previous structures, comprised of an early dilational-extensional phase and a later cavity collapse, partial corrosion and dissolution phase.

The orebody forms have been described as wedge-shaped (Koongarra: Morton, 1977), tabular (McClean: Jagodits *et al.*, 1986), flattened and elongate (Midwest: Ayres *et al.*, 1983), half cylindrical (Key Lake, Collins Bay: Hoeve *et al.*, 1980), and amoeboidal in plan (Ranger: Morton, 1977). The Nabarlek orebody has been described as pods that occur at the intersection of a transcurrent cataclastic zone and horizons of chloritized rocks, and becomes more tabular with depth where the orebody is cut off by a diabase sill (Hegge and Rowntree, 1978).

In summary, it appears that unconformity-type uranium deposits show a strong structural control and consist of pods, lenses, veins, breccia fillings and disseminations.

Alteration. Three distinct alteration episodes and processes associated with unconformity-type uranium deposits have been recognized. In geochronological order they are:

(a) "alteration related to retrogression" of the high-grade metamorphic assemblage, during the waning stages of the Hudsonian Orogeny. This is not extensive and has only been described from a few localities. Ayres *et al.* (1983) mention this type of alteration at the Midwest deposit where sericitization and chloritization of cordierite, biotite and feldspars have occurred;

(b) a major episode of **"alteration located at the sub-Athabasca"** level of erosion and weathering, where a paleoweathering soil profile or regolith developed. This paleosoil was preserved from later erosion by being capped by the Athabasca Group sediments. Wallis *et al.* (1984) give a detailed description of the regolithic horizons which developed during this weathering episode. Characteristically, the regolith shows a vertical zonation and is superimposed on the crystalline basement rocks. Wallis *et al.* (1984) describe an upper red division of the paleosoil consisting of hematite capped by a thin (few centimetres) layer of bleached material. The lower half of the regolith is green and is gradational to the overlying red profile and the underlying fresh bedrock. The zonation of the regolith may be better defined as kaolinite at the top and illite and chlorite at the bottom. Hematite is ubiquitous except at the very base of the profile (Hoeve *et al.*, 1980). At the McClean Lake deposit, the thickness of the regolith varies from 7 to 222 m, and is dependent on the underlying lithology, ranging from the thinnest (less than 21 m) over meta-arkose to the thickest (up to 106 m) over the carbonaceous metapelites (Wallis *et al.*, 1986). Clearly the depth of the regolith extends even further over fault zones and channelways.

The genesis of the regolith has been

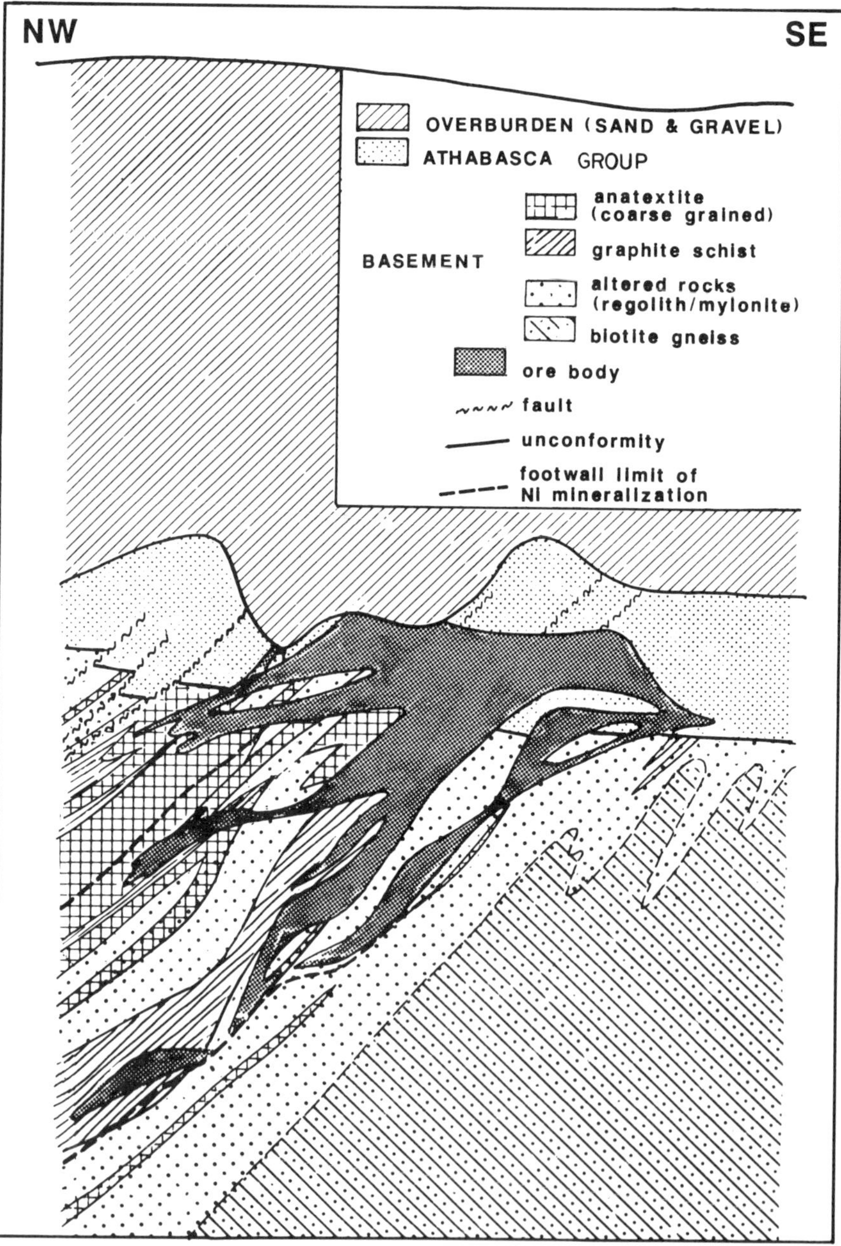

Figure 5 *Generalized geological cross-section of the Deilmann orebody, Key Lake deposit. (After Dahlkamp, 1978).*

debated by various workers and is attributed to two fundamental processes: (1) the regolith represents a paleo-lateritic horizon which formed prior to the sedimentation of the Athabasca Group, or (2) the regolith is a result of diagenesis after the basinal sedimentation.

Wallis *et al.* (1984) assume that the regolith is a paleo-laterite which is the source for quartz, clay and hematite in the Athabasca Group sediments. However, Ramaekers (1983) points out that the diagenesis and intense alteration of the Athabasca Group sediments continued several hundreds of million years after their deposition, if not in fact it is still continuing to date. Hoeve *et al.*, (1980) also emphasize that the red colouration of the Athabasca Group sediments is a consequence of oxidation and post-depositional leaching and, although there are hematite pebbles in the sandstone, the bulk of the hematite in the Athabasca Group sediments is diagenetically induced. Hematite, kaolinite and illite alteration of the post-Athabasca diabase dykes, as well as bleaching (reduction) of the oxidized regolith, also attest to the post-sedimentation alteration (Hoeve *et al.*, 1980).

It is therefore reasonable to assume that although the regolith was formed prior to the formation of the unconformity and deposition of the sandstone, it has been affected by later diagenesis.

Literature on the Australian deposits does not contain sufficient information on the extent and nature of the regolithic horizon; it may be that most of the regolith has been eroded or that it was only poorly developed.

(c) "the alteration directly associated with the mineralization" which overprints the regolith assemblage (Figure 6). In most instances, where mineralization-related alteration (referred to as hydrothermal alteration in the literature) is superimposed upon the basement rocks and the regolith, characterization of various types of alteration and their paragenesis is difficult or impossible. The best descriptions of this alteration have been given where it has affected the lithologically simpler Athabasca Group sandstone. One of the characteristic features of the hydrothermal alteration is that it is far more extensive than the mineralization and therefore provides an excellent exploration target. Fouques *et al.* (1986) point out that the alteration at the Cigar Lake deposit extends up to 100 m below and 200 m above the unconformity, but is limited laterally.

The dominant alteration types consist of chloritization, argillization, carbonatization (commonly dolomitization), silicification, sulphidation and tourmalinization. The intensity of the alteration increases with proximity to better mineralized sections. The present literature does not provide sufficient data on the nature of original textures of the primary lithologies which often have been obliterated by deformation and alteration, but in some cases, as has been documented by Fouques *et al.* (1986) from the Cigar Lake deposit, parts of the basement rocks altered to illite and chlorite have retained recognizable relict textures. In most deposits, an overall zonation may be defined and, depending on the dominant host lithology, one or several of the alteration assemblages predominate. At the Cigar Lake deposit, the central core of the mineralized zone comprises illite, sudoite (Mg-rich chlorite), dravite, and rare phosphates, such as goyazite; in addition, all the carbonaceous material has been removed form the upper basement and has been replaced by siderite and calcite (similar to the Rabbit Lake and the Collins Bay deposit, Hoeve *et al.*, 1980). In contrast with the inner core, 50-100 m below the unconformity, partial replacement by illite is practically the only product. At the Rabbit Lake deposit, Hoeve and Sibbald (1978), describe three types of alteration. An early pre-mineralization type, which is restricted to the high-grade core of the mineralized zone, and consists of dark green Fe-rich chlorite and anatase. The two other types are synchronous with mineralization: a red halo which consists of Mg-rich chlorite, tourmaline (mostly dravite), quartz, anatase and hematite; and a pale green assemblage comprising an assemblage similar to the red halo but lacking the hematite and enriched in pyrite, chalcopyrite, chalcocite and galena. Hoeve and Sibbald (1978) mention that silicification and dolomitization are only locally significant.

A detailed description and paragenesis of the alteration minerals of the Ranger orebodies is discussed by Nash and Frishman (1981). With the aid of x-ray diffraction and microprobe analyses, these workers show that the most extensive alteration is chloritization

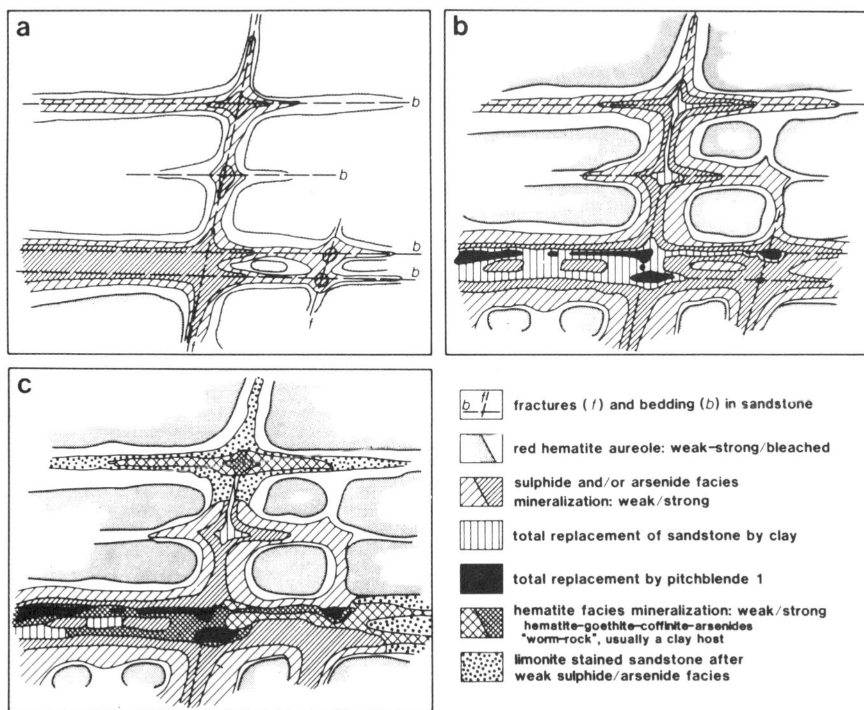

Figure 6 *Alteration and mineralization patterns at the McClean deposits. (After Wallis* et al., *1986).*

(which took place in several episodes), sericitization and argillization (whose distribution is not well known), dolomitization (mostly magnesite and dolomite with calcite totally absent) and formation of apatite and Ti-oxides. Similar assemblages have been reported from Nabarlek and Jabiluka deposits (Hegge and Rowntree, 1978; Binns *et al.*, 1980; Ewers and Ferguson, 1980). Wallis *et al.* (1984) define the alteration package of the McClean deposit based on kaolinite/illite/chlorite ratios. Illite, which is intergrown with various amounts of hematite, exhibits a spectrum of colours, and occurs up to 150 m above the core of the mineralized zone. These authors recognize at least five generations of hematite prior to, and some synchronous with, mineralization and note that it may be replaced by pyrite, siderite and subsequently by limonite. Wallis *et al.* (1984) demonstrate that, in contrast with a deposit such as Key Lake where Mg-rich chloritized rocks are barren and the ore is associated with the Fe-rich chloritic zones (Dahlkamp, 1978), at the McClean deposit, both Fe and Mg-rich chlorites coexist. Bray *et al.* (1987) have dated illite from the alteration halo of the McClean deposit at 1319 ± 3Ma.

In summary, there are three distinct alteration assemblages in this group of deposits, reflecting metamorphic retrogression, a period of weathering and erosion, and mineralization. The mineralization-related alteration is extensive and overprints the two other assemblages.

Mineralization. One of the most important characteristics of this group of ore deposits is that, unlike other types of uranium deposits, the average grade of mineralization is very high. In most deposits, average grades reach several percents and in some deposits highest grades reach tens of percents of U_3O_8 (Hoeve *et al.*, 1980). In addition to U, these deposits may contain concentrations of Ni, Co, Ag, Mo, Cu, Pb, Zn, Bi, Se and As, and less frequently Au and PGE. In some cases, these elements attain economic grades. Examples include gold at Jabiluka II (Figure 7); gold and selenium at D Zone of Cluff Lake; nickel and arsenic at Key Lake, Midwest, Cigar Lake and Dawn Lake; and gold and silver at Collins Bay A zone (Sibbald and Quirt, 1987). The genetic relationships among the metallic concentrations are not well understood. Ewers and Ferguson (1980) find no positive correlation between U and Au in the East Alligator Rivers deposits.

An interesting chemical signature of the mineralization is the presence of many species of solid and gaseous hydrocarbons. Hoeve *et al.* (1980) report carbon dioxide, methane and ethane from Rabbit Lake, Cluff Lake and Collins Bay B zone.

The mineralization has three basic characteristics:

(a) there is a primary (hypogene) and a secondary (supergene) mineral assemblage. The main primary uranium minerals are uraninite and pitchblende (both oxides). The latter shows a variety of textures such as euhedral, subhedral, botryoidal, spherulitic, radial and colloform banding. Primary Ni minerals include rammelsbergite, pararammelsbergite, gersdorffite and millerite. Pyrite, arsenopyrite, galena, sphalerite, chalcopyrite and molybdenite are the usual sulphides.

The secondary (supergene) minerals result from *in situ* oxidation and alteration of primary uranium oxides by low-temperature ground waters (Snelling, 1980). At Nabarlek, Morton (1977) has shown that the depth of the supergene enrichment is about 15 m (at the dry-season water level) and at Koongarra reaches 25 m. In this case, a vast range of new minerals are formed, amongst which, the better known include uranophane and sklodowskite (silicate), torbernite and autunite (phosphate) and various vanadates and sulphates.

(b) the mineral assemblage in the Athabasca Group host rocks differs from that in the basement lithologies. Von Pechmann (1981) shows that the uranium mineralization hosted by the Athabasca sediments consists primarily of sooty pitchblende and coffinite and the Ni occurs mainly as gersdorffite and millerite. In the basement rocks, however, uranium is present as both crystalline and sooty pitchblende and Ni and Co occur as various arsenides. Ayres *et al.* (1983) divide the ore at the Midwest deposit into three types according to their host lithology: (a) the basement ore, which is pitchblende and coffinite; (b) the unconformity ore, which consists of massive and colloform pitchblende; and, (c) the sandstone ore, which is primarily a sooty, fine-grained pitchblende. All three types of uranium ores are accompanied by high concentrations of various Ni-arsenides. At the Dawn Lake deposit, uranium occurs as sooty pitchblende, but in the deeper parts of the mineralized "pods", hosted by the Aphebian metasediments, the texture of the pitchblende becomes botryoidal (Clarke

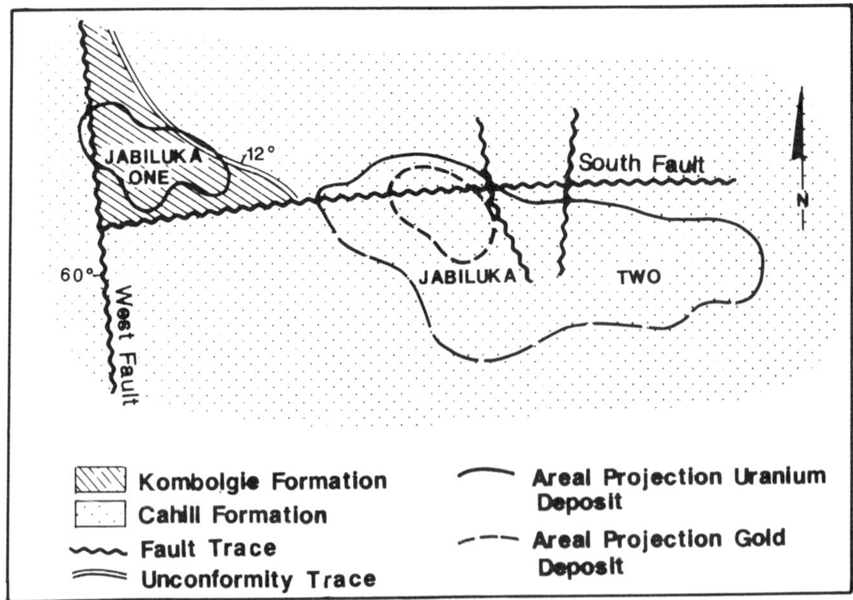

Figure 7 *Plan of Jabiluka I and II deposits; note the overlapping gold mineralization at Jabiluka II. (After Hegge* et al.*, 1980).*

and Fogwill, 1986). This deposit is particularly rich in Co- and Ni-arsenides and chalcopyrite at depth. Hoeve *et al.* (1980) also contrast the lustrous colloform pitchblende in the Athabasca Group ore with the sooty pitchblende and coffinite in the Aphebian metasediments at the Key Lake deposit.

(c) paragenesis and timing of mineralization. Lainé (1986) defines three phases of mineralization at Cluff Lake. Uraninite, brannerite and sulphides formed at 1150-1050 Ma, a second stage of uraninite and sulphides formed at 890-820 Ma and a last phase consisting of pitchblende and hematite formed at 380 Ma. Similarly, Dahlkamp (1978) demonstrates that, at the Key Lake deposit, crystalline pitchblende has been altered to sooty pitchblende and the latter has, in turn, been replaced by coffinite. Fouques *et al.* (1986) compare the paragenesis of the Cigar Lake orebody to Key Lake and Cluff Lake, but emphasize that volumetrically the most important phase was the first generation uraninite. Wallis *et al.* (1986) in subdividing the mineralization at the McClean deposit into three facies, note that early base metal sulphides and Ni-arsenides were contemporaneous with pitchblende, were overprinted by sericite and illite, and precede coffinite, colloform pitchblende and a second generation of Ni-arsenides, all of which are contemporaneous with oxides (Figure 6). Snelling (1980) provides another well-documented example of the paragenesis of the metallic minerals from the Koongarra deposit. In this instance, uraninite is divided into two types on the basis of its calcium and lead contents. The first type, with 1-2% CaO, shows many varieties of textures, whereas the second type, containing 3-5% CaO, has colloform banding and resembles low-temperature mineralogy typical of the sandstone-hosted roll-front type deposits. In this deposit, a vast array of secondary minerals constitute the bulk of the mineralization.

In general, there is consensus that the age of "initial" uranium mineralization in Saskatchewan deposits is relatively consistent from one deposit to another. Hoeve *et al.* (1980) quote U-Pb dates such as 1281 Ma for Rabbit Lake, 1200 Ma for Key Lake, 1330 and 1050 Ma for the D Zone of Cluff Lake and Fryer and Taylor (1984) obtain a Sm/Nd age of 1281 ± 80 Ma for Collins Bay. Trocki *et al.* (1984) obtain an age of 1350 ± 4 Ma (U-Pb) for mineralization at Key Lake and suggest that although many younger ages have been reported from various deposits, no other "specific" time can be determined. Bray *et al.* (1987) obtain an alteration and mineralization age of 1319 ± 3 Ma ($^{40}Ar/^{39}Ar$).

In conclusion, the main features of the mineralization are: (i) a hypogene and a supergene assemblage; (ii) different mineralogy in basement and Athabasca Group host rocks; and, (iii) a multi-phase process with an initial uranium age some 100-150 m.y. younger than the Athabasca Group sediments.

Fluid inclusion and stable isotope data

The extent of data of this nature in the literature is inconsistent. Bray *et al.* (1982, 1984), Pagel *et al.* (1980), Donnelly and Ferguson (1980) and Ypma and Fuzikawa (1980) present data from some of the Saskatchewan and Australian deposits. Most of these workers obtain similar isotopic signatures, the summary of which follows.

Metamorphic fluids. Fluid inclusions from the Cahill metasediments suggest a metamorphic fluid with NaCl and high density CO_2 at 350°C (Ympa and Fuzikawa, 1980). Donnelly and Ferguson (1980) obtain $\delta^{34}S$ values of +2.2‰ for the metamorphic fluids of most of the Pine Creek Geosyncline deposits. Pagel *et al.* (1980) define the metamorphic fluids as carbonic (granulite facies).

Ore-related fluids. Data from samples of ore and alteration minerals show that the ore fluids had different characteristics from the metamorphic fluids. From the Australian deposits, Ympa and Fuzikawa (1980) obtain a very saline fluid with up to 20-30 wt.% $CaCl_2$, 5-10 wt.% $MgCl_2$, 10-20 wt.% NaCl and minor KCl and $FeCl_2$. They estimate a homogenization temperature of 160-110°C. Pagel *et al.* (1980), on the other hand, calculated temperatures ranging from 160-185°C from fluid inclusions from the Rabbit Lake deposit.

Isotopes. Carbon isotope values from the Cahill metapelites (either carbonates or carbonaceous minerals) indicate a sedimentary origin, and suggest an organic derivation (Donnelly and Ferguson, 1980). Bray *et al.* (1984) report that the oxygen isotope values obtained at the McClean deposits do not show a difference between the mineralized and unmineralized rocks. However, through the use of hydrogen isotope data, it is possible to distinguish the illite and chlorite associated with the ore zone from illite in the sandstone and the regolith. Ympa and Fuzikawa (1980), based on their oxygen isotope data from several Australian deposits, suggest a meteoric origin for the ore-forming fluids. Sulphur isotope information (Bray *et al.*, 1982) indicate similar signatures for the ore-related and barren sulphides.

In summary, ore fluids were highly saline (showing many salinity reversals), with an average temperature of 180-200°C. Pagel *et al.* (1980), and Bray *et al.* (1987) compare these fluids to diagenetic brines and point out that their characteristics are different from the earlier metamorphic fluids.

Genesis

As for any group of mineral deposits, debate over the genesis of unconformity-related uranium deposits has continued since the discovery of the first deposit of this type and their recognition as a class.

An examination of the progression of the genetic models developed for these deposits shows that early models, because of lack of sufficient geological and detailed laboratory data, are incomplete and therefore simplistic. Subsequently, when the complexity of the deposits was realized (by access to three-dimensional exposures and larger number of discoveries), the models that developed included all possible geological processes in order to accommodate all features. It is only after many years of careful geological and laboratory documentation that models were produced which involved a single, dominant process capable of explaining most of the features of these deposits. The following is a brief account of this evolutionary path.

The "supergene model". The supergene model was originally proposed by Knipping (1974), and Ruzicka (1975), amongst others. In this case, it is assumed that uranium and other metals present in the Lower Proterozoic rocks, were leached from them by ground and surface waters and were precipitated when and where they encountered a reducing environment such as the carbonaceous metasediments. The timing of this process is presumed to be pre-Athabasca, during the weathering and erosion that resulted in the formation of the regolith.

The supergene model was debated by workers, such as Hoeve and Sibbald (1976, 1978), who demonstrated the inconsistency of many geological observations with the proposed model. For instance, subsequent discoveries have shown that extensive amounts of mineralization occur within the Athabasca Group sediments and that both alteration and mineralization overprint the regolithic alteration. Also, later dating by various workers has proven that initial mineralization post-dates Athabasca Group sedimentation (Bray *et al.*, 1987; Armstrong and Ramaekers, 1985; Bell, 1981, 1985).

The "hypogene model". The sources of the mineralizing structures for the hypogene model range from metamorphogenic/hydrothermal (Hegge and Rowntree, 1978; von Pechmann, 1981; Morton, 1977) to magmatic/hydrothermal (Binns *et al.*, 1980) to a combination of the two. In this model, the source of the fluids is considered to be deep-seated, generated during the metamorphic event that preceded the deposition of the Athabasca Group sediments. Hegge and Rowntree (1978) suggest that during the metamorphism of high-uranium granites (averaging 9.6 ppm U) and metasediments (average of 34 ppm U in the Cahill Formation), uraniferous fluids were generated, as a result of anatectic rejuvenation, migmatization and pegmatitic intrusion; later retrogression caused Mg-metasomatism and uranium deposition. Binns *et al.* (1980) hypothesize that a geological setting such as Jabiluka represents an upwelling centre where radiogenic heat generated from adjacent granites drove a convective cell of circulating metalliferous fluids. These authors suggest the same post-kinematic granitoids as the source for uranium. Fogwill (1981) even considers a mantle-derived source for the Ni-Co-As assemblage. One of the major drawbacks of this model is the fact that it cannot satisfactorily explain the undeniable spatial association of the deposits with the pre-Helikian unconformity.

"Polygenetic, multiphase model". In order to accommodate all the features that are characteristic of the deposits a number of workers have presented a multigenetic model. Dahlkamp (1978), Clarke and Fogwill (1986), McMillan (1977), Fouques *et al.* (1986) and Nash *et al.* (1981) present an involved and complex history of mineralization, which includes the following stages:

(a) Lower to Upper Proterozoic synsedimentary concentrations in carbonaceous sediments at 2200-1900 Ma;
(b) mobilization and further concentration of the existing mineralization during the Hudsonian Orogeny at 1900-1800 Ma;
(c) formation of the regolith, whereby weathering and surface leaching removed some of the mineralization and redeposited a new generation of uranium minerals in fractures and faults of the basement rocks; this event took place at 1800-1350 Ma;
(d) deposition of the Athabasca basin, generating diagenetic fluids and another generation of uranium minerals at 1350-1000 Ma; and,
(e) finally, secondary mineralization during several episodes of uplift and erosion between 1000 and 200 Ma.

One of the major problems with this model is that syngenetic, economically significant deposits have never been identified in these environments (Nash *et al.*, 1981). Hegge and Rowntree (1978) report a background value of 34 ppm uranium in the Cahill Formation. However, Nash *et al.* (1980) remark that these values were not obtained from entirely barren rocks and represent a mixture of mineralized and unmineralized samples. Trocki *et al.* (1984), in discussing the Key Lake deposit, note that a concentration of at least 240 ppm U is necessary in the protore in order to reach the subsequent concentrations. The most problematic aspect of the polygenetic model is the timing of events. Bray *et al.* (1987), Armstrong and Ramaekers (1985), Baadsgaard *et al.* (1984) and Blenkinsop and Bell (1981) present many isotopic ages which clearly demonstrate that the timing of the first stage of mineralization postdates the deposition of the Athabasca sediments.

In summary, the following reasons suggest that evidence to prove the early stages of the model are insufficient: (i) lack of anomalous concentrations in the country rocks; (ii) absence of early mineralized veins; (iii) superposition of the alteration/mineralization assemblage upon the regolith; and, (iv) post-Athabasca, initial mineralization ages.

"Diagenetic model". Hoeve and Sibbald (1976, 1978), Hoeve *et al.* (1980) and Jones (1980), amongst others, are the advocates of the diagenetic model (Figure 8). Sibbald (1985) proposes that interaction of the basement and Athabasca Group rocks through the mixing of chemically contrasting fluids was responsible for precipitation of uranium. Hoeve *et al.* (1980) explain that after the deposition of the Athabasca Group sediments, intraformational

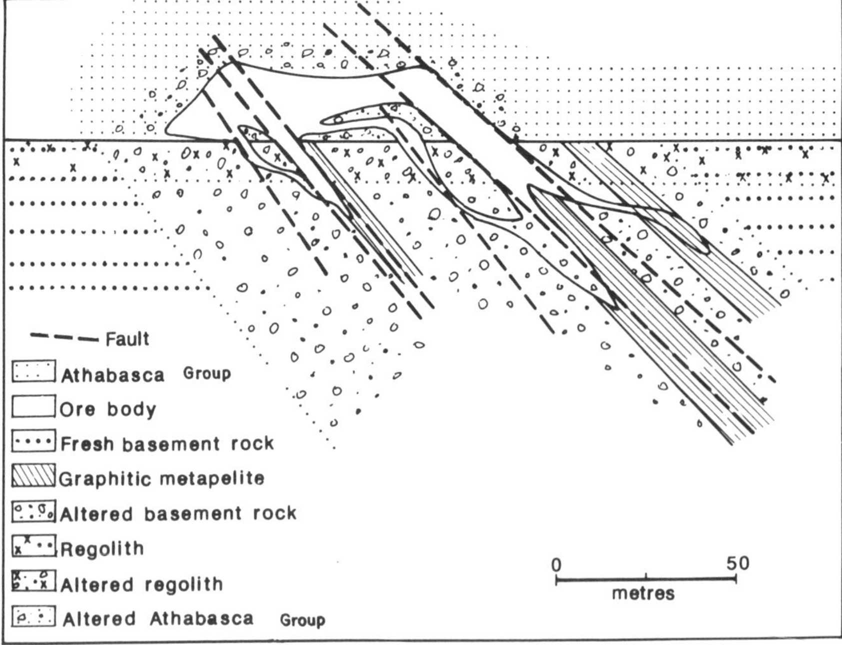

Figure 8 *Schematic cross-section of an orebody showing the general setting of unconformity-related deposits. (After Hoeve and Sibbald, 1978).*

fluids percolated through them to encounter a different physico-chemical environment upon reaching the unconformity and the basement contact. The oxidizing metalliferous fluids were heated during their descent along the geothermal gradient (T = 180-220°C and P = 700 bars from fluid inclusions, Pagel, 1975, 1977) and intersected a reducing environment at the base of the unconformity which could either be the carbonaceous metasediments or methane-bearing fluids ascending faults and fractures in the basement. At this redox front, which is described by Sibbald (1985) and Sibbald and Quirt (1987) as dynamic but basically stationary, metals were released from the fluids and precipitated in favourable structural sites such as the porous regolith, faults, fractures and breccias. Wallis *et al.* (1986) support the diagenetic model and propose a "two-fluid" system (the reductant plume model) to account for the redox variations. Hoeve *et al.* (1980) and Sibbald and Quirt (1987) consider the Athabasca sediments themselves as a possible source of metals, whereby breakdown of feldspars, mafic and heavy minerals could have released uranium and other metals which would have stayed in solution within the oxidizing environment of the sediments.

The strongest evidence in support of this genetic interpretation is the fact that it agrees with the paragenesis of the alteration and mineralization assemblages and the isotopic age data.

Summary and Discussion

The unconformity-related uranium deposits: (i) are hosted by varied lithologies located below, at, and above the unconformity between Lower and Middle Proterozoic rocks; (ii) occur as veins, breccias, and open-space fillings usually associated with reverse and normal faulting; (iii) are associated with three episodes of alteration related to the retrogression of the amphibolite metamorphic event, weathering and erosion, and later hydrothermal activity; (iv) are commonly polymetallic consisting of uranium, V and Mo oxides, Ni, Co, Cu, Zn and Pb sulphides and arsenides, and in some cases native Au; (v) have isotopic signatures which indicate a high salinity ore-forming fluid, ranging in temperature between 160 and 200°C (similar to diagenetic fluids); and, (vi) have initial mineralization ages between 1350-1200 Ma, which are 100-150 m.y. younger than the depositional age of the Middle Proterozoic sediments.

Most of the geological relationships, such as the paragenesis of the alteration and mineralization assemblages as well as the radiogenic and stable isotope data, suggest that, of the models presented, the "diagenetic" model accounts for the majority of the features of this class of deposits. The source of uranium and other metals is still not well established, and as proposed by Hoeve *et al.* (1980), the metals may have derived from the sedimentary basins or they may be from other origins. Hoeve *et al.* (1980) compare the unconformity-related deposits to younger uranium deposits which are hosted by sandstone beds and their siting is controlled by a redox front (hence called roll-front type). In In many respects, the mode of transportation, the site of deposition and the alteration associated with roll-front deposits are comparable with unconformity-type deposits emphasizing that similar ore-forming processes were active during various geological episodes.

Acknowledgements

The author would like to express sincere thanks to V. Ruzicka of the Geological Survey of Canada, E. Craigie of BP Selco, and R.G. Roberts of the University of Waterloo for their review of the manuscript. Special thanks are due to T. Sibbald of Saskatchewan Geological Survey for his extensive critical review from which the manuscript benefited tremendously.

References

Armstrong, R.L. and Ramaekers, P., 1985, Sr isotope study of Helikian sediment and diabase dikes in the Athabasca Basin, northern Saskatchewan: Canadian Journal of Earth Sciences, v. 22, p. 399-407.

Ayres, D.E., Wray, E.M., Farstad, J. and Ibrahim, H., 1983, Geology of the Midwest deposit, *in* Cameron, E.M., ed., Uranium exploration in Athabasca Basin, Saskatchewan,Canada: Geological Survey of Canada, Paper 82-11, p. 33-40.

Baadsgaard, H., Cumming, G.L. and Worden, J.M., 1984, U-Pb geochronology of minerals from Midwest uranium deposit, northern Saskatchewan: Canadian Journal of Earth Sciences, v. 21, p. 642-648.

Beck, L.S., 1976, History of uranium exploration in northern Saskatchewan with special reference to the changing ideas of metallogenesis, *in* Dunn, C.E., ed., Uranium in Saskatchewan: Saskatchewan Geological Society, Special Publication # 3, p. 1-10.

Bell, K., 1981, A review of the geochronology of the Precambrian of Saskatchewan — some clues to uranium mineralization: Mineralogical Magazine, v. 44, p. 371-378.

Bell, 1985, Geochronology of the Carswell area, Northern Saskatchewan, *in* Lainé, R., Alonso, D. and Svab, M., eds., The Carswell Structure Uranium Deposits, Saskatchewan: Geological Association of Canada, Special Paper 29, p. 33-46.

Binns, R.A., Ayres, D.E., Wilmhurst, J.R. and Ramsden, A.R., 1980, Petrology and geochemistry of alteration associated with uranium mineralization at Jabiluka, Northern Territory, Australia, *in* Ferguson, J. and Goleby, A.B., eds., Uranium in the Pine Creek Geosyncline: International Atomic Energy Agency, Vienna, p. 417-438.

Blenkinsop, J. and Bell, K., 1981, Whole rock Rb/Sr dating of selected suites from the Precambrian of Saskatchewan : Saskatchewan Geological Survey, Miscellaneous Report 81-4 p. 25-44.

Bray, C.J., Spooner, E.T.C., Golightly, J.P. and Saracoglu, N., 1982, Carbon and sulphur isotope geochemistry of unconformity-related uranium mineralization, McClean Lake deposit, N. Saskatchewan, Canada: Geological Society of America, Abstracts with Program, v. 14, p. 451.

Bray, C.J., Spooner, E.T.C., Hall, C.M., York, D., Bills, T.M. and Krueger, H.W., 1987, Laser probe $^{40}Ar/^{39}Ar$ and conventional K/Ar dating of illites associated with the McClean unconformity-related uranium deposits north Saskatchewan, Canada: Canadian Journal of Earth Sciences, v. 24, p. 10-23.

Bray, C.J., Spooner, E.T.C. and Longstaffe, F., 1984, Oxygen and hydrogen isotope geochemistry of unconformity-related uranium mineralization, McClean deposits, N. Saskatchewan, Canada: Geological Association of Canada— Mineralogical Association of Canada, Program with Abstracts, v. 9, p. 48.

Burwash, R., Baadsgaard, H. and Morton, R.D., 1962, Precambrian K/Ar dates from the Western Canada sedimentary basin: Journal of Geophysical Research, v. 67, p. 1617-1625.

Clarke, P.J. and Fogwill, W.D., 1986, Geology of the Dawn Lake Uranium deposits, northern Saskatchewan, *in* Evans, E.L., ed., Uranium Deposits of Canada: Canadian Institute of Mining and Metallurgy, Special Volume 33, p. 184-192.

Curtis, L. and Miller, A.R., 1980, Uranium geology in the Amer-Dubawnt-Yathkyed-Baker Lake Region, Keewatin District, N.W.T., Canada, *in* Ferguson, J. and Goleby, A.B., eds., Uranium in the Pine Creek Geosyncline: International Atomic Energy Agency, Vienna, p. 595-616.

Dahlkamp, F.J., 1978, Geologic appraisal of the Key Lake U-Ni deposits, northern Saskatchewan: Economic Geology, v. 73, p. 1430-1449.

de Carle, A.L., 1986, Geology of the Key Lake deposits, *in* Evans, E.L., ed., Uranium Deposits of Canada: Canadian Institute of Mining and Metallurgy, Special Volume 33, p. 170-177.

De Voto, R.H., 1978, Uranium Geology and Exploration: Lecture Notes and References: Colorado School of Mines, Golden, Colorado, 396 p.

Donnelly, T.H. and Ferguson, J., 1980, A stable isotope study of three deposits in the Alligator Rivers uranium field, N.T., *in* Ferguson, J. and Goleby, A.B., eds., Uranium in the Pine Creek Geosyncline: International Atomic Energy Agency, Vienna, p. 397-406.

Ewers, G.R. and Ferguson, J., 1980, Mineralogy of the Jabiluka, Ranger, Koongarra, and Nabarlek uranium deposits, *in* Ferguson, J. and Goleby, A.B., eds., Uranium in the Pine Creek Geosyncline: International Atomic Energy Agency, Vienna, p. 363-374.

Fogwill, W.D., 1981, Canadian and Saskatchewan uranium deposits: compilation, metallogeny, models, exploration, *in* Sibbald, T.I.I. and Petruk, W., eds., Geology of Uranium Deposits: Canadian Institute of Mining and Metallurgy, Special Volume 32, p. 3-19.

Fouques, J.P., Fowler, M., Knipping, H.D. and Schimann, K., 1986, Le gisement d'uranium de Cigar Lake: découverte et caractéristiques générales: Canadian Institute of Mining and Metallurgy, Bulletin, v. 79, #886, p. 70-82.

Fraser, W.J., 1980, Geology and exploration of the Rum Jungle uranium field, *in* Ferguson, J. and Goleby, A.B., eds., Uranium in the Pine Creek Geosyncline: International Atomic Energy Agency, Vienna, p. 287-297.

Fryer, B.J. and Taylor, R.P., 1984, Sm/Nd direct dating of the Collins Bay hydrothermal uranium deposit, Saskatchewan: Geology, v. 12, p. 479-482.

Hegge, M.R., Mosher, D.V., Eupene, G.S. and Anthony, P.J., 1980, Geologic Setting of the East Alligator uranium deposits and prospects, *in* Ferguson, J. and Goleby, A.B., eds., Uranium in the Pine Creek Geosyncline: International Atomic Energy Agency, Vienna, p. 259-272.

Hegge, M.R. and Rowntree, J.C., 1978, Geologic setting and concepts of the origin of main deposits in the East Alligator River Region, N.T., Australia: Economic Geology, v. 73, p. 1420-1429.

Hoeve, J. and Sibbald, T.I.I., 1976, The Rabbit Lake uranium deposit, *in* Dunn, C.E., ed., Uranium in Saskatchewan: Saskatchewan Geological Society, Special Publication #3, p. 331-354.

Hoeve, J. and Sibbald, T.I.I., 1978, On the Origin of Rabbit Lake and other unconformity-type uranium deposits in northern Saskatchewan, Canada: Economic Geology, v. 73, p. 1450-1473.

Hoeve, J., Sibbald, T.I.I., Ramaekers, P. and Lewry, J.F. 1980, Athabasca Basin unconformity-type uranium deposits: a special class of sandstone-type deposits, *in* Ferguson, J. and Goleby, A.B., eds., Uranium in the Pine Creek Geosyncline: International Atomic Energy Agency, Vienna, p. 575-594.

Jagodits, F.L., Betz, J.E., Krause, B.R., Saracoglu, N. and Wallis, R.H., 1986, Ground geophysical surveys over the McClean Uranium deposits, northern Saskatchewan: Canadian Institute of Mining and Metallurgy, Bulletin, v. 79, #886, p. 35-50.

Jones, B.E., 1980, The geology of the Collins Bay uranium deposit, Saskatchewan: Canadian Institute of Mining and Metallurgy, Bulletin, v. 73, #818, p. 84-90.

Knipping, H.D., 1974, The concepts of supergene *versus* hypogene emplacement of uranium at Rabbit Lake, Saskatchewan, Canada, *in* Formation of Uranium Ore Deposits, International Atomic Energy Agency, Vienna, p. 531-548.

Lainé, R.T., 1986, Uranium Deposits of the Carswell Structure, *in* Evans, E.L., ed., Uranium Deposits of Canada: Canadian Institute of Mining and Metallurgy, Special Volume 33, p. 155-169.

Langford, F.F., 1986, Geology of the Athabasca Basin, *in* Evans, E.L., ed., Uranium Deposits of Canada: Canadian Institute of Mining and Metallurgy, Special Volume 33, p. 123-133.

Lewry, J.F. and Sibbald, T.I.I., 1979, A review of pre-Athabasca basement geology in Northern Saskatchewan, *in* Parslow, G.R., ed., Uranium Exploration Techniques: Geological Society of Saskatchewan, Special Publication #4, p. 19-58.

Macdonald, C.C., 1981, Mineralogy and geochemistry of the sub-Athabasca regolith near Wollaston Lake, *in* Sibbald, T.I.I. and Petruk, W., eds., Geology of Uranium Deposits: Canadian Institute of Mining and Metallurgy, Special Volume 32, p. 155-163.

McMillan, R.H., 1977, Metallogenesis of Canadian uranium deposits: a review, Geology, Mining and Extractive Processing of Uranium: Institution of Mining and Metallurgy, Special Publication, p. 43-55.

Morton, R.D., 1977, The Western and Northern Australia Uranium Deposits - Exploration guides or exploration deterrents for Saskatchewan, *in* Dunn, C.E., ed., Uranium in Saskatchewan: Saskatchewan Geological Society, Special Publication #3, p. 211-255.

Nash, J.T., 1978, Uranium Geology in Resource Evaluation and Exploration: Economic Geology, v. 73, p. 1401-1406.

Nash, J.T. and Frishman, D., 1981, Progress report on geologic studies of the Ranger orebodies, Northern Territory, Australia, *in* Sibbald, T.I.I. and Petruk, W., eds., Geology of Uranium Deposits: Canadian Institute of Mining and Metallurgy, Special Volume #32, p. 205-215.

Nash, J.T., Granger, H.C. and Adams, S.S., 1981, Geology and concepts of genesis of important types of uranium deposits: Economic Geology, 75th Anniversary Volume, p. 63-116.

Needham, R.S. and Stewart-Smith, P.G., 1980, Geology of the Alligator Rivers uranium field, *in* Ferguson, J. and Goleby, A.B., eds., Uranium in the Pine Creek Geosyncline: International Atomic Energy Agency, Vienna, p. 233-257.

Page, R.W., Compston, W. and Needham, R.S., 1980, Geochronology and evolution of the Late Archean basement and Proterozoic rocks in the Alligator Rivers uranium field, Northern Territory, Australia, *in* Ferguson, J. and Goleby, A.B., eds., Uranium in the Pine Creek Geosyncline: International Atomic Energy Agency, Vienna, p. 39-68.

Pagel, M., 1975, Détermination des conditions physico-chimiques de la silicification diagénétique des grès Athabasca (Canada) au moyen des inclusions fluides: Académie des Sciences, Comptes Rendus Hebdomadaires des Séances, Série D, v. 280, p. 2301-2304.

Pagel, M., 1977, Microthermometry and chemical analysis of fluid inclusions from the Rabbit Laka uranium deposit, Saskatchewan, Canada, (Abstract): Institution of Mining and Metallurgy, Transactions, v. 86, p. B157.

Pagel, M., Poty, B. and Sheppard, S.M.F., 1980, Contribution to some Saskatchewan uranium deposits mainly from fluid inclusion and isotope data, *in* Ferguson, J. and Goleby, A.B., eds., Uranium in the Pine Creek Geosyncline: International Atomic Energy Agency, Vienna, p. 639-654.

Ramaekers, P., 1976, Athabasca Formation, northeast edge (64L, 74I, 74P); Part I, Reconnaissance Geology, *in* Christopher, J.E. and Macdonald, R., eds., Survey Investigations 1976: Saskatchewan Geological Survey, p. 73-77.

Ramaekers, P., 1983, Geology of the Athabasca Group, NEA/IAEA Athabasca Test Area, *in* Cameron, E.M., ed., Uranium Exploration in Athabasca Basin, Saskatchewan, Canada: Geological Survey of Canada, Paper 82-11, p. 15-25.

Robertson, D.S., Tilsley, J.E. and Hogg, G.M., 1978, The time-bound character of uranium deposits: Economic Geology, v. 73, p. 1409-1419.

Ruzicka, V., 1975, Some metallogenic features of the "D" uranium deposit at Cluff Lake, Saskatchewan: Geological Survey of Canada, Paper 75-1C, p. 279-283.

Ryan, G.R., 1977, Uranium in Australia, *in* Jones, M.J., ed., Geology, Mining and Extractive Processing of Uranium: Institution of Mining and Metallurgy, p. 24-42.

Saracoglu, N., Wallis, R.H., Brummer, J.J. and Golightly, J.P., 1983, The McClean uranium deposits, northern Saskatchewan - discovery: Canadian Institute of Mining and Metallurgy, Bulletin, v. 76, #852, p. 63-79.

Sibbald, T.I.I., 1983, Geology of the crystalline basement, NEA/IAEA Athabasca Test Area, *in* Cameron, E.M., ed., Uranium Exploration in Athabasca Basin, Saskatchewan, Canada: Geological Survey of Canada, Paper 82-11, p. 1-14.

Sibbald, T.I.I., 1985, Geology and genesis of the Athabasca Basin uranium deposits, *in* Summary of Investigations 1985: Saskatchewan Geological Survey, Miscellaneous Report 85-4, p. 133-155.

Sibbald, T.I.I. and Quirt, D., 1987, Uranium deposits of the Athabasca Basin: Saskatchewan Research Council, Publication #R-855-1-G-87, 72 p.

Snelling, A.A., 1980, Uraninite and its alteration products, Koongarra uranium deposit, *in* Ferguson, J. and Goleby, A.B., eds., Uranium in the Pine Creek Geosyncline: International Atomic Energy Agency, Vienna, p. 487-498.

Tremblay, L.P., 1978, Geologic setting of Beaverlodge-type of vein-uranium deposit and its comparison to that of the unconformity-type, *in* Kimberley, M.M., ed., Uranium Deposits: Their Mineralogy and Origin: Mineralogical Association of Canada, Short Course Handbook #3, p. 431-456.

Trocki, L.K., Curtis, D.B., Gancarz, A.J. and Banar, J.C., 1984, Ages of major uranium mineralization and lead loss in the Key Lake Uranium deposit, Northern Saskatchewan, Canada: Economic Geology, v. 79, p. 1378-1386.

von Pechmann, E., 1981, Mineralogy of the Key Lake U-Ni orebodies, Saskatchewan, Canada: Evidence for their formation by hypogene hydrothermal processes, *in* Sibbald, T.I.I. and Petruk, W., eds., Geology of Uranium Deposits: Canadian Institute of Mining and Metallurgy, Special Volume 32, p. 27-37.

Wallis, R.H., Saracoglu, N., Brummer, J.J. and Golightly, J.P., 1984, The geology of the McClean uranium deposits, north Saskatchewan: Canadian Institute of Mining and Metallurgy, v. 77, #864, p. 69-96.

Wallis, R.H., Saracoglu, N., Brummer, J.J. and Golightly, J.P., 1986, The geology of the McClean uranium deposits, northern Saskatchewan, *in* Evans, E.L., ed., Uranium Deposits of Canada: Canadian Institute of Mining and Metallurgy, Special Volume 33, p. 193-217.

Worden, J.M., Cumming, G.L. and Baadsgaard, H., 1981, Geochronology of host rocks and mineralization of the Midwest uranium deposit, northern Saskatchewan, *in* Sibbald, T.I.I. and Petruk, W., eds., Geology of Uranium Deposits: Canadian Institute of Mining and Metallurgy, Special Volume 32, p. 67-72.

Wray, E.M., Ayres, D.E. and Ibrahim, H., 1981, Geology of the Midwest uranium deposits, northern Saskatchewan, *in* Sibbald, T.I.I. and Petruk, W., eds., Geology of Uranium Deposits: Canadian Institute of Mining and Metallurgy, Special Volume 32, p. 54-66.

Wright, G.M., 1967, Geology of the southeastern barren ground, Northwest Territories: Geological Survey of Canada, Memoir 350, 91 p.

Ypma, P.J.M. and Fuzikawa, K., 1980, Fluid inclusion and oxygen isotope studies of the Nabarlek and Jabiluka uranium deposits, Northern Territory, Australia, *in* Ferguson, J. and Goleby, A.B., eds., Uranium in the Pine Creek Geosyncline: International Atomic Energy Agency, Vienna, p. 375-395.

Accepted, as revised, 29 October 1987.
Originally published in *Geoscience Canada* v. 14 Number 4
(December 1987)

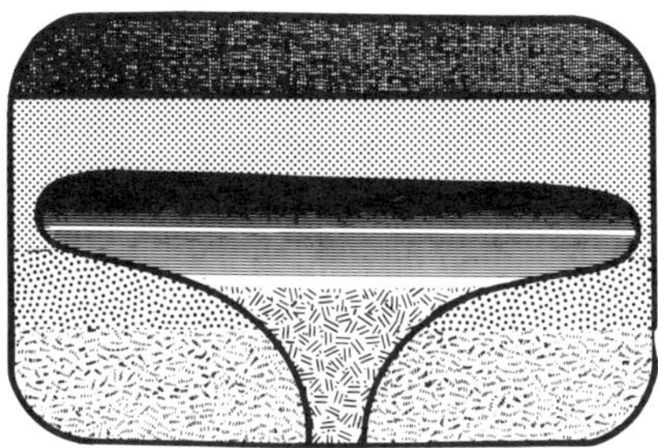

The Platinum Group Element Deposits: Classification and Genesis

A. James Macdonald
Ontario Geological Survey
1028 - 77 Grenville Street
Toronto, Ontario M7A 1W4

Abstract

The platinum group element deposits are sub-divided within a genetic framework; the three principal deposit types are formed by (1) Orthomagmatic, (2) Alluvial or (3) Hydrothermal processes. The Orthomagmatic class can be further subdivided into three subclasses: deposits that formed as a result of (a) magma mixing, (b) contamination of magma by material from an external source, and (c) deuteric fluid activity, *i.e.* flow of fluids derived from the same magma as the intrusive host rocks.

Alluvial deposits include modern placers, which commonly show an association with mafic/ultramafic complexes such as "Alaskan-" or "Alpine-type" intrusions, and paleoplacers, of which the Witwatersrand is the only known example that is sufficiently rich in the platinum group of elements to permit recovery.

Hydrothermal platinum-palladium deposits include those in which a hydrothermal system has been channelled through mafic/ultramafic host rocks from which the precious metals may have been leached. Examples include the New Rambler Mine in Wyoming, the Rathbun Lake occurrence in Ontario, and the Nicholson No. 2 uranium ores in the Beaverlodge area of Saskatchewan. The alkaline suite of porphyry copper deposits comprises the second type of hydrothermal mineralization in which platinum and palladium are significantly concentrated. The huge Kupferschiefer Cu-Ag deposits in Central Europe contain locally significant concentrations of platinum and palladium, associated with redox fronts in carbonaceous shales. At the Coronation Hill deposit, in Australia's Northern Territory, PGE are also spatially associated with carbonaceous material, and with uranium and gold mineralization.

Introduction

The objective of this contribution is to provide a genetic framework in which to consider the several deposit classes of the platinum group of elements (Os, Ir, Ru, Rh, Pt, Pd, or the PGE). A new classification scheme is proposed and the geologic processes responsible for metal concentration in each deposit class are considered in turn. A brief consideration of the geochemistry of the PGE is presented prior to discussion of classification and genetic processes.

Geochemistry of the PGE

The understanding of PGE geochemistry has been constrained until relatively recently by the inability to analyse these elements at levels in which they typically occur in unmineralized rocks, *i.e.*, at the part per billion level. Recent advances, particularly with the aid of radiochemical and instrumental neutron activation analysis, has now removed this obstacle, permitting accurate determination of background PGE contents of most rock types. The most comprehensive survey of PGE geochemistry has been presented by Crocket (1981), from which the bulk of this summary is synthesized.

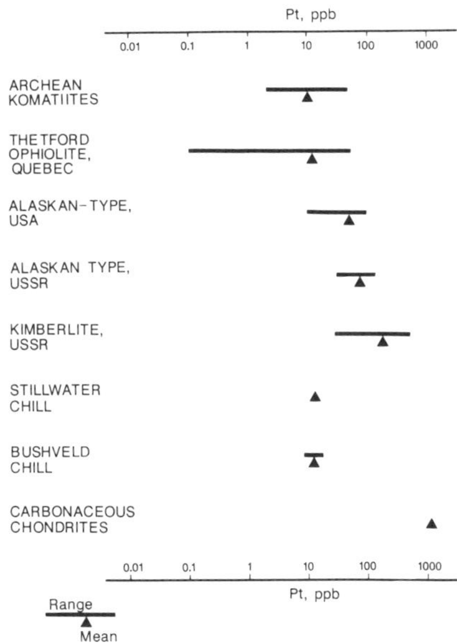

Figure 1 *Pt content of some mafic and ultramafic igneous rocks. Bushveld and Stillwater chills may represent magma compositions. (Data from Crocket, 1981; Campbell* et al.*, 1983; and Davies and Tredoux, 1985).*

The most obvious characteristics displayed by the PGE are their siderophile nature and their affinity for sulphides in mafic and ultramafic rocks. Figure 1 gives typical means and ranges of Pt contents in mafic and ultramafic rocks. The average Pt content of carbonaceous chondrites, on the other hand, is approximately 1000 ppb, attesting to the siderophilic character of Pt. Similar data for the other PGE are presented in Naldrett and Duke (1980), Crocket (1981) and Campbell *et al.* (1983). PGE contents in intermediate and felsic igneous rocks, and in metamorphic and sedimentary rocks are less well known. Crocket (1981) suggests that typical levels for individual PGE in felsic and intermediate rocks are in the sub-ppb range.

As can be seen from Figure 1, the average Pt content of unmineralized mafic and ultramafic rocks is approximately 10 ppb, with a range from 0.1 ppb to 500 ppb. A typical economic PGE deposit may have a mean platinum grade of between 5 and 10 ppm, which is the same order of magnitude as gold deposits. This suggests that the geologic process(es) which are responsible for formation of a PGE deposit involve enrichment factors of approximately one thousand.

Although the sulphide content of igneous rocks hosting orthomagmatic PGE deposits can be highly variable, the precious metals display a ubiquitous association with sulphides, which has led most researchers to suggest that formation of an economically significant deposit is dependent upon the segregation of immiscible sulphides, into which the PGE preferentially partition (*e.g.*, Naldrett and Duke, 1980). For sulphide to become a discrete phase within a mafic/ultramafic magma, the solubility of sulphur in the magma must be exceeded. The precipitation of an immiscible sulphide phase, which is a critical step in the formation of a PGE deposit, may result from one or more of the following processes: (1) assimilation of sulphur from an external source; (2) rapid cooling of the magma; (3) assimilation of silica; (4) blending of two or more disparate magmas; (*e.g.*, Irvine, 1975, 1977; Naldrett and Macdonald, 1980; Todd *et al.*, 1982).

Different types of PGE deposit may form as a result of each of these processes, which are discussed in detail in the following section.

Classification of PGE deposits

Classification of mineral deposits can be approached from several perpectives: (a) associated host rock, *e.g.*, shale-hosted Pb-Zn deposits; (b) mineralogical association, *e.g.*, telluride-gold ores; (c) geographic location, *e.g.*, Alpine-type chromite deposits; (d) archetypal, *e.g.*, "Carlin"-type gold deposits; (e) tectonic setting, *e.g.*, rift-related Cu-Ni deposits; (f) mineralization process, *e.g.*, exhalative massive sulphides. The optimal approach to classification of a mineralization type, such as ores of the platinum group of elements, separates radically different classes of mineralization into a manageable number of categories, without permitting duplication of individual deposits within two or more categories. Classification by the sixth approach given above, (f), based upon the process by which the metals are concentrated, avoids potential duplication which can be imposed by classifications reliant upon associated host rocks, mineralogy, geography, archetype, or tectonic setting. The approach must be

Figure 2 *(opposite page - upper)* *Geological map of the Bushveld Igneous Complex, South Africa. (From Willemse, 1969).*

Figure 3 *(opposite page - middle)* *Geological Map of the Stillwater Igneous Complex, Montana, USA. (From Conn, 1979).*

Figure 4 *(opposite page - lower)* *Simplified stratigraphy and cumulus phases in the Bushveld and Stillwater Igneous Complexes. (Modified from Campbell* et al.*, 1983).*

Table 1 A classification of PGE Deposits

Deposit Class	Examples
1. ORTHOMAGMATIC	
1a) Magma Mixing	UG2 †, Merensky Reef*, Platreef*, BIC J-M Reef *, SIC Great Dyke *, Zimbabwe Lac des Iles *, Canada
1b) Magma Contamination	Noril'sk ‡, USSR Sudbury ‡, Canada Kambalda ‡, Australia Thompson ‡, Canada
1c) Deuteric	Dunite Pipes *, BIC
2. ALLUVIAL	
2a) Placer	Choco *, Colombia Urals *, USSR
2b) Paleoplacer	Witwatersrand ‡, South Africa
3. HYDROTHERMAL	New Rambler ‡, USA Rathbun Lake ‡, Canada Allard Stock ‡, USA Kupferschiefer ‡, Poland Coronation Hill †, Australia

BIC - Bushveld Igneous Complex, South Africa
SIC - Stillwater Igneous Complex, USA
* Primary Product
† Co-product
‡ Bi-product

Table 2 PGE Reserves in some Deposits (Data from Buchanan (1979), Naldrett (1981), Robson (1985) and Naldrett *et al.* (1987))

	Bushveld Complex			Great Dyke *	Sudbury	Noril'sk	Stillwater
	Merensky Reef *	UG2 *	Platreef *				J-M Reef *
Million Tonnes	2160	3700	1700	1679	310	1640	49
Grade (Total PGE + Au, g/t)	8.1	8.71	7.27	4.7	0.9	3.8	22.3
Contained PGE + Au (tonnes)	17496	32227	11900	7890	279	6232	1093
Percentage of total	26.8	49.4	18.2	12.1	0.4	9.6	1.7

* Primary Product

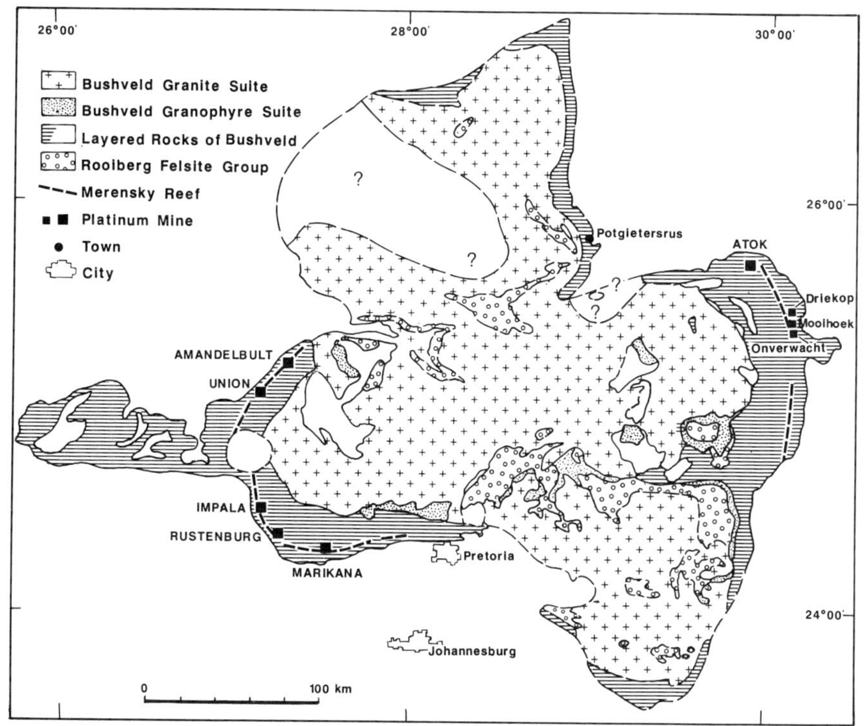

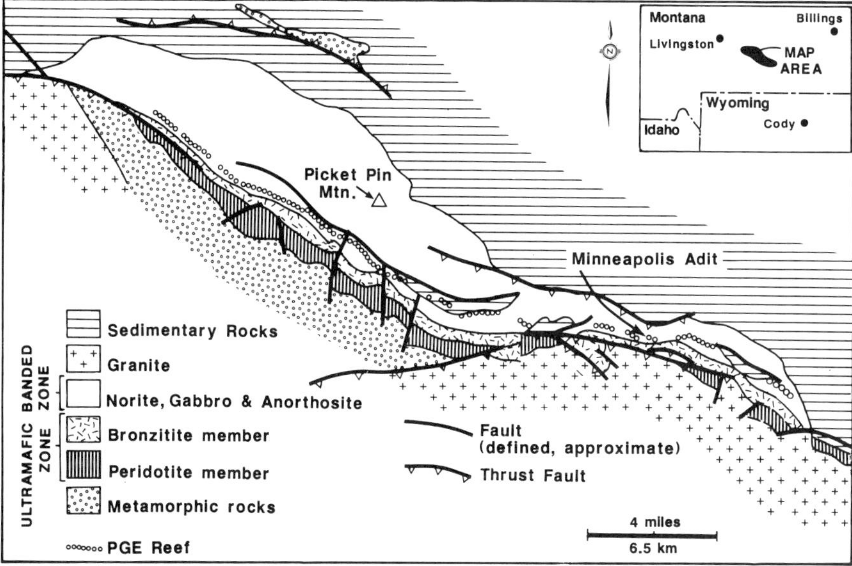

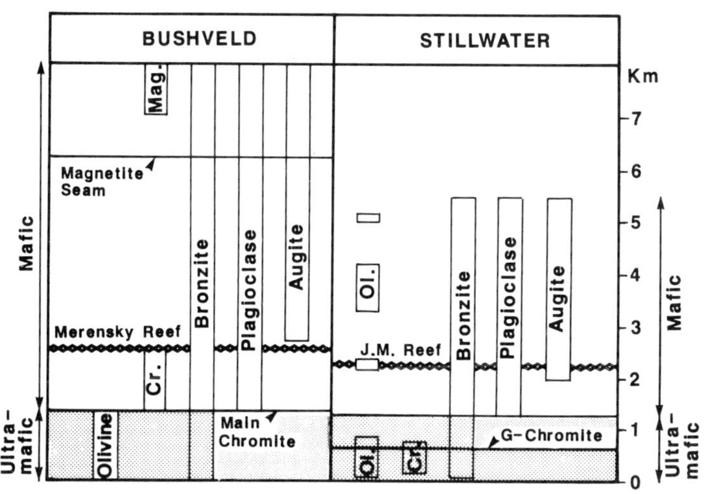

flexible enough to permit modification in response to an increased understanding of the processes in question.

A classification of PGE deposit types is given in Table 1, based upon the inferred process by which the PGE were concentrated, and whether the PGE are the primary product, a co-product or a bi-product.

CLASS 1: Orthomagmatic. The orthomagmatic class contains all those deposits that form solely within the magmatic environment. This includes deposits in which the PGE were concentrated during mixing of more than one magma (Subclass 1a), by contamination of a magma with various different types of material from external sources (Subclass 1b) or by deuteric processes, *i.e.*, involving volatile-rich fluids derived from within the magma chamber from which the host rocks crystallized (Subclass 1c). Some PGE deposits may have formed from a combination of one or more of these orthomagmatic processes.

Table 2 gives reserves for a number of the better known PGE deposits. Past production and published reserves for the two other deposit classes in Table 1 are approximately two orders of magnitude less than those for the orthomagmatic class. This does not preclude, of course, significant contributions in the future from deposit types that do not currently constitute a PGE reserve.

SUBCLASS 1a: Magma Mixing. The data in Tables 1 and 2 underscore the pre-eminent position of Subclass 1a, those deposits whose formation is attributed to mixing of more than one magma. The Merensky Reef (sulphide), UG2 Reef (chromitite) and Platreef (sulphide) in the Bushveld Complex, South Africa, together with the J-M Reef (sulphide) in the Stillwater Complex, Montana, USA, and the Great Dyke of Zimbabwe (sulphide) contain over 90% of reserves within the principal PGE deposits (Table 2). This percentage is likely to increase as a result of further exploration in the Stillwater Complex.

The Bushveld and Stillwater Complexes exhibit considerable morphological, lithological, mineralogical and chemical similarities. The mafic and ultramafic Bushveld Igneous Complex (BIC) is 2.1 Ga, 240 x 400 km in areal extent and approximately 8 km thick (Figure 2), and intrudes 2.26 Ga sediments and lavas (Hamilton, 1977).

The mafic and ultramafic Stillwater Igneous Complex (SIC, Figure 3), at 2.7 Ga (DePaolo and Wasserburg, 1979), intrudes 3.14 Ga sedimentary rocks, which are hornfelsed (Page, 1977; Page and Zientek, 1985), and is overlain unconformably by Phanerozoic sedimentary rocks. The SIC has a strike length of 48 km (Figure 3), and the portion of the complex remaining after erosion is 5.7 km thick. The extent of erosion, and hence original thickness, is unknown. Both intrusions contain prominent lithological layering, permitting the establishment of a simplified stratigraphy, consisting of lower ultramafic zones and upper mafic zones in each complex (Figure 4).

In detail, the BIC has been subdivided into four zones; a Lower Zone of harzburgites, bronzitites and dunites (1200 m), the Critical Zone of bronzitites, norites, anorthosites and chromitites (1400 m), the Main Zone of gabbro, norite and anorthosite (3600 m) and an Upper Zone of magnetite-bearing gabbros, troctolites, olivine diorites and magnetitites (2000 m). Detailed descriptions of BIC geology are given in Von Gruenewaldt (1979) and Von Gruenewaldt *et al.* (1985).

The SIC comprises a Basal Zone (Page and Zientek, 1985), an Ultramafic Zone, and a Banded Zone (McCallum *et al.*, 1980). The Basal Zone (60 m), contains inclusions of hornfelsed sediments within noritic rocks and also includes dykes and sills related to the SIC (Page, 1979). The Ultramafic Zone consists of a lower peridotite (915 m), overlain by a bronzitite (305 m), with at least 13 chromitite groups associated with peridotite. (N.B. - chromitites occur above the ultramafic rocks in the BIC). The Banded Zone of the SIC (4.5 km) contains layers of norite, gabbronorite, anorthosite and minor troctolite and gabbro. The geology of the SIC is described in detail by McCallum *et al.* (1980) and Czamanske and Zientek (1985).

The Merensky Reef, BIC, is a coarse grained to pegmatoidal, feldspathic pyroxenite (*e.g.*, Vermaak, 1976) that stratigraphically overlies an anorthositic/leuconoritic unit, and underlies a finer grained pyroxenite or melanorite (Figure 5). Two chromite seams are usually associated with the Merensky Reef, seperated by less than 1 m of pegmatoidal pyroxenite, although the exact stratigraphy may vary from locale to locale (*e.g.*, refer to sections in Naldrett, 1981). PGE grade correlates with sulphide abundance, which is typically less than 2 wt.%, and highest PGE grades are associated with the chromite seams (Campbell *et al.*, 1983). The thickness of the mineralized unit is variable, but is of the order of 1 m. The reef is exposed around the perimeter of the eastern and western lobes of the BIC (Figure 2), over a strike length of approximately 250 km. The most common PGE-bearing minerals are braggite, cooperite, laurite and Pt-Fe alloy, the latter being intergrown with base metal sulphides (Vermaak and Hendriks, 1976).

The J-M Reef, SIC, is approximately 2 m thick and has been traced along strike for approximately 40 km, although only a portion, in the vicinity of the Minneapolis Adit, is included within published mining reserves of the Stillwater Mining Company.

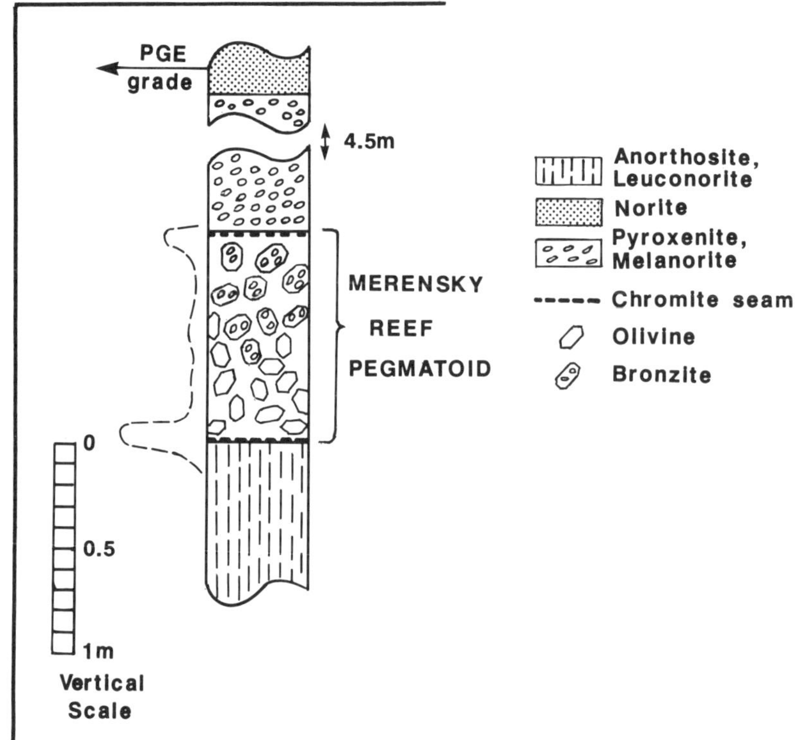

Figure 5 *Stratigraphy in the vicinity of the Merensky Reef, Bushveld Igneous Complex (Naldrett, 1981).*

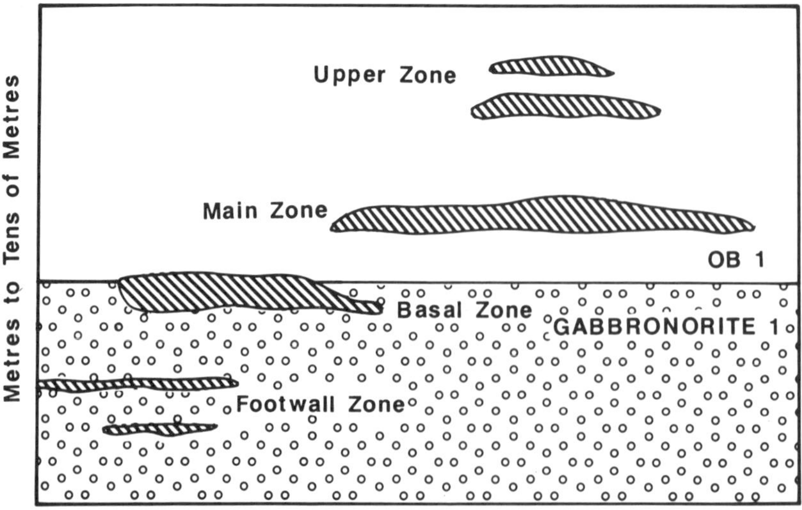

Figure 6 *Zones of mineralization and stratigraphy, J-M Reef, Stillwater Igneous Complex (Raedeke and Vian, 1986).*

Stratigraphic control on mineralization in the J-M Reef is less well constrained than that of the Merensky Reef. Although the J-M Reef is localized generally within the vicinity of the lowest olivine-bearing rock within the Banded Series, known as olivine-bearing Zone 1 (OB1), (McCallum *et al.*, 1980), PGE and sulphides may be found at four different stratigraphic levels in the Minneapolis Adit area (Figure 6, Raedeke and Vian, 1986): (a) in gabbronorite, 10 m below OB1; (b) across the basal contact of OB1; (c) within 3 m of the basal contact, but totally within OB1; (d) within OB1, but well above the basal contact. In the last three zones sulphides are usually disseminated over 1 to 3 metres. Most commonly, mineralization in any specific area occurs only at one stratigraphic level, although locally mineralization may be found at up to three discrete levels. Sixty to seventy percent of the higher grade mineralization is found at levels (b) and (c), as described by Raedeke and Vian (1986).

In a comparison of the Merensky and J-M Reefs, Campbell *et al.* (1983) point out a number of features common to both which must be accounted for by any genetic model: (1) the PGE are in sulphides; (2) the sulphides are at or near the base of cyclic (macrorhythmic) units, marked by the re-appearance of high temperature cumulus phases (chromite + olivine), and reversals in mineral fractionation trends; (3) local discordances in the footwall rocks, known as "potholes", are present below both reefs; (4) development of coarse-grained, pegmatoidal rocks; (5) stratigraphic level; both reefs occur about 500 m above the level at which plagioclase first becomes a cumulus phase (*i.e.* above the level at which the stratigraphy is first dominated by gabbroic, rather than ultramafic, rocks) - Figure 4.

Campbell *et al.* (1983) contend that any model for the origin of these deposits must account for these similarities.

The UG2 chromitite layer, which averages over 0.8 m thick (Naldrett, 1981) is as little as 30 m below the Merensky Reef in the northern sector of the BIC, and as much as 400 m below in the northeastern sector (Figure 2). As seen in Table 2, the PGE resource contained in the UG2 is greater than that of the Merensky Reef. The UG2 is also noteworthy for its higher Rh grades, not only a more valuable metal but also an asset for the manufacture of three-way autocatalysts, which utilize Rh to reduce oxides of nitrogen (Robson, 1985, 1986). The UG2 mineralogy, with a low sulphide, and high chromite content, caused initial metallurgical difficulties which rendered the ore difficult to treat. By 1985 these problems were resolved and at least two of South Africa's principal PGE mining houses, Rustenburg and Western Platinum, now derive a significant proportion of their production from the UG2 Reef (Robson, 1986).

The UG2 Reef is also associated with a 0.5 m thick pegmatoidal, feldspathic bronzite cumulate (*cf.* Merensky and J-M Reefs). PGE in the UG2 Reef are present as sulphides and alloys within the chromitite, principally laurite, cooperite, braggite and Pt-Fe alloy (Gain, 1980). Recently Naldrett *et al.* (1987) have proposed that while sulphides, with associated PGE and oxides, were deposited synchronously with the host rocks, PGE, Ni and Cu tenors were upgraded as a result of the subsolidus loss of FeS.

Much evidence is accumulating to indicate that the PGE-rich reefs in the BIC and SIC are spatially associated with mixing zones of magmas with two distinct compositions. For example, Todd *et al.* (1982) described two igneous suites within the SIC: rocks with an ultramafic (U) affinity, which exhibit a crystallization order olivine, bronzite, plagioclase, augite; the second magma has an anorthositic (A) affinity with crystallization from plagioclase, to olivine, augite and finally bronzite. Todd *et al.* (1982) proposed that the J-M Reef sulphides formed during mixing of the two magma types.

In the same year, Sharpe (1982) described two magma suites in chilled margins of the BIC, one with pyroxenitic affinity, the other gabbroic. Sharpe also suggested that mixing of the two magma types may have given rise to host rocks for the UG2 and Merensky Reefs.

Irvine *et al.* (1983) refined the magma mixing hypothesis in terms of double-diffusive convection of the two magmas (U and A) for both the Bushveld and Stillwater Complexes. Their principal conclusions are that the J-M, Merensky and UG2 Reefs owe their existence to the extraordinary composition of the Stillwater and Bushveld U-type magmas, which display an olivine-boninitic affinity, *i.e.*, rich in SiO_2 (52-56%), MgO (12-16%), Cr (800-2000 ppm) and incompatible elements (*e.g.*, 20-50 ppm Rb, 150-400 ppm Zr), and, most importantly, exceptionally rich in the PGE (up to 100 ppb total PGE + Au). The A-type magmas are closer to tholeiitic basalt in composition (SiO_2 48-50%, MgO 8-10%), with low Cr and incompatible elements. Note that, interestingly, the specialized magma with ultramafic affinity actually has a higher silica content than the gabbroic magma. For further discussion of these magmas, refer to the review by Von Gruenewaldt *et al.* (1985).

Irvine *et al.* (1983) also point out that perhaps one of the most significant characteristics of the U-type liquids is their low sulphur content. The ore reefs mark the first occurrence within the stratigraphy of the BIC and SIC, with the exception of basal zones in contact with country rock, that prominent sulphide precipitation — or, sulphide liquation, to emphasize that initial sulphide formation is as liquid droplets, rather than solid crystals — occurred on a sustained basis, permitting scavenging of available PGE by sulphides. The scavenging is a function of the strong partitioning of the PGE in favour of sulphide relative to silicate phases (Naldrett and Duke, 1980). If sulphide liquation had occurred earlier in the development of the U-type liquids, then magmatic enrichment of the PGE to the observed levels would not be possible.

Sulphide precipitation in the magmatic environment, whilst being perhaps the single most important phenomenon in localizing and concentrating PGE, is a poorly understood process. A number of schemes have been proposed to account for the formation of sulphide droplets in a silicate magma: (1) simple cooling of the magma (Haughton *et al.*, 1974); (2) mixing of two magmas, one primitive, the other more evolved (Irvine, 1977); (3) addition of silica by contamination (Irvine, 1975; Naldrett and Macdonald, 1980), discussed further under Subclass 1b.

As much of the gross lithological, mineralogical and chemical data for the Merensky and J-M Reefs, outlined above, are suggestive of mixing two disparate magmas, much effort has been undertaken to relate sulphide-bearing PGE mineralization directly to the hypothesized magma mixing event.

Campbell *et al.* (1983) presented an elegant model to relate efficiency of mixing of two magmas with stratigraphic level in the BIC and SIC. Their fluid dynamic model emphasizes the control on mixing exerted by density contrasts between the magma that is resident in the chamber and the incoming, new magma pulse. Density of the resident magma will vary in sympathy with fractionation, temperature and composition. If a new pulse is more dense, it will spread along the floor of the chamber, with minimal mixing (Figure 7a). A dense magma jet with considerable momentum will spurt into the chamber before falling back to the floor, with only minor mixing (Figure 7b). If the magma pulse is lighter, it will rise through the resident magma to the roof of the chamber (Figure 7c), again with little mixing. If the new magma pulse has the same density as some portion of the magma chamber, it will spread across the chamber at the level in question, mixing with the resident magma (Figure 7d). Campbell *et al.* (1983) demonstrate that in magma chambers which are crystallizing plagioclase — a low density mineral — the density of the remaining magma increases with fractionation. The gradual density increase of the resident magma eventually attains and exceeds that of a new magma pulse (Figure 8). This density crossover point is suggested by Campbell *et al.* (1983) to occur at the approximate stratigraphic levels of the Merensky and J-M Reefs.

Irvine (1977) and Irvine and Sharpe (1986) have also proposed that magma mixing is responsible for the formation of chromitite horizons, with attendant PGE, in layered igneous intrusions. The magma mixing model alone is capable of accounting for formation of both sulphide and oxide phases, but may also involve components of the other two hypotheses mentioned above (cooling and silica contamination): the jet of hot, incoming magma (Figure 7d), upon mixing with the cooler, resident magma, will itself cool, perhaps sufficiently to promote sulphide liquation. In addition, the mixing of two magmas with disparate chemical compositions can also cause sulphide to precipitate, if one has a greater silica content than the other (Irvine, 1975). The chemistry of this process is described more fully in the discussion of the next Subclass, 1b.

While the points of similarity between PGE mineralization in the BIC and SIC are best explained by orthomagmatic processes, with evidence from detailed petrography, geochemistry and radiogenic isotopes (see, for example, Von Gruenewaldt *et al.*, 1985), the same data, when combined with additional observations, suggest that some modifications to the magma-mixing model may be required. These observations include: (a) the close spatial association between PGE and pegmatoidal rocks; (b) the locally transgressive nature of the reef mineralization, often associated with "potholes" (*e.g.*, Viljoen and Hieber, 1986; Turner *et al.*, 1985); (c) the coexistence of hydrous silicates, such as amphiboles, micas, talc and serpentine, often Cl-rich (*e.g.*, Johan and Watkinson, 1985; Mathez *et al.*, 1985); (d) graphite, present in "stockworks" beneath the J-M Reef (Volborth and Housley, 1984) or within potholes in the Merensky Reef (Ballhaus and Stumpfl, 1985, 1986); (e) CO_2-CH_4-H_2O-bearing and chloride-rich fluid inclusions in quartz from pegmatoids in the Merensky Reef (Ballhaus and Stumpfl, 1986).

At least four, fundamentally different hypotheses have been proposed to account for the pothole features which form depressions up to several hundred metres in length and 25 m deep into the footwall stratigraphy below both the Merensky and J-M Reefs. The PGE mineralization also sympathetically drapes down into the pothole depressions (Stumpfl and Ballhaus, 1986). The potholes are proposed to be slump structures over voids in the footwall (Cousins, 1964; Leeb-du Toit, 1986). On the other hand, Campbell (1986) suggests that during magma mixing (*e.g.*, Figure 7d) plumes of convecting magma pass downward towards the accreting crystal pile, and induce partial erosion where the down-convecting jet intersects the crystal pile.

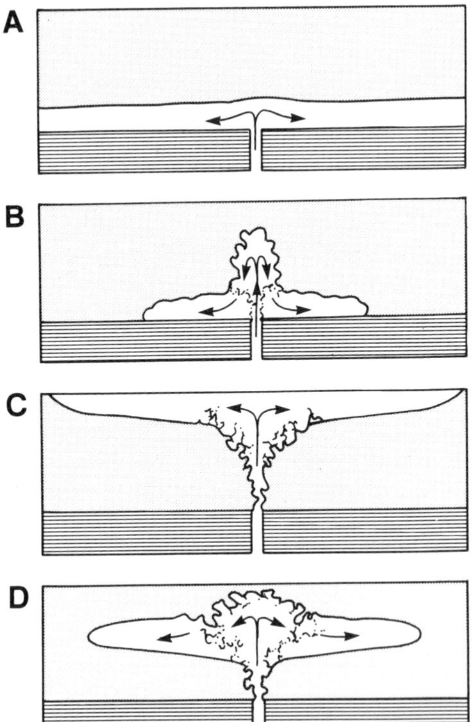

A

B

C

D

Figure 7 *Magma injection plumes into resident magma, with the effect of varying density difference. See text for details. (After Campbell* et al., *1983).*

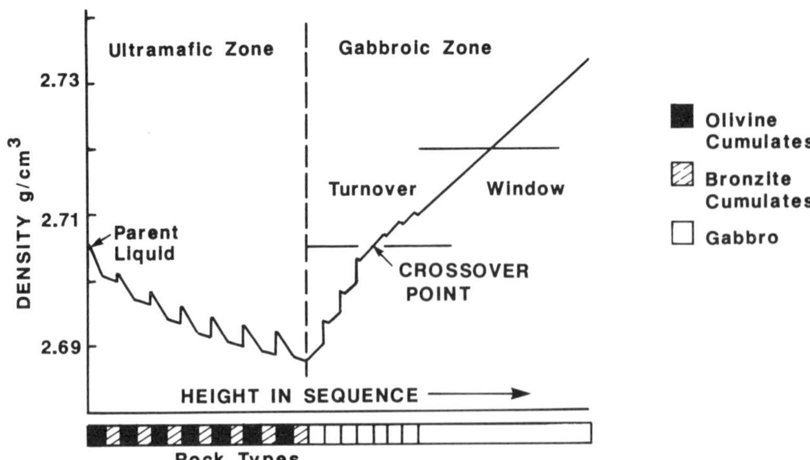

Figure 8 *Density variation of magma during sequential injection of primitive melts. Each cyclic unit is assumed to mark the introduction of a new pulse of magma into the chamber equal to 25% of the volume of the magma in the chamber. The turnover window is the density range in the fractionated magma for which magma turnover is possible. (After Campbell* et al., *1983).*

A third proposed origin for potholes is based upon the "de-watering" hypothesis, where volatile-bearing, intercumulus liquid is displaced upward during compaction of the lower stratigraphic pile. The upward percolation of fluids could have discharged into the magma as a fumarole. Perhaps the potholes represent the egress zones of this fumarolic activity, and it has been suggested that the Merensky Reef sulphides were up-graded by PGE carried by the upward percolating fluids (Vermaak, 1976; Von Gruenewaldt, 1979). Instead of the fluids being derived through compaction of the lower portions of the magma chamber, Ulmer *et al.* (1981) propose that the fluids were released through dewatering of the underlying Transvaal sediments.

Stumpfl and Ballhaus (1986), in proposing a fourth hypothesis, reject both the magmatic erosion (*e.g.*, Campbell, 1986) and fumarolic models (*e.g.*, Ulmer *et al.*, 1981) and emphasize the presence of stratigraphically-controlled mineralization and alteration that is present within the stratigraphy around the margins of the potholes in the BIC. Turner *et al.* (1985) have noted similar features in the Minneapolis Adit sector of the J-M Reef, SIC. Turner *et al.* describe "pothole margin mineralization ... concentrated in layered footwall gabbro rocks on pothole margins adjacent to the contact with the Reef package and along favourable layers marginal to pothole depressions". The PGE mineralization has replaced certain horizons, with fluids passing from the J-M Reef at the floor of the pothole into the marginal rocks by a diffusion mechanism. Stumpfl and Ballhaus (1986) suggest that pothole morphology and mineralogy reflect localized, elevated volatile contents during crystallization of the magma that gave rise to the lithological units that define the potholes.

Previous understanding of the complex geometry of the Merensky Reef had been hampered by aspects of confidentiality surrounding the platinum mining industry in South Africa (Viljoen and Hieber, 1986). Recently, however, a number of articles (*e.g.*, Viljoen and Hieber, 1986; Farquar, 1986; Mossom, 1986) have described the widespread, ubiquitous presence of complex pothole (depression) and koppie (elevation) morphologies of the Merensky Reef, with transgressive geometries of both lithological units and mineralization, much as seen in the J-M Reef, SIC (*cf.* Raedeke and Vian, 1986; Turner *et al.*, 1985). The recent Bushveld descriptions emphasize an interesting association between the locations of potholes, their associated replacement pegmatites and large-scale fracture trends, possibly reflecting a regional tectonic control on these features.

A third style of PGE mineralization in the BIC is located in the Potgietersrus Limb (Figure 2). The Platreef mineralization is found at the basal contact of the Bushveld Complex, and is hosted by rocks generally considered to be part of the Critical Zone (Van der Merwe, 1976), the same general stratigraphic level within the BIC at which the Merensky Reef is located. Disseminated sulphides are present sporadically within a feldspathic pyroxenite and harzburgite over a strike length of 60 km, with a thickness of up to 200 m. Less is known about this style of mineralization than the Merensky Reef, as mining commenced on the Platreef in 1926 and terminated shortly thereafter. Recently the area has seen renewed interest, re-evaluation and detailed study. Wagner (1929) gave the original geological description, providing the basis for further work by Van der Merwe (1976), Mostert (1982), Gain and Mostert (1982), Buchanan *et al.* (1981), Cawthorn *et al.* (1985) and Barton *et al.* (1986). Until the work of Cawthorn *et al.* (1985), most authors related mineralization to contamination or hybridization of the Platreef host rocks by wallrock, mainly dolostone from the adjacent Transvaal sediments. On the basis of geochemical and isotopic constraints, however, Cawthorn *et al.* (1985) and Barton *et al.* (1986) developed an hypothesis for contamination of the mafic magma by rheomorphic influx of a felsic magma, probably generated by partial melting of the immediate floor rocks, banded tonalite gneisses; these felsic fluids mixed with the BIC magma, and may have promoted sulphide liquation and PGE scavenging. Barton *et al.* (1986) point out, however, that formation of an immiscible sulphide liquid predated contamination of the BIC magma by rheomorphic felsic magmas, and suggest that initial sulphide liquation was in response to a new influx of BIC magma, a process similar to that described for the Merensky Reef. It appears possible, therefore, that the mineralization was modified subsequently by processes related to influx of a felsic magma. As understanding of the Platreef mineralization evolves, the popularity of the sediment-contamination theory is waning, with the bulk of the new evidence indicating that blending of ultramafic, mafic and felsic magmas appears to have been responsible for the observed features. Mixing in the Potgietersrus limb is, therefore, a more complicated process than that invoked for the Merensky and J-M Reef mineralization. In the latter, specialized magmas of ultramafic and mafic affinities — possibly differentiates from the same parent magma, in a master chamber at depth — have mixed. In the Platreef, however, an additional magmatic component of melted country rock gave rise to felsic magmas which entered the Bushveld magma chamber, to mix with the mafic/ultramafic magmas therein.

Cabri and Naldrett (1984) suggested that the Lac des Iles deposit in Northern Ontario, Canada, is a Platreef-type of PGE deposit, based upon mineralogy and metal contents. Recent geological data support this hypothesis (Macdonald, 1987). Several sulphide zones with anomalous PGE grades (up to 1000 ppb total PGE + Au) are found in gabbroic rocks contaminated by felsic magmas, which formed by rheomorphic melting of country rock. Higher grades of PGE mineralization at Lac des Iles, up to 15,000 ppb PGE + Au, are related to intrusion of an ultramafic liquid into, and blending with, a more evolved, iron-rich, anorthositic gabbro. The orthomagmatic mineralization experienced considerable remobilization due to influx of deuteric fluids, locally concentrating the noble metals into transgressive gabbro pegmatites, which may carry up to 37,000 ppb PGE + Au.

The Great Dyke in Zimbabwe represents a highly significant PGE resource, estimated to contain at least approximately 8000 tonnes of PGE (as reported in Naldrett *et al.*, 1987), or approximately 50% of the estimated Merensky Reef reserve (Table 2). Worst (1960), Wilson and Wilson (1981), Wilson (1982) and Prendergast (1987a,b) described the geology of the Great Dyke, which is a complex of four, sub-parallel, narrow, layered mafic-ultramafic intrusions. The lower ultramafic series consists of chromitite, dunite, harzburgite, olivine-bronzitite, bronzitite and websterite, with at least fourteen cycles of magma replenishment. The overlying mafic sequence consists of olivine gabbro, norite and gabbronorite in three cycles. Little is available in the literature on the stratiform Main Sulphide Zone (MSZ) of the Great Dyke, although it has been known and explored for over 60 years. The MSZ, unlike the Merensky and J-M Reefs, is hosted by ultramafic rocks, just below the contact between the Main Websterite of the ultramafic series and the lowermost olivine gabbro of the mafic series (Prendergast, 1987a,b). The MSZ is hosted by bronzitite, several metres stratigraphically below the Main Websterite, and exhibits three principal features: (1) considerable stratigraphic continuity; (2) characteristic vertical metal profiles; (3) systematic lateral variations in width of mineralization, sulphide content, intermetallic ratios and PGE content when normalized to 100% sulphide (Prendergast, 1987a,b). Prendergast (1987b) considers that the MSZ mineralization of the Great Dyke resulted from a combination of magma replenishment, convective overturn, and magma cooling.

With the resurgence of interest in the PGEs during the latter half of this decade, much new information on the MSZ is anticipated, as a result of not only academic research possibilities offered by the recent availability of diamond drill core that penetrated the MSZ, but also by aggressive exploration and development by the many mining companies with interests in Zimbabwe.

SUBCLASS 1a: Summary. While various competing hypotheses continue to support, on the one hand, magma mixing and, on the other, deuteric concentration of PGE to form the Merensky/UG2/J-M Reefs, the following observations are generally accepted: (a) Mixing of two discretely different magmas (U and A) occurred at the stratigraphic levels occupied by the Reefs; (b) Effects of a conspicuous deuteric component within the magma are manifest in the Reefs' host rocks as hydrous silicates, pegmatoidal layers and cross-cutting pegmatite dykes, locally abundant graphite in potholes and complex fluid inclusions trapped in quartz.

It is also generally agreed (dissenters include Ulmer *et al.*, 1981) that the fluids responsible for the variable textures and compositions in the Reefs' host rocks are deuteric in origin; that is, the fluids are derived from within the parent magma, retaining the integrity of the orthomagmatic hypothesis (Table 1). Until conclusive evidence is available to the contrary, the optimal working hypothesis for the origin of the UG2, Merensky and J-M Reef mineralization invokes magma mixing of two specialized magmas, with some degree of deuteric effect upon (a) silicate crystallization, and (b) subsequent, local redistribution of the PGE-bearing sulphides.

In other deposit types whose formation is attributed to mixing of magmas, such as the Platreef and Lac des Iles, contamination by rheomorphic, felsic melts appears to have modified both magma chemistry and the geometry of mineralized zones. Mineralization containing higher grades of PGE are, however, related to blending of mafic and ultramafic magmas; simple mixing of felsic and gabbroic magmas does *not* appear to be a critical factor in the formation of economically significant PGE mineralization in this deposit type.

SUBCLASS 1b: Magma Contamination. The second subclass of orthomagmatic, PGE-bearing sulphide deposits are found within mafic/ultramafic intrusions and extrusions in which sulphur saturation has been achieved by addition to the magma of external material, *i.e.*, wallrock contamination.

In 1959, Sullivan suggested that Ni deposits in the Manitoba Nickel Belt, Canada, formed by "sulphurization", the reaction of sulphur, from an external source, with metals (*e.g.*, Fe, Ni) in the magma. Naldrett (1966) proposed a similar process for Fe-Ni sulphides in the Porcupine District of Ontario.

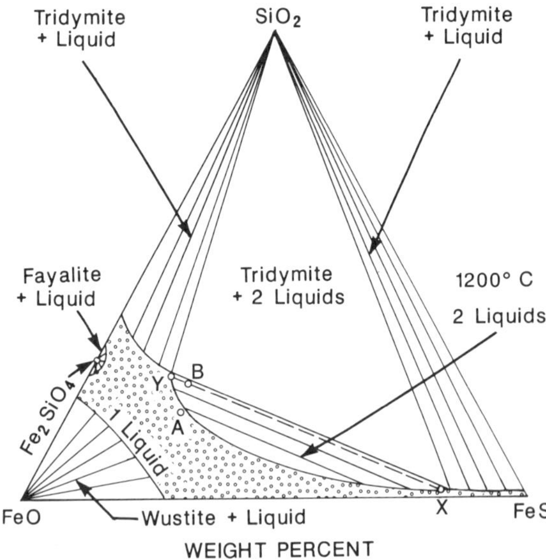

Figure 9 *The effect of silica contamination upon an FeO-FeS melt. See text for details. (After Naldrett and Macdonald, 1980).*

Irvine (1975) described an elegant model for the origin of stratiform chromite and sulphide layers in large, layered igneous intrusions, where addition of silica from an external source can initiate precipitation of chromite and/or liquation of sulphide. In 1977, however, Irvine also demonstrated how the same phenomena may result from the mixing of two magmas, discussed in the previous section.

The Sudbury deposit in Ontario is Canada's principal PGE resource (Table 1). For a recent synthesis of Sudbury geology, see Pye *et al.* (1984). A number of features of the Sudbury ores and host rocks led Naldrett and Macdonald (1980) to propose that contamination of the mafic magma by relatively silica-rich wallrock was the cause of sulphide liquation. These features include (a) the high percentage of normative quartz, co-existing with pyroxenes and plagioclase whose composition is relatively primitive, reflecting little differentiation; (b) numerous, partially digested inclusions in the host rocks (Stevenson and Colgrove, 1968); (c) the relatively high $^{87}Sr/^{86}Sr$ ratio of 0.706 for the noritic host rock (Gibbins and McNutt, 1975); (d) the relative lack of cyclical units and modal layering, when compared with mafic intrusions of similar size, suggesting a viscous magma, which is consistent with elevated silica content. The silica-contamination model is demonstrated in Figure 9. The FeO-SiO_2 side of the ternary is equivalent to silicate melts and the FeS corner is equivalent to sulphide magmas. A magma at a composition indicated by Point A is a one phase, homogeneous FeO-SiO_2-FeS liquid. If SiO_2 is added such that the composition changes to that of, for example, Point B, the magma now consists of a two-phase, silicate-rich liquid (Y) and co-existing sulphide liquid (X).

The Noril'sk-Talnakh-Kharaelakh Ni-Cu-PGE camp in Siberia, is the Soviet Union's premier source of the PGEs (Table 1). The geology of the Noril'sk area is described by Smirnov (1966), Bazunov (1976), Glaskovsky *et al.* (1977), and is summarized by Naldrett and Macdonald (1980). The Ni-Cu-PGE mineralization has been described recently by Godlevsky and Likhachev (1986). Sulphur isotopic data for the Noril'sk area ($\delta^{34}S$ = +7 to +10‰) are consistent with sulphur assimilation from sulphate-bearing, Devonian evaporites that were intruded by the Permian and Triassic gabbro-dolerite host rocks for the mineralization (Godlevsky and Grinenko, 1963; Kovalenko *et al.*, 1975).

Komatiite-associated Ni-Cu sulphide deposits, such as at Kambalda, Australia, and Thompson, Manitoba, are important hosts for sulphide deposits with significant PGE contents. At Kambalda, for example, the ores contain 2.96% Ni, 0.22% Cu, 0.07% Co, 8.09% S, 1141 ppb PGE, 339 ppb Au and 1170 ppb Ag (Hudson and Donaldson, 1984). Until recently, all components of these ores were generally considered to be solely magmatic in origin, modified by metamorphism and deformation (*e.g.*, Groves *et al.*, 1979; Barrett *et al.*, 1977; Marston and Kay, 1980). Lesher and Groves (1986), on the other hand, suggest that the sulphur is externally derived and present arguments that the sulphide ores exsolved from komatiitic magmas *in situ*, in response to assimilation of country rocks during emplacement. They point out that the sulphide deposits show a marked spatial association with sulphidic sediments.

SUBCLASS 1b: Summary. Contamination by wallrock is an effective process for initiating sulphide liquation, whether the contaminant be sulphur itself, silica in the form of assimilated sediments, or as a granitic melt. Sulphide deposits may be expected, therefore, at contacts between mafic/ultramafic bodies and their country rocks, where chemical interaction between the two has occurred. The PGE tenor of this type of sulphide deposit is a function of (a) the amount of magma available for PGE scavenging by sulphide, and (b) the PGE content of the magma.

SUBCLASS 1c: Deuteric Processes. While the deuteric component of mineralization in the Merensky, UG2 and J-M Reefs is still controversial, some PGE deposits are contained within features that demonstrably transect stratigraphy within the layered intrusions. Perhaps the most well-known of these are the dunite, or hortonolite (olivine with 50-70% fayalite) pipes. The pipes are commonly contained within envelopes of olivine dunite and olivine-bronzite-plagioclase pegmatoid (Wagner, 1929). These bodies have not provided large tonnages: the largest is the Driekop Pipe, up to 25 m in diameter, and mined through a vertical depth of 200 m (Schiffries, 1982). Grades, on the other hand can be spectacularly high, up to 2000 ppm total PGE, although average mining grades are substantially less: the Onverwacht Pipe averaged approximately 10 ppm, the Driekop Pipe contained 6 ppm total PGE (Wagner, 1929).

Where a dunite pipe cuts across an easily identifiable stratigraphic marker, such as a chromitite layer at the Onverwacht Pipe (Wagner, 1929) disrupted blocks of chromitite are found within the pipe, more or less at their projected stratigraphic level (Figure 10). The layered rocks around a pipe locally form collapse structures, with layering becoming subvertical at the pipe contact (Schiffries, 1982).

The dunite pipes do not reflect intrusion of an olivine-rich magma; rather, the presence of chromitite slabs within pipes suggests an infiltration metasomatism, or replacement phenomenon (Stumpfl and Rucklidge, 1982; Schiffries, 1982). Schiffries (1982) proposes that a channelled, high-temperature chloride solution is responsible for reacting with the noritic host rock, resulting in a dunite. The replacement process involves desilication of pyroxene, and production of an iron-rich olivine (Schiffries, 1982).

Wagner (1929) noted that not all dunite pipes are mineralized. Stumpfl and Rucklidge (1982) draw attention to the presence of graphite and pegmatoidal textures, seen also in the Merensky Reef, implying that there may be some relation between pipe and reef formation.

Desilication of pyroxene may be a common phenomenon in other intrusions. An example may be the classic "pillowed troctolite" exposure in the Stillwater Complex, described by Hess (1960), containing elliptical masses of plagioclase-olivine rock (troctolite) in an anorthositic matrix. According to a model proposed by McCallum *et al.* (1977), the troctolite "clumps" are the vestiges of original gabbro (cumulus plagioclase and augite) included within anorthosite; the original augite has been replaced by olivine. Boudreau and McCallum (1986) describe transgressive sulphide zones on Picket Pin Mountain in Anorthosite Subzone II, the thickest anorthosite member (600 m) of the Stillwater Complex, approximately 3 km stratigraphically higher than the J-M Reef. The cross-cutting sulphide zones pass upward into conformable sulphide-bearing zones (1% pyrrhotite, pentlandite and chalcopyrite), that contain approximately 800 ppb total PGE + Au. Boudreau and McCallum (1986) infer that the transgressive sulphide zone reflects a fluid conduit for upward percolating fluids to feed a stratabound sulphide deposit. Perhaps the Picket Pin mineralization can be considered to be an intra-magma chamber, exhalative deposit!

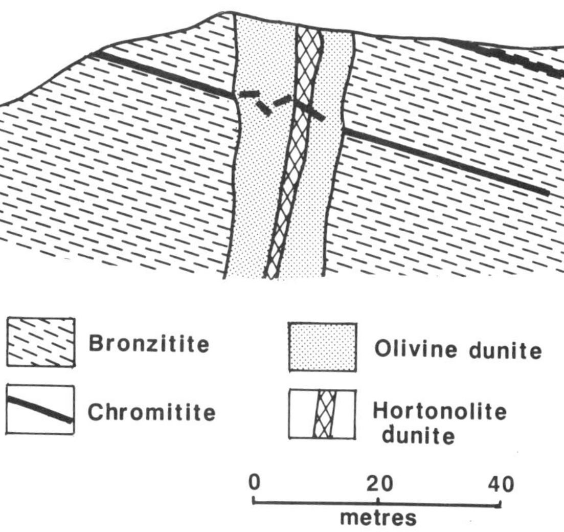

Figure 10 *Dunite pipe transecting stratigraphy, Onverwacht, Bushveld Igneous Complex. Note rafts of chromitite within pipe at approximately same stratigraphic level as chromitite seam outside pipe. (After Stumpfl and Rucklidge, 1982).*

SUBCLASS 1c: Summary. Deuteric and sub-solidus phenomena may be significant processes, especially in large magma chambers, causing considerable modification to original magmatic mineralogy, chemistry, and morphology of lithologies and mineralization. Alteration and replacement features are beginning to receive considerable attention within mafic and ultramafic intrusions, such as the Bushveld and Stillwater complexes. PGE-bearing sulphide mineralization can, it appears, form within this deuteric, late-stage period in the development of a mafic/ultramafic magma body. To date, however, only relatively small mineralized bodies, such as the dunite pipes, are considered to owe their genesis to solely deuteric processes. An understanding of the effects of late-magmatic fluids is, however, crucial to explain fully all features observed in the much more voluminous "reef"-style mineralization.

CLASS 2: Alluvial Deposits

SUBCLASS 2a: Placer Deposits. Platinum was "discovered" by Spanish Conquistadors in Colombia in the mid-16th Century. A detachment of troops moving along a drainage system in northwest Colombia, found gold and grains of an unknown, heavy, silvery mineral in the banks of a river that, subsequently, became known as the Rio Platino del Pinto. As the unknown metal had an extremely high melting point, it was regarded as useless and a hinderance to the refining of gold, and was contemptuously dubbed "platino", meaning silver of poor quality. The Choco district of Colombia has been a source of alluvial PGE since this time, although has long since ceased to be significant, contributing approximately 0.5% of supply to the Western world (Robson, 1985).

The Goodnews Bay alluvial deposits in Alaska yielded just over 3 tonnes of PGE between 1927 and 1982, and is currently under re-evaluation (Robson, 1986). This deposit, like many other alluvial placer deposits such as those in Colombia, is associated with "Alaskan-type" ultramafic complexes (Naldrett, 1981). Similar placers are present down stream from the Tulameen ultramafic/mafic complex in British Columbia (Racevic and Cabri, 1976). The weathering process enriched the Pt/(Pt + Pd) ratio in the placer, with respect to the source, as palladium is more easily dissolved and removed in the weathering environment (Naldrett, 1981). "Alpine-type" ultramafic bodies are also commonly the source for placer PGE, such as in the Ural Mountains of the USSR (Mertie, 1969). Interestingly, placer PGE in the Urals are also associated with "Alaskan-type" zoned mafic/ultramafic complexes (Naldrett, 1981).

SUBCLASS 2b: Paleoplacers. The Witwatersrand paleoplacer in South Africa, discovered in 1892, comprises PGE-bearing auriferous-uraniferous conglomerates within the Witwatersrand Supergroup. Pretorius (1975) presents a comprehensive summary of the geological environment of mineralization in the Witwatersrand conglomerate.

Although the PGE are only a trace constituent of the Witwatersrand (less than 10 ppb) the PGE are collected as a bonus during extraction of gold from crushed ore. Annual production of the PGE, since 1921, was in the range of 150-210 kg until 1965, after which time production figures were not released, for security reasons (Reimer, 1979). It is estimated, however, that annual production of the PGE in recent years has risen to 350 kg, giving a total of about 12 tonnes from this one deposit, or 10% of the PGE produced from conventional placer sources (Reimer, 1979).

CLASS 2: Summary. Placer deposits of the PGE are spatially associated with mafic/ultramafic complexes, most commonly of the "Alaskan-" or "Alpine-type" of mafic/ultramafic intrusions. The only known paleoplacer PGE deposit of economic significance, the Witwatersrand, may have resulted from greenstone belt erosion, although arguments still rage as to the origin of other constituents, such as gold (*cf.* Skinner and Merewether, 1986).

CLASS 3: Hydrothermal PGE Deposits. Hydrothermal PGE deposits form from epigenetic fluids, whose compositions are, as yet, not well constrained, in a wide variety of geological environments; including: (1) associated with shear zones cutting mafic/ultramafic host rocks; (2) associated with alkaline porphyry copper-precious metal deposits; (3) in late diagenetic flow of metal-bearing brines in carbonaceous sediments. This deposit type is generally rich in platinum and palladium, with respect to the other PGE (Naldrett, 1981). Hydrothermal PGE deposits have yet to contribute significantly to global PGE production, although this may be a function of lack of exploration for this deposit type.

SUBCLASS 3a: Shear Zone-related. The Rathbun Lake PGE-bearing Cu-Ni sulphide deposit is located within the Wanapitei gabbronorite intrusion in Central Ontario (Dressler, 1982). Fluids focussed in sheared rocks are inferred to have been formed a secondary silicate assemblage of chlorite, sericite, quartz, epidote and biotite (Rowell and Edgar, 1986). Massive and disseminated chalcopyrite and pyrite are the principal base metal sulphides. Grab samples average 9% Cu, 0.2% Ni, 20.8 ppm Pd, 9.7 ppm Pt and 3.1 ppm Au. The ratio (Pt + Pd)/(Os + Ir + Ru) of >1000, much higher than for the orthomagmatic J-M Reef (250) and Merensky Reef (12.5) (*cf.* Naldrett, 1981), is taken as supportive evidence for a hydrothermal origin (Rowell and Edgar, 1986). The massive sulphide body at Rathbun Lake is, however, of insignificant size, measuring 14 m x 0.3 m to 0.6 m wide. Unfortunately, the occurrence is no longer exposed and its association with a fault can only be inferred (Rowell and Edgar, 1986). This leaves open the possibility that the deposit may be the result of orthomagmatic, deuteric activity, as suggested by Dressler (1982).

The New Rambler PGE-bearing Cu deposit is located within Precambrian rocks of the Medicine Bow Mountains, Southern Wyoming (McCallum and Orback, 1968). The host rocks are metagabbroic; other spatially associated lithologies include upper amphibolite metamorphic grade leucogneiss, and medium-grained, equigranular granite, with associated quartz monzonite, aplite and pegmatite, and a large luxullianite (tourmaline-rich) body exposed on the northern margin of the stock. Major minerals in the granite are albite, quartz, microcline, shorl, muscovite and garnet (McCallum *et al.*, 1976).

Mineralization at New Rambler is associated with both quartz-carbonate-sulphide fissure veins and large sulphide "pods", up to 20 m in horizontal diameter and 15 m in depth. The pods contain little quartz and carbonate, but considerable, finely banded, cryptocrystalline, jasperoid, or more correctly, chrysocolla. The pods are considered to be, at least in part, of supergene origin; the copper sulphide-bearing veins, on the other hand, are probably primary, hypogene features (McCallum and Orback, 1968).

The copper ores contain high Pd and Pt grades (average 75 ppm and 4 ppm, respectively), contrastingly low grades of the other PGE (<50 ppb in total), and are considered by McCallum *et al.* (1976) to be solely hydrothermal origin, there being no evidence for remobilization of a magmatic sulphide protore. They consider that faults and shear zones, related to the Mullen Creek-Nash Fork Shear Zone (Houston *et al.*, 1968) may have acted as fluid conduits for the hydrothermal solutions. Again, as at Rathbun Lake, geological investigation of this deposit is hampered by the fact that exposure of the mineralization is not available as mining terminated after a fire in 1918. Investigation of the few outcrops on the property, and study of material from the dump by this author and B.O. Dressler (Ontario Geological Survey) in 1985, reveals intensely hydrothermally altered gabbroic rocks, which are not, however, highly strained. The abundance of felsic dykes, pegmatites and veins cutting gabbroic rock, may indicate that a hydrothermal system related to the felsic intrusion was responsible for hypogene mineralization. High Pt and Pd tenors may either reflect an original hydrothermal fluid content, or, alternatively, be a function of hypogene and/or supergene leaching from the mafic host rocks.

Recently, Hulbert (1986) has described PGE-bearing, uraniferous veins in the Beaverlodge area of Northern Saskatchewan. Veins at, for example, the Nicholson Number 2 deposit, are spatially associated with a mafic intrusion. Hulbert (1986) reports that PGE grades appear to decrease with increasing distance from the mafic rocks, which may therefore, have been the source from which hydrothermal fluids leached PGE.

SUBCLASS 3b: Alkaline Porphyry Deposits. The alkaline suite of porphyry copper deposits has been known for some time to be enriched in precious metals. Kesler (1973), for example, proposed a porphyry copper-gold subclass of the porphyry copper family and Finch *et al.* (1983) noted, in addition, a significant enrichment of PGE. Werle *et al.* (1984) described the syenitic Allard Stock, an epizonal Laramide intrusion at the southwestern end of the Colorado Mineral Belt, 25 km northwest of Durango, Colo. A porphyry copper/precious metals deposit is hosted within the Allard Stock, containing typically 0.6% Cu, 2-20 ppm Ag, 10-200 ppb Au, 3 ppm Mo and 100-170 ppb total PGE. Werle *et al.* (1984) conclude that fractionation of magma and volatiles in the parent magma chamber of the Allard Stock concentrated H_2O, CO_2 (up to 10% CO_2 is present in the syenitic rocks, and CO_2-bearing fluid inclusions are present in both host rocks and veins), F and S. The volatile-, base and precious metal-bearing fluid phase was responsible for breccia and vein mineralization, precipitating metallic minerals, calcite, fluorite and quartz.

SUBCLASS 3c: Late Diagenetic/ Epithermal. In contrast to Subclasses 3a and 3b, the hydrothermal Kupferschiefer stratabound Cu-Ag (Pb-Zn) deposits show no spatial nor temporal relationship with igneous activity. Kupferschiefer mineralization is at the contact between Lower Permian volcanic rocks and rift-filling red beds, and Upper Permian marine carbonates, evaporites and red beds, with the largest and richest deposits being found in Southwest Poland. Most recent studies have indicated that the Kupferschiefer mineralization is epigenetic, deposited from late diagenetic, convecting fluids (Jowett, 1986).

At the Lubin and Polkawice Mines in Poland, Kucha (1982) described Pt-rich (>10 ppm) shales that were mapped along strike lengths in excess of 1.5 km. Maximum values of >200 ppm Pt have been found over 50 m strike lengths, associated with the contact zone between white, oxidized sandstone, and black, reduced shale, containing thucholite, ketones, kerogen, phenols, tertiary alcohols and aromatic hydrocarbons. Kucha (1982) suggests that the precious metals are concentrated in the shale, during late diagenetic fluid flow, by a process of auto-oxidation and desulphurization of the organic matter. Kucha (1982) further suggests that the PGE may act as catalysts in this reaction.

A second interesting polymetallic, PGE-bearing deposit, not obviously related to felsic intrusions, has been described recently at Coronation Hill, in the Northern Territory of Australia (Needham and Stewart-Smith, 1987). The Coronation Hill Mine is one of a suite discovered in the South Alligator Valley district, mined for uranium in the mid-1950s to 1960s. The uranium deposit is hosted by highly faulted conglomerate, altered volcanic rocks and carbonaceous schist. Needham and Stewart-Smith (1987) interpret the deposit to be allied to the class of epigenetic, sandstone-hosted uranium deposits, with uranium-enriched felsic volcanic rocks as the metal source, sandstone beds as fluid conduit, and carbonaceous schist acting as reductant. The precious metal mineralization (PGE + Au) is finely disseminated in felsic volcanic rocks, altered to quartz, sericite and chlorite, immediately to the east of the open cut from which uranium was produced. The gold is very fine grained, not visible to the naked eye, and also not associated with any prominent quartz veining. Only minor uranium mineralization is found directly associated with the precious metals. Felsic dikes locally intrude the host rocks. A joint venture between BHP Minerals Limited, Noranda Pacific Limited and EZ Industries Limited discovered the precious metal mineralization in 1985 and report that of 177 two-metre core assays, 163 gave gold values greater than 1 g/t, averaging 0.61 g/t Pt, 1.15 g/t Pd and 7.72 g/t Au (Noranda Pacific Limited, Quarterly Report, December 31, 1985). It is presently unclear whether the spatial association between precious metal and uranium mineralization is purely fortuitous. The exploration companies believe that the precious metal mineralization is epithermal in nature, related to sub-aerial, acid volcanism.

CLASS 3: Summary. While hydrothermal PGE deposits are less well understood than the more extensively described orthomagmatic deposits, at least three distinct categories are currently identified: (a) those associated with epigenetic fluid flow in mafic/ultramafic rocks, (b) precious metal-enriched (PGE, Au, Ag), porphyry copper deposits, of alkaline affinity, (c) carbonaceous, shale-hosted deposits, related to diagenetic fluid flow. As research progresses, other subclasses of the hydrothermal PGE deposit type will undoubtedly be identified.

Implications For Exploration

The current domination of global PGE markets by deposits of the orthomagmatic class, suggests that, based upon "track record", this deposit type remains the optimal exploration target. Subclass 1a is currently the most desirable target within the orthomagmatic deposit class, as the PGE are a primary product, rather than a biproduct of less economic, base-metal mining. Utilizing an understanding of Bushveld and

Stillwater mineralization, the following criteria can be regarded as positive indicators for mineralization: (1) large, compositionally stratified, mixed mafic and ultramafic, layered intrusive complex; (2) zones of mixing between liquids of different U- and A-type affinities (Todd *et al.*, 1982; Irvine *et al.*, 1983), within the mafic portions of the intrusive complex, rather than the lower ultramafic zone. *Nota bene* the MSZ in the Great Dyke is an exception, located within ultramafic host rocks, several metres below the contact with the mafic series; (3) sulphides, typically only 1-2%, in zone of mixing.

Magma mixing in the BIC and SIC has been documented in the context of chemical and isotopic signatures (*e.g.*, Irvine *et al.*, 1983; Sharpe *et al.*, 1986). Field expressions of the proposed mixing phenomena are complex (*e.g.*, Turner *et al.*, 1985) inasmuch as the geology within the zones of mixing can be highly variable on all scales, contrasting with the relative uniformity of layered rocks stratigraphically above and below the mineralized units. Detailed mapping of the Minneapolis Adit in the Stillwater Complex (Figure 3) reveals such complexity on a small scale that certain packages of rocks are mapped collectively as "Mixed Rock" (Bow *et al.*, 1982). Mixed Rock is an complex igneous breccia, with a matrix of plagioclase cumulate gabbro, containing 20-40% amoeboidal olivine, associated with clusters of pegmatoidal, poikilitic crystals of bronzite and augite. The rock also contains boulders of olivine cumulate, plagioclase-olivine cumulate (troctolite) and plagioclase cumulate.

Macdonald (1985, 1987) has mapped rocks of similar complexity hosting PGE mineralization at Lac des Iles, Canada. In contrast with the relatively monotonous gabbroic rocks that comprise the bulk of the intrusive complex hosting significant PGE at Lac des Iles, the immediate host rocks are highly complex: a wide (500 m) zone of texturally and compositionally variable lithologies — from fine grained to pegmatoidal, pyroxenite, norite, gabbro, quartz gabbro to anorthosite — is cored by an igneous breccia, up 100 m wide that is spatially associated with the higher grades of PGE + Au. The breccia has a gabbroic matrix, with clasts, up to in excess of 3 m in long dimension, of gabbro, pegmatoidal gabbro, anorthosite and pyroxenite. The breccia is cut by gabbro pegmatite dikes, locally containing up to 37,000 ppb total PGE + Au. Mineralization at Lac des Iles is spatially, and probably genetically, related to a pyroxenite that has intruded and locally blended with the anorthositic gabbro host (Macdonald, 1987).

While deposits that formed in response to magma mixing exhibit evidence of at least a partial role played by deuteric fluids, it is by no means certain that this is an essential component of the mineralization process; suffice to say that deuteric effects are not considered deleterious to economic potential.

The renewed interest in PGE mineralization, driven by political and market forces, may result in the discovery of potentially economic concentrations of the PGE in unconventional geologic settings. For example, the Kupferschiefer-style mineralization was not widely recognized until 1982. The PGE-black shale connection is worthy of further investigation in, for example, the Central African Copper Belt and the shale-hosted copper deposits of Oklahoma, *etc.* The relatively recent (1985) discoveries of PGE mineralization associated with apparently epithermal gold mineralization at Coronation Hill, in Australia's Northern Territory, suggests that the numerous deposits of this type, in a similar geological environment, warrant a closer inspection for PGE content.

The association of PGE deposits with CO_2-rich, alkaline igneous intrusions is also in need of further description. The deposit type may have to be expanded or redefined to include carbonatite bodies: the pyroxenite-syenite-carbonatite complex at Phalaborwa, in northeastern South Africa, for example, in addition to supporting apatite and vermiculite mines, contains one of the most efficient open pit copper mines in the world, producing 100MT of ore per annum, grading 0.48-0.57% Cu. The composition of recoverable precious metals is dependent largely upon copper content (typical concentrate contains 45% ± 15 Cu). The following composition of anode slimes is typical: 3.5 kg/t Au, 85 kg/t Ag, 0.45 kg/t Pt, 0.55 kg/t Pd and 0.025 kg/t Rh (Verwoerd, 1986).

Conclusions

Deposits of the platinum group of elements form in three fundamental environments: (a) as an integral component of orthomagmatic processes, including mixing of magmas, wallrock contamination of magmas and by deuteric processes; (b) by alluvial concentration of degraded source material — most modern placers being derived from mafic/ultramafic complexes of the "Alaskan-" or "Alpine-type"; (c) by hydrothermal transport, related either to leaching of PGE from mafic/ultramafic host rocks, or to alkaline porphyry copper and/or epithermal precious metal mineralization, or to late-diagenetic fluid flow in sedimentary basins.

Acknowledgements

G. Atkinson is thanked for drafting the diagrams. An earlier version of this manuscript was greatly improved following careful readings by H.R. Williams, J.M. Duke, A.J. Naldrett, J.F.H. Thompson and R.G. Roberts, to whom I am most grateful. This paper is published with the permission of the Director, Ontario Geological Survey.

References

Ballhaus, C.G. and Stumpfl, E.F., 1985, Occurrence and petrological significance of graphite in the Upper Critical Zone, western Bushveld Complex, South Africa: Earth and Planetary Letters, v. 75, p. 58-68.

Ballhaus, C.G. and Stumpfl, E.F., 1986, Sulfide and platinum mineralization in the Merensky Reef: evidence from hydrous silicates and fluid inclusions: Contributions to Mineralogy and Petrology, v. 94, p. 193-204.

Barrett, F.M., Binns, R.A., Groves, D.I., Marston, R.J. and McQueen, K.G., 1977, Structural history and metamorphic modification of Archean volcanic-type nickel deposits, Yilgarn Block, Western Australia: Economic Geology, v. 72, p. 1195-1244.

Barton, J.M., Cawthorn, R.G. and White, J., 1986, The role of contamination in the evolution of the Platreef of the Bushveld Complex: Economic Geology, v. 81, p. 1096-1104.

Bazunov, E.A., 1976, Development of the main structures of the Siberian Platform: history and dynamics: Tectonophysics, v. 36, p. 289-300.

Boudreau, A.E. and McCallum, I.S., 1986, Investigations of the Stillwater Complex: III. The Picket Pin Pt/Pd deposit: Economic Geology, v. 81, p. 1953-1975.

Bow, C., Wolfgram, D., Turner, A., Barnes, S., Evans, J., Zdepski, M. and Boudreau, A., 1982, Investigations of the Howland Reef of the Stillwater Complex, Minneapolis Adit area: stratigraphy, structure and mineralisation: Economic Geology, v. 77, p. 1481-1492.

Buchanan, D.L., 1979, Platinum metal production from the Bushveld Complex and its relationship to world markets: Bureau for Mineral Studies, University of Witwatersrand, Johannesburg, Report No. 4, 31 p.

Buchanan, D.L., Nolan, J., Suddaby, P., Rouse, J.E., Viljoen, M.J. and Davenport, J.W.J., 1981, The genesis of sulphide mineralisation in a portion of the Potgietersrus limb of the Bushveld complex: Economic Geology, v. 76, p. 568-579.

Cabri, L.J. and Naldrett, A.J., 1984, The nature of the distribution and concentration of platinum-group elements in various geologic environments: 27th International Geological Congress, Moscow: Proceedings of Mineralogy, v. 10, p. 17-46.

Campbell, I.H., 1986, A fluid dynamic model for the potholes of the Merensky Reef: Economic Geology, v. 81, p. 1118-1125.

Campbell, I.H., Naldrett, A.J. and Barnes, S.J., 1983, A model for the origin of the platinum-rich sulfide horizons in the Bushveld and Stillwater Complexes: Journal of Petrology, v. 24, p. 133-165.

Cawthorn, R.G., Barton, J.M. and Viljoen, M.J., 1985, Interaction of floor rocks with the Platreef on Overysel, Potgietersrus, Northern Transvaal: Economic geology, v. 80, p. 988-1006.

Conn, H.K., 1979, Discovery and evaluation of the Johns-Manville platinum-palladium prospect, Stillwater Complex, Montana, U.S.A.: Canadian Mineralogist, v. 17, p. 463-468.

Cousins, C.A., 1964, The platinum deposits of the Merensky Reef, *in* Haughton, S.H., The geology of some ore deposits in Southern Africa, vol II: Geological Society of South Africa, Johannesburg, p. 225-238.

Crocket, J.H., 1981, Geochemistry of the Platinum Group Elements, *in* Cabri, L.J., ed., Platinum Group Elements: Mineralogy, Geology, Recovery: Canadian Institution of Mining and Metallurgy, Special Volume 23, p. 47-64.

Czamanske, G.K. and Zientek, M.L., 1985, The Stillwater Complex, Montana: Geology and Guide: Montana Bureau of Mines and Geology, Special Publication 92, 396 p.

Davies, G. and Tredoux, M., 1985, The platinum group element and gold contents of the marginal rocks and sills of the Bushveld Complex: Economic Geology, v. 80, p. 838-849.

DePaolo, D.J. and Wasserburg, G.J., 1979, Sm-Nd age of the Stillwater Complex and the mantle evolution curve for neodymium: Geochimica et Cosmochimica Acta, v. 43, p. 999-1008.

Dressler, B.O., 1982, Geology of the Wanapitei Lake area, District of Sudbury: Ontario Geological Survey, Report 213, 131 p. plus maps 2450, 2451.

Farquar, J., 1986, The Western Platinum Mine, *in* Anhaeusser, C.R. and Maske, S., eds., Mineral Deposits of Southern Africa: The Geological Society of South Africa, p. 1135-1142.

Finch, R.J., Ikramuddin, M., Mutschler, F.E. and Shannon, S.S., 1983, Precious metals in alkaline suite porphyry copper systems, western North America: Geological Society of America, Abstracts with Programs, v. 15, p. 572.

Gain, S.B., 1980, The geology and PGE distribution in the Upper Group chromitite layers at Maandagshoek 254, KT, Eastern Bushveld Complex: Institute for Geological Research on the Bushveld, Research report 22, 24 p.

Gain, S.B. and Mostert, A.B., 1982, The geological setting of the platinoid and base metal mineralisation in the Platreef of the Bushveld Complex in Drenthe, north of Poitgietersrus: Economic Geology, v. 77, p. 1395-1404.

Gibbins, W.A. and McNutt, R.H., 1975, The age of the Sudbury Nickel Irruptive and the Murray Granite: Canadian Journal of Earth Science, v. 12, p. 1970-1989.

Glaskovsky, A.A., Gorbunov, G.I. and Sysoev, F.A., 1977, Deposits of nickel, *in* Smirnov, V.I., ed., Ore Deposits of the USSR, v. II.

Godlevsky, M.N. and Grinenko, L.N., 1963, Some data on the isotopic composition of sulfur in the sulfides of the Noril'sk deposit: Geochemistry, v. 1, p. 35-41.

Godlevsky, M.N. and Likhachev, A.P., 1986, Types and distinctive features of ore-bearing formations of copper-nickel deposits, *in* Friedrich, G.H., Genkin, A.D., Naldrett, A.J., Ridge, J.D., Sillitoe, R.H. and Vokes, F.M., eds., Geology and metallogeny of copper deposits: Springer Verlag, p. 124-134.

Groves, D.I., Barrett, F.M. and McQueen, K.G., 1979, The relative roles of magmatic segregation, volcanic exhalation and regional metamorphism in the generation of volcanic-associated nickel ores of Western Australia: Canadian Mineralogist, v. 17, p. 319-336.

Hamilton, J., 1977, Sr isotope and trace element studies of the great dike and Bushveld mafic phase and their relation to early Proterozoic magma genesis in Southern Africa: Journal of Petrology, v. 18, p. 24-52.

Haughton, D.R., Roeder, P.L. and Skinner, B.J., 1974, Solubility of sulfur in mafic magmas: Economic Geology, v. 69, p. 451-467.

Hess, H.H., 1960, Stillwater igneous complex, Montana - a quantitative mineralogical study: Geological Society of America Memoir 80, 230 p.

Houston, R.S., *et al.* (approximately 20 others), 1968, A regional study of rocks of Precambrian age in that part of the Medicine Bow Mountains lying in Southeastern Wyoming - with a chapter on the relationship between Precambrian and Laramide structure: Geological Survey of Wyoming, Memoir No. 1, 167 p.

Hudson, D.R. and Donaldson, M.J., 1984, Mineralogy of platinum group elements in the Kambalda nickel deposits, Western Australia, *in* Buchanan, D.L. and Jones, M.J., eds., Sulphide deposits in mafic and ultramafic rocks: Institute of Mining and Metallurgy, p. 55-61.

Hulbert, L., 1986, Platinum group element environments in the Churchill Province of Northern Saskatchewan and Manitoba: Geological Survey of Canada, Current Activities Forum, Program with Abstracts, Paper 87-8, p. 5.

Irvine, T.N., 1975, Crystallisation sequences in the Muskox intrusion and other layered intrusions- II. Origin of chromitite layers and similar deposits of magmatic ores: Geochimica et Cosmochimica Acta, v. 39, p. 991-1020.

Irvine, T.N., 1977, Origin of chromitite layers in the Muskox intrusion: A new interpretation: Geology, v. 5, p. 273-277.

Irvine, T.N., Keith, D.W. and Todd, S.G., 1983, The J-M platinum-palladium Reef of the Stillwater Complex, Montana: II. Origin by double-diffusive convective magma mixing and implications for the Bushveld Complex: Economic Geology, v. 78, p. 1287-1334.

Irvine, T.N. and Sharpe, M.R., 1986, Magma mixing and the origin of stratiform oxide ore zones in the Bushveld Complex: Geocongress '86, Extended Abstracts, p. 599-602.

Johan, Z. and Watkinson, D.H., 1985, Significance of a fluid phase in platinum-group-element concentration: Evidence from the Critical Zone, Bushveld Complex: Canadian Mineralogist, v. 23, p. 305-306.

Jowett, E.C., 1986, Genesis of Kupferschiefer Cu-Ag deposits by convective flow of Rotliegende brine during Triassic rifting: Economic Geology, v. 81, p. 1823-1837.

Kesler, S.E., 1973, Copper, molybdenum and gold abundances in porphyry copper deposits: Economic Geology, v. 68, p. 106-112.

Kovalenko, V.A., Gladyshev, G.D. and Nosik, L.P., 1975, Isotopic composition of sulphide sulphur from deposits of Talnakh ore node in relation to their selenium content: International Geology Review, v. 17, p. 725-736.

Kucha, H., 1982, Platinum-group metals in the Zechstein copper deposits, Poland: Economic Geology, v. 77, p. 1578-1591.

Leeb-du Toit, A., 1986, The Impala Platinum Mines, *in* Anhaeusser, C.R. and Maske, S., eds., Mineral Deposits of Southern Africa: The Geological Society of South Africa, p. 1091-1106.

Lesher, C.M. and Groves, D.I., 1986, Controls on the Formation of komatiite-associated nickel-copper sulphide deposits, *in* Friedrich, G.H., Genkin, A.D., Naldrett, A.J., Ridge, J.D., Sillitoe, R.H. and Vokes, F.M., eds., Geology and metallogeny of copper deposits: Springer Verlag, p. 43-62.

Macdonald, A.J., 1985, The Lac des Iles platinum group metals deposit,, Thunder Bay District, Ontario; Ontario Geological Survey, Miscellanaeous Paper 126, p. 235-241.

Macdonald, A.J., 1987, PGE mineralisation and the relative importance of magmatic and deuteric processes: field evidence from the Lac des Iles deposit, Ontario, Canada: GEOPLATINUM '87, Milton Keynes, U.K., April 23-24, 1987, abstract.

Marston, R.J. and Kay, B.D., 1980, The distribution, petrology and genesis of nickel ores at the Juan Complex, Kambalda, Western Australia: Economic Geology, v. 75, p. 546-565.

Mathez, E.A., Boudreau, A.E. and McCallum, I.S., 1985, Apatite and biotite from the Stillwater and Bushveld Complexes and the nature of hydrothermal activity: Canadian Mineralogist, v. 23, p. 308.

McCallum, I.S., Raedeke, L.D. and Mathez, E.A., 1977, Stratigraphy and petrology of the Banded Zone of the Stillwater Complex, Montana: Eos, v. 58, p. 1245.

McCallum, I.S., Raedeke, L.D. and Mathez, E.A., 1980, Investigations of the Stillwater Complex: Part I. Stratigraphy and structure of the banded zone: American Journal of Science, v. 280-A, p. 59-87.

McCallum, M.E., Loucks, R.R., Carlson, R.R., Cooley, E.F. and Doerge, T.A., 1976, Platinum metals associated with hydrothermal copper ores of the New Rambler Mine, Medicine Bow Mountains, Wyoming: Economic Geology, v. 71, p. 1429-1450.

McCallum, M.E. and Orback, C.J., 1968, The New Rambler copper-gold-platinum district, Albany and Carbon counties: Wyoming Geological Survey Preliminary Report 8, 12 p.

Mertie, J.B., 1969, Economic Geology of the Platinum Metals; United States Geological Survey, Professional Paper 630, 120 p.

Mossom, R.J., 1986, The Atok Platinum Mine, *in* Anhaeusser, C.R. and Maske, S., eds., Mineral Deposits of Southern Africa: The Geological Society of South Africa, p. 1143-1154.

Mostert, A.B., 1982, The mineralogy, petrology and sulphide mineralization of the Plat Reef north-west of Potgietersrus, Transvaal, Republic of South Africa: South Africa Geological Survey, Bulletin 72, 48 p.

Naldrett, A.J., 1966, The role of sulphurization in the genesis of iron-nickel sulphide deposits of the Porcupine District, Ontario: Canadian Institute of Mining and Metallurgy, Transactions, v. 69, p. 147-155.

Naldrett, A.J., 1981, Platinum-Group Element Deposits, *in* Cabri, L.J., ed., Platinum-Group Elements: Mineralogy, Geology, Recovery: Canadian Institute of Mining and Metallurgy, Special Volume 23, p. 197-231.

Naldrett, A.J., Cameron, G., von Gruenewaldt, G. and Sharpe, M.R., 1987, The formation of stratiform PGE deposits in layered intrusions, *in* Parsons, I., ed., Origins of Igneous Layering: D. Reidel Publishing Company, Amsterdam, p. 313-397.

Naldrett, A.J. and Duke, J.M., 1980, Platinum metals in magmatic sulphide ores: Science, v. 208, p. 1417-1424.

Naldrett, A.J. and Macdonald, A.J., 1980, Tectonic setings of some Ni-Cu sulfide ores: their importance in genesis and exploration, *in* Strangway, D.W., ed., The Continental Crust and its Mineral Deposits: Geological Association of Canada, Special Paper 20, p. 633-657.

Needham, R.S. and Stewart-Smith, P.G., 1987, Coronation Hill U-Au Mine, South Alligator Valley, Northern Territory: an epigenetic sandstone-type deposit hosted by debris-flow conglomerate: BMR Journal of Australian Geology and Geophysics, v. 10, p. 121-131.

Page, N.J., 1977, Stillwater Complex, Montana: rock succession, metamorphism and structure of the complex and adjacent rocks: United States Geological Survey, Professional Paper 999, 79 p.

Page, N.J., 1979, Stillwater Complex, Montana - Structure, mineralogy and petrology of the basal zone with emphasis on the occurrence of sulphides: United States Geological Survey, Professional Paper 1038, 69 p.

Page, N.J. and Zientek, M.L., 1985, Petrogenesis of the metamorphic rocks beneath the Stillwater Complex: Lithologies and structures, *in* Czamanske, G.K. and Zientek, M.L., eds., The Stillwater Complex, Montana: Geology and guide: Montana Bureau of Mines and Geology, Special Publication 92, p. 55-69.

Prendergast, M.D., 1987a, The PGE-Rich Main Sulphide Zone of the Great Dyke, Zimbabwe, GEOPLATINUM '87, Milton Keynes, U.K., April 23-24, 1987, poster abstract.

Prendergast, M.D., 1987b, The PGE-Rich Main Sulphide Zone of the Great Dyke, Zimbabwe, unpublished manuscript, 27 pages, plus 1 table, 2 plates, and 8 figures.

Pretorius, D.A., 1975, The depositional environment of the Witwatersrand gold fields: a chronologic review of speculations and observations: Minerals Science and Engineering, v. 7, p. 18-47.

Pye, E.G., Naldrett, A.J. and Giblin, P.E., 1984, eds., The geology and ore deposits of the Sudbury Structure: Ontario Geological Survey, Special Volume 1, 603 p.

Racevic, D. and Cabri, L.J., 1976, Mineralogy and concentration of Au- and Pt-bearing placers from the Tulameen River Area in British Columbia: Canadian Institute of Mining and Metallurgy, v. 69, #770, p. 111-119.

Raedeke, L.D. and Vian, R.W., 1986, A three-dimensional view of mineralization in the Stillwater J-M Reef: Economic Geology, v. 81, p. 1187-1195.

Reimer, T.O., 1979, Platinoids in auriferous Proterozoic conglomerates of South Africa: Evaluation of existing data: Neues Jahrbuch fř Mineralogie Abhandlung, v. 135, p. 287-314.

Robson, G.G., 1985, Platinum 1985: Johnson Matthey Public Limited Company, London, 72 p.

Robson, G.G., 1986, Platinum 1986: Johnson Matthey Public Limited Company, London, 52 p.

Rowell, W.F. and Edgar, A.D., 1986, Platinum-group element mineralization in a hydrothermal Cu-Ni sulfide occurrence, Rathbun Lake, Northeastern Ontario: Economic Geology, v. 81, p. 1272-1277.

Schiffries, C.M., 1982, The petrogenesis of a platiniferous dunite pipe in the Bushveld Complex: infiltration metasomatism by a chloride solution: Economic Geology, v. 77, p. 1439-1453.

Sharpe, M.R., 1982, Noble metals in the marginal rocks of the Bushveld Complex: Economic Geology, v. 77, p. 1286-1295.

Sharpe, M.R., Evensen, N.M. and Naldrett, A.J., 1986, Sm/Nd and Rb/Sr isotopic evidence for liquid mixing, magma generation and contamination in the eastern Bushveld Complex (abstract), *in* GEOCONGRESS '86, Johannesburg July 7-11, 1986.

Skinner, B.J. and Merewether, P., 1986, Genesis of Witwatersrand ores: evidence *vs.* prejudice: GEOCONGRESS '86, Extended Abstracts, Johannesburg, July, 1986, p. 81-84.

Smirnov, M.F., 1966, The Noril'sk nickeliferous intrusions and their sulfide ores: Nedra Press, Moscow [in Russian].

Stevenson, J.S. and Colgrove, G.L., 1968, The Sudbury Irruptive: some petrogenetic concepts based upon recent field work: Proceedings 23rd International Geological Congress, v. 4, p. 27-35.

Stumpfl, E.F. and Ballhaus, C.G., 1986, Stratiform platinum deposits: New data and concepts: Fortschritte der Mineralogie, v. 64, p. 205-214.

Stumpfl, E.F. and Rucklidge, J.C., 1982, The platiniferous dunite pipes of the Eastern Bushveld: Economic Geology, v. 77, p. 1419-1431.

Sullivan, C.J., 1959, The origin of massive sulphide ores, Candian Institute of Mining and Metallurgy, Bulletin, v. 52, p. 613-619.

Todd, S.G., Keith, D.W., Le Roy, L.W., Schissel, D.J., Mann, E.L. and Irvine, T.N., 1982, The J-M platinum-palladium Reef of the Stillwater Complex, Montana: 1. Stratigraphy and Petrology: Economic Geology, v. 77, p. 1454-1480.

Turner, A.R., Wolfgram, D. and Barnes, S.J., 1985, Geology of the Stillwater County sector of the J-M Reef, including the Minneapolis Adit, *in* Czamanske, G.K. and Zientek, M.L., eds., The Stillwater Complex, Montana: Geology and Guide: Montana Bureau of Mines and Geology, Special Publication 92, p. 210-230.

Ulmer, G.C., Elliot, W.C. and Gold, D.P., 1981, Origin of Merensky Platinum: Redox geochemistry of fortuitous geophysics: Third International Platinum Symposium, Pretoria, South Africa, abstract, p. 52.

Van der Merwe, M.J., 1976, The layered sequence of the Potgietersrus limb of the Bushveld Complex: Economic Geology, v. 71, p. 1337-1351.

Vermaak, C.F., 1976, The Merensky Reef - thoughts on its environment and genesis: Economic Geology, v. 71, p. 1270-1298.

Vermaak, C.F. and Hendriks, L.P., 1976, A review of the mineralogy of the Merensky Reef with specific reference to new data on precious metal mineralogy: Economic Geology, v. 71, p. 1244-1269.

Verwoerd, W.J., 1986, Mineral deposits associated with carbonatites and alkaline rocks, *in* Anhaeusser, C.R. and Maske, S., eds., Mineral Deposits of Southern Africa: The Geological Society of South Africa, p. 2173-2191.

Viljoen, M.J. and Hieber, R., 1986, The Rustenburg Section of Rustenburg Platinum Mines Limited, with reference to the Merensky Reef, *in* Anhaeusser, C.R. and Maske, S., eds., Mineral Deposits of Southern Africa: The Geological Society of South Africa, p. 1107-1134.

Volborth, A.A. and Housley, R.M., 1984, A preliminary description of complex graphite, sulphide, arsenide, and platinum group element mineralization in a pegmatoid pyroxenite of the Stillwater Complex, Montana, USA: Tschermaks Mineralogische und Petrographische Mitteilungen, v. 33, p. 213-230.

Von Gruenewaldt, G., 1979, A review of some recent concepts of the Bushveld Complex with particular reference to the sulphide mineralization: Canadian Mineralogist, v. 17, p. 233-256.

Von Gruenewaldt, G., Sharpe, M.R. and Hatton, C.J., 1985, The Bushveld Complex: Introduction and Review: Economic Geology, v. 80, No. 4, p. 803-812.

Wagner, P.A., 1929, The platinum deposits and mines of South Africa, Republished 1973 by C. Struik (Pty) Ltd., Cape Town, R.S.A., 338 p.

Werle, J.L., Ikramuddin, M. and Mutschler, F.E., 1984, Allard Stock, La Plata Mountains, Colorado - an alkaline rock-hosted porphyry copper-precious metal deposit: Canadian Journal of Earth Sciences, v. 21, p. 630-641.

Willemse, J., 1969, The geology of the Bushveld Complex, the largest repository of magmatic ore deposits in the world, *in* Wilson, H.D.B., ed., Magmatic Ore Deposits: Economic Geology, Monograph 4, p. 1-22.

Wilson, A.H., 1982, The geology of the Great Dyke, Zimbabwe: the ultramafic rocks: Journal of Petrology, v. 23, p. 240-292.

Wilson, A.H. and Wilson, J.F., 1981, The Great Dyke, *in* Hunter, D.R., ed., Precambrian of the Southern Hemisphere: Elsevier, Amsterdam, p. 572-578.

Worst, J.F., 1960, The Great Dyke of Southern Rhodesia: Southern Rhodesia Geological Survey Bulletin, No. 47, 234 p.

Accepted, as revised, 27 June 1987.
Originally published in *Geoscience Canada* v. 14 Number 3 (September 1987)

Magmatic Segregation Deposits of Chromite

J.M. Duke
Geological Survey of Canada
601 Booth Street
Ottawa, Ontario K1A 0E8

Introduction

Chromite is a vital industrial commodity. Its importance derives mainly from the fact that it is the only ore mineral of chromium which is an essential constituent of stainless and some other steels as well as of certain non-ferrous alloys. Chromite is also an industrial mineral that is used in its natural form in the manufacture of refractory bricks for furnace linings and as a foundry sand. It is the raw material for the production of chromium chemicals used in a variety of applications, such as electroplating, leather tanning, pigments and dyes. The suitability of chromite for various applications is determined, in part, by its chemical composition. For example, until recently, a relatively high chromium to iron ratio was required for metallurgical-grade chromite (*e.g.*, Cr/Fe ≥ 2:8:1). However, advances in smelting technology have meant that this is no longer a critical factor, and ferrochrome is now produced routinely from chromite ores with substantially lower ratios (*e.g.*, Cr/Fe ~ 1.5:1).

Canadian requirements for chromium materials are met entirely from imports. Chromite ores and concentrates for use in refractory applications account for roughly one-half of Canadian consumption in terms of weight. The balance is largely in the form of ferrochrome for use in the iron and steel industry, whereas chromium chemicals make up a relatively small portion of imports. Chromite has been identified as a strategic mineral in many of the industrialized nations that have no domestic production capacity because chromite supply is dominated by a few countries and is perceived to be subject to disruption, and chromium or chromite is irreplaceable in certain applications, and, in others, substitution would impose penalties of increased cost or decreased performance.

Chromite is mined almost exclusively from massive to heavily disseminated segregations in ultramafic or mafic igneous rocks. Eluvial and alluvial deposits derived by the erosion of such rocks account for a minor proportion of total production. Hard rock chromite ores are usually assigned to one of two classes on the basis of deposit geometry, as well as the petrologic character and tectonic setting of their host rocks. *Stratiform* deposits are sheet-like accumulations of chromite that occur in layered ultramafic to mafic igneous intrusions. The intrusions are typically of Precambrian age and occur in intracratonic terranes. *Podiform* deposits are irregular, but fundamentally lenticular, chromite-rich bodies that occur within alpine peridotite or ophiolite complexes (Thayer, 1964). This terminology is not entirely satisfactory because certain important deposits have characteristics of both classes. Nevertheless, it is used widely in the literature and I have retained it here. Podiform deposits currently account for about 55% of world production, with the most important deposits in the USSR, Albania, Philippines, Turkey and India. There are relatively few productive stratiform deposits but, although they are marginally less important than the podiform ores in terms of total production, they account for more than 98% of the world chromite reserves. In this brief review, I will attempt to summarize the important geological characteristics of stratiform and podiform chromite ore deposits and present some of the models that have been proposed for their genesis.

Characteristics of Stratiform Chromite Deposits

The intrusions that host the important stratiform chromite deposits occur within intracratonic terranes. In some cases, they intrude granitic or gneissic basement rocks, and, in others, they intrude supracrustal rocks which rest upon sialic basement. The intrusions fall into two broad categories with respect to morphology. The first includes essentially tabular bodies that were emplaced as horizontal sill-like intrusions in which the igneous layering is generally conformable to the floor (*e.g.*, Stillwater, Kemi, Bird River). The second group are the funnel-shaped intrusions in which the igneous layering dips at a shallow angle toward the centre, giving a synclinal cross-section (*e.g.*, Muskox, Great Dyke, Bushveld).

The chromite deposits typically comprise thin, laterally extensive chromite-rich layers in the lower parts of the intrusions. Despite local irregularities, the chromite-rich layers are generally conformable to, and form an integral part of, the igneous layering that characterizes such intrusions. The individual layers of massive chromite (chromitite) range from less than 1 cm to more than 1 m in thickness, but their lateral dimensions are measured in kilometres or tens of kilometres. Orebodies may comprise a single chromitite layer or a number of closely spaced layers separated by ultramafic rock which contains disseminated chromite.

Typical chromitites are composed of from about 50% to more than 95% fine-grained (~0.2 mm) cumulus chromite with interstitial olivine, orthopyroxene, plagioclase, clinopyroxene or their alteration products. Brown mica is a common, albeit minor, constituent. The chromite tends to be euhedral to subhedral against interstitial silicates, but shows evidence of mutual interference during crystal growth against other chromite grains (Figure 1a).

Cameron and Desborough (1969) have pointed out that there is a tendency to stress the similarities of vertical distribution and petrologic association of chromitites in layered intrusions when, in fact, the differences are quite striking. Therefore, the most important productive stratiform chromite deposits are described briefly below. These are the Bushveld Complex of South Africa, the Selukwe and Great Dyke deposits of Zimbabwe, the Kemi deposit in Finland and the Campo Formoso and Jacurici Valley deposits of Brazil. Also described are two non-producing deposits that are important strategic resources in the North American context: the Stillwater Complex of Montana and the Bird River Sill in Manitoba.

Bushveld Complex. The mafic rocks of the Bushveld Complex extend over an oval-shaped area about 480 km by 380 km in the Kaapvaal Craton in South

Africa. The layered rocks crop out in three crudely arcuate bands referred to as the eastern, western and northern Bushveld, which are superficially similar, but differ considerably in detail. It is believed that the Complex was emplaced as seven shallow, partly overlapping, conical intrusions, which eventually coalesced into three larger magma chambers corresponding to the eastern, western and northern segments (von Gruenewaldt, 1979; Vermaak and von Gruenewaldt, 1981). The mafic rocks have been dated at 2095 ± 24 Ma by the Rb-Sr method (Hamilton, 1977).

The Bushveld Complex includes a sequence of layered igneous rocks that locally exceeds 9000 m in thickness, and which has been divided into the marginal, lower, critical, main and upper zones. A complete description of the igneous stratigraphy is not warranted here, but some of the salient features are summarized in Figure 2. The lower part of the sequence is predominantly ultramafic in character with bronzitites and harzburgites being the principal rock types in the lower zone and the lower subzone of the critical zone. The base of the critical zone is currently defined by a sharp increase in the abundance of postcumulus plagioclase in bronzitite. The base of the upper subzone of the critical zone marks the main cumulus appearance of plagioclase, and the overlying rocks in the layered series are generally mafic in character and include norites, anorthosites, gabbronorites, and minor gabbro and pyroxenite. The base of the upper zone corresponds to the cumulus appearance of magnetite.

Chromite is a cumulus mineral in parts of the lower zone but is not ubiquitous (Cameron, 1978). Chromite-rich rocks and, in particular, chromitite seams are largely restricted to the critical zone. Cameron and Desborough (1969) defined numerous "chromitic intervals" within the critical zone where chromite is present in amounts in excess of 1%. The rock types within the "chromitic intervals" have chromite as the only cumulus mineral (*i.e.*, chromitite), or chromite accompanied by bronzite, plagioclase, olivine, bronzite and olivine, or bronzite and plagioclase.

There are many tens of chromitite layers in the critical zone, but the total number is uncertain due to the difficulty

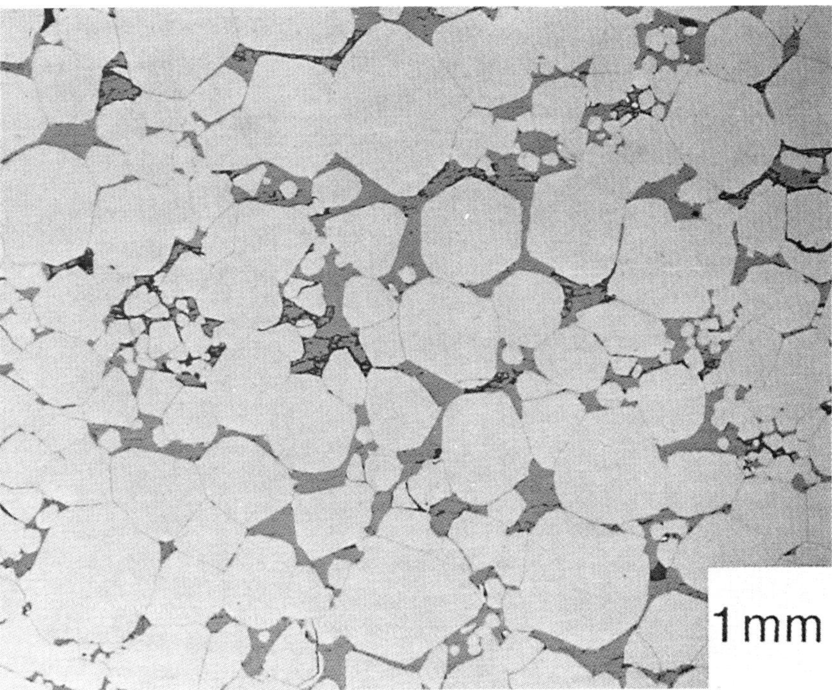

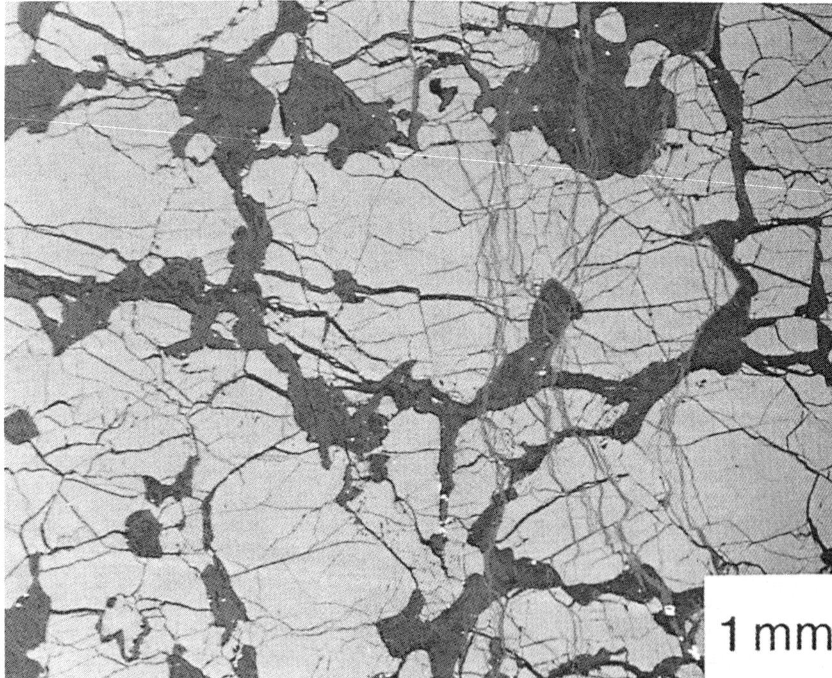

Figure 1 *Contrasting chromite textures in stratiform and podiform ores.*
(a) *Fine-grained, massive chromite from the A chromitite zone of the Stillwater Complex, Montana. The cumulus chromite exhibits euhedral to subhedral boundaries against intercumulus silicate minerals (dark grey) but shows effects of mutual interference during crystal growth against other chromite grains.*
(b) *Coarse-grained, massive chromite from the Sterrett Mine, St. Cyr, Quebec. Note the extremely irregular shapes of the chromite anhedra.*

in correlating between sections. Individual seams may be a single chromitite layer or several closely-spaced layers separated by layers or "partings" of pyroxenite or anorthosite (*e.g.*, Cameron and Desborough, 1969). The chromitite seams range in thickness from less than 1 cm to more than 2 m with the majority falling toward the lower end of the range.

Two seams in the lower subzone of the critical zone account for the bulk of the chromite production and reserves. The LG6 seam is continuous over a strike length of 70 km in the western Bushveld (where it is also called the Magazine or Main seam) with an average thickness of 0.8 m. In the eastern Bushveld, the LG6 is known as the Steelpoort seam and has a strike length of 90 km. It has a thickness of 0.6-1.3 m where mined (Buchanan, 1979). A second seam is mined in the eastern Bushveld: this is the F chromitite which has a strike length of 35 km and a thickness of 1.3 m. The ore reserves of the LG6 seam to a depth of 300 m have been estimated at 752 million tonnes, and those of the F chromitite are 312 million tonnes (Buchanan, 1979). Average grades range from 46.0% to 47.6% Cr_2O_3 and Cr/Fe ratios are about 1.6:1.

Great Dyke. The Great Dyke of Zimbabwe is 530 km in length and averages about 6 km in width. It is, in fact, not a dyke at all, but rather a series of four, coalescing, funnel-shaped bodies which are, from north to south, the Musengezi, the Hartley, the Selukwe and the Wedza Complexes (Worst, 1958). Each complex consists of a lower ultramafic-layered series and an upper mafic-layered series. In the Hartley Complex, the largest of the four, more than 2100 m of ultramafic rocks and 900 m of mafic rocks are observed. The layering dips gently inward at the margins of each complex and flattens out toward the centre, giving the bodies a synclinal form in cross-section. Each of the complexes is thought to represent an intrusive centre, and Hamilton (1977) reported a Rb-Sr age of 2461 ± 16 Ma.

Worst (1958) observed that the ultramafic series comprises a number of cyclic units and Wilson (1982) has outlined 14 such units in the Hartley Complex. The ideal sequence within each unit is, from the base upward, chromitite, dunite, harzburgite, olivine bronzitite, bronzitite (Figure 3); however, cumulus bronzite is absent toward the base of the layered series, whereas dunite and chromitite are missing from some of the upper units. The top of the ultramafic series is marked by websterite in each complex and this is overlain by mafic rocks.

Chromite is a widespread cumulus mineral in the olivine-bearing rocks of the ultramafic series in amounts less than about 5%, but it is only rarely observed in bronzitite. The chromitites, however, generally contain in excess of 90% cumulus chromite with minor cumulus olivine and postcumulus orthopyroxene (Wilson, 1982). Wilson has observed 11 chromitite seams in the Hartley Complex which range in thickness from 2 to 12 cm. The chromitite differs from that in most other stratiform deposits with respect to the coarse grain size (>2 mm) of much of the contained chromite.

Stillwater Complex. The Stillwater Complex is a steeply dipping, tabular intrusion which crops out over a strike length of 48 km and a width of up to 5.5 km. The intrusion was probably emplaced as a horizontal sheet at about 2701 Ma (DePaolo and Wasserburg, 1979), but was subsequently faulted and tilted into its present configuration. The exposed part of the intrusion has been subdivided from the base upward into the Basal Zone, the Ultramafic Zone, the Banded Zone and the Upper Zone (Jones *et al.*, 1960). However, the upper part of the intrusion is unconformably overlain by early Paleozoic sedimentary rocks, and so the original nature of the upper contact and the total thickness are unknown.

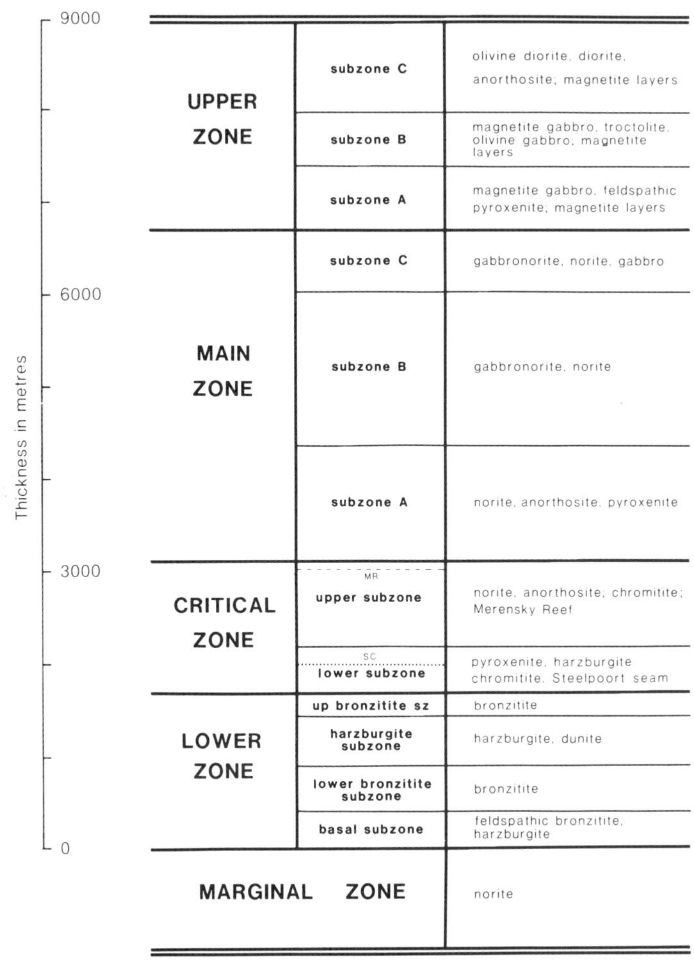

Thickness in metres: 9000 – 6000 – 3000 – 0

Zone	Subzone	Rocks
UPPER ZONE	subzone C	olivine diorite, diorite, anorthosite; magnetite layers
	subzone B	magnetite gabbro, troctolite, olivine gabbro; magnetite layers
	subzone A	magnetite gabbro, feldspathic pyroxenite; magnetite layers
MAIN ZONE	subzone C	gabbronorite, norite, gabbro
	subzone B	gabbronorite, norite
	subzone A	norite, anorthosite, pyroxenite
CRITICAL ZONE	MR upper subzone	norite, anorthosite; chromitite; Merensky Reef
	SC lower subzone	pyroxenite, harzburgite chromitite, Steelpoort seam
LOWER ZONE	up bronzitite sz	bronzitite
	harzburgite subzone	harzburgite, dunite
	lower bronzitite subzone	bronzitite
	basal subzone	feldspathic bronzitite, harzburgite
MARGINAL ZONE		norite

Figure 2 *Informal subdivision of the Bushveld Complex, South Africa (after Vermaak and von Gruenewaldt, 1981) which alone accounts for more than one-third of world chromite production and two-thirds of reserves. The approximate stratigraphic positions of the Steelpoort chromite seam and the Merensky platinum reef are shown.*

It is the Ultramafic Zone that is of interest in the present context. The Zone averages 1070 m in thickness and has been subdivided into a lower Peridotite Member and an upper Bronzitite Member (Jackson, 1961). The Peridotite Member is made up of as many as 15 cyclic units which are characterized by the upward sequence poikilitic harzburgite (olivine cumulate), granular harzburgite (olivine-bronzite cumulate) and bronzitite (bronzite cumulate). These cyclic units, which range from 3 to 381 m in thickness (Page, 1977), are illustrated in Figure 3. The lower poikilitic harzburgite parts of at least 13 of these cyclic units contain "chromitite zones" that range in thickness from less than 2 cm to almost 4 m (Jackson, 1968). The lower half of each "chromitite zone" is essentially massive chromite, whereas the upper half is composed of alternating layers of chromitite and olivine chromitite. The "reserves" of the most important deposit, the Mouat Mine, are reported to be 4 milion tonnes grading 22.5% Cr_2O_3 with Cr/Fe ratio of about 1.6:1 (Kingston *et al.*, 1970).

Kemi. A number of layered intrusions about 2436 million years old occur at the unconformable contact between Archean basement gneisses and overlying Karelian schists in northern Finland (Piirainen, 1978; Piirainen *et al.*, 1974). Four of these intrusions (Kemi, Koitelainen, Penikat and Tornio) host stratiform chromite deposits, but only those at Kemi are currently being mined.

The Kemi intrusion is a steeply dipping, lenticular body at least 15 km in length and 2 km across at its widest point. It comprises a lower ultramafic zone and an upper mafic zone which are of similar average thickness. The ultramafic rocks, which have undergone widespread alteration, include dunite, peridotite and pyroxenite as well as chromitite, whereas the mafic zone consists mainly of gabbro and norite with minor anorthosite and pyroxenite (Kujanpää, 1980). Layering is well developed in both zones, but it is not known whether any cyclic units occur; a conformable chromitite layer occurs in the lower part of the ultramafic zone. Over much of the intrusion, it is a few centimetres to a few metres thick; however, over a strike-length of 4.5 km in the widest part of the intrusion, there are several successive swellings where the layer attains thicknesses of 30 to 90 m (Kujanpää, 1980). It is these swellings that constitute the orebodies, eight of which have been identified. The main chromitite layer has peridotite or its altered equivalent (serpentinite or talc-carbonate rock) in both the footwall and hangingwall. The peridotite above the ore contains numerous thin chromitite layers, and is overlain, in turn, by pyroxenite and more peridotite. Much of the ore is brecciated, particularly in the lower parts of the principal orebodies, but igneous textures are typically preserved and there is little evidence of plastic deformation. It seems likely that the remarkable thickening of the chromitite was largely a magmatic phenomenon rather than a result of tectonism.

The Kemi ore reserves are reported to be about 50 million tonnes averaging 26% Cr_2O_3 with a Cr/Fe ratio of 1.55:1. Although the grade is low in comparison with the deposits in southern Africa, the extraordinary width of the orebodies makes them suitable for mining by relatively inexpensive open pit methods, and deposits of the Kemi type should constitute an attractive exploration target.

Selukwe. The chromite deposits of the Selukwe schist belt were the first to be discovered in what is now Zimbabwe

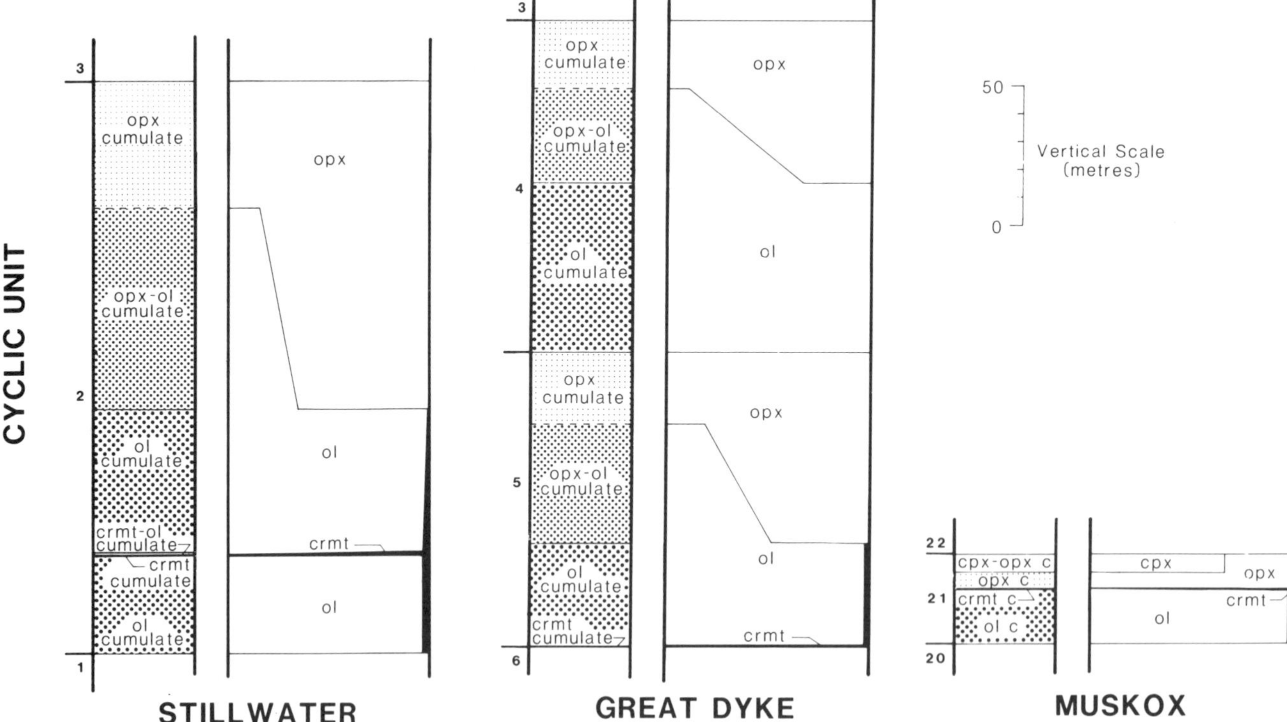

Figure 3 *Position of chromite layers within the cyclic units of the Great Dyke (Wilson, 1982), Stillwater Complex (Jackson, 1968), and Muskox intrusion (Irvine and Smith, 1969).*

and probably were the world's most important single source of chromite during the first half of this century. The Selukwe ultramafic complex, not to be confused with the Selukwe Complex of the Great Dyke (see above), is a highly deformed stratiform intrusion that lies within the Sebakwian Group in the Rhodesian craton and is about 3500 million years old (Cotterill, 1979). The ultramafic succession is up to 1000 m thick, and includes dunite, chromitite and pyroxenite. Cotterill (1979) reports the presence of cyclic units in which dunite is overlain successively by olivine pyroxenite and pyroxenite. Chromitite layers, where present, occur in the olivine cumulates at or near the contact with pyroxenite.

The Selukwe deposits have certain features in common with ores occurring in ophiolite complexes and are often referred to as podiform deposits in the literature. The individual orebodies have an elongate, lenticular form and average about 300 m in length, 60 m in depth and 10 m in width (Cotterill, 1969). Moreover, the grain size in much of the massive ore is very coarse (up to 4 mm), and is thus similar to that in podiform ores.

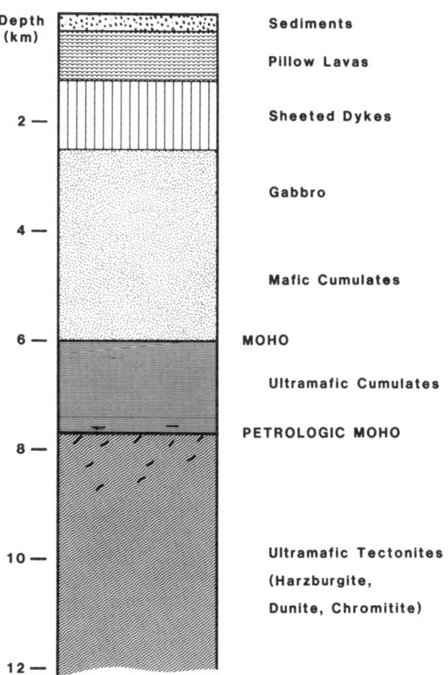

Figure 4 *Ideal ophiolite succession. Podiform chromite bodies tend to be most abundant in the uppermost part of the tectonite unit but also occur at depth and in the lowermost cumulates.*

Campo Formoso and Jacurici Valley. The Campo Formoso chromite deposits are located 350 km northwest of Salvador in the State of Bahia, Brazil, and have been an important source of chromite since 1962. The deposits occur within a metaperidotite body at least 40 km long and up to 1.1 km wide (de Deus *et al.*, 1982). The ultramafic rocks are underlain by granulites of the Archean Caraiba Group and are uncomformably overlain by younger quartzites and phyllites of the Jacobina Group. The ultramafic rocks are intruded by the 2000 Ma Campo Formoso granite which defines the minimum age for the deposits. The ultramafic complex has been metamorphosed to greenschist facies and little remains of primary textures and minerals. However, Hedlund *et al.* (1974) believed that the body consisted mainly of peridotite and chromitite with minor pyroxenite and gabbro layers. Chromite occurs in massive chromitite, in net-textured olivine chromitite and disseminated in metaperidotite. Hedlund *et al.* (1974) described a measured section at the Coitezeiros A Mine where five layers of chromitite 0.3-1.4 m thick occur over a stratigraphic interval of 10 m. Measured, indicated and inferred ore reserves exceed 55 million tonnes. Much of this is low grade (15-20% Cr_2O_3), but there is a significant proportion of massive ore which averages 38% Cr_2O_3 with Cr/Fe = 1.9:1 (de Deus *et al.*, 1982).

The Jarurici Valley chromite district is about 80 km east of Campo Formoso. The chromite deposits occur in ultramafic-mafic sills which intrude granulites, gneisses and migmatites of the Archean Pedra Vermelha Complex (de Deus and Viana, 1982). The sills crop out discontinuously over a strike length of 70 km and have a maximum thickness of 300 m. In the Medrado sill, the layered sequence comprises, from the base upward, harzburgite, chromitite, harzburgite, bronzitite and norite. The chromitite layer is 7 m thick and is made up of 80% cumulus chromite with intercumulus diopside making up the balance. Measured, indicated and inferred reserves amount to 3 million tonnes grading from 33 to 42% Cr_2O_3 with Cr/Fe in the range of 1.3 to 2.4.

Bird River Sill. The chromite deposits of the Bird River Sill in the Lac du Bonnet area of southeastern Manitoba have been explored periodically since their discovery in 1942, but have never been mined. The sill occurs within supracrustal rocks of the Archean Rice Lake Group. The sill and its host rocks have been folded about a broad anticlinal axis, resulting in dips of 65° to 90°. The sill has been longitudinally segmented by numerous faults, with the segments having an aggregate strike length of about 25 km. It averages 1 km in thickness and comprises, from the base upward, a Marginal Group composed mainly of olivine gabbro, a Layered Ultramafic Series, and a Layered Gabbro Series (Trueman, 1971, 1980). Trueman (1980) described the igneous stratigraphy at the Chrome Property where the layered succession is 580 m thick. The Layered Ultramafic Series of the Chrome Property is 200 m thick, and is dominated by dunite toward the base and peridotite in the upper part. A total of 21 chromitite layers occur in the Series and most of these are grouped toward the top. The peridotite becomes somewhat feldspathic just beneath the main chromitite layers. Cumulus plagioclase occurs in the uppermost Layered Ultramafic Series, and the contact with the overlying Layered Gabbro Series is described as transitional. The latter series includes gabbro, anorthositic gabbro and anorthosite with minor granophyre.

Characteristics of Podiform Chromite Deposits

The ideal ophiolite succession comprises, in descending order, marine sediments, pillowed basaltic lavas, sheeted diabase dykes, noncumulate and cumulate mafic rocks, ultramafic cumulates and ultramafic tectonites (Figure 4). In most cases, the succession has been dismembered by faulting and one cannot be sure that all alpine peridotites are in fact ophiolitic. Most ophiolites are clearly allochthonous and are believed to represent transported fragments of oceanic lithosphere. The ultramafic cumulates are composed of dunite, wehrlite and clinopyroxenite that accumulated at the floor of a relatively shallow crustal magma chamber. The tectonites are mainly foliated harzburgite with pockets of dunite, and represent depleted mantle which has undergone partial melting and from which a basaltic melt fraction has been extracted. The contact between the ultramafic cumulates and tectonites is the petrologic Moho, or the boundary between crust and mantle.

The chromite deposits occur almost exclusively within the ultramafic rocks of the ophiolite succession. They are most abundant in the tectonites, but also occur in the lowermost parts of the cumulate sequences. Dickey (1975) indicates that there is a tendency for the abundance of podiform deposits to increase toward the top of the tectonite. The immediate host rocks of the chromite are commonly dunitic; thus, the deposits occurring within harzburgitic tectonites are normally within an envelope of dunite a few centimetres to a few metres thick.

The morphology of podiform chromite deposits is irregular and unpredictable. Thayer (1964, p. 140) observed that "Although many podiform chromite deposits have shapes that defy description, the majority might be characterized as tabular lenses, irregular pencils, or combinations of these two basic forms". Individual pods range from a few kilograms to several million tonnes in size, but bodies exceeding one million tonnes are rare. An ore deposit normally comprises a number of discrete pods. For example, the Kavak Mine in Turkey includes 21 orebodies that total about 2 million tonnes of ore with an average grade of 28 to 30% Cr_2O_3 (Ergunalp, 1980). The podiform bodies are commonly, but not always, conformable to the foliation or layering of the host rocks. Cassard *et al.* (1981) divided the New Caledonian deposits into discordant, subconcordant and concordant types depending upon their orientation with respect to the penetrative fabric of the enclosing peridotite. They also noted that concordant orebodies are more deformed and more regular in shape than discordant ones.

The podiform chromite ores are texturally diverse. Disseminated chromite in the ultramafic cumulates is very similar in grain shape and size to that observed in stratiform intrusions, whereas that in the tectonite often occurs as large, elongated anhedra (Greenbaum, 1977). Heavily disseminated chromite in both the ultramafic cumulates and tectonites often displays typical cumulate textures (Thayer, 1969), rhythmic layering (Greenbaum, 1977) and graded layers. The chromite in massive ore tends to occur as coarse-grained, interlocking anhedra; grain sizes on the order of 5-10 mm are not uncommon (Figure 1b). Perhaps the most striking feature of podiform ores is the remarkable nodular texture which is characterized by loosely-packed, ellipsoidal chromite nodules 5-20 mm in diameter in a dunite matrix. Thayer (1969) found the nodules to be aggregates of smaller grains (up to 3 mm), and Greenbaum (1977) observed that the cores of some nodules from the Troodos Complex consisted of an intergrowth of skeletal chromite and secondary silicate minerals. In any event, the nodular texture is widely viewed to be a primary magmatic feature (Thayer, 1969; Greenbaum, 1977; Bergath and Weiser, 1980).

Johan and his co-workers have documented the occurrence of a fascinating suite of inclusions in chromite from podiform deposits in Cyprus, New Caledonia, Oman and Saudi Arabia (*e.g.*, Johan *et al.*, 1982). Among the silicate mineral inclusions are olivine, ortho- and clinopyroxene, paragonite, sodic mica, albite and jadeite. Equilibration temperatures deduced from the co-existing pyroxenes are 1050 $\pm$ 20°C. Aqueous fluid inclusions are ubiquitous, and indicate rather high temperatures of formation, and may imply a role for a fluid phase in the precipitation of chromite from the magma.

Chemical Composition of Chromite

Chrome spinel is a solid solution having the general formula $(Mg, Fe^{2+})(Cr, Al, Fe^{3+})_2O_4$ and natural chromites are characterized by a wide range of chemical composition. Chromite composition is, to a large extent, indicative of the deposit type. The variation of trivalent cations is depicted in Figure 5. In podiform deposits, the proportion of Fe^{3+} in chromite is very low and the main differences in composition reflect the $Cr \leftrightarrow Al$ substitution. The chromite in stratiform ores generally contains more Fe^{3+} than does podiform chromite, and both Fe^{3+} and Al tend to increase with decreasing Cr. The $Mg/(Mg+Fe^{2+})$ ratio of chromite in podiform deposits does not vary greatly, but does show a slight negative correlation with $Cr/(Cr+Al)$, whereas the $Mg/(Mg+Fe^{2+})$ ratio in stratiform chromite varies over a comparatively wide range (Figure 6). A further compositional distinction between podiform and stratiform chromites is the TiO_2 content, which is generally less than 0.3 wt.% in the former, but is greater in the latter (Dickey, 1975).

The composition of chromite varies systematically within individual stratiform deposits, although the variation is not in the same sense in every deposit that has been studied. Cameron (1977) has documented a systematic cryptic variation of chromite composition across 26 chromitite layers in the Bushveld critical zone whereby the $Mg/(Mg+Fe^{2+})$, $Cr/(Cr+Al)$, and Cr/Fe ratios decrease upward through the sequence. Similar trends are suggested by Worst's (1958) data for the Great Dyke. Jackson (1968) reported that the Cr/Fe ratio increases upward toward the middle of the sequence of chromitites in the Stillwater and then decreases toward the top. However, the ratio decreases upward from the base in individual chromitite layers, and there is also a significant lateral variation in the relative proportions of Fe^{2+} and Fe^{3+} along many of the layers.

Systematic compositional variation has also been documented within some podiform chromite deposits. Neary and Brown (1979) reported a small upward decrease in the Cr/Fe ratio of chromite within a single chromitite lens in the Al'Ays ophiolite complex in Saudi Arabia. Brown (1980) has shown that both the Cr/Fe and $Cr/(Cr+Al)$ ratios of chromite increase significantly as a function of the depth of the chromitite body beneath the petrologic Moho across a thickness of 6.5 km of ultramafic tectonite in the Oman ophiolite.

Origin of Chromite Deposits

Stratiform Deposits

The chromite-rich layers that constitute stratiform ores are igneous cumulates, and their formation is simply a special case of the larger problem of the origin of layered igneous rocks. For many years, cumulate rocks were widely believed to represent accumulations formed by the settling of crystals to the floor of a magma chamber under the influence of gravity, and igneous layering was largely explained by processes analogous to those responsible for the layered sedimentary rocks. Thus, mechanical sorting through the action of magmatic currents, or differential settling of minerals of different density or grain size, was invoked by various workers as being at least partly responsible for stratiform chromite accumulations (*e.g.*, Cameron and Emerson, 1959; Irvine and Smith, 1969).

Jackson (1961) observed that certain features of the layering of the Ultramafic Zone of the Stillwater Complex are not consistent with crystal settling, and proposed that crystal nucleation and growth occurred in a stagnant zone very close to the floor of the magma chamber. A dramatic change in the perception of igneous layering has occurred in recent years, and it is now believed that, although crystal settling and deposition from magmatic currents are important in certain circumstances, layering is probably more often the result of crystallization *in situ* at the floor, roof or walls of the magma chamber (Campbell, 1978; McBirney and Noyes, 1979; Rice, 1981).

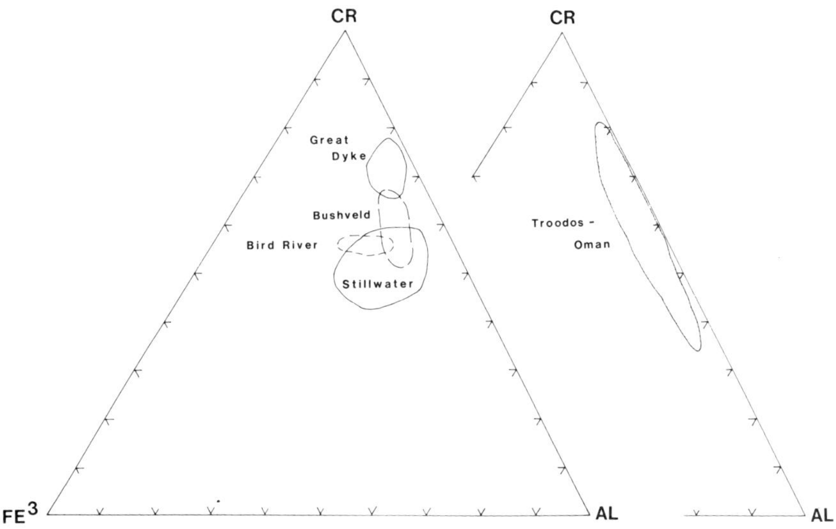

Figure 5 *Proportions of trivalent cations in chromite from certain stratiform (left) and podiform deposits. (Data from Worst, 1958; Cameron, 1977; Gait, 1964; Jackson, 1969; Greenbaum, 1977; and Brown, 1980).*

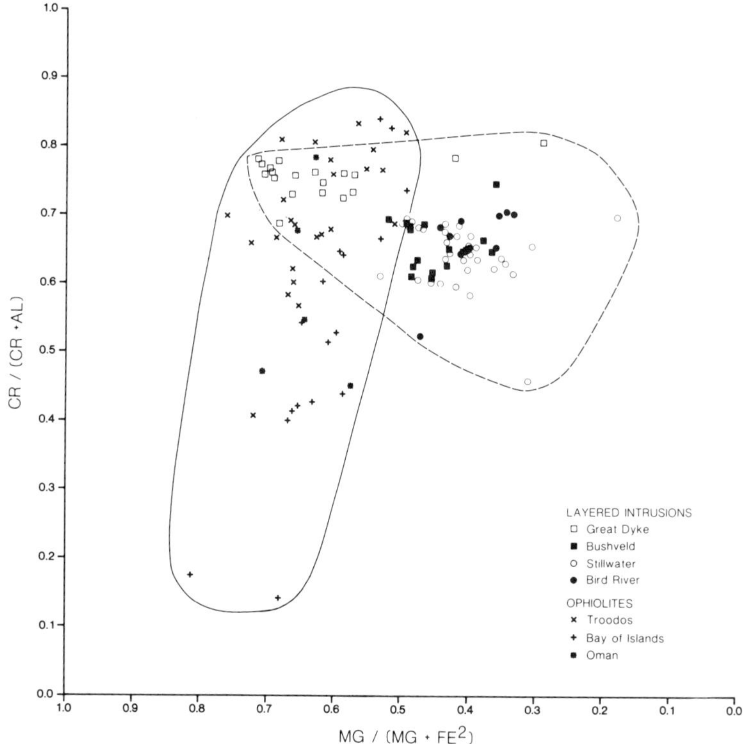

Figure 6 *Variation of Mg/(Mg + Fe^{2+}) and Cr/(Cr + Al) ratios in chromite from certain stratiform and podiform deposits. Data from same sources as Figure 5.*

Irvine (1970) demonstrated that the sequence of appearance of the principal cumulus minerals in layered mafic rocks, that is, olivine, plagioclase and the pyroxenes, is largely in accord with experimentally determined phase relationships and the magma compositions inferred for the various intrusions. The occurrence of chromitite layers in which chromite is the only cumulus mineral is problematical in this context because, as pointed out by Irvine and Smith (1969), chromite is not expected to be alone on the liquidus during the normal course of fractional crystallization. Most of the stratiform chromite ores occur in sequences involving olivine, orthopyroxene, and chromite as cumulus minerals, and the problem is elucidated by reference to the schematic phase diagram devised by Irvine (1975) and presented here as Figure 7.

Although chromite is commonly an early crystallization product of mafic and ultramafic magmas, it is seldom the earliest phase. The intrusions of interest typically have abundant olivine cumulates in their lowermost parts, and it follows that the compositions of their parent magmas would plot in the olivine liquidus field in Figure 7, at a point such as A. Fractional crystallization of olivine from such a magma would drive the liquid composition toward the olivine-chromite cotectic (B), at which point the two phases would co-precipitate and the liquid composition would move along the cotectic toward the distribution point (D). At the distribution point, both olivine and chromite disappear and orthopyroxene appears as the sole crystalline phase. The proportions of olivine and chromite separating at any point along the olivine-chromite cotectic are given by the intersection of the tangent to the cotectic with the olivine-chromite join. For realistic magma compositions, in which the solubility of chromite is very low, the chromite will comprise, at most, a few percent of the solid fraction. This is in accord with the typical proportions of disseminated chromite that are observed in dunites and peridotites.

In order to form a chromitite, the magma composition must lie in the chromite field, and one of two things must happen: either the magma composition must change such that it enters the primary liquidus field of chromite, or the position of the olivine-chromite (or orthopyroxene-chromite) boundary must shift toward the olivine-silica join. Exactly how the shift in magma composition or phase boundary is accomplished is the key to the origin of stratiform chromite deposits, and a number of possible mechanisms have been proposed.

Ulmer (1969) reviewed the abundant experimental data that demonstrated that increasing the oxygen fugacity greatly enlarged the liquidus field of the Mg-Fe-Al spinel minerals in mafic melts and suggested that the liquidus of chrome spinel would be similarly affected. Cameron and Desborough (1969) and Ulmer (1969) proposed that the "chromitic intervals" in the critical zone of the Bushveld Complex represented periodic increases in the ambient oxygen fugacity which caused chromite to precipitate alone or with various proportions of cumulus silicates. The experiments of Hill and Roeder (1974) confirmed that oxygen fugacity had an important effect on the liquidus relations of chromite in basaltic systems, but also led Cameron (1977) to conclude that the cryptic variation of chromite composition in the Bushveld Complex is inconsistent with the hypothesis that increasing oxygen fugacity led to the precipitation of chromite. Moreover, in a broader context, it seems likely that the oxygen fugacity which prevails in a large body of magma is, to a large extent, internally buffered and is not subject to the rapid, but spatially uniform, fluctuations implied by the rhythmically layered chromitites.

Cameron (1977, 1980) proposed that the formation of the Bushveld chromitites could reflect changes in the total pressure under which crystallization occurred. Osborn (1978) demonstrated that the liquidus field of Mg-Fe-Al spinel in a simplified basalt system expanded with increasing pressure, and Cameron inferred that the effect on the stability of chromite would be qualitatively similar. Thus, Cameron suggested that the Bushveld magma composition was in the bronzite field, but close to the chromite-bronzite cotectic or the chromite-bronzite-olivine triple point during the accumulation of the lower subzone of the critical zone, and that the intermittent appearance of chromite represents shifts in the phase boundaries in response to changes in total pressure. Slight changes in total pressure might occur as a result of tectonic activity or the addition or removal of batches of magma to or from the magma chamber. This model is appealing from the standpoint that the pressure variations would be laterally uniform.

Irvine (1975) proposed that salic contamination of a magma, for example, by assimilation of the roof rocks of an intrusion, could bring about chromite saturation. This derives from the curvature of the olivine-chromite cotectic (Figure 7), which means that addition of silica-rich material to a liquid on the cotectic at, for example, C would move its composition toward the SiO_2 apex and into the liquidus field of chromite to a point such as E. The liquid would then crystallize chromite alone, and its composition would move back toward the cotectic at G, at which point, the normal crystallization path would be resumed. Such a sequence of events could result in a crystallization order such as that observed in the cyclic units in the Stillwater Complex (Figure 3). Assimilation of a larger proportion of salic material, such that the liquid composition was driven toward the chromite-orthopyroxene boundary rather than the chromite-olivine boundary, could lead to cyclic units of the Muskox type.

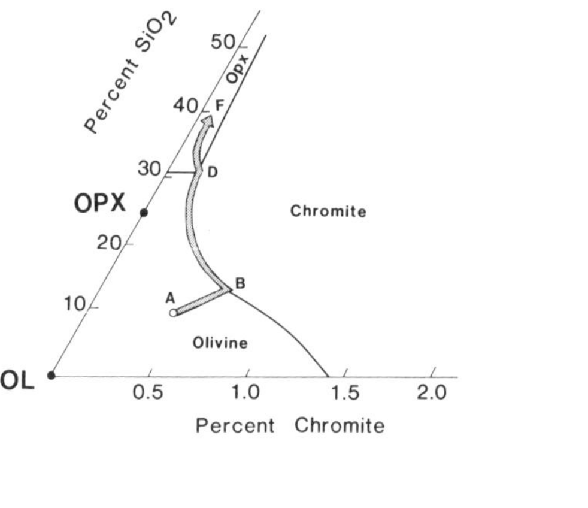

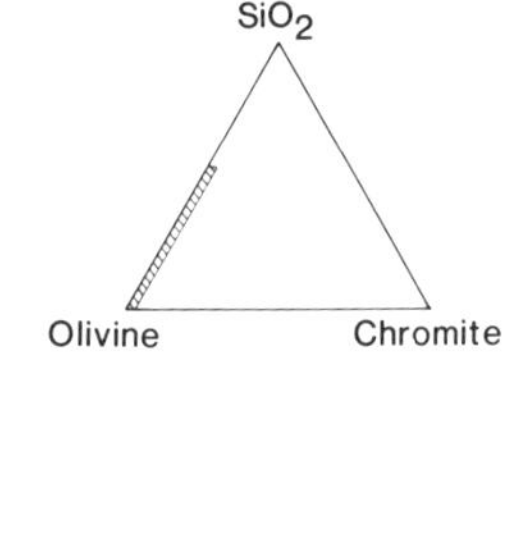

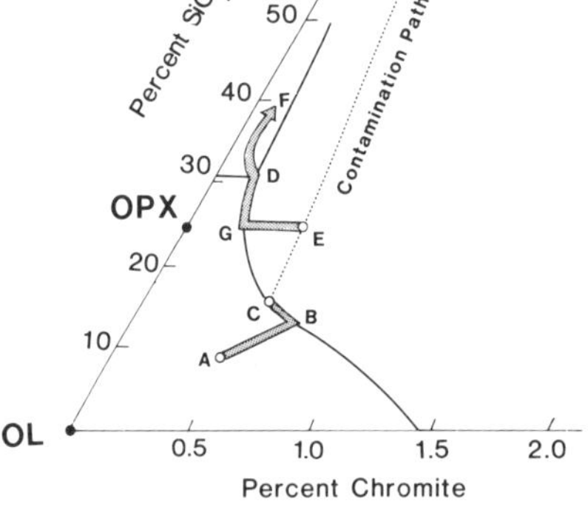

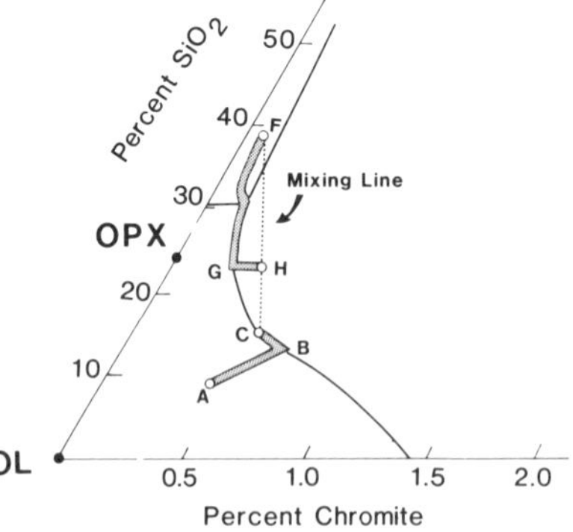

Figure 7 *Schematic phase diagram of the system olivine-chromite-silica. (After Irvine, 1975, 1977, with modifications). Point* ***D*** *is a distribution point and the normal crystallization path (heavy arrow) of a magma of initial composition* ***A*** *leads to the cumulate sequence olivine, olivine + chromite, orthopyroxene (upper left). Contamination of a liquid on the olivine-chromite cotectic (e.g., at* ***C****) with salic material results in the sequence olivine, olivine + chromite, chromite, olivine + chromite, orthopyroxene (lower left). A similar modification of the cumulate sequence arises if a liquid on the cotectic mixes with a more evolved liquid (e.g., at* ***F****). Note that the proportions are weight percentages and that the horizontal scale is greatly expanded.*

Irvine (1977) subsequently concluded that, although the salic contamination model was sound in principle, it would be very difficult to achieve the necessary amount and uniformity of contamination in nature. He proposed an alternative hypothesis, based on the same phase relationships, which he considered more likely to occur in nature. Accordingly, suppose a magma chamber contained a liquid that had an initial composition on, or close to, the olivine-chromite cotectic at C, and that the liquid had followed a normal fractionation path, with the result being that it evolved to composition F in the orthopyroxene field. If an influx of primitive magma of composition C then entered the chamber and mixed with the evolved liquid, the resulting mixture could lie within the liquidus field of chromite at H and bring about the precipitation of chromite alone. I consider this to be one of the more satisfying models of chromite genesis, since multiple injections of magma are very plausible in many intrusions, and differences in the proportions and compositions of the liquids that are mixed would allow a variety of crystallization sequences.

Podiform Deposits

Although many characteristics of podiform chromite deposits may be attributed to deformation, the common occurrence of cumulate textures in the ores suggests that they formed initially by magmatic segregation processes fundamentally similar to those that produced stratiform deposits. The analogy seems particularly appropriate for podiform deposits within the ultramafic cumulate rocks of ophiolite successions which have been generated by fractional crystallization of basaltic magma in chambers at the base of the oceanic crust. The precipitation of chromite from such magmas is subject to the same phase equilibria constraints as those outlined above for stratiform deposits, but the environment of crystal accumulation is undoubtedly rather different. It is unlikely that the chromitites once formed layers that were continuous over many kilometres as in the stratiform deposits which were subsequently broken up by tectonic activity. The discontinuous nature of the podiform deposits is regarded as a reflection of the magmatic environment. Greenbaum (1977), noting the occurrence of chromite in isolated deposits of limited lateral extent, suggested that the conditions necessary for the precipitation of massive chromitite (*i.e.*, a chromite cumulate rather than an olivine-chromite cumulate) prevailed only locally within the magma chamber. Thayer (1969) interpreted various structural and textural features of podiform ores as indicative of deformation at magmatic temperatures. Because ophiolitic magmatism occurs at active spreading centres, the magma chamber is subject to tectonic disturbance throughout its evolution, and it is likely that the partially consolidated cumulate pile would undergo slumping, flowage and other processes which would disrupt primary cumulate features.

Diverse explanations have been offered for the occurrence of chromite deposits within the ultramafic tectonite which represents the refractory residua of partial melting. It might be argued that the dunite and chromitite bodies within the residual harzburgite are simply pockets of more advanced partial melting. Dickey *et al.* (1971) found that chrome diopside melts incongruently to chromite + liquid, and Dickey and Yoder (1972) suggested that chromite formed in this way could be aggregated into pods by winnowing as magma percolates through the partially molten rock. However, such an origin is not in accord with the presence of cumulate textures. As noted above, the chromitites are typically enveloped by dunite, and are most abundant near the contact of the harzburgite with the overlying cumulates. Accordingly, it has been suggested that they are outliers of the cumulate succession that were emplaced in the tectonite either by gravitational sinking of dense autoliths (Dickey, 1975) or by infolding of the lowermost cumulate layers (Greenbaum, 1977). Thayer and Lipin (as quoted in Anonymous, 1979, p. 9) report that neither hypothesis is consistent with field relations or the composition of chromite in some ophiolites which they have studied, and conclude that the chromite deposits are indigenous to the harzburgite. Similarly, the systematic variation of chromite composition with depth below the petrologic Moho and spatial distribution of chromitite bodies documented by Brown (1980) in the Oman ophiolite are not consistent with an origin either involving introduction from the overlying cumulates or as a refractory residuum from partial melting. On balance, the evidence supports the view that the chromitites and associated dunites within the ultramafic tectonite are magma chambers in which olivine and chromite accumulated during fractional crystallization of ascending basaltic liquids. Neary and Brown (1979) described these as "mini-chambers", on the order of 0.5-1 km in extent, that are transitional between "grain boundary partial melting" below and the "well defined" magma chamber within the overlying oceanic crust. They clearly do not represent chambers where the ascending magmas came to rest and crystallized as a closed system: if this were the case, significant volumes of gabbroic rocks would be present. Rather, these small magma chambers were stopover points *en route* to the crust. Lago *et al.* (1982) have taken a somewhat similar view in suggesting that the precipitation and accumulation of chromite took place in small, steeply inclined cavities that were essentially widenings of the conduits through which the magmas ascended. However, it would not appear that this hypothesis is consistent with textures and cryptic variation which suggest accumulation of chromite in horizontal layers in many occurrences.

Conclusion

The chromite deposits described here are undoubtedly magmatic segregation deposits that have formed through the precipitation and accumulation of chromite from mafic or ultramafic magmas. Moreover, the geological settings in which chromite deposits occur are reasonably well defined, albeit in rather broad terms. Because geophysical and geochemical methods have enjoyed only limited success in chromite prospecting, rational exploration should depend heavily upon a good understanding of the origin of chromite ores. However, in my opinion, we do not yet understand the details of the ore-forming processes sufficiently well to allow formulation of a definitive genetic model. There is a critical need for theoretical and experimental studies of the factors that control the precipitation of chromite in a range of magma compositions. The many theories of the origin of layering in igneous rocks must be tested. Careful field and laboratory investigations of many known chromite deposits are needed. When such research has been carried out, we may be in a position to adopt a genetic model that will be of use in chromite exploration. In the meantime, exploration criteria will be empirical in nature and based, in large part, on analogy with known chromite deposits.

Acknowledgements
R.I. Thorpe and R.F. Emslie of the Geological Survey of Canada, and D.H. Watkinson of Carleton University provided constructive criticism of the manuscript. Brian Williamson assisted in the preparation of the diagrams. This review was undertaken as part of Geological Survey of Canada Project 800025.

References

Anonymous, 1979, Geological Survey Research 1979: United States Geological Survey Professional Paper 1150.

Brown, M., 1980, Textural and geochemical evidence for the origin of some chromite deposits in the Oman ophiolite, *in* A. Panayiotou, ed., Ophiolites—Proceedings of an International Ophiolite Symposium, Cyprus 1979: Geological Survey Department Cyprus, 781 p.

Bergath, K. and Weiser, Th., 1980, Primary features and genesis of Greek podiform chromite deposits, *in* A. Panayiotou, ed., Ophiolites—Proceedings of an International Ophiolite Symposium, Cyprus 1979: Geological Survey Department Cyprus, 781 p.

Buchanan, D.L., 1979, Chromite production from the Bushveld Complex: World Mining, v. 32, No. 10, p. 97-101.

Cameron, E.N., 1977, Chromite in the central sector of the Eastern Bushveld Complex, South Africa: American Mineralogist, v. 62, p. 1082-1096.

Cameron, E.N., 1978, The Lower Zone of the Eastern Bushveld Complex in the Oliphants River Trough: Journal of Petrology, v. 19, p. 437-462.

Cameron, E.N., 1980, Evolution of the Lower Critical Zone, Central Sector, Eastern Bushveld Complex, and its chromite deposits: Economic Geology, v. 75, p. 845-871.

Cameron, E.N. and Desborough, G.A., 1969, Occurrence and characteristic of chromite deposits - Eastern Bushveld Complex: Economic Geology, Monograph 4, p. 23-40.

Cameron, E.N. and Emerson, M.E., 1959, The origin of certain chromite deposits in the eastern part of the Bushveld Complex: Economic Geology, v. 54, p. 1151-1213.

Campbell, I.H., 1978, Some problems with cumulus theory: Lithos, v. 11, p. 311-323.

Cassard, D., Rabinovitch, M., Nicholas, A., Moutte, J., Leblanc M. and Prinzhofer, A., 1981, Structural classification of chromite pods in southern New Caledonia: Economic Geology, v. 76, p. 805-831.

Cotterill, P., 1969, The chromite deposits of Selukwe, Rhodesia: Economic Geology, Monograph 4, p. 154-186.

Cotterill, P., 1979, The Selukwe schist belt and its chromite deposits: Geological Society of South Africa, Special Publication 5, p. 229-245.

de Deus, P.B. and Viana, J., 1982, Jacurici Valley chromite district: International Symposium on Archean and Early Proterozoic Geologic Evolution and Metallogenesis, Abstracts and Excursions, p. 97-107.

de Deus, P.B., Viana, J., Duarte, P.M. and de Queiroz, W.J.A., 1982, Campo Formoso chromite district: International Symposium on Archean and Early Proterozoic Geologic Evolution and Metallogenesis, Abstracts and Excursions, p. 107-114.

DePaolo, D.J. and Wasserburg, G.J., 1979, Sm-Nd age of the Stillwater Complex and the mantle evolution curve for neodymium: Geochimica et Cosmochimica Acta, v. 43, p. 999-1008.

Dickey, J.S., Jr., 1975, A hypothesis of origin for podiform chromite deposits: Geochimica et Cosmochimica Acta, v. 39, p. 1061-1074.

Dickey, J.S., Jr. and Yoder, H.S., Jr., 1972, Partitioning of chromium and aluminium between clinopyroxene and spinel: Carnegie Institution of Washington, Yearbook, v. 71, p. 384-392.

Dickey, J.S., Jr., Yoder, H.S., Jr. and Schairer, J.F., 1971, Chromium in silicate-oxide systems: Carnegie Institution of Washington, Yearbook, v. 70, p. 118-122.

Ergunalp, F., 1980, Chromite mining and processing at Kavak mine, Turkey: Institute of Mining and Metallurgy, Transactions, v. 89, p. A179-A184.

Gait, R.J., 1964, The Mineralogy of Chrome Spinels of the Bird River Sill, Manitoba: Unpublished M.Sc. thesis, University of Manitoba, 64 p.

Greenbaum, D., 1977, The chromitiferous rocks of the Troodos ophiolite complex, Cyprus: Economic Geology, v. 72, p. 1175-1194.

Hamilton, J., 1977, Sr isotope and trace element studies of the Great Dyke and Bushveld mafic phase and their relation to early Proterozoic magma genesis in Southern Africa: Journal of Petrology, v. 18, p. 24-52.

Hedlund, D.C., Moriera, J., Pinto, A., da Silva J. and Souza, G., 1974, Stratiform chromitite at Campo Formoso, Bahia, Brazil: United States Geological Survey, Journal of Research, v. 2, p. 551-562.

Hill, R. and Roeder, P., 1974, The crystallization of spinel from basaltic liquid as a function of oxygen fugacity: Journal of Geology, v. 82, p. 709-730.

Irvine, T.N., 1967, Chromian spinel as a petrogenetic indicator, Part 2. Petrologic applications: Canadian Journal of Earth Sciences, v. 4, p. 71-103.

Irvine, T.N., 1970, Crystallization sequences in the Muskox intrusion and other layered intrusions. I. Olivine-pyroxene-plagioclase relations: Geological Society of South Africa, Special Publication 1, p. 442-476.

Irvine, T.N., 1975, Crystallization sequences in the Muskox intrusion and other layered intrusions. II. Origin of chromitite layers and similar deposits of other magmatic ores: Geochemica et Cosmochemica Acta, v. 39, p. 991-1020.

Irvine, T.N., 1977, Origin of chromitite layers in the Muskoka intrusion and other stratiform intrusions: A new interpretation: Geology, v. 5, p. 273-277.

Irvine, T.N. and Smith, C.H., 1969, Primary oxide minerals in the layered series of the Muskox intrusion: Economic Geology, Monograph 4, p. 76-94.

Jackson, E.D., 1961, Primary Textures and Mineral Associations in the Ultramafic Zone of the Stillwater Complex, Montana: United States Geological Survey, Professional Paper 358, 106 p.

Jackson, E.D., 1968, The chromite deposits of the Stillwater Complex, Montana, *in* Ridge, J.D., ed., Ore Deposits of the United States, v. 2: American Institute of Mining Engineers, 1880 p.

Jackson, E.D., 1969, Chemical variation in coexisting chromite and olivine in chromite zones of the Stillwater Complex: Economic Geology, Monograph 4, p. 41-71.

Johan, Z., Le Bel, L., Robert, J.L. and Volfinger, M., 1982, Role of reducing fluids in the origin of chromite deposits from ophiolitic complexes: Geological Association of Canada—Mineralogical Association of Canada, Program with Abstracts, v. 7, p. 58.

Jones, W.R., Peoples, J.W. and Howland, A.L., 1960, Igneous and tectonic structures of the Stillwater Complex, Montana: United States Geological Survey, Bulletin 1071-H, p. 281-340.

Kingston, G.A., Miller, R.A. and Carillo, F.V., 1970, Availability of U.S. Chromium Resources: United States Bureau of Mines, Information Circular 8465, 23 p.

Kujanpää, J., 1980, Geology of the Kemi chromite deposits, *in* Hakli, T.A., ed., Precambrian Ores of Finland: Geological Survey of Finland, 48 p.

Lago, B.L., Rabinowicz, M. and Nicholas, A., 1982, Podiform chromite ore bodies: a genetic model: Journal of Petrology, v. 23, p. 103-125.

McBirney, A.R. and Noyes, R.M., 1979, Crystallization and layering of the Skaergaard intrusion: Journal of Petrology, v. 20, p. 487-554.

Neary, C.R. and Brown, M.A., 1979, Chromites from the Al'Ays Complex, Saudi Arabia and the Semail Complex, Oman: Institute of Applied Geology, Kingdom of Saudi Arabia, Bulletin 3, p. 193-205.

Osborn, E.F., 1978, Change in phase relations in response to change in presssure from 1 atm to 10 kbar for the system Mg_2SiO_2-iron oxide-$CaAl_2Si_2O_8$-SiO_2: Carnegie Institution of Washington, Yearkbook, v. 77, p. 784-790.

Page, N.J., 1977, Stillwater Complex, Montana: Rock Succession, Metamorphism, and Structure of the Complex and Adjacent Rocks: United States Geological Survey, Professional Paper 999, 79 p.

Piirainen, T., 1978, General geology and metallogenic features of Finland: Metallogeny of the Baltic Shield Symposium, Helsinki 1978, Excursion Guide, p. 1-19.

Piirainen, T., Hugg, R., Isohanni, M. and Juoperri, A., 1974, On the geotectonics and ore forming processes in the basic intrusive belts of Kemi-Suhanko, and Syote-Naranka-Vaara, Northern Finland: Geological Society of Finland, Bulletin, v. 46, p. 93-104.

Rice, A., 1981, Convective fractionation: a mechanism to provide cryptic zoning macrosegregation, layering, crescumulates, banded tuffs, and explosive volcanism in igneous processes: Journal of Geophysical Research, v. 86, p. 405-417.

Thayer, T.P., 1964, Geologic features of podiform chromite deposits, *in* Woodtli, R., ed., Methods of Prospection for Chromite: Organization for Economic Cooperation and Development, Paris, 244 p.

Thayer, T.P., 1969, Gravity differentiation and magmatic re-emplacement of podiform chromite deposits: Economic Geology, Monograph 4, p. 132-146.

Trueman, D.L., 1971, Petrological, Structural, and Magnetic Studies of a Layered Basic Intrusion, Bird River Sill, Manitoba: Unpublished M.Sc. thesis, University of Manitoba, 67 p.

Trueman, D.L., 1980, Stratigraphy, Structure and Metamorphic Petrology of the Archean Greenstone Belt at Bird River, Manitoba: Unpublished Ph.D. thesis, University of Manitoba, 151 p.

Ulmer, G.C., 1969, Experimental investigations of chromite spinels: Economic Geology, Monograph 4, p. 114-131.

Vermaak, C.F. and von Gruenewaldt, G., 1981, The Bushveld Complex Excursion Guide: Geocongress '81, 3rd International Platinum Symposium, 62 p.

von Gruenewaldt, G., 1979, A review of some recent concepts of the Bushveld Complex, with particular reference to sulfide mineralization: Canadian Mineralogist, v. 17, p. 233-256.

Wilson, A.H., 1982, The geology of the "Great Dyke", Zimbabwe: the ultramafic rocks: Journal of Petrology, v. 23, p. 240-292.

Worst, B.G., 1958, The differentiation and structure of the Great Dyke of Southern Rhodesia: Geological Society of South Africa, Transactions and Proceedings, v. 61, p. 283-358.

MS received 3 December 1982
Originally published in
Geoscience Canada v. 10 Number 1 (March 1983)

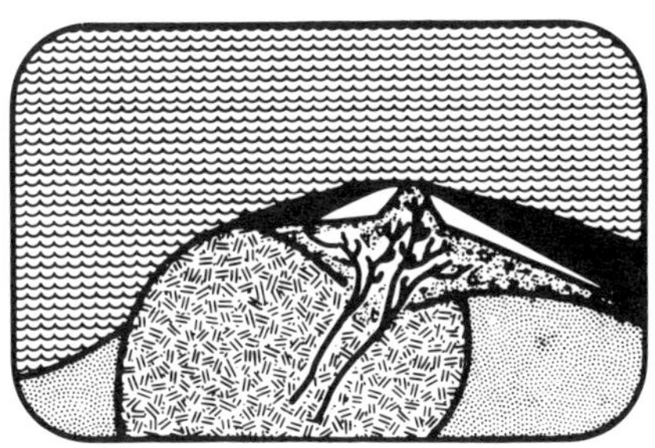

Volcanogenic Massive Sulphide Deposits Part 1: A Descriptive Model

John W. Lydon
Geological Survey of Canada
601 Booth Street
Ottawa, Ontario K1A 0E8

Introduction

Viewed from both the economic and scientific perspectives, volcanogenic massive sulphide deposits occupy a unique position of importance among mineral deposit types. Economically, deposits of this type are a major source of copper, zinc, lead, silver and gold, and a range of by-products including tin, cadmium, antimony and bismuth. For example, of the total production from Canadian mines during the period 1977-78, about 30% of the Cu, 63% of the Zn, 27% of the Pb, 58% of the Ag and 8% of the Au was obtained from volcanogenic massive sulphide deposits. From the scientific point of view, problems related to the characterization and genesis of these deposits have attracted the attention of geoscientists more than those of any other deposit type, with the result that in the last twenty-five years more than five thousand articles directly related to volcanogenic massive sulphide deposits have been published. Over the last few years, the flames of scientific, and to some extent economic and political, interest have been further fanned by the discovery of high temperature (350°C) hydrothermal vents on spreading ridges of the eastern Pacific Ocean that are actively precipitating metal sulphides with many similarities to volcanogenic massive sulphide deposits.

With such a history of scientific interest in the subject, even this relatively brief and generalized account of volcanogenic massive sulphide deposits must to some extent constitute a review of the literature, but at the same time, because of its brevity, must be highly selective in its content.

An ore deposit model can be considered to consist of two components. One of these is a descriptive model which embodies those features of the geological settings, morphology, chemistry, mineralogy, zoning, *etc.*, judged to be characteristic of the deposit type. The descriptive model therefore can be considered to be an idealized example of the deposit type scaled down in degree of detail and complexity from the actual examples on which it is based. The other is a genetic model, which attempts to give a rational and consistent explanation of the characteristics of the deposit type in terms of known or postulated geological processes.

This article (Part 1) deals only with a descriptive model for volcanogenic massive sulphide deposits. The next article (Part 2) will discuss aspects of genetic models for this ore deposit type. In both articles, emphasis is on generalization and the use of Canadian examples. For more detailed and extensive reviews of the descriptive and genetic models, the reader is referred to Franklin *et al.* (1981); Ohmoto and Skinner (1983); Klau and Large (1980); Finlow-Bates (1980); Sangster and Scott (1976); Solomon (1976); Gilmour (1976); Lambert and Sato (1974); Hutchinson (1973); and Sangster (1972).

Terminology

Volcanogenic massive sulphide deposits belong to the larger class of concordant massive sulphide deposits that includes all massive or semi-massive sulphide deposits formed by the discharge of hydrothermal solutions into the seafloor. Although a complete spectrum of types is represented (Gilmour, 1976), the great majority of individual concordant massive sulphide deposits are readily classified into one of two major groups defined by the chemical, mineralogical, morphological, textural, grade and tonnage characteristics of the deposit itself. However, in naming these major groups, emphasis has been placed on the most common lithologies of the host rocks. Over one part of the spectrum are the *sedimentary-exhalative, sediment-hosted,* or *shale-hosted* stratiform massive sulphides, which include such famous deposits as Sullivan, Broken Hill, Mt. Isa and Rammelsberg. Over the other part of the spectrum are the *volcanogenic, volcanic-associated, volcanic-hosted,* or *volcanophile massive sulphides*, which are the subject of this article. Many authors correctly object to the term *volcanogenic massive sulphides* because it implies that the deposits themselves are an integral part of the volcanic process, which does not appear to be the case. Rather, they seem to be the product of a specialized type of hydrothermal system that is only occasionally developed in a submarine volcanic environment. However, the term *volcanogenic massive sulphide* has such wide usage, it is debatable whether it should be dropped. For convenience, the acronym VMS will be used here.

Geological Setting and Distribution

VMS deposits typically, if not exclusively, occur within geological domains which can be defined by the presence of submarine volcanic rocks. Although the immediate host rocks to the deposits are most commonly of direct volcanic origin, such as lava or pyroclastic rocks, or of indirect volcanic origin, such as volcaniclastic rocks, other sedimentary marine lithologies with no volcanic affiliation, such as shales or greywackes, are by no means rare.

There does not appear to be any preferred geotectonic environment for VMS deposits except that, like the submarine volcanic rocks themselves, they are more commonly formed near plate margins (Sillitoe, 1973; Sawkins, 1976). Thus VMS deposits are found at divergent plate margins (ophiolite-associated deposits) which may reflect mid-ocean ridges or spreading back arc basins (*e.g.*, deposits of Cyprus and Baie Verte area, Newfoundland); at convergent plate margins in island arcs or continental margins (*e.g.*, Kuroko deposits of Japan and Spanish-Portuguese Pyrite Belt); associated with intra-plate oceanic islands (Aggarwal and Nesbitt, 1984); and, of course, in more enigmatic plate tectonic environments such as those represented by Archean greenstone belts.

It is also evident from the above that, since plate tectonic environments are most commonly diagnosed by the petrochemistry of the associated igneous rocks, VMS deposits are not confined to any particular petrochemical type of volcanic rock (Klau and Large, 1980). It has been suggested that there is a preferential association of VMS deposits with the most differentiated phases of a calc-alkaline magma (*e.g.*, Sangster and Scott, 1976; Solomon, 1976). In some cases, the importance of calc-alkaline host rocks may have been exaggerated due to confusion over the distinction between real calc-alkaline trends and "pseudo-calc-alkaline" trends superimposed on tholeiitic rocks by the laterally extensive hydrothermal alteration commonly associated with areas of VMS deposits (MacGeehan and MacLean, 1980). In order to scientifically test whether or not VMS deposits are preferentially related to a specific petrochemical type, their spatial distribution should be normalized to number of occurrences per unit area of outcrop of each petrochemical type of volcanic rock. To the author's knowledge, this has never been done.

There also does not appear to be any preferred time distribution for VMS deposits, which range in age from about 3500 Ma in the Pilbara Block of Australia to the modern sulphide deposits of the East Pacific Rise. Hutchinson (1973) pointed out that there are definite age span groupings for VMS deposits which, as Sangster (1980a) observed for Precambrian deposits of North America, correspond to periods of deposition of thick, supracrustal accumulations, and should not be considered as marking unique metallogenic phenomena. Therefore, again, it is more than likely that these age-frequency peaks would disappear, if the number of deposits were normalized to area of volcanic outcrop of given age ranges.

However, there is no doubt that within submarine volcanic domains of the same age and petrochemical type, there is a strongly preferred spatial distribution of VMS deposits. For example, 83 economic VMS deposits are known in 2650-2730 Ma volcanic belts of the Canadian Shield, but only two are known in compositionally similar volcanic belts of the same age in Australia (Franklin *et al.*, 1981). Similarly, on a smaller scale within a single volcanic belt or domain, VMS deposits are not evenly distributed. For example, out of the nine essentially similar volcanic complexes identified by Goodwin and Ridler (1970) in the Abitibi Belt, only four contain significant VMS deposits. Characteristically, within most volcanic domains VMS deposits tend to occur in spatial groups or clusters, separated from one another by lithologically similar rocks that may contain only a few, isolated, small VMS deposits. Sangster (1980b) calculated that the average area occupied by a typical cluster was about 850 square kilometres, equivalent to a circular area of about 32 km in diameter, and that it contained an average of 12 deposits and 94 million tonnes of ore.

Within each cluster, most of the deposits tend to occur within a single stratigraphic interval, which occupies only a fraction of the total stratigraphic interval occupied by the host volcanic edifice as a whole. This most productive stratigraphic interval is often referred to as the *favourable horizon*, and it is particularly evident in the Noranda (Spence and de Rosen-Spence, 1975), Matagami (Roberts and Reardon, 1978; MacGeehan, 1978) and Bathurst (Davies, 1980) areas of Canada and the Green Tuff Belt of Japan (Lambert and Sato, 1974). Figure 1 illustrates the case in the Noranda area, where it can be seen that a large proportion of the deposits occur close to a single stratigraphic horizon. This concentration of VMS deposits in such a small stratigraphic interval is even more remarkable if one takes into account that below the stratigraphic interval illustrated in Figure 1, there are more than 16,000 m^3 of both felsic and mafic volcanic rocks that contain no evidence of VMS deposits (Franklin *et al.*, 1981).

Within the stratigraphic confines of the favourable horizon and the lateral confines of a cluster, the localization of individual VMS lenses seems to be strongly related to structural controls in the substrate and positive or negative topographical features of the ocean floor (see below). Sangster (1972) and Scott (1978) noted that the distribution of deposits in the Noranda and Hokuroku mining camps, respectively, appeared to be controlled by sets of linear fractures. Solomon (1976) deduced that 50% of VMS deposits are spatially associated with felsic volcanic rocks, the VMS deposits themselves show a propensity of association with rhyolite domes or felsic fragmental rocks. Knuckey (1975) suggested that many of the individual ore lenses of the Millenbach Mine, Noranda, as well as the rhyolite domes and their feeder dykes, with which some, but not all, ore lenses are spatially related,

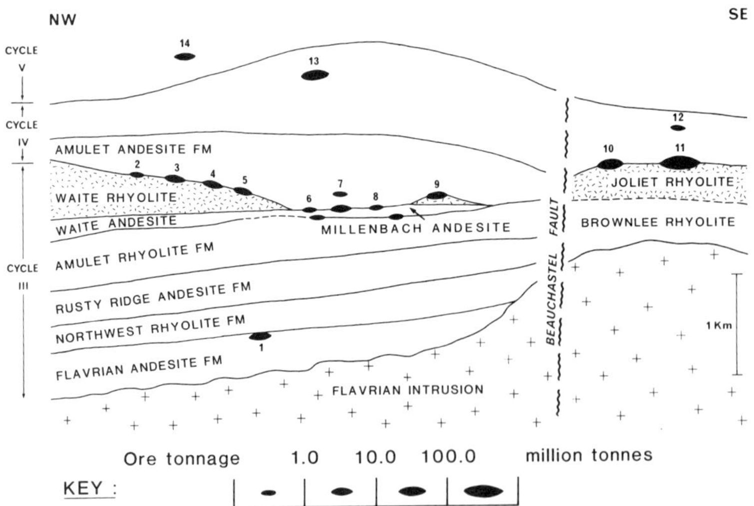

Figure 1 *A schematic composite section of the Noranda massive sulphide district, showing stratigraphic relationships in the upper part of the Blake River Group and the stratigraphic positions of major VMS deposits. (After Spence and de Rosen-Spence, 1975; Knuckey* et al., *1982; Knuckey and Watkins, 1982). Cycles refer to the andesite-rhyolite volcanic cycles of Spence and de Rosen-Spence (1975).*
Key to deposits: ***1*** *- Corbet;* ***2*** *- Vauze;* ***3*** *- Norbec;* ***4*** *- E. Waite;* ***5*** *- O. Waite;* ***6*** *- Amulet C and F;* ***7*** *- Amulet Upper A and Lower A;* ***8*** *- L. Dufault No. 1;* ***9*** *- Millenbach;* ***10*** *- Quemont;* ***11*** *- Horne;* ***12*** *- Delbridge;* ***13*** *- W. MacDonald (Mine Gallen);* ***14*** *- Mobrun*

are associated with synvolcanic faults with vertical displacements. In a more general perspective, Hodgson and Lydon (1977) proposed that many VMS deposits are associated with the fracture systems produced by resurgent calderas or subvolcanic intrusions.

Cumulatively, the evidence cited above does not suggest a simple relationship between VMS deposits and submarine volcanism *per se*. If VMS deposits were simply a direct and integral product of submarine volcanism, then possibly a more or less uniform stratigraphic distribution of the deposits, combined with a preferred spatial association with volcanic vents, could be expected for every submarine volcanic edifice of a specific petrochemical type. Instead, there is obviously a very selective distribution of VMS deposits which, regardless of the petrochemical magma type, occur only in a minority of submarine volcanic edifices. Furthermore, within these mineralized edifices VMS deposits tend to be relatively common. In contrast with the lack of distinct control by petrochemical type of volcanism, the spatial relationship of the deposits to synvolcanic faults, rhyolite domes or topographic depressions, caldera rims or subvolcanic intrusions suggests that the deposits are closely related to particular hydrologic, topographic and geothermal features of the ocean floor that only infrequently combine to give the specific configuration that is necessary to form VMS deposits.

Architecture of VMS Deposits

The idealized architecture of a VMS deposit is depicted in Figure 2. The typical deposit consists of a concordant lens of massive sulphide, composed of 60% or more sulphide minerals (Sangster and Scott, 1976), that is stratigraphically underlain by a discordant stockwork or stringer zone of vein-type sulphide mineralization contained in a pipe of hydrothermally altered rock. The upper contact of the massive sulphide lens with hanging wall rocks is usually extremely sharp, but the lower contact is usually gradational into the stringer zone. A single deposit or mine may consist of several individual massive sulphide lenses and their underlying stockwork zones. The conventional interpretation is that the stockwork zone represents the near-surface channelways of a submarine hydrothermal system and the massive sulphide lens represents the accumulation of sulphides precipitated from the hydrothermal solutions on the sea floor above and around the discharge vent. The characteristics that are illustrated in Figure 2 represent the simplest configuration, in which no complications, such as syndepositional slumping of the massive ore away from its stockwork zone, are taken into consideration.

The morphology of a single massive lens ranges from that of a steep-sided cone to that of a tabular sheet. The

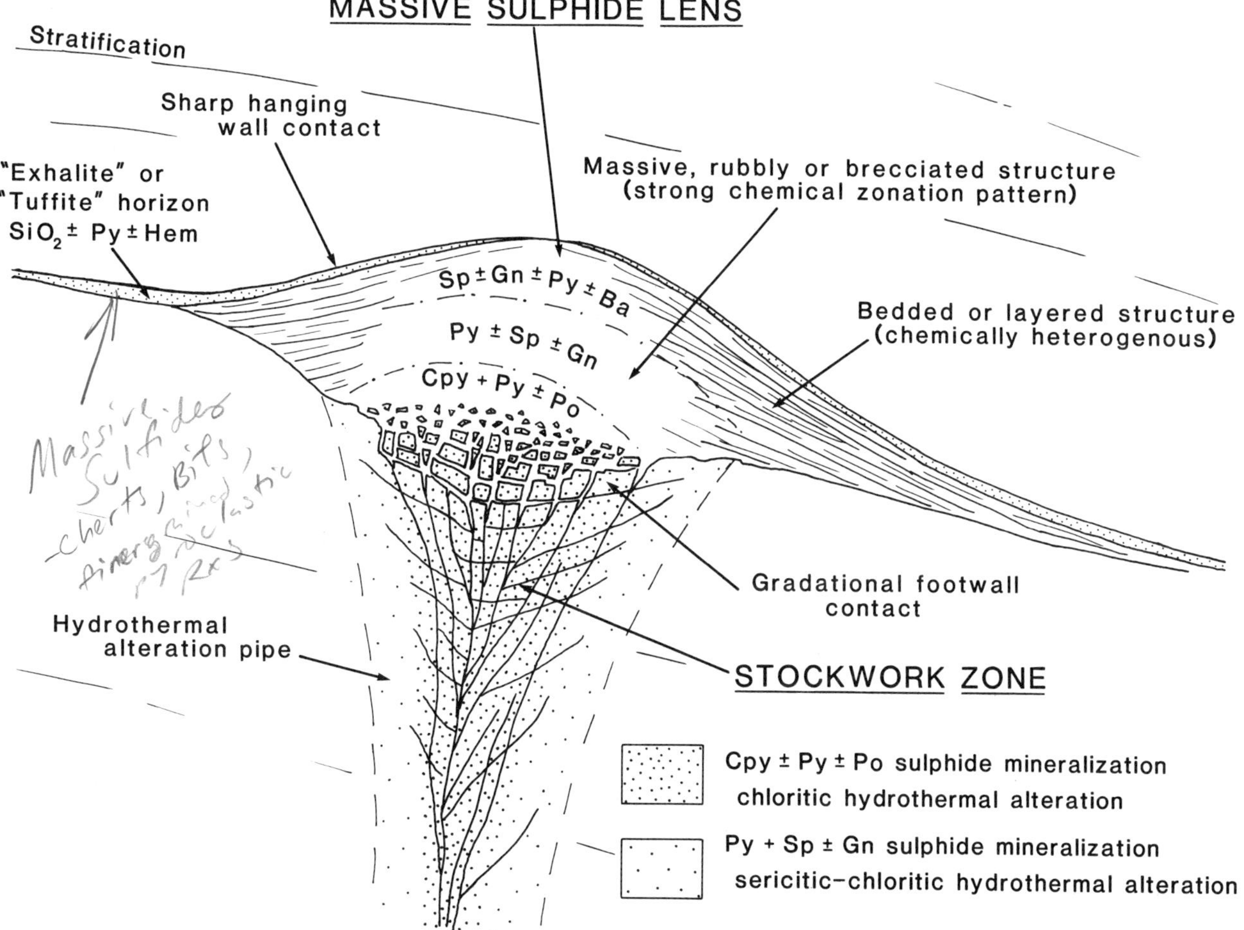

Figure 2 *Essential characteristics of an idealized volcanogenic massive sulphide deposit.*

majority of cone-shaped deposits appear to have accumulated on the top or flanks of a positive topographic feature, such as a rhyolite dome in the case of the Millenbach deposit, whereas the majority of sheet-like deposits, such as the Brunswick No. 12 deposit, appear to have accumulated in topographic depressions. Most Canadian deposits have undergone penetrative deformation, so that in extreme cases, a deposit may have been stretched into a pencil-shaped body, with the stockwork zone transposed to a position of apparent lateral conformity with the massive sulphide lens.

By far the most common sulphide mineral in the massive sulphide lens is pyrite. Pyrrhotite, chalcopyrite, sphalerite, galena and, more rarely, sulphosalts and bornite comprise the subordinate sulphide species. The most common non-sulphide metallic minerals include magnetite, hematite and cassiterite. Of the gangue or associated minerals that may occur as co-precipitates with the sulphides are quartz, chlorite, barite, gypsum and carbonates.

Textures and structures of the ores are very variable. In the least metamorphosed deposits, the massive ore is usually a fine-grained mosaic of sulphide grains which increase in coarseness with increasing metamorphic grade. In the cone-shaped deposits, massive, rubbly or brecciated textures tend to predominate in the central part of the lens, whereas silt- to boulder-sized sulphide fragments form an apron of clastic sulphide rock that often exhibits spectacular sedimentary structures around the periphery of the lens. Regularly layered or laminated sulphides are more typical of the sheeted deposits. Textures and structures of the most metamorphosed and deformed massive sulphide lenses are more aptly described as gneissose.

Perhaps the most diagnostic features of VMS deposits are the pronounced zonations of chemistry, mineralogy and textures of the ores and the metasomatic changes to the host rock within the hydrothermal alteration pipe. The most obvious and consistent of these zonation patterns is the systematic decrease in chalcopyrite/(sphalerite + galena) ratios, or more simply and conveniently, the Cu/Zn ratio, upward and outward from the core of the alteration pipe and the base of the massive sulphide lens (Figure 2). Of the other metallic minerals, pyrrhotite, magnetite and bornite (if present) tend to be concentrated in the core of the stockwork zone and central basal part of the massive sulphide lens, corresponding to the zone of highest Cu/Zn ratios. Barite, when present, generally occurs with the greatest sphalerite and galena concentrations in the outermost zone of the massive sulphide lens. Pyrite, though generally ubiquitous throughout the sulphide zonation pattern, tends to achieve its maximum modal relative proportion where sphalerite becomes predominant over chalcopyrite. In many cases, a thin, bedded, pyritic or hematitic, siliceous *exhalite* or *tuffite* horizon forms a veneer over the top of the sulphide mound and extends as a stratigraphic marker laterally away from the deposit. This sedimentary horizon is thought to largely represent chemical precipitation from the waning stages of hydrothermal activity during volcanic quiescence. In some deposits, there is a spatial association of magnetite-hematite iron formation (*e.g.*, Bathurst, New Brunswick area) or manganese oxide formation (*e.g.*, Iberian pyrite belt), with clusters of VMS deposits. These widespread metal oxide sediments usually occur somewhat stratigraphically higher than the favourable horizon, but the exact genetic relationship between the sulphide and oxide deposits is not clearly understood.

Hydrothermal Alteration of Host Rocks

Within and surrounding the stockwork zone, there is generally a pronounced zonation in the intensity and type of metasomatism produced by the hydrothermal alteration of the host rocks, which to some extent corresponds to the abundance of the sulphide veining. The most detailed documentation of metasomatic effects has been recorded for deposits of the Abitibi Belt. At the Millenbach (Riverin and Hodgson, 1980; Knuckey *et al.*, 1982) and Corbet (Knuckey and Watkins, 1982) deposits of the Noranda area, the alteration pipes consist of inner chloritized cores surrounded by sericitized peripheries. The chloritic core is characterized by major additions of iron and magnesium and by depletions of calcium, sodium and silicon, reflecting the destruction of the feldspar component of the original felsic or mafic volcanic rock during the process of chloritization. Potassium tends to be depleted in the chloritic zone but is enriched in the surrounding sericitic zone. Otherwise, the metasomatic changes of the sericitic zone are less intense gradational continuations of those of the chloritic core zone. This gradational change, reflecting mainly a change in intensity of metasomatism from the "fresh" footwall rocks inward to the core of the alteration pipe, gives the impression that the complete zonation pattern is due to a single metasomatic gradient imposed by the ore-forming solution. In this connection, it is interesting to note that Riverin and Hodgson (1980) pointed out that the same metasomatic zonation pattern occurs in the alteration selvages of individual sulphide veins in the stockwork zone of the Millenbach deposit. An extreme form of this metasomatic progression may be the talc-actinolite assemblage described by Roberts and Reardon (1978) at the Mattagami Lake deposit, which these authors suggest was formed as a result of the removal of a major amount of Al by the hydrothermal solutions.

A similar configuration of a sericitic halo to a chloritic core is reported for other deposits in the Abitibi Belt (Lickus, 1965; Sakrison, 1966; Spitz and Darling, 1973, 1975). Elsewhere in the world, chloritization of the central part of the hydrothermal alteration pipe is the most common form of alteration, and for many deposits this core is surrounded by an outer zone of potassium enrichment, though the main potassic mineral is not necessarily sericite (*e.g.*, Walford and Franklin, 1982; Rui, 1973; Lydon, 1984). In deposits that have been metamorphosed above the stability of chlorite, the major magnesium and iron addition to the alteration pipe is reflected by a cordierite-anthophyllite assemblage in the core zone, as in the Flin Flon-Snow Lake (Froese, 1969; Whitmore, 1969; Walford and Franklin, 1982) and Manitouwadge (Pye, 1960; James *et al.*, 1978) areas of Canada and the Kvikne mines of Norway (Morton, 1972).

A strong silica metasomatism is characteristic of the upper part of the alteration pipe of some deposits, particularly the Kuroko deposits of Japan (Shirozo, 1974; Urabe *et al.*, 1983) and the VMS deposits of Cyprus (Lydon, 1984). Silica and chlorite also form the predominant non-sulphide fraction of the massive sulphide lens, and occur as cross-cutting veins, as a matrix cement to sulphide grains or as distinct lithological lenses. This silica and chlorite, and in some cases even talc (Costa *et al.*, 1983; Aggarwal and Nesbitt, 1984), represent hydrothermal minerals whose components were supplied directly by the hydrothermal solutions. In considering metasomatic processes, these hydrothermal precipitates must be distinguished from similar minerals whose modal increase is due to the redistribution of components of the host rocks during hydrothermal alteration (*e.g.*, the release of quartz during chloritization of a plagioclase feldspar).

Generalizations about wall-rock alteration beyond this point do not seem warranted at this stage of scientific accomplishment, as the documentation and understanding of wall-rock alteration associated with VMS deposits is one of the areas that requires much more research. There are many documented examples in which the alteration mineralogy of the stockwork alteration pipe clearly departs from the "normal" chloritic core and sericitic periphery. At the Mattabi Mine, for instance, the main alteration minerals of the feeder pipe are siderite, chloritoid and andalusite, with chlorite and sericite occuring only as irregularly distributed pods (Franklin *et al.*, 1975), but in this case, the footwall succession itself is unusual in that it consists of a thick sequence of epiclastic rocks containing 10-15% dolomite.

A relatively recent development that has important consequences for mineral exploration is the increasing recognition that in several mining areas a laterally widespread zone of rock alteration occurs stratigraphically below the favourable horizon. In the Mattabi-Sturgeon Lake area, the footwall epiclastic horizon appears to have been depleted in sodium over an outcrop area some 8 km long and 1 km wide (Franklin *et al.*, 1975). In the Snow Lake area, a conformable alteration zone at least 2 km long and several hundreds of metres thick is marked by a staurolite- and chlorite-bearing rhyolite, reflecting a regional sodium depletion and magnesium and/or iron enrichment (Walford and Franklin, 1983). In the Mattagami Lake area, MacGeehan (1978) and MacGeehan and MacLean (1980) concluded that all but the uppermost part of the footwall basaltic sequence had undergone a regional alteration involving addition of silica and sodium, and depletion of iron, magnesium, calcium, titanium, zinc and copper, resulting in secondary albite and epidote-quartz mineral assemblages. These authors noted that the elements mobilized by this regional alteration were exactly those that were enriched in the stockwork alteration zone of the Garon Lake deposit. In the Noranda area, Gibson *et al.* (1983) found that the Amulet Rhyolite Formation, which occurs immediately below the most favourable ore horizon, in large part actually consists of mafic flows and flow breccia that have undergone regional alteration consisting of intense silicification in its upper part and epidote-quartz and chlorite-sericite alteration at progressively lower horizons. Gibson (1979) noted that the alteration involved the mobilization of aluminum, iron, magnesium, titanium and zinc, which are those elements that have been enriched in the alteration pipes of the Millenbach deposit.

This continuity of the stockwork alteration pipe into zones of regional alteration serves to complicate the definition and understanding of different metasomatic events. In the Kuroko deposits, Shirozo (1974) and Ijima (1974) distinguished only a single quartz-sericite-chlorite alteration zone immediately underlying the deposits and enclosing the stockwork zone, but described surrounding concentric alteration zones, defined on clay and zeolite assemblages, that continue into the hanging walls of the deposits. Hanging wall alteration has been described for other deposits. With the present state of knowledge, it is not possible to make a generalized statement as to whether most of this hanging wall alteration is due to the continuation of the same ore-forming hydrothermal activity after deposition of the hanging wall rocks to the main sulphide body, as in the case of the Corbet deposit (Knuckey and Watkins, 1982), or whether it is due to anomalous effects of normal groundwater circulation in the vicinity of the geochemically and mineralogically anomalous zones of an ore deposit during its long post-depositional burial history, as in the case of some deposits in Cyprus (Lydon, 1984).

Classification

Most proposals for the classification of VMS deposits have placed emphasis on the geotectonic setting or immediate footwall lithology of the deposits, rather than on characteristics of the deposits themselves. Sillitoe (1973) distinguished those deposits that formed at spreading centres and usually had high Cu/Zn ratios from those formed in island arc or continental margin settings that usually had relatively elevated concentrations of Pb, Zn, Ag and Ba. Sawkins (1976) recognized three main types of VMS deposits: (1) Kuroko-type, occurring in felsic, calc-alkaline volcanic sequences of Archean to Tertiary age at sites of plate convergence in ocean areas; (2) Cyprus-type, occurring in low-potassium basaltic volcanic rocks in the upper part of ophiolite complexes at sites of plate spreading; and (3) Besshi-type, occurring in clastic sediments and mafic volcanics but with no clear-cut plate tectonic setting. However, in general, plate tectonic theory does not account for many of the genetic aspects of mineral deposits (Sangster, 1979).

Klau and Large (1980) used host rock lithologies as a basis for classifying VMS deposits into deposits associated with: (1) felsic volcanic rocks in Archean greenstone belts; (2) post-Archean calc-alkaline and tholeiitic volcanic sequences; and (3) mafic volcanic rocks. They noted that deposits of similar chemical characteristics overlapped these three divisions. Hutchinson (1973) proposed a three-fold classification of VMS deposits based on their major ore element chemistry: (1) Zn-Cu-types occurring in fully differentiated magmatic suites of tholeiitic and calc-alkaline affinities and predominantly of Archean age; (2) Pb-Zn-Cu-types in intermediate to felsic calc-alkaline volcanic rocks and predominantly of Phanerozoic age; and (3) Cu-types in poorly differentiated ophiolitic or tholeiitic suites of Phanerozoic age. Solomon (1976) used a very similar classification of (1) Zn-Pb-Cu-types; (2) Zn-Cu-types; and (3) Cu types, the terminology reflecting the order of relative abundance of the major ore metals in the deposits.

Classification of VMS deposits according to their major ore-element chemistry seems to be the most satisfactory method. Using VMS deposits of the four major metallogenic provinces of the Abitibi Belt, Bathurst-Newcastle area, New Brunswick, Norwegian Caledonides and the Green Tuff Belt, Japan (Table 1) as a representative sampling of the VMS class as a whole, Figures 3 and 4 show that two major groups can be distinguished on the basis of bulk Zn/(Zn + Pb) ratios of the deposits. These two groups are termed the Zn-Pb-Cu type and Cu-Zn type, respectively, the terminology reflecting the major ore metal associations. These two groups are not readily obvious from a visual inspection of a Cu-Zn-Pb ternary diagram showing only the distribution of bulk compositions of individual deposits (Figure 3a), but are easily distinguished if each point is weighted in terms of the tonnes of ore metal contained in the deposit and the points contoured for tonnes of contained ore metal per unit area of the diagram (Figure 3b). The same is true for histograms showing the frequency-distribution of Zn/(Zn + Pb) ratios of individual deposits (Figure 4a) compared with a similar diagram showing the amount of contained metal per Zn/(Zn + Pb) ratio division (Figure 4b). These diagrams show that, by far, most of the ore is contained in VMS deposits that have a bulk Zn/(Zn + Pb) ratio of either between 0.70 and 0.80 (the Zn-Pb-Cu type) or greater than 0.95 (the Cu-Zn type). Deposits outside of these limits contain only a small proportion of the total ore, and hence can be considered to be unusual. An apparent merging of the two groups is to be expected for several reasons, not the least being that because of the strong ore metal zonation characteristic of the deposit type, incomplete preservation, or only partial economic viability of the original complete geological mineral deposit, would bias the metal ratios of the economic ore deposit that have been used in constructing the diagrams.

This classification of VMS deposits into just two major compositional groups is thus similar to that proposed by Hutchinson (1973) and Solomon (1976), except that a Cu-type is not recognized and no affiliation to any particular geotectonic setting, volcanic petrochemical

Table 1 Average grade and tonnage data for volcanogenic massive sulphide deposits of selected areas.

Area	Dominant Deposit Type	Number of Deposits	Average Grade and Tonnage						
			Cu (%)	Zn (%)	Pb (%)	(*)	Ag (g/t)	Au (g/t)	million tonnes
Abitibi Belt, Canada	Cu-Zn	52	1.47	3.43	0.07	(47)	31.9	0.8	9.2
Norwegian Caledonides	Cu-Zn	38	1.41	1.53	0.05	(0)	na	na	3.5
Bathurst, N.B., Canada	Zn-Pb-Cu	29	0.56	5.43	2.17	(28)	62.0	0.5	8.7
Green Tuff Belt, Japan	Zn-Pb-Cu	25	1.63	3.86	0.92	(7[a])	95.1	0.9	5.8

Notes: (*) Number of deposits for which data available to calculate average Ag and Au grades
[a] Average Ag and Au grade for Green Tuff Belt is average of 27 individual lenses from 7 different deposits

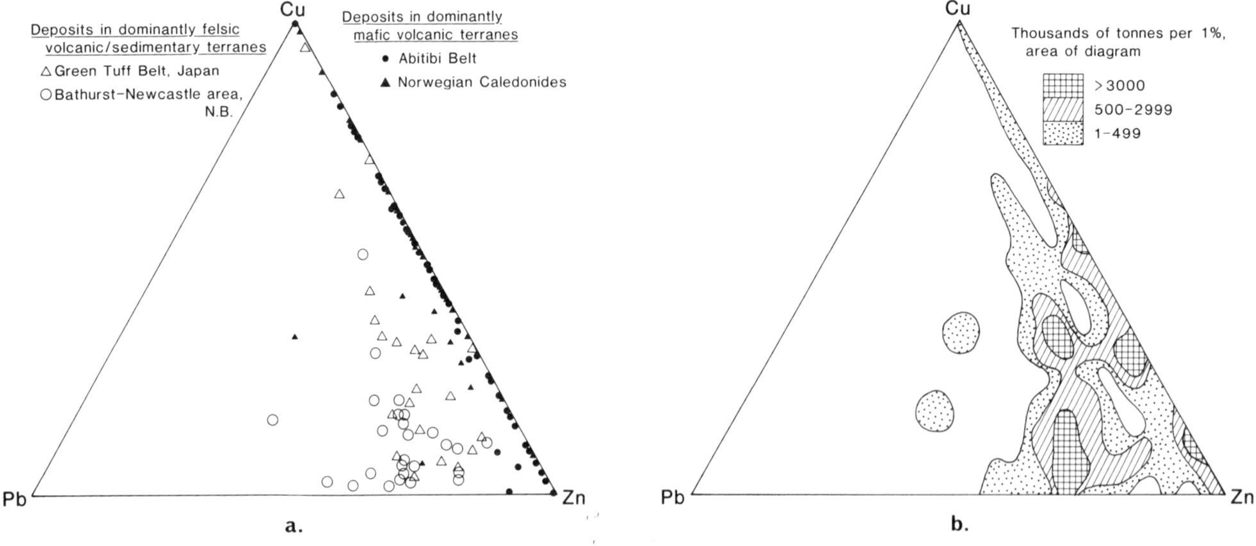

Figure 3 *Cu-Zn-Pb ternary diagrams for bulk compositions of VMS deposits of Abitibi Belt, Bathurst area, New Brunswick, Norwegian Caledonides and Green Tuff Belt, Japan, as reported in the literature.*
(a) *Individual mines.*
(b) *Data of Figure 3a contoured for thousands of tonnes of contained Cu + Zn + Pb per 1% area of plot.*

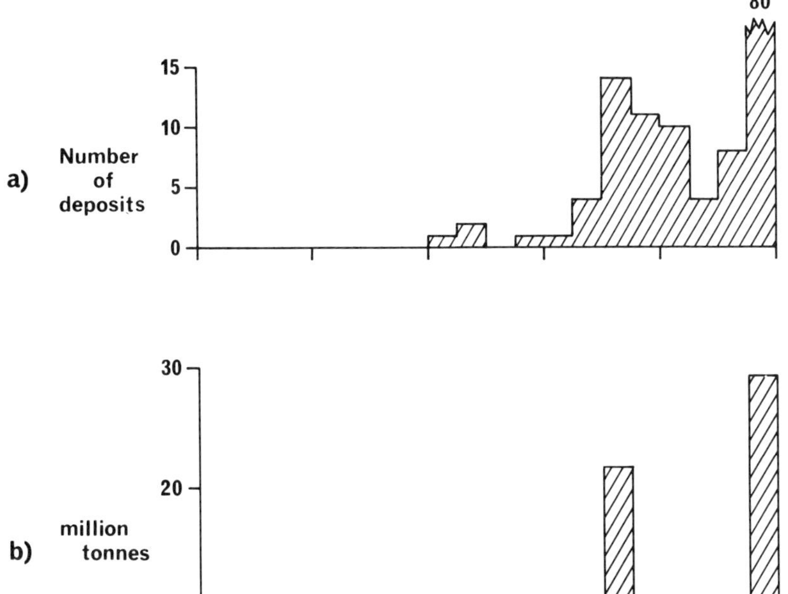

suite, or age is assigned to either group. The elimination of the Cu-type is quite justified when it is noted that most deposits which plot near the Cu apex of the Cu-Pb-Zn ternary diagram owe their position more to the economics of mining than geology when, especially for low grade deposits, the less valuable zinc and lead contents may have been ignored in reporting reserves. For example, none of the Cyprus deposits, often regarded as prime examples of the Cu-type, are devoid of sphalerite, and in some cases Zn contents greatly exceed those of Cu (Lydon, 1984). Most of the deposits described as Cu-types should be considered to belong to the Cu-Zn group.

That this division of VMS deposits into two major groups based on Zn/(Zn + Pb) ratios reflects a natural grouping is emphasized by the high degree of correlation with other features of the deposits. For example, barite as a significant gangue mineral is confined to deposits of Zn-Pb-Cu type; Au/Ag ratios of the Cu-Zn type are usually higher than those of the Zn-Pb-Cu type (Table 1); sulphur isotope ratios of the two types clearly form two populations, with ratios of the Cu-Zn type generally being significantly lighter (Figure 5).

Volcanogenic Massive Sulphides

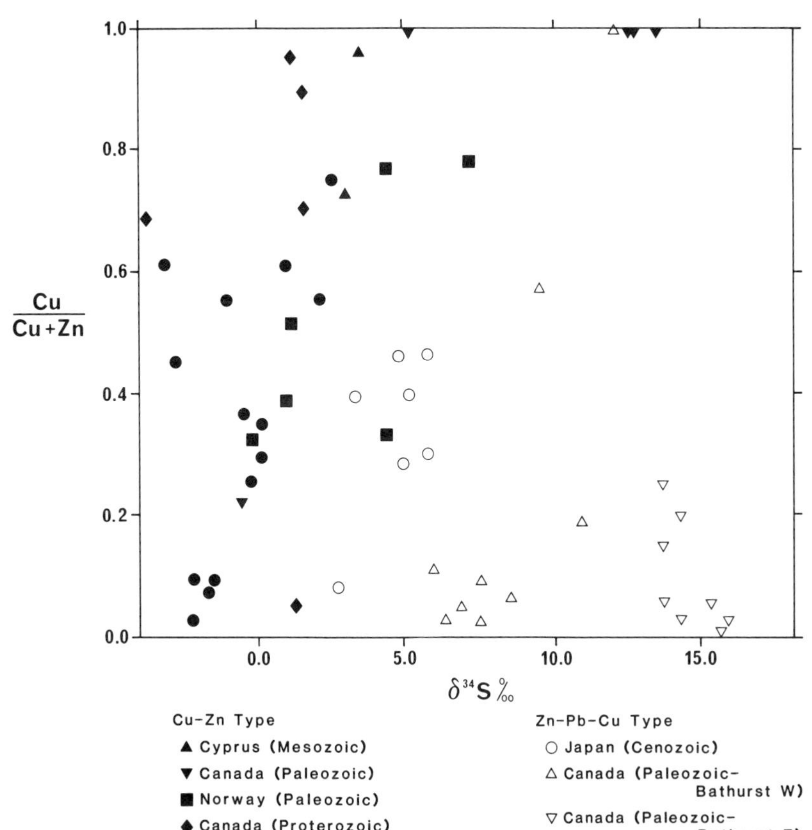

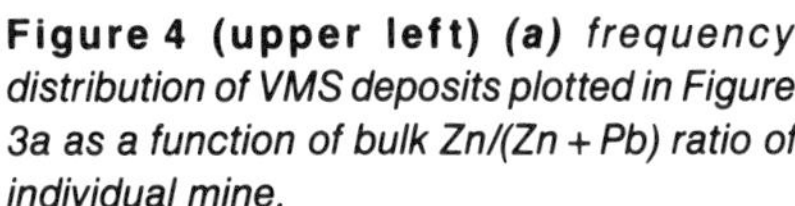

Figure 4 (upper left) ***(a)*** *frequency distribution of VMS deposits plotted in Figure 3a as a function of bulk Zn/(Zn + Pb) ratio of individual mine.*
(b) *Cumulative tonnes of ore metals contained in VMS deposits plotted in Figure 3a as a function of bulk Zn/(Zn + Pb) ratio of individual mine.*

Figure 5 (lower left) *Average sulphur isotope ratios of pyrite from the massive sulphide lenses of VMS deposits plotted against the bulk Cu/(Cu + Zn) ratio of the deposit, showing a clear separation of the two major compositional groups. Sulphur isotope data compiled from the literature. A minimum of three samples from each deposit was used to derive the arithmetic average for the sulphur isotope ratio.*

Perhaps the most significant correlation, especialy from the point of view of genetic modelling, is the strong relationship between the deposit type and the dominant lithology for 1-5 km stratigraphically below the deposit. Deposits of the Cu-Zn type occur where the dominant regional footwall lithology is mafic volcanic rocks or their direct sedimentary derivatives, whereas deposits of the Zn-Pb-Cu type occur where the regional footwall succession is composed of dominantly felsic volcanic rocks or mica/clay-bearing sedimentary rocks. Obviously, a corollary to this statement is that deposits of the same mining camp are of the same type. For example, non-intrusive rocks of the Abitibi Belt are comprised of about 70% mafic volcanic rocks, 24% sedimentary rocks (dominantly clastics derived from volcanic rocks), and 6% felsic volcanic rocks (Goodwin and Ridler, 1970). Of the 52 major VMS deposits that occur in the Abitibi Belt, 50 are clearly of the Cu-Zn type (Figure 3). Similarly, in the Bathurst-Newcastle area of New Brunswick, planimetric measurements of a 1:250,000 scale map (Map NR-3, New Brunswick Department of Natural Resources, 1974) indicates that non-intrusive lithologies of the Tetagouche Group are composed of 53% sediments, 32% felsic volcanic rocks and 15% mafic volcanic rocks. All the VMS deposits of the area are of the Zn-Pb-Cu type. The significance of this correlation will be discussed in Part 2.

Conclusion

Volcanogenic massive sulphide deposits are synvolcanic accumulations of sulphide minerals that occur in geological domains characterized by submarine volcanic rocks. The deposits are composed of iron sulphides with subordinate amounts of chalcopyrite, sphalerite and galena. Typically, a deposit consists of a stratiform lens of massive sulphide containing the bulk of the mineralization and a discordant zone of stockwork type sulphide mineralization within hydrothermally altered rocks of the stratigraphic footwall.

The deposits are not confined to a single plate tectonic environment, to a particular petrochemical type of volcanism, or to any particular geological time span. They are not essential products of submarine volcanism, but are the result of special hydrologic, geothermal and topographic conditions of the ocean floor.

The typical economic deposit may consist of several individual massive sulphide lenses or stockwork zones and contains 1-10 million tonnes of ore with an average grade of 2-10% Cu + Zn + Pb. The largest deposits contain in excess of 100 million tonnes of ore. Deposits tend to occur in clusters forming a mineralized district on average 32 km in diameter, and, within each cluster, individual deposits tend to occur within a single stratigraphic interval.

Characteristically, VMS deposits show a pronounced zonation of ore, gangue and hydrothermal alteration minerals outward and upward from the core of the stockwork zone and the base of the massive sulphide lens. VMS deposits are best classified into just two major groups, called the Cu-Zn type and Zn-Pb-Cu type, respectively, which reflects the associations of major ore metals and other geological characteristics.

Acknowledgements

R.I. Thorpe, D.F. Sangster and A. Galley of the Geological Survey of Canada, and R.G. Roberts of the University of Waterloo are thanked for their suggestions and comments which improved the quality of the manuscript. R.D. Lancaster and K. Nguyen drafted the diagrams and the GSC Word Processing Centre typed the manuscript.

References

Aggarwal, P.K. and Nesbitt, B.E., 1984, Geology and geochemistry of the Chu Chua massive sulphide deposit, British Columbia: Economic Geology, v. 79, p. 815-825.

Costa, U.R., Barnett, R.L. and Kerrich, R., 1983, The Mattagami Lake Mine Archean Zn-Cu sulphide deposit, Quebec: hydrothermal coprecipitation of talc and sulphides in a seafloor brine pool - evidence from geochemistry, $^{18}O/^{16}O$, and mineral chemistry: Economic Geology, v. 78, p. 1144-1203.

Davies, J.L., 1980, Geology of the Bathurst-Newcastle area, Northern New Brunswick: Geological Association of Canada Field Guide 16.

Finlow-Bates, T., 1980, The chemical and physical controls on the genesis of submarine exhalative orebodies and their implications for formulating exploration concepts: A review. 4: Geologisches Jahrbuch, v. 40, p. 131-168.

Franklin, J.M., Kasarda, J. and Poulsen, K.H., 1975, Petrology and chemistry of the alteration zone of the Mattabi massive sulfide deposit: Economic Geology, v. 70, p. 63-79.

Franklin, J.M., Lydon, J.W. and Sangster, D.F., 1981, Volcanic-associated massive sulfide deposits: Economic Geology, 75th Anniversary Volume, p. 485-627.

Froese, E., 1969, Metamorphic rocks from the Coronation mine and surrounding area: Geological Survey of Canada, Paper 68-5, p. 55-77.

Gibson, H.L., 1979, Geology of the Amulet Rhyolite Formation, Turcotte Lake section, Noranda area, Quebec: Unpublished M.Sc. thesis, Carleton University, Ottawa, 154 p.

Gibson, H.L., Watkinson, D.H. and Comba, C.D.A., 1983, Silicification: Hydrothermal alteration in an Archean geothermal system within the Amulet Rhyolite Formation, Noranda, Quebec: Economic Geology, v. 83, p. 954-971.

Gilmour, P., 1976, Some transitional types of mineral deposits in volcanic and sedimentary rocks, *in* Wolf, K.H., ed., Handbook of Stratabound and Stratiform Ore Deposits: Elsevier, Amsterdam, v. 1, p. 111-160.

Goodwin, A.M. and Ridler, R.H., 1970, The Abitibi orogenic belt, *in* Baer, A.J., ed., Basins and Geosynclines of the Canadian Shield: Geological Survey of Canada, Paper 70-40, p. 1-30.

Hodgson, C.J. and Lydon, J.W., 1977, Geological setting of volcanogenic massive sulphide deposits and active hydrothermal systems: some implications for exploration: Canadian Institution of Mining and Metallurgy, Bulletin, v. 70, p. 95-106.

Hutchinson, R.W., 1973, Volcanogenic sulfide deposits and their metallogenic significance: Economic Geology, v. 68, p. 1223-1246.

Iijima, A., 1974, Clay and zeolitic alteration zones surrounding Kuroko deposits in the Hokuroku District, northern Akita, as submarine hydrothermal-diagenetic alteration products: Mining Geology Japan, Special Issue 6, p. 267-289.

James, R.S., Grieve, R.A.F. and Pauk, L., 1978, The petrology of cordierite-anthophyllite gneisses and associated mafic and pelitic gneisses at Manitouwadge, Ontario: American Journal of Science, v. 278, p. 41-63.

Klau, W. and Large, D.E., 1980, Submarine exhalative Cu-Pb-Zn deposits - a discussion of their classification and metallogenesis: Geologisches Jahrbuch, v. 40, p. 13-58.

Knuckey, M.J., 1975, Geology of the Millenbach copper-zinc deposit, Noranda, Quebec, Canada: Society of Economic Geologists—American Institute of Mining and Metallurgical Engineers, Annual General Meeting, Feb. 1975.

Knuckey, M.J., Comba, C.D.A. and Riverin, G., 1982, Structure, metal zoning and alteration at the Millenbach deposit, Noranda, Quebec, *in* Hutchinson, R.W., Spence, C.D. and Franklin, J.M., eds., Precambrian Sulphide Deposits, Geological Association of Canada, Special Paper 25, p. 256-295.

Knuckey, M.J. and Watkins, J.J., 1982, The geology of the Corbet massive sulphide deposit, Noranda district, Quebec, Canada, *in* Hutchinson, R.W., Spence, C.D. and Franklin, J.M., eds., Precambrian Sulphide Deposits: Geological Association of Canada, Special Paper 25, p. 297-317.

Lambert, T.B. and Sato, T., 1976, The Kuroko and associated ore deposits of Japan: A review of their features and metallogenesis: Economic Geology, v. 69, p. 1215-1236.

Lickus, R.J., 1965, Geology and geochemistry of the ore deposits at the Vauze mine, Noranda district, Quebec: Unpublished Ph.D. thesis, McGill University.

Lydon, J.W., 1984, Some observations on the mineralogical and chemical zonation patterns of volcanogenic massive sulphide deposits of Cyprus: Geological Survey of Canada, Paper 84-1A, p. 611-616.

MacGeehan, P.J., 1978, The geochemistry of altered volcanic rocks at Matagami, Quebec: a geothermal model for massive sulphide genesis: Canadian Journal of Earth Sciences, v. 15, p. 551-570.

MacGeehan, P.J. and MacLean, W.H., 1980, Tholeiitic basalt-rhyolite magmatism and massive sulphide deposits at Matagami, Quebec: Nature, v. 283, p. 153-157.

Morton, R.D., 1972, A discussion: Sulphide mineralization and wall rock alteration at Rodhammeren mine, Sur-Trondelag, Norway: Norske Geologisk Tidsskrift, v. 52, p. 313-315.

Ohmoto, H. and Skinner, B.J., 1983, eds., The Kuroko and related volcanogenic massive sulfide deposits: Economic Geology, Monograph 5, 604 p.

Pye, E.G., 1960, Geology of the Manitouwadge area: Ontario Department of Mines, v. 66, part 8, 1957, 114 p.

Riverin, G. and Hodgson, C.J., 1980, Wall-rock alteration at the Millenbach Cu-Zn mine, Noranda, Quebec: Economic Geology, v. 75, p. 424-444.

Roberts, R.G. and Reardon, E.J., 1978, Alteration and ore forming processes at Mattagami Lake Mine, Quebec: Canadian Journal of Earth Sciences, v. 15, p. 1-21.

Rui, I., 1973, Geology and structure of the Rostvangen sulphide deposit in the Kvikne district, central Norwegian Caledonides: Norske Geologisk Tidsskrift, v. 53, p. 433-442.

Sakrison, H.C., 1967, Chemical studies of the host rocks of the Lake Dufault mine, Quebec: Unpublished Ph.D. thesis, McGill University.

Sangster, D.F., 1972, Precambrian volcanogenic massive sulphide deposits in Canada: a review: Geological Survey of Canada, Paper 72-22, 44 p.

Sangster, D.F., 1979, Plate tectonics and mineral deposits: A view from two perspectives: Geoscience Canada, v. 6, p. 185-188.

Sangster, D.F., 1980a, Distribution and origin of Precambrian massive sulphide deposits of North America, *in* Strangway, D.W., ed., The Continental Crust and its Mineral Deposits: Geological Association of Canada, Special Paper 20, p. 723-740.

Sangster, D.F., 1980b, Quantitative characteristics of volcanogenic massive sulphide deposits: 1. Metal content and size distribution of massive sulphide deposits in volcanic centres: Canadian Institution of Mining and Metallurgy, Bulletin, v. 73, p. 74-81.

Sangster, D.F. and Scott, S.D., 1976, Precambrian, strata-bound, massive Cu-Zn-Pb sulphide ores of North America: *in* Wolf, K.H., ed., Handbook of Stratabound and Stratiform Ore Deposits: Elsevier Scientific Publishing Co., Amsterdam, v. 6, p. 130-221.

Sawkins, F.J., 1976, Massive sulphide deposits in relation to geotectonics, *in* Strong, D.F., ed., Metallogeny and Plate Tectonics: Geological Association of Canada, Special Paper 14, p. 221-240.

Scott, S.D., 1978, Structural control of the Kuroko deposits of the Hokuroku district, Japan: Mining Geology Japan, v. 28, p. 301-311.

Scott, S.D., 1980, Geology and structural control of Kuroko-type massive sulphide deposits, *in* Strangway, D.W., ed., The Continental Crust and its Mineral Deposits: Geological Association of Canada, Special Paper 20, p. 705-722.

Shirozo, H., 1974, Clay minerals in altered wall rocks of the Kuroko-type deposits: Society of Mining Geologists Japan, Special Issue 6, p. 303-311.

Sillitoe, R.H., 1972, Formation of certain massive sulphide deposits at sites of sea-floor spreading: Institute of Mining and Metallurgy, Transactions, v. 81, p. B141-B148.

Sillitoe, R.H., 1973, Environments of formation of volcanogenic massive sulfide deposits: Economic Geology, v. 68, p. 1321-1325.

Solomon, M., 1976, "Volcanic" massive sulphide deposits and their host rocks - a review and an explanation, *in* Wolf, K.H., ed., Handbook of Stratabound and Stratiform Ore Deposits: Elsevier, Amsterdam, v. 2, p. 21-50.

Spence, C.D. and de Rosen-Spence, A.F., 1975, The place of sulphide mineralization in the volcanic sequence at Noranda, Quebec: Economic Geology, v. 70, p. 90-101.

Spitz, G. and Darling, R., 1973, Pétrographie de roches encaissantes du gîsement cuprifère de Louvem: Canadian Journal of Earth Sciences, v. 10, p. 760-777.

Spitz, G. and Darling, R., 1975, The petrochemistry of Altered Volcanic Rocks surrounding the Louvem Deposit: Canadian Journal of Earth Sciences, v. 12, p. 1820-1849.

Urabe, T., Scott, S.D. and Hattori, K., 1983, A comparison of footwall-rock alteration and geothermal systems beneath some Japanese and Canadian volcanogenic massive sulphide deposits, *in* Ohmoto, H. and Skinner, B., eds., The Kuroko and Related Volcanogenic Massive Sulphide Deposits: Economic Geology, Monograph 5, p. 345-364.

Walford, D.C. and Franklin, J.M., 1982, The Anderson Lake Mine, Snow Lake, Manitoba, *in* Hutchinson, R.W., Spence, C.D. and Franklin, J.M., eds., Precambrian Sulphide Deposits: Geological Association of Canada, Special Paper 25, p. 481-523.

Whitmore, D.R.E., 1969, Geology of the Coronation copper deposit: Geological Survey of Canada, Paper 68-5, p. 37-54.

Received 29 June 1984.
Originally published in
Geoscience Canada v. 11 Number 4
(December 1984)

Volcanogenic Massive Sulphide Deposits Part 2: Genetic Models

John W. Lydon
Geological Survey of Canada
601 Booth Street
Ottawa, Ontario K1A 0E8

Introduction

Since Oftedahl (1958) revived the idea that volcanogenic massive sulphide (VMS) deposits were formed on the sea floor, more articles have been published on the genesis of VMS deposits than any other deposit type. This article is intended as a summary of current opinions on the genesis of VMS deposits, rather than a review of the progression of scientific thought by which these opinions have been reached (see reviews by Sangster, 1972; Sangster and Scott, 1976; Solomon, 1976; Franklin *et al.*, 1981 for this background). Current ideas on the topic are highly influenced by studies of hydrothermal systems and sulphide accumulations at modern submarine spreading centres. Following this research trend, this article places emphasis on discussion of these studies and their applicability to understanding the genesis of ancient VMS deposits.

A genetic model for a mineral deposit type is an attempt to give a rational and consistent explanation of the characteristics of the deposit type in terms of physical and chemical processes. Characteristics of VMS deposits were described in Part 1 of this article (Lydon, 1984a). Discussion here will be limited to those processes and related geological aspects that have been of the most concern in the recent literature and include: (i) processes of sulphide accumulation which form the massive sulphide lens and which in turn control the morphology, texture, mineralogical zonation, etc. of the sulphide body; (ii) processes in the immediate footwall of proximal VMS deposits that give rise to the hydrothermal alteration pipe and stockwork ore; (iii) reasons for fluid flow in the hydrothermal system, which in turn help explain the geological setting and distribution of VMS deposits; and (iv) origin and nature of the hydrothermal fluids, which in turn determine the source of the ore constituents and the chemistry of the sulphide ores.

GSC Contribution No. 29187

Accumulation of sulphides

Historical perspective. The debates of the 1950s and 1960s resulted in a consensus that VMS deposits were formed on the sea floor by the accumulation of sulphides precipitated from hydrothermal fluids. The main evidence in support of this opinion came from observations on well-preserved proximal-type deposits (*i.e.*, those deposits which formed immediately around a hydrothermal vent, Plimer, 1978; Franklin *et al.*, 1981), and, as illustrated by figure 9 and figure 2 of Lydon (1984a), includes:

(i) The presence of sedimentary structures within the massive sulphide lens of which graded-bedding, cross-bedding, and the intercalation of fragmental with laminated sulphide lithologies are the most convincing. The descriptive papers in Ishihara (1974) illustrate classic examples from the Kuroko deposits of Japan.

(ii) The stratigraphically upper contact of the massive sulphide lens is sharp and conformable with hanging wall rocks.

(iii) The massive sulphide lens is commonly laterally continuous with an horizon of chemical and/or clastic metalliferous sediment. Examples are the Key Tuffite of the Matagami area (Roberts, 1975) and the C contact exhalative sediment of the Noranda area (Gibson *et al.*, 1983), which are both pyritic tuffaceous cherts.

(iv) The massive sulphide lens is commonly rooted in a discordant footwall hydrothermal alteration pipe, interpreted to represent the uppermost part of the hydrothermal conduit. The upward termination of the conduit at the horizon of concordant sulphides indicates that ore-forming hydrothermal activity was confined to the time interval between deposition of the footwall rocks and the deposition of the hanging wall rocks.

Sato (1972) postulated that a hydrothermal solution discharged into seawater would behave in one of three ways, depending on its initial temperature and density and its subsequent degree of mixing with seawater (Figure 1):

Type I or highly saline hydrothermal solutions whose densities are greater than cold seawater at all degrees of mixing.

Type II solutions that, although initially less dense than seawater, at some stage in the mixing process pass through a density maximum which is greater than that of cold seawater.

Type III solutions that are initially less dense than cold seawater and which remain buoyant at all degrees of mixing.

Following the prevalent perception of the time that any deposition on the sea floor must follow the laws of sedimentary superposition, Sato (1972) suggested that each of his three ore solution types would give rise to deposits with a characteristic morphology. Type I solutions would give rise to brine pool deposits that characteristically have a tabular or sheet-like morphology (see figure 9 for morphology types). Type II ore solutions would give rise to deposits with a conical or mound-like morphology, which is typical of proximal deposits. Precipitates from his Type III ore solutions would be dispersed by the buoyant plume and would only form a sedimentary veneer of plume fall-out of large lateral extent.

Despite the large volume of literature on the genesis of VMS deposits, relatively few papers have addressed the problem of the mechanisms of sulphide accumulation. Prior to the discovery of the mound-chimney sulphide deposits on the East Pacific Rise (EPR) (Francheteau *et al.*, 1979; Hekinian *et al.*, 1980), mechanisms by which sulphides accumulated were most frequently visualized as analogous to those for the Red Sea brine pool metalliferous sediments (Miller *et al.*, 1966; Degens and Ross, 1969) or were based on the buoyancy models of Sato (1972), whereby sulphides were precipitated from the hydrothermal fluids after discharge from the vent and accumulated as heavy gelatinous muds (Sangster, 1972). For example, Hutchinson and Searle (1971) interpreted the VMS

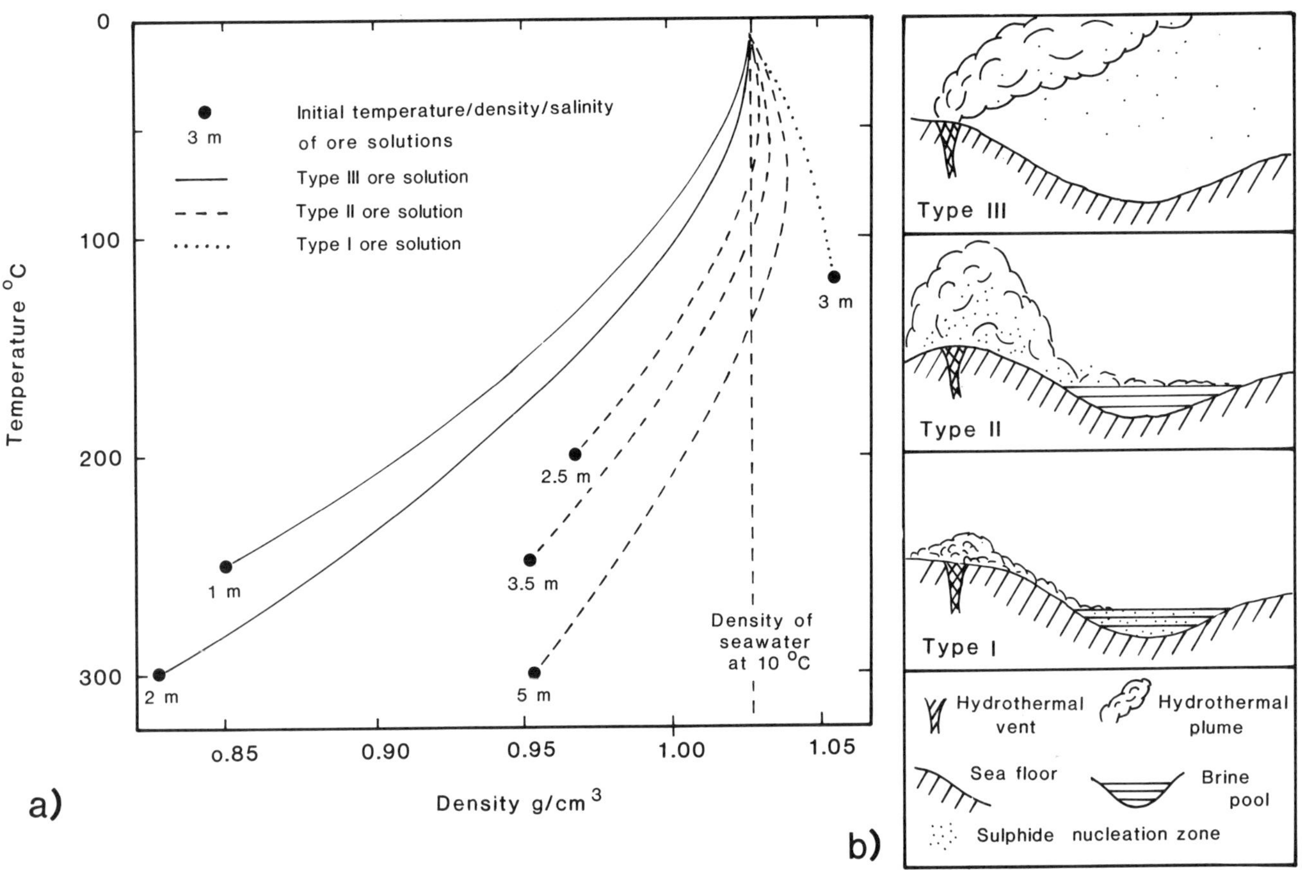

Figure 1 (a) (upper left) *Density-temperature relationships of hot NaCl solutions mixed with seawater at 10 °C (after Sato, 1972) illustrating the definition of Sato's (1972) three types of hydrothermal solutions.* **(b) (upper right)** *Schematic representation of the buoyancy behaviour of and sulphide precipitation from Sato's (1972) three types of hydrothermal solutions.*

Table 1 **The compositions (in mg·g⁻¹) of some natural metalliferous hydrothermal solutions compared with modern seawater. References: *Salton Sea:* Helgeson (1968); *Red Sea:* Shanks and Bischoff (1977); *Kuroko:* Pisutha-Arnond and Ohmoto (1983) (maximum concentrations in fluid inclusions); *East Pacific Rise (EPR), Seawater:* Janecky and Seyfried (1984) (EPR, 21°N, average); *Southern Juan de Fuca Ridge (SJFR):* Von Damm and Bischoff (1987) (SJFR, Plume vent).**

	Salton Sea	Red Sea	Kuroko	EPR	SJFR	Seawater
Temp (°C)	340	60	320	350	224	2
pH	5.5	5.5	4.5	3.5	3.2	8.0
Na	53000	92600	17500	9800	18300	10790
K	16500	1870	5000	1000	2020	395
Ca	28800	5150	4400	860	3860	413
Mg	10	764	510	< 1	< 1	1280
SiO_2	400	60	?	960	1400	10
Cl	155000	156000	40400	17335	38640	19355
SO_4	10	840	?	< 1	-50	2745
H_2S	30	?	?	221	63	< 1
$\sum CO_2$	500	140	8800	282	?	103
Fe	2000	81	6	100	1045	< 1
Mn	1370	82	?	34	197	< 1
Zn	500	5.4	3	7	59	< 1
Cu	3	0.3	5	1	< 0.1	< 1
Pb	80	?	3	< 1	?	< 1
Ba	250	0.9	?	13	?	< 1

deposits of Cyprus to be brine pool muds, whereas it appears to be implicit in genetic models of the Japanese Kuroko deposits by Kajiwara (1973), Sato (1973), Large (1977) and Urabe and Sato (1978) that the sulphides accumulated as fall-out from hydrothermal plumes of Sato's (1972) Type II hydrothermal solutions. Experimental studies of buoyant hydrothermal plumes (Turner and Gustafson, 1978; Solomon and Walshe, 1979a) also carried with them the assumption that sulphides accumulated from plume fall-back and formed mounds of sulphide gel around the vent (Solomon and Walshe, 1979a).

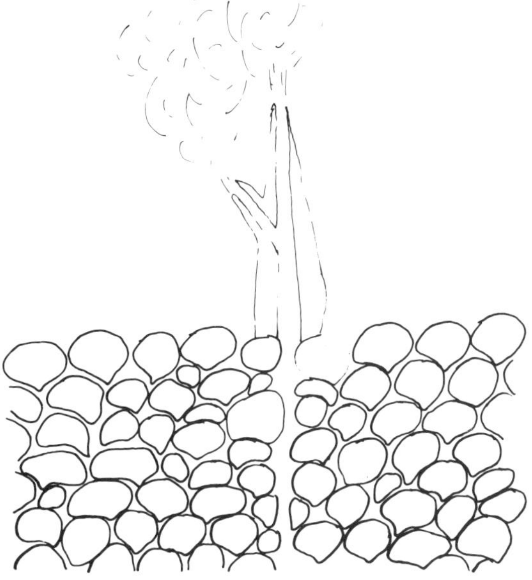

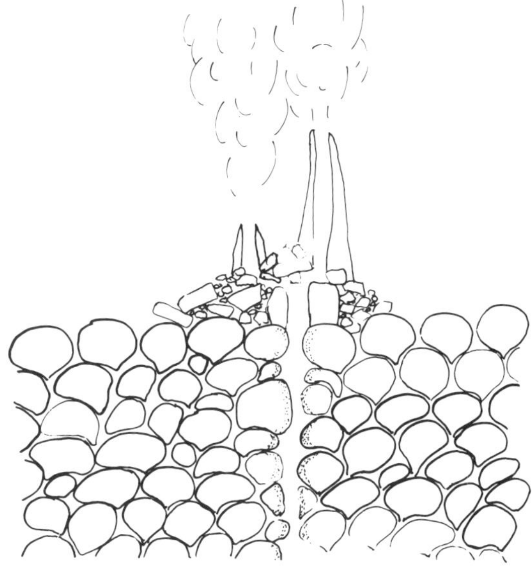

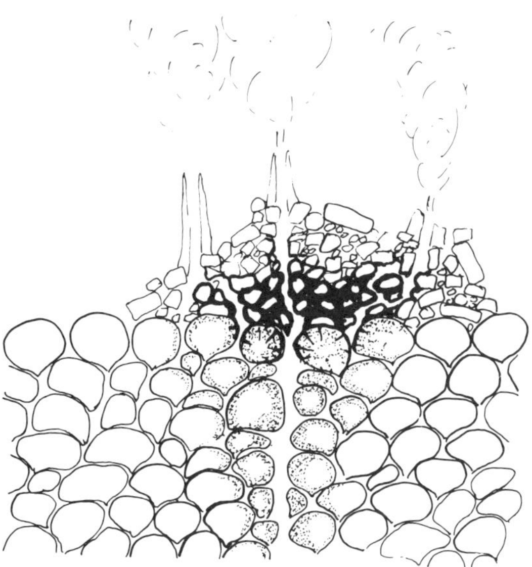

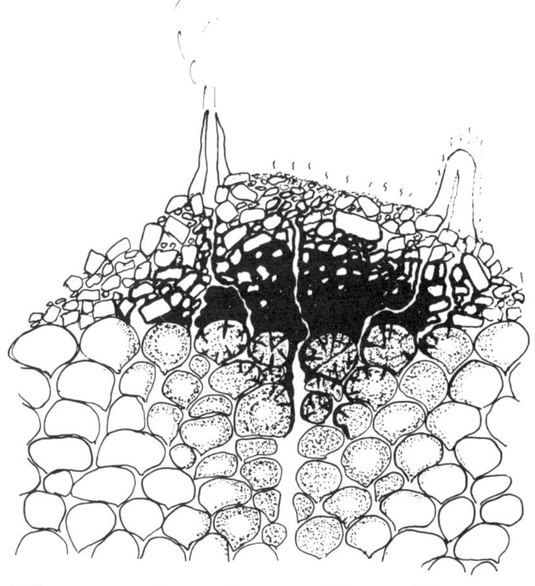

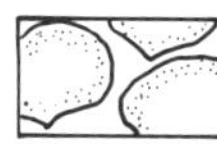

Figure 2 *Schematic representation of the growth of a modern mound-chimney sulphide deposit. (Based on the descriptions by Hekinian and Foquet, 1985; Goldfarb* et al., *1983; Jonasson* et al., *1986).*

In contrast, studies of modern mound-chimney deposits emphasize that sulphides in proximal deposits accumulate as rigid edifices. It is not being suggested that studies of the modern sulphide deposits by themselves have completely revolutionized geological thought on the accumulation of VMS deposits, because many of the current concepts, including the rigidity of proximal sulphide lenses, were previously widely appreciated by at least the mineral exploration community (*e.g.*, Simmons *et al.*, 1973; Spence, 1975). What these studies on modern sulphide deposits have accomplished is to provide, by documentation and observation in a natural experimental laboratory, the scientific confirmation of processes that already existed as concepts in the minds of exploration geologists, and to bring new details of the phenomena and processes, together with the concepts, into the mainstream of scientific literature.

Modern submarine hydrothermal activity. Aspects of modern sulphide deposits at submarine spreading centres have been reviewed by Rona (1984) and Franklin (1986). As of 1984, with only about 1% of the 50,000 km length of oceanic ridges having been explored at the detail necessary to detect hydrothermal activity, about 60 hydrothermal fields had been discovered (Rona, 1984). The majority of known sites are in the Eastern Pacific, where most vents occur in the median valley of the ridge crest and the hydrothermal fluids emanate directly from a volcanic (mainly basaltic) substrate. At some sites, notably those of the Guaymas Basin in the Gulf of California, Middle Valley at the northern end of the Juan de Fuca Ridge and the Escanaba Trough adjacent to the Gorda Ridge, hydrothermal discharge is through a sediment cover as much as 500 m thick. In addition to the median valley of the ridge crest, seamounts have also proved to be a favourable setting for high temperature hydrothermal discharge and sulphide deposition (Lonsdale *et al.*, 1982; Hekinian and Fouquet, 1985; CASM, 1985).

The temperature of the hydrothermal fluids ranges from 400°C at some vents to just a few degrees above ambient seawater at others. The highest temperature fluids are considered to be pristine hydrothermal fluids (*i.e.*, fluids which have equilibrated with reservoir rocks at pressure-temperature conditions close to the maximum values of the hydrothermal system), whereas those at lower temperatures are usually mixtures of pristine hydrothermal fluid and pore water (mainly seawater) that has become entrained into the hydrothermal conduit in the subsurface. The hydrothermal fluids (Table 1) have salinities up to twice that of modern seawater (3.2 wt.% NaCl), and, as predicted by Sato (1972) for his Type III solutions, form buoyant plumes on discharge. The Red Sea brines, with salinities seven times that of seawater, discharge at temperatures of over 200°C (Schoell, 1976), and, like Sato's (1972) Type I solutions, collect as brine pools in the deepest parts of the axial valley. Recent evaluation (Turner and Campbell, 1987) of the problem addressed by Sato (1972) suggests that all high temperature (>300°C) fluids will be initially buoyant in seawater, but if their salinities are greater than about 7 wt.% NaCl will show reversing buoyancy on mixing with seawater. This means that all the vent fluids of the Eastern Pacific that have been studied to date are of Sato's Type III solutions, but visual observations (Lonsdale *et al.*, 1982) and data from fluid inclusions (Delaney and Cosens, 1982; Stakes and Vanko, 1986) suggest that fluids with salinities high enough to form reversing plumes like Sato's Type II solutions do in fact exist in the modern oceanic crust.

Active sulphide deposition on the ocean floor is confined to high temperature (>200°C but more commonly >300°C) vents. Sulphide deposits formed from buoyant hydrothermal fluids are characterized by sulphide chimneys that rise from a basal mound of sulphides (Figure 2). The most-studied area to date is the East Pacific Rise (EPR) at 21°N (Spiess *et al.*, 1980), where a typical mound-chimney combination has been

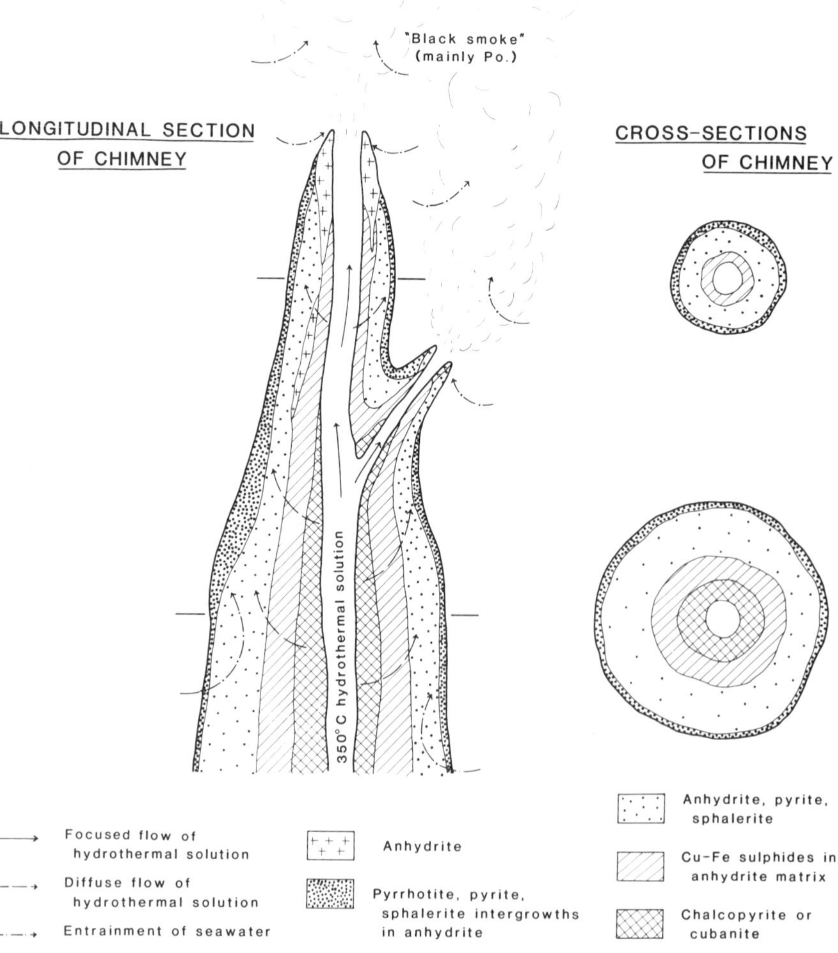

Figure 3 *Characteristics of mineral zonation of a modern "black smoker" sulphide chimney. (After Haymon and Kastner, 1981; Haymon, 1983).*

estimated to contain about 1000 tonnes of metal (Rona, 1984) or about 2000 tonnes of sulphide. At 13°N on EPR, more than 80 separate sulphide deposits have been recognized along a 20 km segment of the axial graben, but it has been estimated that collectively they do not contain more than 20,000 tonnes of sulphide (Hekinian and Fouquet, 1985). Perhaps the largest mound-chimney deposit discovered to date occurs in the TAG hydrothermal field on the Mid-Atlantic Ridge at 26°N, where a single deposit has been estimated to contain 4.5 million tonnes of sulphide (Rona *et al.*, 1986). Current unpublished speculation is that sulphide deposits of sediment-covered segments of ridges at Middle Valley (Davis *et al.*, 1987) and Escanaba Trough (Zierenberg *et al.*, 1986) may be significantly larger. Deposits of the mound-chimney type have only been grab-sampled, which does not allow an accurate estimate of their ore grades (see Rona, 1984). The largest sulphide deposits known at present on the modern ocean floor are the metalliferous sediments of the Red Sea brine pools. The Atlantis II deposit consists of about 100 million tonnes of sulphide/oxide sediment containing 2% Zn, 0.4% Cu, 90 $g \cdot t^{-1}$ Ag, 0.8 $g \cdot t^{-1}$ Au (Blissenbach and Nawab, 1982) and about 0.1% Pb (Lowell and Rona, 1985).

Perhaps the most unexpected discovery with regard to modern submarine hydrothermal vents is the associated prolific biota and their food-chain based on chemosynthetic bacteria (Grassle, 1983; Jannasch, 1983). Megafauna of the hydrothermal vents are characterized by vestimentiferan tube-worms and giant clams in the Eastern Pacific (Desbruyeres and Laubier, 1983), by shrimps and anemones in the mid-Atlantic (Rona *et al.*, 1986), and by large gastropods in the Manus Basin, Papua New Guinea (Both *et al.*, 1986). Traces of these megafauna, notably tube casts of vestimentifera, are preserved as textural relicts in chimney sulphides.

Accumulation of modern sulphide chimneys and mounds. Growth of a sulphide chimney (Figure 3) is initiated by the precipitation of anhydrite in the narrow vertical zone of high thermal gradient between the buoyant hydrothermal fluid and surrounding cold seawater (Haymon, 1983). Anhydrite precipitates from seawater at this thermal interface because its solubility decreases with increasing temperature. Modern seawater precipitates anhydrite when it is heated (without evaporation) to 130°C (Haymon and Kastner, 1981). The porous wall or collar of anhydrite that forms at the thermal interface continues to grow upward around the hydrothermal jet for as long as high temperature flow is sustained. Although some calcium may be contributed by the hydrothermal fluids, sulphur isotope analyses indicate that sulphate of the anhydrite is derived from seawater (Styrt *et al.*, 1981).

Most of the hydrothermal fluid flows upward along the central conduit of the anhydrite collar and discharges into the surrounding seawater. However, a small proportion (< 1%) of the hydrothermal fluid flows through the porous anhydrite wall. In doing so, the fluid descends a steep physico-chemical gradient from high temperature (> 300°C), acidic (~ 3.5), reduced ($H_2S >> SO_4$) conditions on the inside of the wall, to conditions approaching the low temperature (2°C), alkaline (~ pH 7.8), oxidized ($SO_4 >> H_2S$) nature of ambient seawater on the outside of the wall. Minerals precipitated from the hydrothermal fluid due to this physicochemical gradient are trapped in the pore spaces of the anhydrite wall, so that, as the anhydrite collar on top of the chimney extends upward, the walls of the lower part of the chimney gradually thicken by the precipitation of sulphides in the interior portions and anhydrite in the exterior portions. A mature chimney structure has a characteristic concentric zonation from chalcopyrite (± isocubanite ± pyrrhotite) on the inside, via a zone dominated by pyrite, sphalerite and wurtzite in an anhydrite matrix, to an exterior zone of anhydrite with minor sulphides, amorphous silica, barite, magnesium hydroxysulphate-hydrate (MHSH), and talc (Spiess *et al.*, 1980; Haymon and Kastner, 1981; Haymon, 1983; Oudin, 1981, 1983; Goldfarb *et al.*, 1983; Tivey and Delaney, 1986).

Mathematical modelling indicates that the zonation is predominantly due to the temperature decrease across the wall, rather than to a chemical gradient, and that the cooling is probably brought about more by mixing of the hydrothermal fluid with cold seawater than by conductive heat exchange (Janecky and Seyfried, 1984; Bowers *et al.*, 1985). In detail, sulphide precipitation in the chimney walls is a complex process, in which reversals of the pressure gradient across the wall, due to the dynamics of high velocity flow within an irregular hydrothermal conduit, plays a dominant role in causing oscillation of the thermal and chemical profiles (Tivey, 1986). Mineral precipitation in any small volume of chimney wall may therefore consist of intermixed mineral assemblages that collectively reflect a wide range of physicochemical conditions. In the longer term, mineral zone boundaries migrate outward with time, so that lower temperature assemblages are replaced by higher temperature assemblages (Haymon, 1983; Goldfarb *et al.*, 1983). Anhydrite is eventually dissolved by the sulphate-free hydrothermal fluids in the interior zones of the chimneys, and by cold seawater in the exterior zones as the chimney cools (Haymon and Kastner, 1981), leaving a sulphide assemblage in which opaline silica is the dominant gangue mineral (Oudin, 1983; Tivey and Delaney, 1986).

The vertical growth of a sulphide chimney has been measured at 8 cm per day (Hekinian *et al.*, 1983) and has been estimated to be as much as 30 cm per day (Goldfarb *et al.*, 1983). Ultimately the chimney becomes mechanically unstable and collapses to form a mound of chimney talus, on which chimney growth begins again (Figure 2). Repetition of this process eventually gives rise to a mound which completely covers the original vent orifice and causes diversion of hydrothermal flow to multiple discharge sites on the mound. Individual chimneys formed at the different discharge points evolve at different rates, and on the same mound "white smokers", "snowballs" and "dead" chimneys may co-exist with the "black smoker" vents (Speiss *et al.*, 1980). "White smokers", which precipitate white clouds of barite, silica and minor pyrite (Rona, 1984), are chimneys in which hydrothermal precipitation has sealed open orifices and decreased the permeability of the chimney walls. The lower flux of hydrothermal fluid through these low permeability walls allows more intra-chimney cooling by seawater entrainment and by heat conduction, so that the fluid exit temperatures, in the range 100-330°C, are lower than those of black smoker chimneys. In the eastern Pacific at least, these less active chimneys sustain the most prolific growth of tube worms, which may become so densely packed as to form a spherical mass of white tubes ("snowballs"). Pseudomorphs of the worm tubes are a common textural feature in chimney sulphides (Speiss *et al.*, 1980; Haymon and Kastner, 1981; Oudin and Constantinou, 1984).

The growth of the basal mound (Figure 2), which can be envisaged as the coalescence of adjacent chimneys into larger edifices (Goldfarb *et al.*, 1983; Hekinian and Fouquet, 1985; Tivey and Delaney, 1986; Alt *et al.*, 1987), is fundamentally due to the same principles which govern chimney growth. The major difference is that in mound growth a horizontal blanket of chimney talus carries out the function of the vertical porous anhydrite wall in chimney growth. Defocussing of hydrothermal discharge by the mound of chimney talus over the original vent orifice promotes advective/conductive cooling and sulphide/silica precipitation in the outer part of the mound (Tivey and Delaney, 1986). These hydrothermal precipitates decrease the permeability of the mound in much the same way as a white smoker chimney evolves, and by constricting fluid channelways through cementation of chimney talus forms a surface crust of low permeability (Goldfarb *et al.*, 1983). In constraining fluid escape, the low-permeability crust causes the high-temperature fluids to circulate within the mound, which leads to the upward migration of isotherms within the mound and consequently to the replacement of previous sulphides by higher temperature mineral assemblages. The creation of new fluid channelways by hydraulic, seismic or tectonic fracturing of the mound in turn initiates the growth of new chimneys, and mound growth continues by repetition of the process for as long as high temperature hydrothermal flow is sustained.

This model for mound growth is supported by several lines of evidence. The fact that high temperature hydrothermal fluids circulate within the mound can be demonstrated by rupturing of the mound crust (Goldfarb *et al.*, 1983), or breaking open inactive chimneys (Hekinian *et al.*, 1983; Von Damm *et al.*, 1985), which results in the immediate emission of "black smoker" fluids. That cooling of some of the hydrothermal fluid takes place within the mound is indicated by the differences in exit temperatures of fluids at different orifices of the same mound. That sulphide precipitation takes place *within* the mound-chimney edifice is demonstrated by the loss of metal sulphide components in the effluent from white smokers that co-exist on the same mound as black smokers.

Seafloor sulphides are chemically very unstable in modern seawater, and rapidly oxidize upon cessation of hydrothermal activity to form ochreous deposits dominated by hydrated iron oxides (Hekinian *et al.*, 1980; Haymon and Kastner, 1981; Alt *et al.*, 1987). Unless the sulphide deposits are quickly covered by a volcanic flow, they are unlikely to be preserved as sulphides in the geological record.

Principles of chimney-mound accumulation. The accumulation of sulphides from modern black smoker vents is an extremely inefficient process. It is estimated that more than 99% of the metal carried by hydrothermal fluids (that have not mixed with seawater or otherwise cooled in the sub-surface) is dispersed in the water column by black smoker plumes (Rona, 1984), and eventually becomes incorporated into distal marine sediments (*e.g.*, Boström and Peterson, 1966; Boström, 1983). Metalliferous components from the hydrothermal fluid have been detected in the water column at distances of more than 750 km from its source (Rona, 1984).

The vertical flow rates of hydrothermal fluids that exit at temperatures of 350°C from single vents with orifice diameters of 2.7-4.7 cm have been measured as 1.0-2.4 $m \cdot s^{-1}$ (Converse *et al.*, 1984). The most common size of sulphide particles in the black smoker plume is 1-3 μm (Mottl, 1986) and these have settling rates of about 2.0×10^{-5} $m \cdot s^{-1}$. Thus, only in the absence of ocean currents could there be any appreciable accumulation of sulphides in the vicinity of the hydrothermal vent (Cathles, 1983; Converse *et al.*, 1984; Campbell *et al.*, 1984), a prediction in keeping with the scarcity of plume fall-back sediment in the vicinity of modern black smoker vents, and a confirmation of Sato's (1972) prediction for his Type III hydrothermal solutions.

The near-vent accumulation of large quantities of sulphides in the modern submarine environment would therefore appear to depend upon the paradoxical requirement that a mound of sulphide be developed over the vent. Once established, the mound has the capacity to grow by the precipitation of sulphides within the mound and by the accumulation of chimney talus on its surface for as long as high temperature hydrothermal flow is maintained. The role of the porous anhydrite collar in initiating sulphide accumulation in the first place is therefore of paramount importance in understanding the accumulation of sulphides from buoyant sea-floor hydrothermal fluids.

As can be easily demonstrated in the laboratory by injecting dilute sulphuric acid into a beaker of calcium chloride solution, anhydrite forms a rigid, though fragile, mass of interconnected acicular and tabular crystals, even when it nucleates rapidly. It is presumed that the anhydrite collar of modern hydrothermal vents precipitates in a similar fashion since it commonly exhibits similar acicular and tabular crystal habits (see mineral texture descriptions by Haymon and Kastner, 1981). This porous network of anhydrite crystals serves three main functions in promoting the accumulation of sulphides:

(1) It impedes the upward flow of any buoyant hydrothermal fluid that enters its latticework. It thus promotes near-vent degeneration of the hydrothermal fluid (*e.g.*, mixing with entrained seawater; cooling by conduction) and consequently near-vent nucleation of sulphides. The lower velocity of the fluid within the anydrite wall permits a higher proportion of precipitated sulphides to remain at the nucleation site rather than being mechanically transported in suspension by the rapid movement of the hydrothermal jet of an open vent.

(2) It provides a substrate upon which hydrothermal precipitation can take place by crystal growth.

(3) It acts as a filter in trapping sulphide particles suspended in the hydrothermal fluid. Its filtering capacity may be expected to increase with time, as the diameter of fluid passage ways decrease due to hydrothermal precipitation.

The common denominator to these three functions is diffuse hydrothermal discharge through a porous medium. It is this defocussing of hydrothermal discharge by a porous barrier that is the main factor in maximizing the cooling of the hydrothermal fluid close to the sea floor and the enhancement of near-vent accumulation of precipitates.

Sulphide accumulation in VMS deposits. The general similarities (*e.g.*, mineralogy, ore mineral zonation, wall rock lithologies) between modern mound-chimney sulphide edifices and ancient proximal VMS deposits are so strong, that drawing direct analogy between the two is almost unavoidable. The discovery of chimney fragments and fossil worm tubes in ophiolitic VMS deposits of Cyprus (Oudin and Constantinou, 1984) and Oman (Haymon *et al.*, 1984) certainly suggests that the mode of accumulation of sea-floor sulphides, at least as far back as the Cretaceous, involved mound-chimney edifices much like the modern ones. Furthermore, the size and inferred tonnage of sulphide mounds discovered recently on the Mid-Atlantic Ridge (Rona *et al.*, 1986; Honnorez *et al.*, 1986) indicate

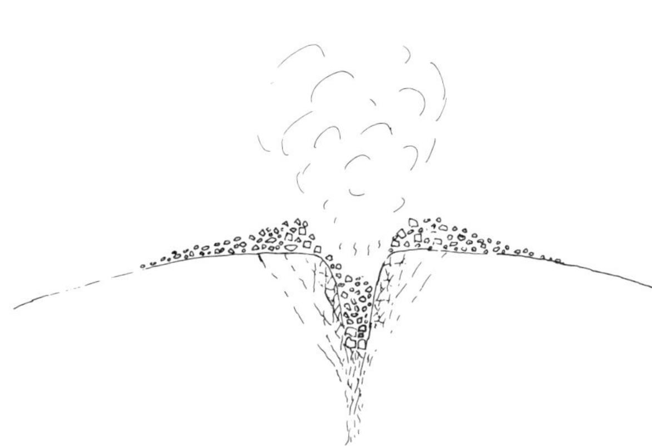

1. Initial hydrothermal eruption.
Accumulation of eruption breccia over hydrothermal vent.

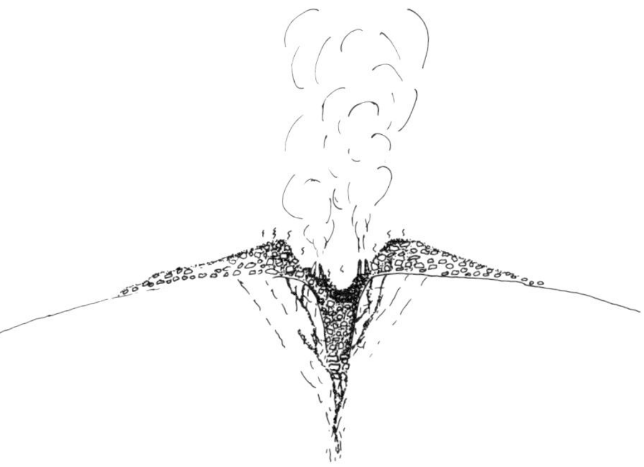

2. Decrease of eruption breccia permeability by intraclast precipitation of sulphides etc.
Initiation of surface sulphide accumulation by crust formation, plume fallback, and chimney talus.

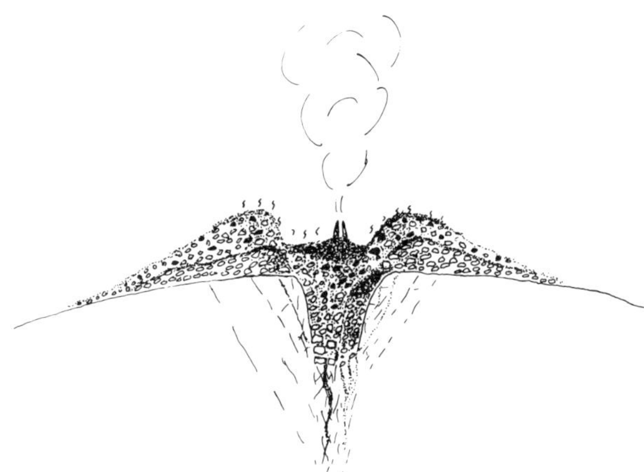

3. Recurrence of hydrothermal eruption.
Hydrothermal eruption breccia contains sulphide clasts.

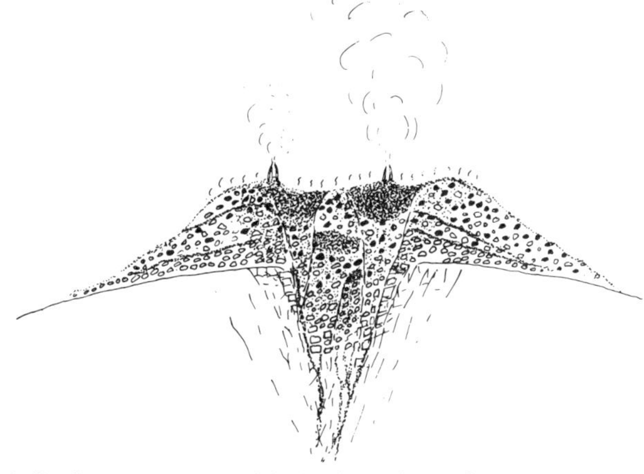

4. Further recurrence of hydrothermal eruption.
As the mound grows:
Increase in the proportion of sulphide clasts in eruption:
Hydrothermal discharge becomes increasingly diffuse:
Recrystallization and redistribution of sulphides within the mound.

LEGEND:

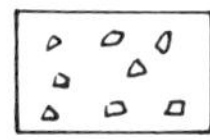

Lithic clasts

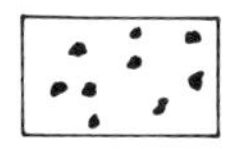

Sulphide clasts

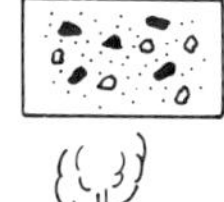

Breccia with sulphide infilling of interstices

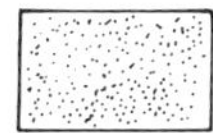

Massive sulphide,
chimney talus, plume fallback,
surface crust, fine particulates fluidized from mound interstices.

Focused discharge from sulphide chimney

Diffuse discharge

SCALE: 100 meters

Figure 4 *Schematic representation of the growth of a proximal VMS deposit formed by a recurrently eruptive hydrothermal vent. The final stage of the progression is illustrated by Figure 5.*

that deposits with physical dimensions comparable to an average ancient VMS deposit (Lydon, 1984a) can be accumulated by the apparently inefficient sulphide-depositing hydrothermal process at modern sea-floor spreading centres.

Fluid inclusion studies of a variety of proximal deposits, particularly from Cyprus (Spooner and Bray, 1977) and Japan (Pisutha-Arnond and Ohmoto, 1983 and references therein), indicate that ore solutions for VMS deposits typically had maximum temperatures of 300±50°C and salinities up to twice those of modern seawater. These temperature-salinity relationships indicate that the ore fluids would be highly buoyant in seawater if they reached the sea floor in an unmodified state (Solomon and Walshe, 1979a, b) and would behave in a similar fashion to the fluids of the modern high-temperature vents.

As discussed above, the main prerequisite for the accumulation of a proximal type VMS deposit is the establishment of a porous barrier over the hydrothermal vent which causes diffuse discharge of the hydrothermal fluids and initiates the accumulation of sulphides. Although the interference between adjacent plumes from a multiple-orifice vent would allow the accumulation of a high proportion of precipitates (Solomon and Walshe, 1979a, b), plume fallback itself (Eldridge *et al.*, 1983) would not appear to be a prime candidate for initiating the growth of a sulphide mound, as this would mean the effect preceding the cause.

Recognizing the beneficial qualities of a porous anhydrite wall in modern sulphide chimneys for promoting initial sulphide accumulation, Campbell *et al.* (1984) suggested that some volcanogenic massive sulphide deposits form beneath an anhydrite cap. As realized by the authors, the weakness in this suggestion is that, with the exception of the Kuroko deposits, anhydrite/gypsum is rare in ancient VMS deposits, even considering that much of the original anhydrite may have been dissolved by digestion in later ore fluids or by aging in cold seawater on the ocean floor, as is the case with modern sulphide mound deposits (Tivey and Delaney, 1986). The typical siting of Sekko ore (anhydrite/gypsum) below and/or adjacent to sulphide ore (*e.g.*, Lambert and Sato, 1974; Hirabayashi, 1974; Shikazono *et al.*, 1983) does suggest a genetic link between the two ore types in the case of the Kuroko deposits, presumably involving the mixing of ore fluids with seawater at an early stage of hydrothermal activity (Sato, 1972, 1973; Shikazono *et al.*, 1983), and the anhydrite may indeed have served as an initial, largely sub-surface, porous blanket which caused subsequent diffuse hydrothermal discharge at the surface.

For the majority of proximal VMS deposits, however, a layer of fragmental rocks over the hydrothermal vent is a much more likely candidate to have fulfilled the role of the porous anhydrite wall in modern sulphide chimneys rather than a cap of anhydrite itself. Fragmental rocks have long been recognized as being an intimate associate of VMS deposits (*e.g.*, Gilmour, 1965; Boldy, 1968) and have been affectionately termed "millrock" (Sangster, 1972). By their definition of having a footwall stockwork zone, proximal VMS lenses invariably repose on breccias or

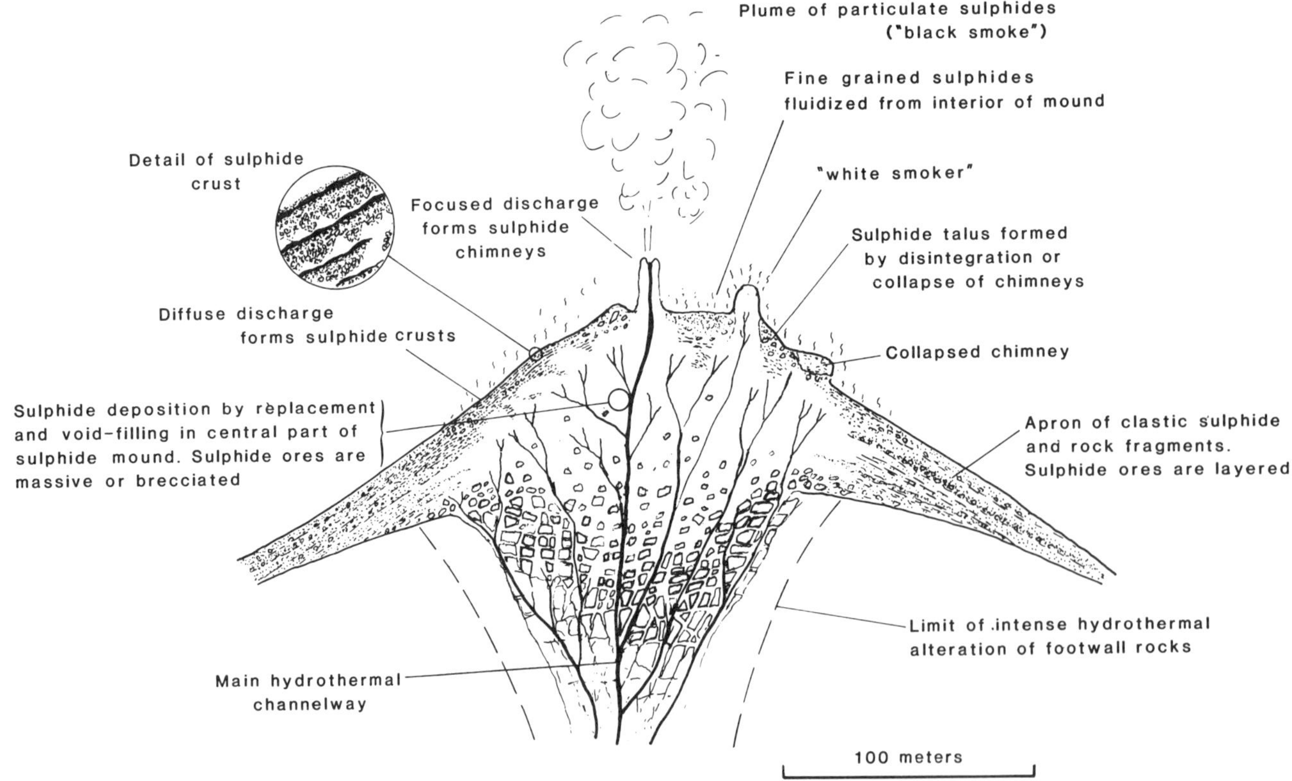

Figure 5 *Schematic representation of some processes of sulphide accumulation in a proximal VMS deposit.*

brecciated rock. The fragmental rocks may have had a magmatic (Burnham, 1983), phreatic (Horikoshi, 1969; Clark, 1983), or hydrothermal (Henley and Thornley, 1979; Lydon, 1986) explosive origin or be talus accumulations (Hekinian and Bideau, 1986).

Of these possible diverse origins to the mound or blanket of breccia within which sulphide accumulation is initiated, perhaps the most common one is hydrothermal eruption, caused by early stage explosive discharge of the ore-forming hydrothermal system (see Discussion). Consistent with this interpretation, the stockwork zone represents a hydrothermal breccia pipe, formed by repeated episodes of hydraulic fracturing and by the concomitant precipitation of hydrothermal minerals in fractures and hydrothermal alteration of the clasts. The volume expansion required by the dilatant hydraulic fracturing is accommodated by the upward displacement of the fragmented rocks of the breccia pipe and their eruption at the surface (Lydon and Galley, 1986). This gradual upward displacement of breccia clasts accounts for the gradational nature (in terms of sulphide:wall rock ratio) of the lower contact of the massive sulphide lens, the downward decrease in sulphide grade in the stockwork zone, and at least in part for the polymictic clastic apron surrounding the massive sulphide lens described in Part 1 of this article (Lydon, 1984a).

Barton (1978) recognized that much of the ore deposition of Kuroko deposits took place by episodes of open space filling and replacement that alternated with episodes of fracturing and slumping of the sulphide ores. Further elaboration of his study and concepts (Eldridge *et al.*, 1983) resulted in a model for ore deposition that is similar in essential details to that shown for the modern mound-chimney deposits in Figure 2. Salient points of the model by Eldridge *et al.* (1983) are that, once initiated, the sulphide mound itself acts as a porous medium over the vent to promote diffuse hydrothermal discharge, and that the "semi-sealed outer sulfide rind" (Eldridge *et al.*, 1983) or the "thin thermal and mechanical blanket" (Barton, 1978) are equivalent to the "low permeability crust" of modern sulphide mounds (Goldfarb *et al.*, 1983) in allowing advective, convective or conductive cooling of the ore fluid within the sulphide mound and the consequent sub-surface precipitation of sulphides. Thus, just as with modern mound-chimney edifices, the Kuroko deposits are envisaged to have accumulated both by the addition of clastic sulphide from above (plume fall-out, chimney talus) and by open space infilling and replacement within the mound.

Essentially the same general conclusions were reached by independent and earlier studies of the Archean deposits of the Noranda area (Simmons *et al.*, 1973; Spence, 1975; Riverin and Hodgson, 1980; Knuckey *et al.*, 1982). Although the concepts derived from the Noranda examples have been used in exploration strategy since the early 1960s (Gilmour, 1965), they have not been so elaborately discussed in the literature as those derived from the modern mound-chimney or the Kuroko deposits. A major difference in emphasis has been the role of explosive volcanic and hydrothermal activity, considered important by the "Noranda" school but not, with some exceptions (*e.g.*, Horikoshi, 1969; Clark, 1971, 1983), by the mainstream "Kuroko" or "modern chimney-mound" schools.

In summary, and with reference to Figure 4 and Figure 5, the model for the accumulation of ancient proximal VMS deposits which incorporates most salient points of current thought visualizes:

(1) The discharge of hydrothermal fluid from a focussed vent into a pile of fragmental rocks, which most commonly is a blanket or mound of lithic hydrothermal eruption breccia produced by the initial explosive discharge of the ore-forming hydrothermal system, or in some cases and more fortuitously, may be pre-existing magmatic, phreatic or talus breccia.

(2) Defocussing of hydrothermal flow by the fragmental pile which enhances advective, conductive or adiabatic cooling of the hydrothermal fluid within the mound and thus the precipitation of sulphides and other hydrothermal minerals. Fine particulates fluidized from the interior may accumulate at the mound surface.

(3) As channelways within the mound become more restricted by hydrothermal precipitation, hydrothermal flow becomes more diffuse and, in turn, the efficiency of sulphide precipitation within the mound increases. Formation of a low permeability crust by conductive or advective cooling of the mound exterior allows only diffuse, low velocity discharge across the mound surface which maximizes the rate of accumulation of hydrothermal particulates formed by quenching in the overlying water column.

(4) Renewed episode of fracturing (hydraulic or otherwise), and perhaps hydrothermal eruption, allows discharge of hydrothermal fluids from major new channelways as focussed buoyant plumes which form new sulphide chimneys but also disperses dissolved components in the overlying water column. Reversing buoyancy plumes would minimize this dispersal of suspended sulphide particles.

(5) Subsequent repetition of Stages 2-4 leads to growth of the mound by the internal creation and filling of fractures, by crust formation, and by the surface accumulation of eruption products, chimney talus and fall-out of plume particulates.

Massive sulphide lenses formed by these mechanisms are most likely to be conical or mound-like in morphology, with maximum lateral extent:thickness ratios (aspect ratio) typically in the range 3:1 to 5:1. Downslope transportation of sulphides as slump breccia flows or slide sheets (Schermerhorn, 1970) initiated by gravitational instability or hydraulic lifting (Eldridge *et al.*, 1983; Cathles, 1983) might be expected to produce transported deposits with different morphologies (Lydon, 1984a, fig. 9).

Deposits with a sheet-like morphology (aspect ratio > 10:1) are more likely to be formed by processes that more closely follow the laws of sedimentary superimposition than do the proximal mound deposits. The most obvious case where this mode of accumulation occurs is in a brine pool formed by high density ore solutions (*i.e.*, Sato's (1972) Type I solutions). Turner and Gustafson (1978) described situations in which initially buoyant plumes that are discharged into a submarine depression may pond below a picnocline (*i.e.*, a variant of Sato's Type II solutions). Since the ponding of the hydrothermal effluent prevents not only the dispersion of plume particulates but also insulates them from the oxidizing effects of normal seawater, sulphide accumulation under these circumstances is more efficient than in the proximal mound type of accumulation. It is thus perhaps significant that the largest VMS deposits, such as Kidd Creek (Coad, 1985) and Brunswick No.12 (van Staal and Williams, 1984) tend to have high aspect ratios.

Zonation in the massive sulphide lens

Mineralogical and chemical zonation.

The increase in Zn:Cu ratio upward and outward from the core of the massive sulphide lens is one of the most definitive characteristics of VMS deposits. The similarity of the ore mineral assemblages and the polarity of their zonation (from the core of the sulphide structure to the seawater interface) to those of modern sulphide chimneys suggests similar controls on sulphide precipitation and deposition in both cases.

Based on evidence from ancient VMS deposits, it has been suggested that the characteristic zonation patterns reflected the effects of local physicochemical gradients on mineral precipitation from, and alteration by, an ore fluid (*e.g.*, Riverin and Hodgson, 1980; Knuckey *et al.*, 1982), rather than changes in the composition or temperature of the pristine ore fluid with time (*e.g.*, Solomon and Walshe, 1979a). This view is supported by sophisticated computer models of mineral deposition from aqueous solutions similar in composition to the EPR 21°N vent fluids (Janecky and Seyfried, 1984; Bowers *et al.*, 1985; Reed, 1983). Furthermore, this computer modelling indicates that it is indeed the progressive local cooling of the solutions, rather than other physicochemical changes such as dilution, oxidation, or pH change (*e.g.*, *cf.* Large, 1977), that causes the Cu-dominant to Zn-dominant zonation. The only constraints on reproducing the sequential sulphide mineral assemblages that mark the zonation are that in the initial solution:

(i) the concentrations of Fe, Cu and Zn are within one or two orders of magnitude of one another; (ii) the metals are in solution dominantly as chloride complexes; (iii) molality dissolved H_2S > molality total dissolved metal; (iv) activity H_2S >> activity SO_4 at chemical eqilibrium (*i.e.*, the solutions are in a very reduced oxidation state).

The reason for this invariability in the sequential Cu-dominant to Zn-dominant sulphide assemblages produced by the cooling of polymetallic, reduced, acidic, sulphurous aqueous solutions is due to the relative solubilities of chalcopyrite and sphalerite as a function of temperature under these conditions (illustrated in Figure 6). The solubility of chalcopyrite in a chloride-bearing aqueous solution in equilibrium with an iron sulphide is dominantly a function of temperature. If 0.1 mg · kg^{-1} Cu in solution is taken as the minimal concentration necessary to precipitate chalcopyrite, then the lower temperature limit for chalcopyrite deposition is above 300°C for solutions from EPR 21°N vents (Janecky and Seyfried, 1984), but may be as low as 150°C in solutions with higher chloride concentrations, lower P_{O_2}, higher metal to sulphide ratio and higher Cu:Fe ratios (*e.g.*, Large, 1977). The solubility of sphalerite (and galena) in chloride solutions is much less temperature dependant than that of chalcopyrite (Figure 6), and significant concentrations of Zn and Pb in solution invariably extend to lower temperatures than does that of Cu under comparable physicochemical conditions. The degree of separation of Zn (and Pb) from Cu in sequential precipitation along the cooling path of a reduced solution depends upon the degree of undersaturation of Zn (and Pb) in the initial solution. The relative position of maximum iron sulphide deposition in the cooling sequence is dependant on the concomitant pH and P_{O_2} profile.

Contours of Cu:Zn ratios in a proximal VMS deposit can therefore be taken to represent an approximate average isotherm pattern (Figure 7). It is only approximate because, as noted above for modern sulphide chimneys, mineral zone boundaries probably oscillated with time, and the overall growth pattern entails the upward and outward expansion of isotherms within the sulphide mound, resulting in the replacement of sphalerite by chalcopyrite in the core and the relative concentration of zinc in the mound crust, chimneys and plume fallback (Eldridge *et al.*, 1983). The preservation of a consistent Cu:Zn zonation pattern implies that a proximal sulphide mound was essentially a rigid

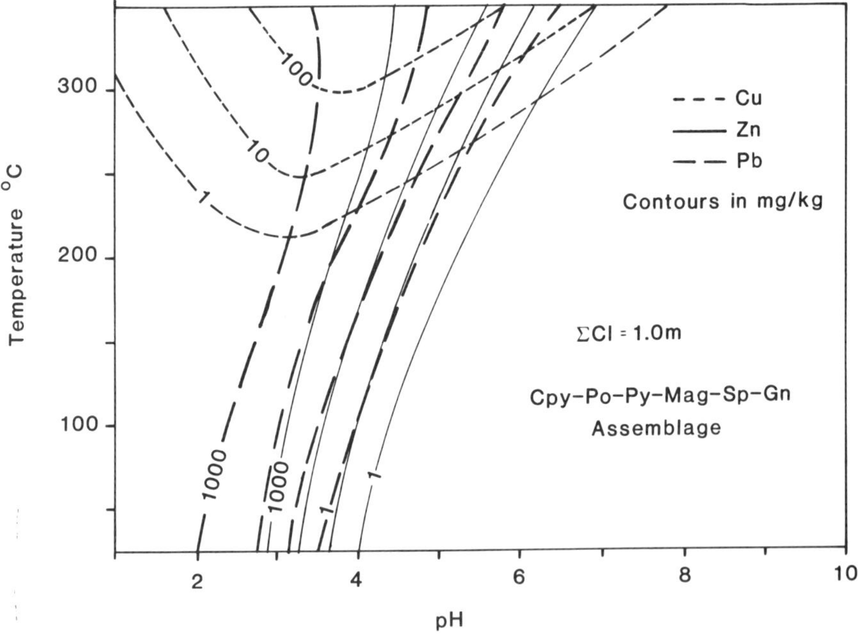

Figure 6 *Sulphide solubility patterns due to chloride complexing. Calculated total metal concentrations (in mg · kg^{-1}) in a 1 m NaCl solution in equilibrium with a chalcopyrite-pyrite-pyrrhotite-magnetite-sphalerite-galena assemblage. Note that the sulphur content of the solution ranges from tt 1 mg · kg^{-1} at low pH and low temperature to > 10,000 mg · kg^{-1} at high pH and high temperature. The isothermal decrease in chalcopyrite solubility with decreasing pH at low pH is due to the decrease in activity of Cl$^-$ as the relative proportion of chloride complexed with metals increases. (Data from Helgeson, 1969; and Crerar and Barnes, 1976).*

edifice. The argument for this conclusion, supported by the steep-sided morphology of conical mounds (Simmons *et al.*, 1973), is that since the zonation is due to a chemical evolution of the fluid, a porous rigid framework is necessary to support and maintain the spatial configuration of precipitates from the fluid.

Little is known of gangue mineral zonation in VMS deposits. In proximal deposits, silica appears to be a pervasive cement throughout the mound (*e.g.*, Costa *et al.*, 1983). The precipitation of silica from solution requires cooling without dilution, such as by conductive or adiabatic cooling (Janecky and Seyfried, 1984). Barite, when present, is concentrated in the upper parts of the deposit and seawater sulphate appears to be involved in its precipitation (*e.g.*, Watanabe and Sakai, 1983; Kowalik *et al.*, 1981). Silicates may be added to the mound by fluidization or eruption from the footwall, lithic fall-out from above, or by hydrothermal precipitation. Talc, formed by the mixing of ore fluid with seawater (Lonsdale *et al.*, 1980; Costa *et al.*, 1983; Aggarwal and Nesbitt, 1984; Bowers *et al.*, 1985) may be the most common primary hydrothermal silicate.

Sulphides accumulated by mechanical processes, including the clastic apron of proximal mounds and deposits with a high aspect ratio morphology, tend to lack a pronounced or consistent zonation. Sheet-like or layered deposits (*e.g.*, Brunswick No.12, Rosebery) generally have high aggregate (Zn + Pb)/Cu ratios and probably largely consist of the lower temperature sulphide precipitates of plume or brine-pool fall-out (Solomon and Walshe, 1979a; Green *et al.*, 1981). Sediments containing a significant proportion of iron oxides, manganese oxides, or silica, including "tuffaceous exhalites" such as the Tetsusekiei associated with the Kuroko deposits, are sometimes spatially associated with the sulphide deposits. These sediments may reflect marginal or terminal oxidation of brine pools as in the Red Sea (*e.g.*, Pottorf and Barnes, 1983), distal products of the progressive oxidation of a buoyant hydrothermal plume (Large, 1977; Kalogeropoulos and Scott, 1983), or products of contemporaneous or later low temperature hydrothermal discharge.

Evidence for the sea-floor oxidation of VMS deposits is scarce, which is surprising considering the rapidity with which sulphides of modern sea-floor sulphide mounds are oxidized. The most popular examples cited for sea-floor oxidation of ancient VMS deposits are the ochres of some deposits of Cyprus (Constantinou and Govett, 1972, 1973), but even in these cases it appears that at least some of the ochres were formed by post-burial oxidation of sulphides by circulating ground water (Lydon, 1984c). The general absence of oxidized cappings to VMS deposits may imply that one of the conditions for preservation (and perhaps accumulation) of VMS deposits is insulation from the oxidizing effects of normal seawater. This insulation may be provided by anoxic bottom conditions, developed during times of ocean stratification or induced by the ponding of sulphurous ore solutions, or by almost instantaneous burial of the sulphide deposit. Another possibility is that burial metamorphism, in tending to equilibrate the diverse mineral assemblages of the primary deposit, may have resulted in sulphidation of oxidation products.

The primary zonation pattern is susceptible to post-burial modification. In the Cyprus deposits, the primary Zn:Cu zonation is locally reversed by the deposition of secondary copper sulphides from oxidized ground water at the redox boundary between sulphide ore and unmineralized lavas (Lydon, 1984c). Mobilization of sulphides during penetrative deformation and metamorphism can result in stockwork-like veins and also distort the primary zonation patterns (van Staal and Williams, 1984).

Textural zonation. Due to the facility with which sulphide minerals anneal, recrystallize and re-equilibrate, only rarely is any semblance of the complex and intricate microscopic textural detail of a pristine VMS deposit preserved in a deposit that has undergone even lower greenschist metamorphism. Only in the youngest VMS deposits, such as those of Cyprus, Oman and Japan, are microscopic textures preserved that are comparable to those of modern mound-chimney sulphides.

However, in the least metamorphosed and deformed ancient VMS deposits, primary macroscopic textures can often be discerned, especially where the form of the texture is marked by a radical change in bulk chemical composition (*e.g.*, lithic clast in sulphide matrix; sphalerite vein in massive pyrite). The general correlation that the pyrrhotite-, pyrite- or chalcopyrite-rich core of a proximal massive sulphide lens tends to be massive or show brecciated textures, whereas the sphalerite- or pyrite-rich outer parts of the lens tend to preserve banded, bedded or clastic textures and structures, is directly

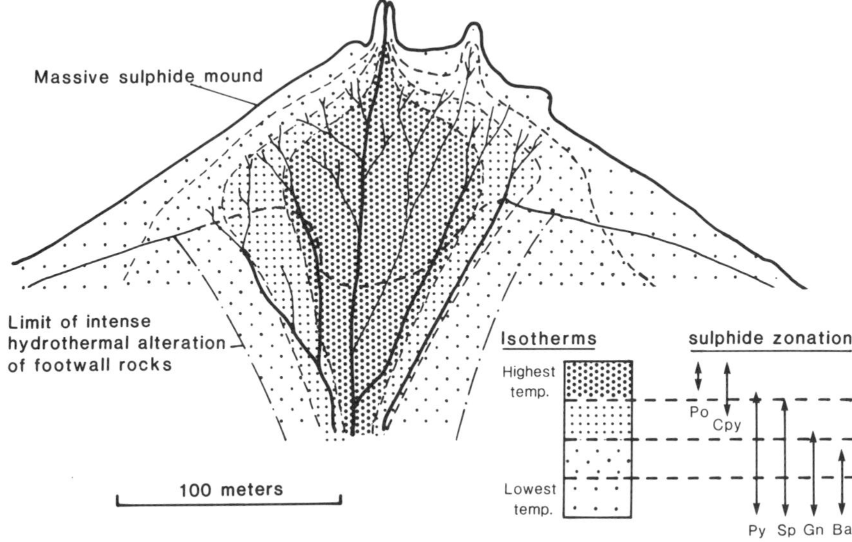

Figure 7 *Idealized sulphide zonation of the proximal VMS deposit illustrated in Figure 5 showing the direct relationship between sulphide zonation and average maximum isotherms if the ore metals were present in the ore fluid dominantly as chloride complexes.*

related to the mechanisms of sulphide accumulation described above. Ore accumulation in the core is mainly by deposition in fractures and channelways from circulating fluids that are contained by the outer low permeability crust. The sustained high temperature fluid circulation in the core leads to annealing and chemical homogenization even while the deposit is on the sea floor. The outer part of the mound, accumulated by crust formation or clastic processes, is not so pervasively subjected to this high temperature reworking of the sulphides, and hence tends to retain its original inhomogeneities.

Due to the tendency for obliteration of primary textures by metamorphism, the significance of many remnant textural features of ancient VMS deposits has probably gone unrecognized. For example, monomineralic patches and bands of sphalerite in massive pyritic ore probably represent the infilling of hydrothermal channelways and fractures, respectively, within the sulphide mound (see some comparative photographs in Lydon, 1984b).

Zonation of alteration pipe

There is consensus that the footwall hydrothermal alteration pipes of proximal VMS deposits represent the conduits for the upward flowing ore fluids (Franklin *et al.*, 1981), but probably a lack of consensus on generalization of its characteristic features (Lydon *et al.*, 1984). A composite of the various features reported for VMS alteration pipes is shown schematically in Figure 8.

Studies of the least metamorphosed Archean Cu-Zn deposits (see summaries in Franklin *et al.*, 1981) show that in terms of silicate mineralogy the alteration pipes are characterized by chloritic cores and sericitic margins, though as Riverin and Hodgson (1980) point out, in some cases the chloritic core may top out upward so that the alteration assemblage immediately below the massive ore is dominantly a sericitic facies (usually sericite-chlorite-quartz) (Figure 8). A magnesium enrichment in the core of the alteration pipe can often be demonstrated (*e.g.*, Riverin and Hodgson, 1980; Knuckey and Watkins, 1982; Walford and Franklin, 1982). A similar configuration occurs in the alteration pipes of Cu-Zn deposits in Cyprus where most commonly chloritic cores are surrounded by illite or illite-smectite peripheries (Clark, 1971; Lydon and Galley, 1986). However, in some cases, such as the Pitharokhoma deposit (Richards *et al.*, in press), illitic alteration dominates the entire alteration pipe. At the Mathiati deposit, at least, magnesium enrichment is confined to a narrow zone around the periphery of the chloritic part of the alteration pipe (Lydon and Galley, 1986), which may be analogous to a narrow zone of high-Mg chlorite, approximately coinciding with the transition between chalcopyrite-dominant and sphalerite-dominant stockwork mineralization, observed by the author at the Archean Millenbach deposit.

The same range of features can be observed in the Cu-Zn-Pb type of deposit. For example, chloritic alteration with magnesium enrichment occurs beneath the La Zarza deposit in Spain (Strauss *et al.*, 1981) whereas quartz-sericite dominates the alteration below Kuroko deposits (Shirozo, 1974). Chlorite which occurs outside of the quartz-sericite zone below Kuroko deposits is ubiquitous in footwall rocks for at least several kilometres away from the sulphide mineralization (Date *et al.*, 1983), and probably should not be considered to indicate a "reverse zonation" of alteration (Franklin *et al.*, 1981, p. 607; Urabe *et al.*, 1983) on the scale at which hydrothermal alteration pipes are usually defined.

The chlorite to sericite (or illite) zonation has been interpreted to represent a decreasing thermal gradient (Riverin and Hodgson, 1980; Lydon and Galley, 1986). This interpretation is consistent with the sulphide zonation pattern of the alteration pipe which characteristically consists of a relatively chalcopyrite-rich core and sphalerite-rich margins. This pattern, as discussed above, is diagnostic of an outward-decreasing thermal gradient. It is also consistent with experimental seawater-basalt interaction, which indicates that the transition between potassium loss from and potassium addition to basalt occurs in the range 150-200°C (Seyfried and Bischoff, 1979).

If the thermal gradient from core to margin of the alteration pipe is due to subsurface mixing of the ore fluid with ambient waters at the margin of the

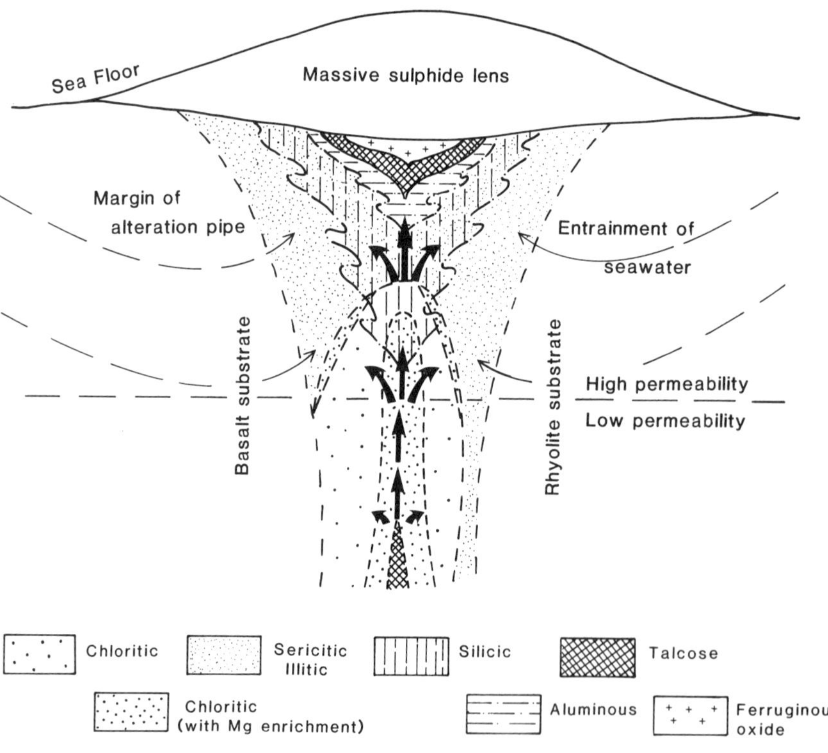

Figure 8 *Composite representation of the various alteration assemblages that have been reported for alteration pipes of VMS deposits. See text for explanation and discussion.*

hydrothermal plume (Lydon and Galley, 1986), then the ratio of sericite (or illite) to chlorite reflects the average degree of mixing that has taken place at that site. If this is correct, then the upward culmination of the chloritic zone depicted in Figure 8 reflects the tendency of hydrothermal plumes to collapse upward due to the entrainment of subsurface waters in a permeable substrate. The low temperature thermal springs of the Galapagos Ridge (Edmond *et al.*, 1979) are thought to represent the discharge of such a collapsed hydrothermal plume, its original sulphide load having been deposited in the sub-surface to give a stockwork deposit analogous to that encountered by DSDP Hole 504B (Alt *et al.*, 1986).

Potassium added to the rock in the form of alteration sericite could have been derived from seawater and the ore fluid (Lydon and Galley, 1986) and/or from that released by the process of chloritization in the core of the alteration pipe (Riverin and Hodgson, 1980). Consistent with experimental seawater-basalt interaction (*e.g.*, Hajash, 1975; Mottl and Holland, 1978; Seyfried and Bischoff, 1979) and the magnesium-free nature of modern pristine hydrothermal fluids emanating from basalts (von Damm *et al.*, 1985), magnesium added in the uppermost or peripheral parts of the alteration pipe is most likely contributed by entrained seawater (Roberts and Reardon, 1978; Lydon and Galley, 1986). The preferential addition of magnesium in the core of the alteration pipe and in sulphide vein selvages implies a derivation from the ore fluids (Riverin and Hodgson, 1980), which likely means either a minor amount of seawater Mg having been entrained into the hydrothermal fluid at depth or else the pristine ore fluids contained Mg and were therefore generated in a feldspar-free lithology (*e.g.*, completely chloritized/epidotized basalt, argillaceous sediments or ultramafic rocks).

Immediately below the massive sulphide lens, the ubiquitous chloritic and/or sericitic alteration (or precursor/successor assemblages) may be supplanted or superimposed by hydrothermal mineral assemblages of contrasting composition. The range includes siliceous (*e.g.*, Shirozo, 1974; Lydon and Galley, 1986), aluminous (*e.g.*, Nilsson, 1968; Walford and Franklin, 1982), carbonate (*e.g.*, Franklin *et al.*, 1975); Deptuck *et al.*, 1982), talcose (Roberts and Reardon, 1978; Aggarwal and Nesbitt, 1984), and ferruginous oxide (*e.g.*, Knuckey and Watkins, 1982; Richards and Boyle, 1986) assemblages. Although Figure 8 implies that these particular mineral assemblages are confined to the uppermost parts of the alteration pipe, this is only a speculative interpretation. Metamorphism, structural complexity and lack of access to alteration pipes away from economic grades of mineralization in most VMS deposits has not allowed studies that unequivocally relate these mineral assemblages to the near-surface part of the hydrothermal discharge vents.

Some of the iron oxide in ferruginous units immediately below the massive sulphide lens originated as early accumulations of oxidized plume fall-out prior to the establishment of anoxic bottom conditions and the stabilization of sulphides (Lydon and Galley, 1986; Richards and Boyle, 1986). Similarly, accumulation of talc may have occurred at the sea floor during the initial discharge of silica-saturated hydrothermal solutions into magnesium-bearing seawater (Costa *et al.*, 1983; Aggarwal and Nesbitt, 1984). Talc-actinolite may have also formed just beneath the massive sulphide lens due to rock alteration by the shallow entrainment of magnesium-bearing seawater into the hydrothermal conduit (Roberts and Reardon, 1978). Boiling of the hydrothermal fluids, causing cooling of the ore solutions without dilution by entrained seawater, would lead to the enhanced precipitation of silica (Lydon and Galley, 1986), and, particularly when the solutions contained dissolved CO_2, other minerals which are dependent on acidity for their solubility (Drummond and Ohmoto, 1985). Adiabatic cooling could also give rise to aluminous mineral assemblages (Riverin and Hodgson, 1980) as could high acidity created by oxidation of sulphurous ore solutions.

Based on a study of fluid inclusions in the upper part of the stockwork zone of Kuroko deposits, Pisutha-Arnond and Ohmoto (1983) demonstrated a waxing and a waning temperature of the hydrothermal system. Time-dependent zonation is difficult to separate from space-dependent zonation in most VMS deposits. This difficulty, combined with the complexities of mineral assemblage genesis noted above, emphasize the caution necessary before drawing generalizations on the nature of the ore fluid and the hydrothermal history of the VMS deposit based on observations in the upper part of an alteration pipe.

Types of hydrothermal systems

As illustrated in Figure 9, a specific hydrothermal system may be a composite of various end-member types in which both the hydrothermal fluids and the energy for fluid circulation are derived from diverse sources. Discussion here is limited to the three main models that have been applied to the genesis of VMS deposits.

Convection cell models. The existence of hydrothermal convection cells in modern oceanic crust is well established. First postulated to explain the discrepancy between the measured and theoretical heat flow of oceanic ridges (Palmason, 1967; Lister, 1972), and later predicted by mathematical models of the heat budget (Wolery and Sleep, 1976; Crane and Normark, 1977), they were confirmed by the discovery of high temperature hydrothermal vents at oceanic ridges (Francheteau *et al.*, 1979). Supported by the reality of the phenomenon, a convection cell model is currently the most popular hydrodynamic model for the formation of VMS deposits.

Spooner and Fyfe (1973) suggested that waters which penetrate and react with oceanic crust could form VMS deposits on their return to the sea floor. This convection cell model was first specifically applied to the VMS deposits of Cyprus (Heaton and Sheppard, 1977; Spooner, 1977). The fundamental concept in this model is that subsurface waters, dominantly of seawater origin, are caused to convect by a magmatic heat source and leach the ore components from the rocks along their flow path. In its application to specific areas, the postulated magmatic heat source is tailored to suit the local geological setting of the VMS deposit under discussion. Suggested heat sources have included rhyolite domes or plugs (*e.g.*, Ohmoto and Rye, 1974), sub-volcanic sills (*e.g.*, Campbell *et al.*, 1981), felsic plutons (*e.g.*, Cathles, 1983), and, of course, spreading ridge magma chambers of both ancient ophiolites (*e.g.*, Spooner, 1977) and the modern oceanic lithosphere (*e.g.*, Lowell and Rona, 1985).

The convection cell scenario has been mathematically modelled by a variety of techniques employing a spectrum of basic assumptions. However, all calculations point to the conclusion that if the ore fluids contain the same concentration of base metals as pristine black smoker fluids of the EPR (*i.e.*, 10 ppm Zn + Cu) there is not enough heat in a reasonably sized magmatic body to form an average sized VMS deposit (about 6 million tonnes sulphide).

Cathles (1978, 1983) modelled Kuroko hydrothermal systems on the basis of assuming a magmatic heat source at an initial temperature of 700°C and hydrothermal flow through a porous medium. He found that even if the ore fluid contained 1000 ppm Cu, the average sized rhyolite plug commonly associated with Kuroko deposits was several orders of magnitude too small to convect sufficient fluid to supply the amount of copper in an average Kuroko deposit (Cathles, 1978). To form the 4.5 million tonnes of base metal resources of the Hokuroku basin from >300°C fluids containing 100 ppm Cu + Zn + Pb would require an intrusive dyke 40 km long, 3.25 km high and 1.3 km wide, provided that all the dissolved metal accumulated at the sea floor discharge site (Cathles, 1983). The deposits could have formed in less than 5,000 years as a result of intrusive pulses in a spreading environment (Cathles, 1983). Similarly, with a minimum requirement of 78 km^3 of felsic magma to heat enough >300°C fluid containing 100 ppm Cu + Zn to form the Noranda deposits, the composite Flavrian "granite" was permissibly the heat source (Cathles, 1983).

Using a discrete fracture model for hydrothermal convection at mid-oceanic ridges, Cann and Strens (1982), Cann *et al.* (1985), Strens and Cann (1986) concluded that heat stored in solid rock within 1 or 2 km of the surface would not allow the formation of a VMS deposit. They found that discharge above 300°C could be sustained only for 10^2 years, which is comparable to the estimated 10^1-10^2 years for modern black smoker vents (Macdonald *et al.*, 1980). Cann and his co-authors suggested that the heat must be extracted from a convecting magma chamber and calculated that ore fluids at 350°C containing 115 ppm Fe requires the crystallization of 30 km^3 of basaltic magma to form a 3 million tonne iron sulphide deposit. The deposit could accumulate in 4000 years assuming 70% accumulation of the sulphides at the sea floor hydrothermal vent.

Using thermal balance considerations, Lowell and Rona (1985) also concluded that the heat content of permeable rocks in oceanic crust was insufficient to form a 3 million tonne sulphide deposit assuming 100 ppm metal concentration in the hydrothermal fluid and 100% efficiency of sulphide accumulation at the hydrothermal vent. However, the same deposit could be formed under the same assumptions, if the heat was extracted from the roof of a vigorously convecting magma chamber. As with the models of Strens and Cann cited above, the essence of this model is a twinned convective system, in which heat is supplied to the base of a

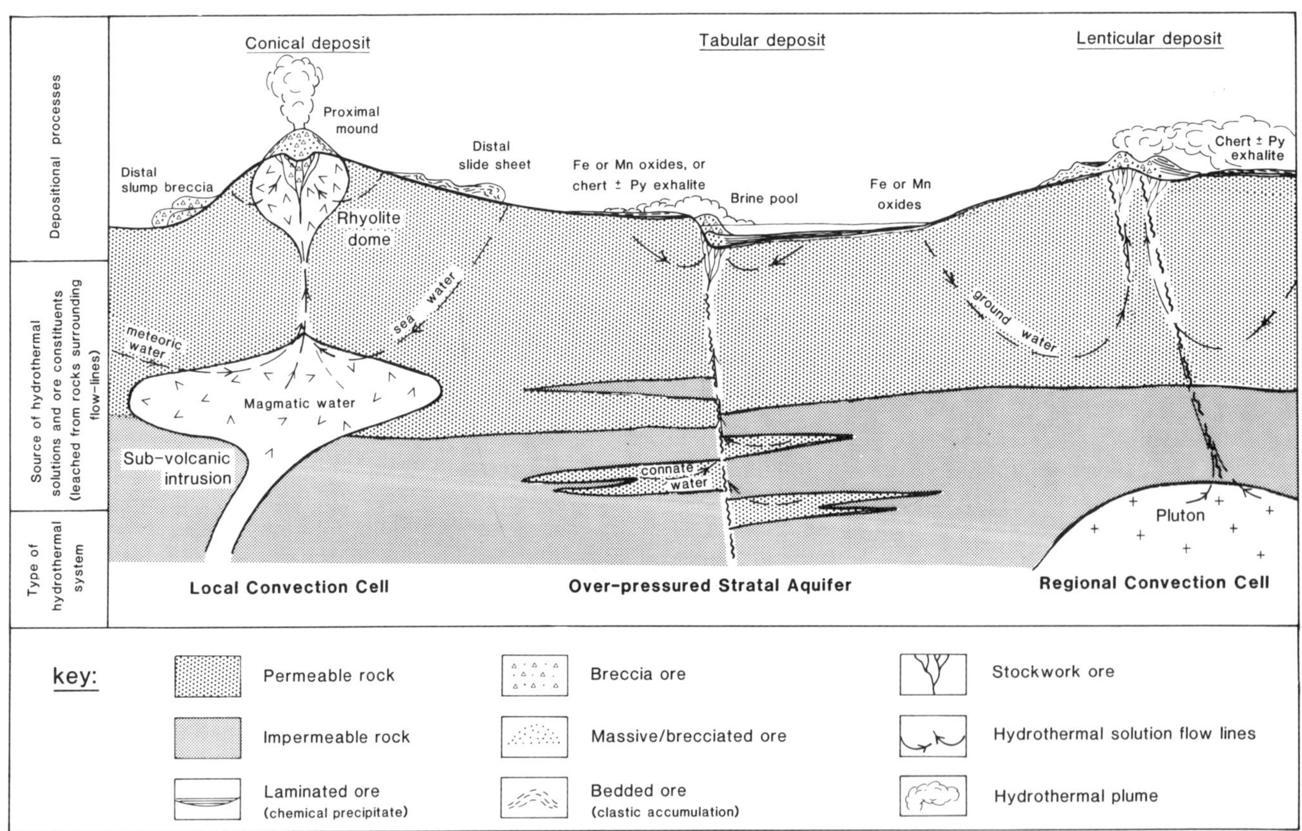

Figure 9 *Schematic representation of some major features of genetic models for VMS deposits. Note that the portrayed linkages between various factors represent only some of the many possible combinations,* e.g., *a brine pool deposit can be formed by the discharge of a local convection cell driven by a cooling pluton.*

hydrothermal convection cell by a magmatic convection cell across a thin-plated gabbro at the roof of the magma chamber. A positive feature of this model is that since the heat exchange site does not migrate with time, the spatial configuration of the hydrothermal convection cell may remain stable for an extended period.

It is apparent from the above selection of convection cell models, that enormous volumes of magma, whether crystallized or not, are required to form an average VMS deposit. The numbers quoted are close to the best-case scenarios, in that it is assumed that all the ore metals transported by high-temperature fluids are precipitated as distinct sulphide deposits. Taking into account the low probability of high temperature discharge being confined to just a few vents and the inefficiency of sulphide accumulation from a buoyant hydrothermal plume, as demonstrated by the modern black smoker vents, it would seem that to account for the formation of the larger VMS deposits by discharge from hydrothermal convection cells, the ore fluids would have had to contain considerably more than 100 ppm total metal.

Magmatic hydrothermal models. There has been the suggestion, but little corroboration, that the ore fluids for VMS deposits are derived from the volatiles of magmas. For example, Urabe and Sato (1978) suggested a magmatic origin for ore fluids for Kuroko deposits largely on the grounds of spatial association with rhyolite domes and weaknesses with alternative models. Bryndzia *et al.* (1983) favoured a magmatic component in Kuroko ore fluids to explain elevated salinities (up to 1.9 times greater than seawater) in fluid inclusions, whereas Sawkins and Kowalik (1981) expressed a similar sentiment for ore fluids of the Buchans orebodies based on considerations of lead isotopes and the lead budget. The perception that VMS deposits can be related to particular igneous suites, particularly the most differentiated products of a calc-alkaline magma (Sangster, 1972; Solomon, 1976), may also be construed as inference of a magmatic affiliation. Although the magmatic model does not seem to have current support, perhaps the concept should not be dismissed. The geochemical data of Perfit and Fornari (1983) and Perfit *et al.* (1983), for example, could be interpreted to reflect the loss of large quantities of sulphur and base metals during a high p_{H_2O} stage of magmatic fractionation.

Stratal aquifer model. Although this model has been proposed for VMS deposits (Hodgson and Lydon, 1977; Gibson *et al.*, 1983), it has found much more favour in genetic models for the sedimentary-exhalative class of sulphide deposits (*e.g.*, Walker *et al.*, 1977; Badham, 1981; Lydon, 1983; Sawkins, 1984; Lydon, 1986). In a sense this is informative, because as Gilmour (1976) has pointed out, the only major difference between some volcanogenic massive sulphide deposits and some sedimentary-exhalative deposits is the lithology of their host rocks. If sedimentary-exhalative sulphide deposits can be formed without the involvement of large volumes of magmas, then it would seem logical that some VMS deposits may have been formed in a similar way.

In the stratal aquifer model, it is visualized that the ore fluids originate as the pore waters of a porous rock unit (the "aquifer") which have been prevented from migrating during burial and compaction by an overlying impermeable barrier (the "cap-rock"). Progessive burial causes heating of the pore fluids along the geothermal gradient and an increase in pore-fluid pressure above hydrostatic pressure. Depending upon the mechanical strength of the cap-rock, pore-fluid pressure may approach or exceed lithostatic pressure. Eventual hydraulic or mechanical fracturing of the cap-rock, perhaps triggered by tectonic activity (*e.g.*, Sibson *et al.*, 1975), allows the upward release of the over-pressured pore waters along fracture zones. Once cross-stratal permeability is established, a short-lived successor convection cell may be initiated, which is driven by the heat stored in rocks of the aquifer.

A unique feature of the stratal aquifer model is that it allows the surface expulsion of very large quantities of fluid within a short time. A modest modern example of the phenomenon may be the expulsion of about 10^{10} kg of saline ground water within a few months of the 1966 initiation of the Matsushiro swarm earthquakes (Tsuneishi and Nakamura, 1970). An attraction of the model is its minimal energy requirements. Since a cap-rock is by definition also a thermal insulator, an over-pressured hydrothermal aquifer actually conserves heat that would otherwise be lost by conduction and convection. Only the normal conductive heat flow of a volcanically active area is required to heat the pore fluids of the aquifer to more than 300°C within 1 km of the surface (McNitt, 1970). In a convection cell model, much of the heat content of a magmatic body is transferred at low temperature during the waxing and waning stages and by peripheral hydrothermal flow. Another positive feature is the minimal water:rock ratio of the stratal aquifer system, which in turn maximizes the capacity of the hydrothermal solution to attain high concentrations of metal by the leaching of the aquifer rocks (see below). The model is also geologically very realistic in terms of the volume requirements of the aquifer. For example, the 4.7×10^{13} kg ore fluid containing 100 ppm ore metal estimated by Cathles (1983) to be necessary to form the Noranda deposits could be contained by a 300 m thick aquifer of 15% porosity occupying an area of 900 km^2, which is about the thickness and area (Cathles, 1983) occupied by the semi-conformable alteration zone underlying the deposits (Gibson *et al.*, 1983). The increasing recognition of semi-conformable alteration zones stratigraphically below the favourable ore horizon in different VMS districts of Canadian greenstone belts (*e.g.*, Franklin *et al.*, 1975; MacGeehan and MacLean, 1980; Gibson *et al.*, 1983; Bailes *et al.*, 1987) may be evidence that the stratal aquifer model has wide applicability.

As noted near the beginning of this article, the largest modern sulphide mounds may be those of the Middle Valley and Escanaba Trough areas, where turbidite sediments infilling the axial valleys cap the underlying basalt substrate. It may not be coincidence that the seismic profile of Middle Valley (Davis *et al.*, 1987, fig. 5) is remarkably similar to figure 12 of Lydon (1986) which, in illustrating a genetic model for the Irish sedimentary-exhalative deposits, depicts 500 m of argillaceous sediments capping arenaceous sediments of a fault-controlled trough. The lithologies may be different in the two areas, but the thermal and hydrodynamic roles they play are probably identical.

Distribution of VMS deposits. There is no obvious advantage to any one of the above genetic models in explaining the areal distribution of VMS deposits. The hydrothermal discharge vents are most commonly localized by fracture systems (Gilmour, 1965; Sangster, 1972;

Scott, 1978) which serve only to focus the hydrothermal discharge and have no direct bearing on the mechanisms by which the hydrothermal fluids were generated. The tendency of VMS deposits to occur in clusters can perhaps be most readily rationalized in terms of a convection cell model or a magmatic model, in which the cluster represents the annular distribution of hydrothermal discharge above a cooling pluton (Cathles, 1983). However, the same distribution patterns can also be explained by the stratal aquifer model, noting that the volcanic stratigraphy of many VMS clusters can be interpreted in terms of a resurgent caldera (Hodgson and Lydon, 1977; Ohmoto, 1978; Ohmoto and Takahashi, 1983), that the range of caldera diameters is comparable to that of VMS clusters (Sangster, 1980), and that collapse calderas represent one of the most favourable environments in volcanic terrains for the development of stratal hydrothermal reservoirs (Hodgson and Lydon, 1977).

The stratal aquifer model probably best explains the phenomenon of "the favourable horizon". A single tectonic pulse can initiate synchronous discharge from different geopressured aquifers over a relatively wide area. Because high volume hydrothermal flow from a geopressured aquifer would be expected to be a single episode of relatively short duration, ore deposition over the entire mining district would likely be restricted to a narrow stratigraphic interval. However, this argument is not definitive, because as the calculations of Cathles (1983) indicate in the Noranda case, the Flavrian pluton could have driven the convective circulation of sufficient quantities of high-temperature fluid to have formed the VMS deposits in just a few thousand years, if the permeability of the pluton and the regional country rocks were within a permissible range. Certainly, insofar as interpretation based on zircon U-Pb ages allow, the Flavrian pluton is co-eval with volcanic rocks that contain the favourable horizon at Noranda (Mortensen, 1987). The preference of the stratal aquifer over the convection cell model to explain the favourable horizon phenomenon depends on the time interval that the favourable horizon represents (*i.e.*, whether it represents perhaps 10^1-10^3 years *versus* 10^3-10^5 years, respectively).

A difficulty with the convection cell model is that, if hydrothermal convection is an integral part of submarine volcanism, as the evidence from the modern ocean floor and the predictions based on mathematical models suggest, then every submarine volcanic centre would be expected to have its attendant hydrothermal convection cells. The model in its simplest form therefore does not appear to rationally explain, for example, the distribution of VMS deposits in the Abitibi Belt, where, as described in Part 1 of this article (Lydon, 1984a), of the nine volcanic complexes identified only four contain VMS deposits. Perhaps in its application to the genesis of VMS deposits of significant size, the convection cell model requires qualification as to a minimum size and maximum depth of the heat source, such as large sub-volcanic sills (Campbell *et al.*, 1981), and a specific range to the regional permeability of the volcanic pile that the plutonic heat source intrudes (Cathles, 1983).

Generation of ore fluids

The consensus is that the hydrothermal fluid that emanates from modern submarine spreading ridges is seawater that has reacted with basaltic rocks as it is circulated through oceanic crust by a convection cell. The model is supported by experimental seawater-basalt interaction (*e.g.*, Hajash, 1975; Mottl, 1983a, b; Seyfried and Bischoff, 1981), computer modelling (Reed, 1983) and observations on sea-floor basalts (Humphris and Thompson, 1978; Thompson, 1983).

The chemical evolution of the seawater (Table 1), as it is heated in the presence of basalt, is dominated in the early stages by the removal of sulphate, which precipitates as anhydrite due to temperature increase. The major cationic exchange is dominated by the loss of magnesium to the basalt in exchange for calcium. Magnesium is also removed above 250°C as MHSH (Bischoff and Seyfried, 1978) causing high acidity in the fluid. High acidity may also be created by epidote formation and by sodium fixation during retrograde cooling when silica activities are high (Seyfried and Janecky, 1985; Shanks and Seyfried, 1987). Potassium is lost to basalt at low temperature, but leached from it above 150°C (Seyfried and Bischoff, 1979). The result is a sulphurous Na-Ca-Cl fluid containing ppm concentrations of Fe, Mn, Zn and Cu (Table 1).

Oxygen and hydrogen isotope data for hydrothermal minerals and for water from fluid inclusions in minerals associated with VMS deposits (*e.g.*, Ohmoto and Rye, 1974; Heaton and Sheppard, 1977; Beaty and Taylor, 1982; Pisutha-Arnond and Ohmoto, 1983) allow the interpretation that the hydrothermal fluids were seawater, possibly containing a component of magmatic and/or meteoric water, that had undergone isotope exchange by reaction with rocks. The elevated salinities of ore fluids generated within oceanic crust can be explained by a variety of processes. Rock hydration reactions by consuming water will concentrate chloride in the pore fluid. For example, a rock of 4% porosity absorbing 4 wt.% water by clay alteration would increase the salinity of the pore fluid by a factor of 2.7 (Cathles, 1983). Phase separation of the NaCl-H_2O fluid under high temperature and pressure conditions produces a brine with salinities two to three times that of seawater (Bischoff and Pitzer, 1985). This chloride brine may be formed and discharged during active hydrothermal venting or may collect in the sub-surface to form a brine reservoir that is tapped later (Von Damm and Bischoff, 1987; Gallinatti, 1984). Ore fluids for VMS deposits associated with sedimentary rocks could attain high salinities simply by the dissolution of evaporites, as in the case of the Red Sea brines, or by a process of ion filtration by argillaceous lithologies (Graf, 1982). Similarly, ore fluids in basaltic rocks could possibly achieve high salinities by the dissolution of a speculative $Fe_2(OH)_3Cl$ phase formed by experimental seawater-basalt reaction at low water: rock ratios (Seyfried *et al.*, 1986).

The $\delta^{34}S$ values of modern sulphide mound-chimney complexes range from about 1.5 to 4 ‰ (per mil) (Styrt *et al.*, 1981; Zierenberg *et al.*, 1984) and the sulphur is interpreted to be derived mainly from the basaltic substrate, with a smaller component derived from reduced seawater sulphate previously deposited as anhydrite (Shanks and Seyfried, 1987). Sulphur isotope values of ancient Cu-Zn VMS deposits have a similar range (Lydon, 1984a, fig. 5), which seemingly indicates that the sulphur in these deposits was likewise largely derived from the footwall mafic volcanic succession. Sangster (1976) noted that average sulphur isotope values for Phanerozoic deposits showed a sympathetic variation with co-eval seawater sulphate. The trend is almost entirely due to the Zn-Pb-Cu type of VMS deposit, which suggests that either these deposits derived their sulphur from footwall sediments, from reduced sulphur in anoxic bottom waters of a stratified ocean, or by direct reduction (bacterial ?) of seawater sulphate. The reason for the weak correlation between decreasing $\delta^{34}S$ values and decreasing Cu/Zn ratios (Lydon, 1984a, fig. 5) is not clear, but the same trend is also apparent from the core to the flanks of the massive sulphide lens in some deposits (*e.g.*, Kajiwara, 1971).

Other isotopic tracers, including strontium (Honma and Shuto, 1979; Farrell and Holland, 1983) and lead (Doe and Zartman, 1979; Thorpe *et al.*, 1981), also point to the ore components being derived by interaction between the ore fluid and rocks stratigraphically below the ore horizon. (Note that this does not mean the immediate footwall to the deposit, but in most cases to a depth of 0.5 to 3 km or more below the co-eval sea floor).

The most popular explanation for the aggregate metal ratios of VMS deposits is that they reflect the trace metal composition of the source rocks (*e.g.*, Hutchinson, 1973; Solomon, 1976; Ohmoto *et al.*, 1983). Most VMS deposits in Archean greenstone belts and Phanerozoic ophiolites, where the dominant lithologies at the time of ore deposition were of mafic composition, are of the Cu-Zn type. In contrast, the substrate to deposits of the Zn-Pb-Cu type (*e.g.*, Kuroko deposits of Japan, Bathurst district of New Brunswick, Buchans district of Newfoundland, Pyrite Belt of Spain) consists dominantly of felsic volcanic rocks and/or argillaceous sediments. It has long been recognized that the overall range of Cu:Zn:Pb ratios of VMS deposits is about the same as that for rocks (Wilson, 1953; Wilson and Anderson, 1959). The argument that the paucity of lead in the Cu-Zn type of VMS deposits is due to the low abundance of lead in mafic rocks is widely accepted. The composition of modern sulphide mound-chimney deposits seems to support this explanation, in that the average Zn:Pb ratios for samples from deposits resting directly on basalts is >60:1 (Rona, 1984). The abundance of lead in the Zn-Pb-Cu type of VMS is ascribed to its derivation from rocks with higher lead contents, notably felsic volcanics and sediments. Again, this view is supported by the compositions of samples from modern sulphide deposits overlying sediments of the Guaymas Basin and Escanaba Trough, which have Zn:Pb ratios as low as 3:1 (data from Hannington *et al.*, 1986; Koski, 1987). Thus, there is empirical evidence that sulphide accumulations with the ore element ratios characteristic of both the Cu-Zn and the Zn-Pb-Cu type of VMS deposits can be formed from acidic chloride solutions, and a factor in the difference between the two is the nature of the source region lithologies (note that the evidence from the Pb-rich modern sulphide deposits is that some of the ore components may be derived from the upper 500 m of the footwall succession).

Discussion

There is no doubt that the study of modern submarine hot springs and sulphide deposits, together with experimental work on seawater-rock interaction, has rapidly advanced the understanding of the genesis of VMS deposits. There also seems to be no doubt that both the modern sulphide deposits and ancient VMS deposits are products of the same fundamental processes. The main question remaining is the extent to which the analogy can be applied.

One of the main differences between modern and ancient VMS deposits is that the average amount of sulphide in the modern deposits is about two orders of magnitude smaller than in ancient VMS deposits though, as Lowell and Rona (1985) pointed out, there is overlap in the range. The reasons for this could be simply that very small VMS deposits are neglected in compilations of VMS statistics because they are not economic to mine and that few large modern deposits have been discovered because of the short exploration history of the modern ocean floor.

It is unlikely that the largest VMS deposits, consisting of more than 100 million tonnes of sulphides (*e.g.*, Kidd Creek, Brunswick No.12, Rio Tinto), were formed from ore fluids with metal concentrations as low as those in the modern vent fluids. The mathematical models indicate that the energy requirements are too large and that it would be unreasonable to expect a single convection cell to maintain the same flow paths for the length of time required to discharge sufficient fluid from a single vent site. The very large sulphide deposits therefore require either higher concentrations of metal in the ore fluid and/or a higher efficiency in the mechanism of sulphide accumulation than occurs in modern mound-chimney deposits.

Metal content of hydrothermal fluids. Although the current consensus is that the ore components for VMS deposits are leached by the ore fluids from the sub-seafloor rock column, little attention has been given to the constraints of the leaching process itself. Lydon (1983, 1986) has pointed out that the trace metals can be considered to occur in rocks either as: (1) labile components, which include those metals that are loosely bound at the surfaces of mineral grains and those held by a grain coating of a material with high adsorption capacities (iron oxide coatings of quartz grains in red sandstones is a good example); or (2) bound components, which occur as diadochic substitutions for essential elements within the crystal structures of rock-forming minerals.

In most lithologies, the greatest proportions of trace elements occur as bound components, and can therefore only be leached during host mineral destruction. The natural destruction of a mineral may be effected either by dissolution or, more commonly, by alteration into secondary minerals. Since the amount of a bound metal that can be leached is proportional to the amount of host mineral destroyed, in cases where mineral destruction or conversion is due to mineral-solution interaction, the amount of metal leached is dependent upon the reactive capacity of the solution. For most natural water-rock interactions, the solution loses its reactive capacity (*i.e.*, chemically equilibrates with the rock) after reacting with just a fraction of its weight of rock, and therefore only low concentrations of metal can be achieved in the leachate. For example (Table 2), the reaction of seawater with a basalt to form smectite, requires a water:rock ratio of 42:1. If 50 ppm metal were leached from the basalt, the final concentration of metal in solution would be

Table 2 Some reactions involving the alteration of basalt to hydrous secondary mineral assemblages. The concentration of trace metal in the product solution is calculated assuming that a total of 50 ppm trace metal is lost from the weight of basalt which reacts with the solution. The minimum seawater:basalt ratio is calculated from the weight of seawater (containing 1350 ppm Mg) required to supply the weight of magnesium and/or water required to supply the reactants.

REACTION 1

0.076 Basalt + 0.963 Mg^{2+} + 2.212 H_2O + 0.084 O_2 → 1.00 Smectite + 0.785 Ca^{2+} + 0.202 Na^+ + 0.024 Fe^{2+} + 0.088 H^+ + 0.587 H_4SiO_4

Minimum seawater : basalt ratio = 42:1 — < 1 ppm trace metal in product solution

REACTION 2

0.104 Basalt + 0.663 Mg^{2+} + 0.155 Na^+ + 1.781 H^+ + 1.187 H_2O + 0.149 O_2 → 1.00 Smectite + 0.50 Albite + 1.260 Ca^{2+} + 0.326 Fe^{2+} + 0.525 H_4SiO_4

Minimum seawater : basalt ratio = 20:1 — < 2 ppm trace metal in product solution

REACTION 3

1.000 Basalt + 0.009 Mg^{2+} + 10.379 H_2O + 0.179 O_2 → 4.90 Albite + 1.604 Chlorite + 2.075 Actinolite + 3.776 Epidote + 1.893 Quartz + 0.009 Ca^{2+}

Minimum seawater : basalt ratio = 1:30 — > 1500 ppm trace metal in product solution

MINERAL FORMULAE

Smectite: $(Na_{.17}Ca_{.09}Mg_{1.73}Fe^{2+}_{.32}Fe^{3+}_{.43}Al_{.39})(Al_{.90}Si_{3.10})O_{10}(OH)_2$ (Seyfried and Bischoff, 1977)

Chlorite: $(Mg_{.51}Fe_{.49})_5\,Al_2Si_3O_{10}\,(OH)_8$ — Approximate average compositions

Actinolite: $Ca_{2.0}\,(Mg_{.60}Fe_{.40})_5\,Si_{8.0}O_{22}\,(OH)_2$ — in altered ocean floor basalts

Epidote: $Ca_{2.0}\,(Fe_{.60}Al_{2.40})\,Si_3O_{12}\,(OH)$ — (Humphris and Thomson, 1978).

Albite: $Na\,Al\,Si_3\,O_8$

Basalt: $Na_{4.90}(Ca_{11.71}Mg_{10.27}Fe^{2+}_{8.83}Fe^{3+}_{1.55})\,Al_{17.18}Si_{49.33}O_{160}$ — Barth unit cell from analysis by Seyfried and Bischoff, 1977

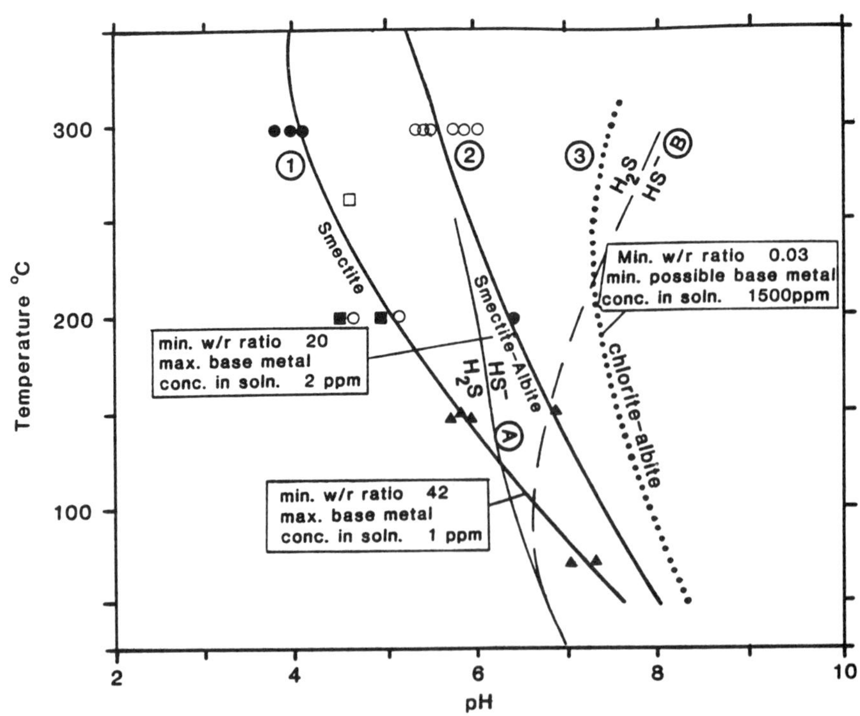

Figure 10 (right) *Summary of experimental data on seawater-basalt interaction showing quenched pH and alteration mineral assemblages for experiments of more than 14 days duration (filled squares: Bischoff and Dickson, 1975; filled circles: Hajash, 1975; filled triangles: Seyfried and Bischoff, 1979; open squares: Seyfried and Bischoff, 1977; open circles: Mottl and Holland, 1978). Bars join pH measurement at the end of 14 days and at the end of the run. Curve for chlorite-albite-(+quartz) calculated assuming 1 m NaCl and an activity of Mg^{2+} equal to that for smectite-albite stability at the same temperature. Reactions, mineral compositions, and leached metal concentrations as shown in Table 2. Boundary between fields of $H_2S(aq)$ dominance and HS^- dominance: Curve A calculated from data of Brewer (1982); Curve B calculated from data of Ellis and Giggenbach (1971).*

about 1 ppm minus the amount incorporated into the secondary alteration mineral assemblage. At the other end of the scale, the conversion of basalt into a chlorite-epidote-actinolite-albite-quartz assemblage (*i.e.*, typical greenschist) requires only hydration. In this case, the reactive component of the solution is the water itself and does not depend upon the supply of dissolved components. The reaction could proceed to completion at a water:rock ratio of 1:30 and result in more than 1500 ppm metal in the pore fluid if 50 ppm metal were leached from the basalt. Interpreted in terms of this leaching constraint and the specific seawater-basalt reactions involved (Table 2 and Figure 10), the 1 to 10 ppm ore metal concentration of both the experimental and the natural ocean ridge hydrothermal fluids most probably constitutes the labile component of the source rocks. If 10 ppm Zn + Cu represents the maximum labile metal available from a basalt for the water:rock ratios of a convection cell, then ore fluids generated within basaltic rocks which require a higher concentration of metal must have derived this additional metal from the bound trace metal component of the source rock.

The most favourable circumstance for leaching the bound metal component of a rock and achieving maximum concentrations of the trace metals in the pore fluids is during thermal or barometric metamorphism of a stratal aquifer. Under these conditions, the mineral-destructive reactions are transformations of one mineral assemblage into another due to changes of temperature and/or pressure. During these mineral-mineral transformations, the bound trace metals have the opportunity to partition into the pore fluid which, in essence, is not a major reactant in the reactions. Since pore fluids in a geopressured stratal aquifer are not continually replaced, leaching of the bound trace metal will continue until the pore fluid becomes saturated with respect to the metals.

Both the reactive capacity and leaching ability of a solution increases with increasing salinity. The reactive capacity of a pore solution is largely due to the cationic concentration and the leaching ability, *via* metal complexing, is due to the anionic concentration.

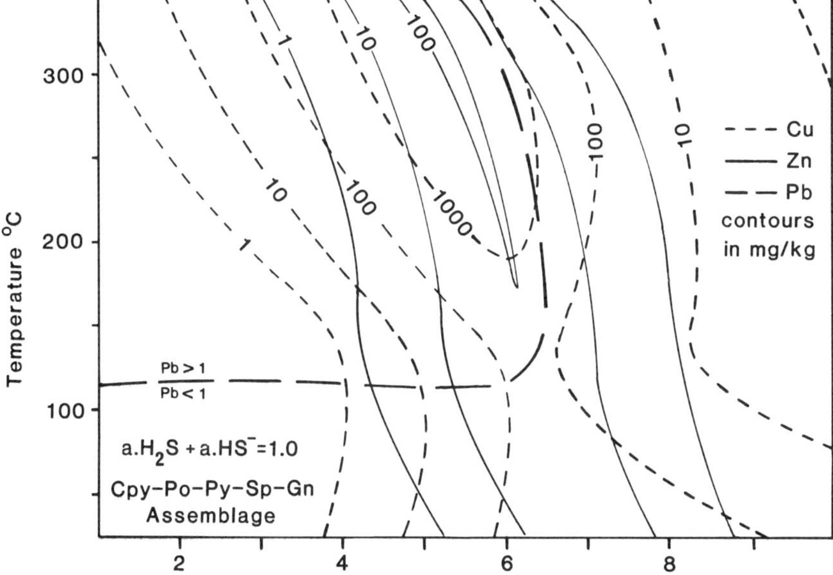

Figure 11 *Sulphide solubilities as bisulphide complexes. Calculated metal concentrations (in mg·kg⁻¹) in a solution in which the total sulphur concentration is 1 molal and which is in equilibrium with a chalcopyrite-pyrite-pyrrhotite-sphalerite-galena assemblage. Metal bisulphide thermodynamic data as listed by or extrapolated from Barnes (1979). Thermodynamic data for aqueous sulphur species from compilation by Brewer (1982).*

It is noteworthy that the most metal-rich modern natural hydrothermal solutions are chloride-rich brines of stratal aquifers (*e.g.*, Salton Sea, oil field brines of Gulf Basin). It is also significant that these brines are saturated with respect to sphalerite and galena (Lydon, 1983; Kharaka *et al.*, 1986) and have Zn:Pb ratios in the range 3:1 to 5:1. These ratios therefore reflect the range of Zn:Pb ratios of solutions saturated with sphalerite and galena when the metals are in solution dominantly as chloride complexes and in the temperature range 100-300°C.

If the ore fluids for most major VMS deposits were saturated with respect to the main ore components prior to commencement of ore accumulation, then the bimodal distribution of Zn:Pb ratios of VMS deposits (Lydon, 1984a, fig. 4) may have an explanation other than the relative abundances of these metals in the source rocks. The great majority of the Zn-Pb-Cu type have Zn:Pb ratios in the range 2.5:1 to 5:1, and this strongly suggests that the pristine ore fluids were chloride-rich solutions saturated with respect to sphalerite and galena, in which the dominant metal species were chloride complexes. In turn, this implies that the ancient ore fluids contained much greater concentrations of Zn and Pb than the undersaturated fluids of modern vents. The Cu-Zn association may also be alternatively explained by metal-saturated solutions if the metals were transported mainly as bisulphide complexes. The solubilities of both chalcopyrite (Crerar and Barnes, 1976) and sphalerite (Barnes and Czamanske, 1967) as bisulphide complexes are about 100 times greater than that of galena (Giordano and Barnes, 1979) under comparable conditions.

A thermodynamic evaluation of the implications of metal transport as bisulphide complexes with regard to the formation of VMS deposits is complicated by the uncertainties in the thermodynamic values for aqueous sulphur species at elevated temperatures. The maximum solubilities of both chalcopyrite and sphalerite as bisulphide complexes occur near the boundary between the fields of H_2S (aq) and HS^- dominance (Figure 11). The data of Ellis and Giggenbach (1971) (the values used most commonly in recent geological literature) indicate that these solubility maxima occur at a pH between 7 and 8 at temperatures above 200°C (Figure 10) and are separated from the field of comparable metal solubilities as chloride complexes at lower pH by a pronounced metal solubility minima. Since, at temperatures above 200°C the solubilities of both chalcopyrite and sphalerite as bisulphide complexes are relatively insensitive to temperature change but very sensitive to the sulphur content of the fluid, it would be difficult to explain the Cu-dominant to Zn-dominant zonation of the Cu-Zn type of VMS deposit if the metals were precipitated directly from bisulphide complexes.

However, the recent assessment and compilation of thermodynamic data for aqueous sulphur species by Brewer (1982) indicates that the boundary between H_2S (aq) and HS^- dominance is located at a much lower pH than reported by Ellis and Giggenbach (1971) (see Figure 10). Brewer's (1982) data base indicates that, providing the solution also contains > 1 molal chloride, the progressive desulphurization of the fluid above 200°C would cause precipitation of much of the copper as chalcopyrite, but zinc would remain in solution as a zinc chloride complex (compare Figure 6 and Figure 11). Only on subsequent cooling of the solution would sphalerite precipitate. Thus, if Brewer's (1982) data base is correct, ore components for the Cu-Zn type of VMS deposits may be transported dominantly as bisulphide complexes (thereby suppressing the mobility of Pb), but the temperature dependent Cu-Zn zonation characteristic of the deposits (which requires precipitation from a solution in which metal chloride complexes are dominant) would still be permissible.

Whether the metals are carried as chloride complexes or bisulphide complexes depends, if the data of Ellis and Giggenbach (1971) are correct, predominatly on the pH of the fluid (and hence the mineral assemblage of the hydrothermal reservoir rocks controlling the pH — see Figure 10 for examples). On the other hand, if Brewer's (1982) data are correct, it will depend predominantly on the Cl:S ratio of the fluid.

There are some supportive arguments for a fundamental dichotomy between the chemical compositions of, and mode of metal transport in, ore fluid that formed the Cu-Zn and Zn-Pb-Cu types of VMS deposits, respectively. Only one will be given here. Barite, which is common in many deposits of the Zn-Pb-Cu type (*e.g.*, Kuroko, Buchans) is characteristically absent from deposits of the Cu-Zn type. Significant mobilization of barium in the natural environment requires reduced acidic conditions, because under oxidized conditions it has a low solubility with respect to barite and under reduced neutral to alkaline conditions it has a low solubility with respect to barium silicates (celsian, cymrite, hyalophane, *etc.*). Because the Ba:Zn ratios of average mafic, felsic and argillaceous rocks are not significantly different, the marked dichotomy of Ba:Zn ratios between deposits of the Cu-Zn type and many of those of the Zn-Pb-Cu type requires an explanation other than the trace element composition of the source rocks. On balance, the most probable explanation is that the ore fluids which formed ancient Cu-Zn VMS deposits were characteristically more alkaline than fluids which formed Zn-Pb-Cu VMS deposits, and this suggests a greater probability that the metals were transported dominantly as bisulphide complexes for the former, but as chloride complexes for the latter type. In the case of barite-free Zn-Pb-Cu deposits (*e.g.*, Brunswick No.12), the absence of barite in ores can be explained by the lack of seawater sulphate due to anoxic bottom conditions, rather than by lack of barium in the ore fluid. However, it is extremely unlikely that the same argument can account for its characteristic absence in the Cu-Zn type of VMS deposits, because this would require that hydrothermal discharge from ancient mafic volcanic complexes occurred only at the time of stratified ocean conditions. The fact that barite is common in the modern "Cu-Zn" type of mound-chimney deposits suggests that the modern hydrothermal fluids are compositionally different, particularly with regard to their pH values, from those fluids which formed ancient Cu-Zn VMS deposits. If this is so, then the ore metal content of modern hydrothermal fluids may also not reflect typical metal concentrations of ore fluids that formed ancient VMS deposits.

Sulphide accumulation. The above discussions, particularly with regard to the size differential between ancient VMS deposits and the modern mound-chimney deposits, have suggested that ore fluids responsible for at least large ancient VMS deposits may have contained many times the concentration of ore metals than do the fluids emanating from modern vents. This, in turn, implies that the ancient ore fluids were more saline than the modern submarine vent fluids of the eastern Pacific, because the latter are close to saturation with respect to at least copper and iron sulphides (Janecky and Seyfried, 1984), and therefore an increase in metal concentration requires an increase in the concentration of the metal-complexing species (*e.g.*, chloride). Comparison of the Juan de Fuca fluids with EPR 21°N fluids (Table 1) may illustrate this point, because in the former, despite their lower temperature (and therefore their lower Cu concentration — see slope of Cu contours Figure 6), the higher chloride concentration allows a higher zinc concentration. Elevated salinities may enhance the efficiency of sulphide accumulation by the mechanisms envisaged by Sato (1972) for the reversing buoyancy plumes of his Type II solutions (Turner and Campbell, 1987). The super-giant member of a VMS cluster (Sangster, 1980) could be explained if a single submarine depression acted as a collector basin for all the bottom-hugging brines discharged within a topographic catchment area that was many times larger than the area of the brine pool itself. This topographic effect could also counteract the dispersal effect of vent migration associated with a long-lived convection cell. It is noteworthy that to explain the unusually heavy oxygen isotope compositions of hydrothermal alteration associated with the Kidd Creek deposit, Beaty and Taylor (1982) suggested that the fluids were derived from an evaporitic basin or sedimentary rocks. Since modern hydrothermal fluids derived from such geological environments are typically highly saline, the involvement of a high-salinity ore fluid in forming the Kidd Creek deposits would also be consistent with the observation made earlier that the high aspect ratio of this deposit suggests deposition from a brine pool.

The origin of the fragmental rock pile

over the hydrothermal vent area, suggested here to be important in the initiation of sulphide accumulation in proximal VMS deposits, has important repercussions for both genetic and exploration models. The association of fragmental rocks and submarine hydrothermal vents could be semi-fortuitous in that there may be no direct genetic link between the two. For example, hydrothermal discharge could be channelled into a pile of pyroclastic rocks by virtue of the ascendant ore fluids following the eruptive magma conduit, or by the ore fluids being channelled into fault scarp talus by following the fractures of the fault system. However, in many cases, there is good evidence that the fragmental rocks (and later fragmental sulphides) are the result of hydrothermal eruption, and an integral part of the ore-forming hydrothermal system. The inverted cone-shape of the brecciated rocks of the stockwork zone, marked by an increase in the degree of brecciation and void creation upward, is typical of a hydrothermal eruption vent (Hedenquist and Henley, 1985). Eruptive fragments of footwall, stockwork and sulphide rocks that occur in the massive sulphide lens (Clark, 1983; Lydon and Galley, 1986) indicate that the ore-forming hydrothermal discharge was recurrently explosive. Consideration of the P-V energy requirements indicates that eruptive hydrothermal discharge is unlikely to take place at water depths in excess of about 1 km if the ore solutions are evolved seawater (Lydon, 1986). Explosive hydrothermal discharge may take place at much greater depths due to water exsolution from a crystallizing magma (Burnham, 1983), but in the context of the genesis of VMS deposits, this mechanism would require that the ore fluids have a substantial magmatic water component.

Eruptive hydrothermal discharge, as a common feature in the genesis of VMS deposits, gives a ready explanation to the phenomenon of the "favourable horizon". This would represent the time interval during which a growing volcanic edifice was at a water depth range of about 1 km to 300 m below the ocean surface. Below 1 km depth, hydrothermal eruption could not form a blanket of eruption breccia, and at depths much shallower than 300 m, the hydrothermal solution would have a maximum temperature of 200°C, which is too low for it to carry sufficient quantities of metal to form a significant VMS deposit. VMS deposits can, of course, form at a greater depth if the porous vent capping is other than a hydrothermal eruption breccia. In this context, it may be significant with regard to the genesis of the Corbet deposit, which occurs stratigraphically below the favourable horizon at Noranda (see Lydon, 1984a, fig. 1), to note that it was formed within a pile of previously erupted pyroclastic and volcaniclastic breccias (Knuckey and Watkins, 1982).

Conclusions

Discovery of sulphide mound-chimney deposits on the modern ocean floor has not only provided corroboration of the hydrothermal exhalative model for the genesis of VMS deposits, but has also provided new impetus and opportunity to achieve an understanding of the ore-forming processes. The most important advances have been derived from the chemical and physical measurements of the ore fluids and the incorporation of this quantitative data into mathematical models of the hydrothermal system and the precipitation of hydrothermal minerals. Detailed documentation of the structure of modern sulphide chimneys and pristine ancient deposits, particularly the Kuroko deposits of Japan, have provided observations against which aspects of the mathematical models can be tested and refined.

If one concept is to be singled out for its impact on our overall appreciation of the genesis of VMS deposits, it must be that the growth of a proximal massive sulphide lens is largely due to sulphide accumulation by open space filling and replacement *within* the lens. It is this intra-mound deposition and reworking under the influence of a conductive and advective thermal gradient that produces the characteristic Cu to Zn zonation.

However, there are still many major problems for which totally satisfactory answers have yet to be given. Of fundamental importance is the nature of the hydrothermal system that most commonly forms VMS deposits. Although convection cells appear to be able to form smaller proximal VMS deposits, it may be unreasonable to suppose that a single convection cell can give rise to the larger ones like Kidd Creek, Brunswick No.12 and Rio Tinto. It may be that at least these larger VMS deposits are genetically more akin to the large-tonnage SEDEX deposits (*e.g.*, Mount Isa, Sullivan, Howards Pass) for which a stratal aquifer model is more appropriate. Similarly, although deposition of sulphides as hard encrustations and infillings within a rigid sulphide mound appears to satisfactorily account for many of the textural features and the mineralogical zonation pattern of many proximal VMS deposits, one cannot ignore the numerous descriptions of gel-like textures and rhythmically layered sulphides in other deposits. Perhaps sulphide oozes, possibly of similar constituency to the Red Sea metalliferous muds, were an important primary component in those VMS deposits that do not so perfectly fit the idealized proximal model described here. Of paramount importance are the compositions of the ore fluids, particularly those for the largest VMS deposits. Elevated salinities of ore fluids certainly increase the probability of forming larger sulphide deposits, both through their increased metal-transporting capacity and their increased tendency to form reversing buoyancy plumes upon discharge at the sea floor which increases the efficiency of accumulation of suspended sulphide particles. Although evolved seawater is the favourite candidate, the nature or cause of its evolution into the optimal ore fluid has not been narrowed down to a single explanation. As recently discussed by Von Damm and Bischoff (1987), the evolution of seawater toward a higher chloride content alone has six plausible explanations, and to evaluate them requires more understanding of sub-surface processes than exists at present. The debate on the generation of ore fluids for VMS deposits is obviously far from reaching a consensus.

Perhaps the major conclusion that emanates from the observations and calculations of the last decade is that it is perhaps a mistake to look for one genetic model to explain all VMS deposits. From observations on modern deposits alone, it is apparent that sulphides can accumulate from buoyant plumes as well as dense brines; they can form on volcanic or on sedimentary substrates; and they can occur in the topographic lows of rift structures or on topographic highs of seamounts. There are therefore few simple "rules of thumb" that are widely applicable for the prognostication of undiscovered VMS deposits. The most important factor in such prognostication is an understanding of the fundamental processes involved in the genesis of VMS deposits, and it is only this knowledge that will allow a prediction of how these processes will express themselves in a particular geological environment.

Acknowledgements
Gwilym Roberts is thanked for his patient encouragement leading to this second part of this article. R.I. Thorpe, C. Jay Hodgson, D.F. Sangster and S.D. Scott are thanked for reviewing the manuscript and providing many valuable suggestions to improve its quality. A. Douma provided much help in drafting of the diagrams.

References

Aggarwal, P.K. and Nesbitt, B.E., 1984, Geology and geochemistry of the Chu Chua massive sulfide deposit, British Columbia: Economic Geology, v. 79, p. 815-825.

Alt, J.C., Honnorez, J., Laverne, C. and Emmerman, R., 1986, Hydrothermal alteration of a 1 km section through the upper oceanic crust, Deep Sea Drilling Project Hole 505B: Mineralogy, chemistry, and evolution of seawater-basalt interaction: Journal of Geophysical Research, v. 91, p. 309-335.

Alt, J.C., Lonsdale, P., Haymon, R. and Muehlenbachs, K., 1987, Hydrothermal sulfide and oxide deposits on seamounts near 21°N, East Pacific Rise: Geological Society of America, Bulletin, v. 98, p. 157-168.

Badham, J.P.N., 1981, Shale-hosted Pb-Zn deposits: products of exhalation of formation waters?: Institution of Mining and Metallurgy, Transactions, v. 90, p. B71-B76.

Bailes, A.H., Syme, E.C., Galley, A., Price, D.P, Skirrow, R. and Ziehlke, D.J., 1987, Early Proterozoic volcanism, hydrothermal activity, and associated ore deposits at Flin Flon and Snow Lake, Manitoba: Geological Association of Canada—Mineralogical Association of Canada, Joint Annual Meeting, Saskatoon, Saskatchewan, Field Trip 1, 95 p.

Barnes, H.L. and Czamanske, G.K., 1967, Solubility and transport of ore minerals, *in* Barnes, H.L., ed., Geochemistry of Hydrothermal Ore Deposits, First Edition: Wiley, New York, p. 334-381.

Barnes, H.L., 1979, Solubilities of ore minerals, *in* Barnes, H.L., ed., Geochemistry of Hydrothermal Ore Deposits, Second Edition: Wiley, New York, p. 404-460.

Barton, P.B., Jr., 1978, Some Ore Textures Involving Sphalerite from the Furutobe Mine, Akita Prefecture, Japan: Mining Geology, v. 28, p. 293-300.

Beaty, D.W. and Taylor, H.P., Jr., 1982, Some petrologic and oxygen isotope relationships of the Amulet Mine, Noranda, Quebec, and their bearing on the origin of Archean massive sulfide deposits: Economic Geology, v. 77, p. 95-108.

Bischoff, J.L. and Dickson, F.W., 1975, Seawater-basalt interaction at 200°C and 500 bars: implications for origin of sea-floor heavy metal deposits and regulation of seawater chemistry: Earth and Planetary Science Letters, v. 25, p. 385-397.

Bischoff, J.L. and Pitzer, K.S., 1985, Phase relations and adiabats in boiling seafloor geothermal systems: Earth and Planetary Science Letters, v. 75, p. 327-338.

Bischoff, J.L. and Seyfried, W.E., 1978, Hydrothermal chemistry of seawater from 25° to 350°C: American Journal of Science, v. 278, p. 838-860.

Blissenbach, E. and Nawab, Z., 1982, Metalliferous sediments of the seabed: The Altlantis-II-Deep deposits of the Red Sea, *in* Borgese, E.M. and Ginsburg, N., eds., Ocean Yearbook 3: University of Chicago Press, p. 77-104.

Boldy, J., 1968, Geological observations on the Delbridge massive sulphide deposit: Canadian Institute of Mining and Metallurgy, Transactions, v. 71, p. 247-256.

Boström, K. and Peterson, M.N.A., 1966, Precipitates from hydrothermal exhalations on the East Pacific Rise: Economic Geology, v. 61, p. 1258-1265.

Boström, K., 1983, Genesis of ferromanganese deposits - diagnostic criteria for recent and old deposits, *in* Rona, P.A., Boström, K., Laubier, L. and Smith, K.L., Jr., eds., Hydrothermal Processes at Seafloor Spreading Centres: Plenum Press, New York, NATO Conference Series IV, v. 12, p. 473-489.

Both, R., Crook, K., Taylor, B., Brogan, S., Chappell, B., Frankel, E., Lui, L., Sinton, J. and Tiffin, D., 1986, Hydrothermal Chimneys and Associated Fauna in the Manus Back-Arc Basin, Papua New Guinea: EOS, v. 67, p. 489-491.

Bowers, T.S., Von Damm, K.L. and Edmond, J.M., 1985, Chemical evolution of mid-ocean ridge hot springs: Geochimica et Cosmochimica Acta, v. 49, p. 2239-2252.

Brewer, L., 1982, Thermodynamic values for desulfurization processes, *in* Hudson, J.L. and Rochelle, G.T., eds., Flue Gas Desulfurization: American Chemical Society, ACS Symposium Series 188, Washington, p. 1-39.

Bryndzia, L.T., Scott, S.D. and Farr, J.E., 1983, Mineralogy, geochemistry, and mineral chemistry of siliceous ore and altered footwall rocks in the Uwamuki 2 and 4 deposits, Kosaka Mine, Hokuroku District, Japan, *in* Ohmoto, H. and Skinner, B.J., eds., Kuroko and Related Volcanogenic Massive Sulphide Deposits: Economic Geology, Monograph 5, p. 507-522.

Burnham, C.W., 1983, Deep submarine pyroclastic eruptions, *in* Ohmoto, H. and Skinner, B.J., eds., Kuroko and Related Volcanogenic Massive Sulphide Deposits: Economic Geology, Monograph 5, p. 142-148.

Burnham, C.W., 1985, Energy Release in Subvolcanic Environments: Implications for Breccia Formation: Economic Geology, v. 80, p. 1515-1522.

Campbell, I.H., Franklin, J.M., Gorton, M.P., Hart, T.R. and Scott, S.D., 1981, The role of subvolcanic sills in the generation of massive sulfide deposits: Economic Geology, v. 76, p. 2248-2253.

Campbell, I.H., McDougall, T.J. and Turner, J.S., 1984, A note on fluid dynamic processes which can influence the deposition of massive sulfides: Economic Geology, v. 79, p. 1905-1913.

CASM (Canadian-American Seamount Expedition), 1985, Hydrothermal vents on an axis seamount of the Juan de Fuca Ridge: Nature, v. 313, p. 212-214.

Cann, J.R. and Strens, M.R., 1982, Black smokers fuelled by freezing magma: Nature, v. 298, p. 147-149.

Cann, J.R., Strens, M.R. and Rice, A., 1985, A simple magma-driven thermal balance model for the formation of volcanogenic massive sulfides: Earth and Planetary Science Letters, v. 76, p. 123134.

Cathles, L.M., 1978, Hydrodynamic Constraints on the Formation of Kuroko Deposits: Mining Geology, v. 28, p. 257-265.

Cathles, L.M., 1983, An Analysis of the Hydrothermal System responsible for Massive Sulfide Deposition in the Hokuroku Basin of Japan, *in* Ohmoto, H. and Skinner, B.J., eds., Kuroko and Related Volcanogenic Massive Sulphide Deposits: Economic Geology, Monograph 5, p. 439-487.

Clark, L.A., 1971, Volcanogenic ores: Comparison of cupriferous pyrite deposits of Cyprus and Japanese Kuroko deposits: Society of Mining Geologists of Japan, Special Issue 3, p. 206-215.

Clark, L.A., 1983, Genetic implications of fragmental ore texture in Japanese Kuroko deposits: Canadian Institute of Mining and Metallurgy, Bulletin, v. 76, p. 105-114.

Coad, P.R., 1985, Rhyolite geology at Kidd Creek - a progress report: Canadian Institute of Mining and Metallurgy, Bulletin, v. 78, p. 70-83.

Constantinou, G., and Govett, G.J.H., 1972, Genesis of sulphide deposits, ochre and umber of Cyprus; Institution of Mining and Metallurgy, Transactions, v. 81, p. B34-B46.

Constantinou, G., and Govett, G.J.H., 1973, Geology, geochemistry, and genesis of Cyprus sulfide deposits: Economic Geology, v. 68, p. 843-858.

Converse D.R., Holland, H.D. and Edmond, J.M., 1984, Flow rates in the axial hot springs of the East Pacific Rise (21°N): implications for the heat budget and the formation of massive sulphide deposits: Earth and Planetary Science Letters, v. 69, p. 159-175.

Costa, U.R., Barnett, R.L. and Kerrich, R., 1983, The Mattagami Lake Mine Archean Zn-Cu sulfide deposit, Quebec: hydrothermal coprecipitation of talc and sulfides in a sea-floor brine pool - evidence from geochemistry, $^{18}O/^{16}O$, and mineral chemistry: Economic Geology, v. 78, p. 1144-1203.

Crane, K. and Normark, W.R., 1977, Hydrothermal activity and crestal structure of the East Pacific Rise at 21°N: Journal of Geophysical Research, v. 82, p. 5336-5348.

Crerar, D.A. and Barnes, H.L., 1976, Ore solution chemistry V: solubilities of chalcopyrite and chalcocite assemblages in hydrothermal solution at 200° to 350°C: Economic Geology, v. 71, p. 772-774.

Date, J., Watanabe, Y. and Saeki, Y., 1983, Zonal Alteration around the Fukazawa kuroko deposits, Akita Prefecture, northern Japan, *in* Ohmoto, H. and Skinner, B.J., eds., Kuroko and Related Volcanogenic Massive Sulphide Deposits: Economic Geology, Monograph 5, p. 365-386.

Davis, E.E., Goodfellow, W.D., Bornhold, B.D., Adshead, J., Blaise, B., Villinger, H. and LeCheminant, G.M., 1987, Massive sulphides in a sedimented rift valley, Northern Juan de Fuca Ridge: Earth and Planetary Science Letters, v. 82, p. 49-61.

Degens, E.T. and Ross, D.A., 1969, eds., Hot brines and recent heavy metal deposits in the Red Sea: Springer-Verlag, New York, 600 p.

Delaney, J.R. and Cosens, B.A., 1982, Boiling and metal deposition in submarine hydrothermal systems: Marine Technology Society Journal, v. 16, p. 62-66.

Deptuck, R., Squair, H. and Wierzbicki, V., 1982, Geology of the Detour Zinc-Copper deposits, Brouillan Township, Quebec, *in* Hutchinson, R.W., Spence, C.D. and Franklin, J.M., Precambrian Sulphide Deposits: Geological Association of Canada, Special Paper 25, p. 319-342.

Desbruyeres, D. and Laubier, L., 1983, Primary consumers from hydrothermal vents animal communities, *in* Rona, P.A., Boström, K., Laubier, L. and Smith, K.L., Jr., eds., Hydrothermal Processes at Seafloor Spreading Centres: Plenum Press, New York, NATO Conference Series IV, v. 12, p. 711-734.

Doe, B.R. and Zartman, R.E., 1979, Plumbotectonics: the Phanerozoic, *in* Barnes, H.L., ed., Geochemistry of Hydrothermal Ore Deposits, Section Edition: Wiley, New York, p. 22-70.

Drummond, S.E. and Ohmoto, H., 1985, Chemical evolution and mineral deposition in boiling hydrothermal systems: Economic Geology, v. 80, p. 126-147.

Edmond, J.M., Measures, C., McDuff, R.E., Chan, L.H., Collier, R., Grant, B., Gordon, L.I. and Corliss, J.B., 1979, Ridge crest hydrothermal activity and the balances of the major and minor elements in the ocean: the Galapagos data: Earth and Planetary Science Letters, v. 46, p. 1-18.

Eldridge, C.S., Barton, P.B., Jr. and Ohmoto, H., 1983, Mineral textures and their bearing on formation of the Kuroko orebodies, *in* Ohmoto, H. and Skinner, B.J., eds., Kuroko and Related Volcanogenic Massive Sulphide Deposits: Economic Geology, Monograph 5, p. 241-281.

Ellis, A.J. and Giggenbach, W., 1971, Hydrogen sulphide ionization and sulphur hydrolysis in high temperature solution: Geochimica et Cosmochimica Acta, v. 35, p. 247-260.

Farrell, C.W. and Holland, D., 1983, Strontium isotope geochemistry of the Kuroko deposits, *in* Ohmoto, H. and Skinner, B.J., eds., Kuroko and Related Volcanogenic Massive Sulphide Deposits: Economic Geology, Monograph 5, p. 302-319.

Francheteau, J., Needham, H.D., Choukroune, P., Juteau, T., and others, 1979, Massive deep-sea sulphide ore deposits discovered on the East Pacific Rise: Nature, v. 277, p. 523-528.

Franklin, J.M., 1986, Volcanogenic massive sulphide deposits - an update, *in* Andrew, C.J., Crowe, R.W.A., Finlay, S., Pennell, W.M. and Pyne, J.F., eds., Geology and Genesis of Mineral Deposits in Ireland: Irish Association for Economic Geolgy, Dublin, p. 49-69.

Franklin, J.M., Kasarda, J. and Poulson, K.H., 1975, Petrology and Chemistry of the Alteration Zone of the Mattabi Massive Sulfide Deposit: Economic Geology, v. 70, p. 63-79.

Franklin, J.M., Lydon, J.W. and Sangster, D.F., 1981, Volcanic-Associated Massive Sulfide Deposits: Economic Geology, 75th Anniversary Volume, p. 485-627.

Gallinatti, B.S., 1984, Initiation and collapse of active circulation in a hydrothermal system at the Mid-Atlantic Ridge, 23°N: Journal of Geophysical Research, v. 89, No.B5, p. 3275-3289.

Gibson, H.L., Watkinson, D.H. and Comba, C.D.A., 1983, Silicification: hydrothermal alteration in an Archean geothermal system within the Amulet Rhyolite Formation, Noranda, Quebec: Economic Geology, v. 78, p. 954-971.

Gilmour, P., 1965, The origin of the massive sulphide mineralization in the Noranda district, Northwestern Quebec: Geological Association of Canada, Proceedings, v. 16, p. 63-81.

Gilmour, P., 1976, Some transitional types of mineral deposits in volcanic and sedimentary rocks, *in* Wolf, K.H., ed., Handbook of Stratabound and Stratiform Ore Deposits: Elsevier, Amsterdam, v. 1, p. 111-160.

Giordano, T.H. and Barnes, H.L., 1979, Ore solution chemistry VI. PbS solubility in bisulphide solutions to 300°C: Economic Geology, v. 74, p. 1637-1646.

Goldfarb, M.S., Converse, D.R., Holland, H.D. and Edmond, J.M., 1983, The genesis of hot spring deposits on the East Pacific Rise, 21°N, *in* Ohmoto, H. and Skinner, B.J., eds., Kuroko and Related Volcanogenic Massive Sulphide Deposits: Economic Geology, Monograph 5, p. 184-197.

Graf, D.L., 1982, Chemical osmosis, reverse chemical osmosis, and the origin of subsurface brines: Geochimica et Cosmochimica Acta, v. 46, p. 1431-1448.

Grassle, J.F., 1983, Introduction to the biology of hydrothermal vents, *in* Rona, P.A., Boström, K., Laubier, L. and Smith, K.L., Jr., eds., Hydrothermal Processes at Seafloor Spreading Centres: Plenum Press, New York, NATO Conference Series IV, v. 12, p. 665-675.

Green, G.R., Solomon, M. and Walshe, J.L., 1981, The formation of the volcanic-hosted massive sulfide ore deposit at Rosebery, Tasmania: Economic Geology, v. 76, p. 304-338.

Hajash, A., 1975, Hydrothermal processes along mid-ocean ridges: an experimental investigation: Contributions to Mineralogy and Petrology, v. 53, p. 205-226.

Hannington, M.D., Peter, J.M. and Scott, S, D., 1986, Gold in sea-floor polymetallic sulfide deposits: Economic Geology, v. 81, p. 1867-1883.

Haymon, R.M., 1983, Growth history of hydrothermal black smoker chimneys: Nature, v. 301, p. 695-698.

Haymon, R.M. and Kastner, M., 1981, Hot spring deposits on the East Pacific Rise at 21°N: preliminary description of mineralogy and genesis: Earth and Planetary Science Letters, v. 53, p. 363-381.

Haymon R.M., Koski, R.A. and Sinclair, C., 1984, Fossils of hydrothermal vent worms from Cretaceous sulfide ores of the Samail Ophiolite, Oman: Science, v. 223, p. 1407-1409.

Heaton, T.H.E. and Sheppard, S.M.F., 1977, Hydrogen and oxygen isotope evidence for sea-water-hydrothermal alteration and ore deposition, Troodos Complex, Cyprus, *in* Volcanic Processes in Ore Genesis: Geological Society of London, Special Publication No.7, p. 42-57.

Hedenquist, J.W. and Henley, R.W., 1985, Hydrothermal eruption in the Waiotapu geothermal system, New Zealand: their origin, associated breccias and relation to precious metal mineralization: Economic Geology, v. 80, p. 1640-1668.

Hekinian, H. and Bideau, D., 1986, Volcanism and mineralization of the oceanic crust on the East Pacific Rise: *in* Gallagher, M.J., Ixer, R.A., Neary, C.R. and Prichard, H.M., eds., Metallogeny of Basic and Ultrabasic Rocks: Institition of Mining and Metallurgy, London, p. 3-20.

Hekinian, R., Fevrier, M., Bischoff, J.L., Picot, P. and Shank, W.C., 1980, Sulfide deposits from the East Pacific Rise near 21°N: Science, v. 207, p. 1433-1444.

Hekinian, R. and Fouquet, Y., 1985, Volcanism and metallogenesis of axial and off-axial structures on the East Pacific Rise near 13°N: Economic Geology, v. 80, p. 221-249.

Hekinian, R., Francheteau, J., Renard, V., Ballard, R.D., Choukroune, P., Cheminee, J.L., Albarede, F., Minster, J.F., Charlou, J.L., Marty, J.C. and Boulegue, J., 1983, Intense hydrothermal activity at the axis of the East Pacific Rise near 13°N: submersible witnesses the growth of sulfide chimney: Marine Geophysical Researches, v. 6, p. 1-14.

Helgeson, H.C., 1968, Geologic and thermodynamic characteristics of the Salton Sea geothermal system: American Journal of Science, v. 266, p. 120-166.

Helgeson, H.C., 1969, Thermodynamics of elevated temperatures and pressures: American Journal of Science; v. 267, p. 729-804.

Henley, R.W. and Thornley, P., 1979, Some geothermal aspects of polymetallic massive sulfide formation: Economic Geology, v. 74, p. 1600-1612.

Hirabayashi, T., 1974, On the Kuroko-type orebodies in the Yokata Mine area, Nishi-Aizu District, Fukushima Prefecture, *in* Ishihara, S., ed., Geology of Kuroko Deposits: Society of Mining Geologists of Japan, Special Issue 6, p. 195-201.

Hodgson, C.J. and Lydon, J.W., 1977, Geological setting of volcanogenic massive sulphide deposits and active hydrothermal systems: some implications for exploration: Canadian Institute of Mining and Metallurgy, Bulletin, v. 70, p. 95-106.

Honma, H. and Shuto, K., 1979, On strontium isotopic ratio of barite from Kuroko-type deposits, Japan: Japanese Association of Mineralogists, Petrologists and Economic Geologists, v. 74, p. 321-325.

Honnorez, J., Detrick, R., Adamson, A., Brass, G., Gillis, K., Humphris, S., Mevel, C., Meyer, P., Petersen, N., Rautenschlein, M., Shibata, T., Staudigel, H., Wooldridge, A. and Yamamoto, K., 1986, Mineralogy and geology of the Snake-Pit hydrothermal sulfide deposit on the Mid-Atlantic Ridge at 23°N, (abstract): EOS, v. 67, p. 1214.

Horikoshi, E., 1969, Volcanic activity related to the formation of the Kuroko-type deposits in the Kosaka District, Japan: Mineralium Deposita, v. 4, p. 321-345.

Humphris, S.E. and Thompson, G., 1978, Hydrothermal alteration of oceanic basalts by seawater: Geochimica et Cosmochimica Acta, v. 42, p. 107-125.

Hutchinson, R.W., 1973, Volcanogenic sulfide deposits and their metallogenic significance: Economic Geology, v. 68, p. 1223-1246.

Hutchison, R.W. and Searle, D.L., 1971, Stratabound pyrite deposits in Cyprus and relations to other sulphide ores, *in* Takeuchi, Y., ed., the Proceedings of the Tokyo-Kyoto IAGOD meeting, 1970: Society of Mining Geologists of Japan, Special Issue 3, p. 198-205.

Ishihara, S., 1974, ed., Geology of Kuroko Deposits: Society of Mining Geologists of Japan, Special Issue 6, 435 p.

Janecky, D.R. and Seyfried, W.E., Jr., 1984, Formation of massive sulfide deposits on oceanic ridge crest: Incremental reaction models for mixing between hydrothermal solutions and seawater: Geochimica et Cosmochimica Acta, v. 48, p. 2723-2738.

Jannasch, H.W., 1983, Microbial processes at deep sea hydrothermal vents, *in* Rona, P.A., Boström, K., Laubier, L. and Smith, K.L., Jr., eds., Hydrothermal Processes at Seafloor Spreading Centres, NATO Conference Series IV, v. 12: Plenum Press, p. 677-709.

Jonasson, I.R., Franklin, J.M. and Embley, R.W., 1986, Nature of alteration zone beneath Galapagos Ridge sulfide mounds, (abstract): EOS, v. 67, p. 1185.

Kajiwara, Y., 1971, Sulfur isotope study of the Kuroko-ores of Shakanai No.1 deposits, Akita Prefecture, Japan: Geochemical Journal, v. 4, p. 157-181.

Kajiwara, Y., 1973, A simulation of the Kuroko type mineralization in Japan: Geochemical Journal, v. 6, p. 193-209.

Kalogeropoulos, S.I. and Scott, S.D., 1983, Mineralogy and geochemistry of tuffaceous exhalites (tetsusekiei) of the Fukazawa Mine, *in* Ohmoto, H. and Skinner, B.J., eds., Kuroko and Related Volcanogenic Massive Sulphide Deposits: Economic Geology, Monograph 5, p. 412-432.

Kharaka, A.S., Maest, T.L., Fries, L.M., Law, L.M. and Corothers, W.W., 1986, Geochemistry of Lead and Zinc in Oil Field Brines: Central Mississippi Salt Dome Basin Revisited, *in* Turner, R.J.W. and Einaudi, M.T., eds., The Genesis of Stratiform Sediment-hosted Lead and Zinc Deposits: Conference Proceedings: Stanford University, p. 181-183.

Knuckey, M.J., Comba, C.D.A. and Riverin, G., 1982, Structure, metal zoning and alteration at the Millenbach deposit, Noranda, Quebec, *in* Hutchinson, R.W., Spence, C.D. and Franklin, J.M., Precambrian Sulphide Deposits: Geological Association of Canada, Special Paper 25, p. 255-295.

Knuckey, M.J. and Watkins, J.J., 1982, The geology of the Corbet massive sulphide deposit, Noranda district, Quebec, Canada, *in* Hutchinson, R.W., Spence, C.D. and Franklin, J.M., Precambrian Sulphide Deposits: Geological Association of Canada, Special Paper 25, p. 297-317.

Koski, R.A., 1987, Geological setting and polymetallic sulphide deposits of the Escanaba Trough, Southern Gorda Ridge (abstract), *in* Recent Hydrothermal Mineralization at Seafloor Spreading Centres: tectonic, petrologic and geochemical constraints, Program with abstracts: Mineral Exploration Research Institute Feb 1987, McGill University, Montreal.

Kowalik, J, Rye, R. and Sawkins, F.J., 1981, Stable isotope study of the Buchans polymetallic sulphide deposits, *in* Swanson, E.A., Strong, D.F. and Thurlow, J.G., eds., The Buchans Orebodies: Fifty Years of Geology and Mining: Geological Association of Canada, Special Paper 22, p. 229-254.

Lambert, I.B. and Sato, T., 1974, The Kuroko and associated ore deposits of Japan: A review of their features and metallogenesis: Economic Geology, v. 69, p. 1215-1236.

Large, R.R., 1977, Chemical evolution and zonation of massive sulfide deposits in volcanic terrains: Economic Geology, v. 72, p. 549-572.

Lister, C.R.B., 1972, On the thermal balance of a mid-ocean ridge: Royal Astronomical Society, Geophysical Journal, v. 26, p. 515-535.

Lonsdale, P.F., Batiza, R. and Simkin, T., 1982, Metallogenesis at seamounts on the East Pacific Rise: Marine Technology Society, Journal, v. 16, p. 54-61.

Lonsdale, P.F., Bischoff, J.L., Burns, V.M. Kastner, M. and Sweeney, R.E., 1980, A high-temperature hydrothermal deposit on the seabed at a Gulf of California spreading center: Earth and Planetary Science Letters, v. 49, p. 8-20.

Lowell, R.P. and Rona, P.A., 1985, Hydrothermal models for the generation of massive sulphide ore deposits; Journal of Geophysical Research, v. 90, p. 8769-8783.

Lydon, J.W., 1983, Chemical parameters controlling the origin and deposition of sediment-hosted stratiform lead-zinc deposits, *in* Sangster, D.F., ed., Sediment-hosted Stratiform Lead-Zinc Deposits: Mineralogical Association of Canada, Short Course Handbook, v. 9, p. 175-250.

Lydon, J.W., 1984a, Volcanogenic massive sulphide deposits Part I: a descriptive model: Geoscience Canada, v. 11, p. 195-202.

Lydon, J.W., 1984b, Some observations on the morphology and ore textures of volcanogenic sulphide deposits of Cyprus: Geological Survey of Canada, Paper 84-1A, p. 601-610.

Lydon, J.W., 1984c, Some observations on the mineralogical and chemical zonation patterns of volcanogenic sulphide deposits of Cyprus: Geological Survey of Canada, Paper 84-1A, p. 611-616.

Lydon, J.W., 1986, Models for the generation of metalliferous hydrothermal systems within sedimentary rocks amd their applicability to the Irish Carboniferous Zn-Pb deposits, *in* Andrew, C.J., Crowe, R.W.A., Finlay, S. Pennell, W.M. and Pyne, J.F., eds., Geology and Genesis of Mineral Deposits in Ireland: Irish Association for Economic Geology, Dublin, p. 555-577.

Lydon, J.W., Franklin, J.M. and Sangster, D.F., 1984, Volcanic-associated massive sulphide, *in* Eckstrand, O.R., ed., Canadian Mineral Deposit Types: A Geological Synopsis: Geological Survey of Canada, Economic Geology Report 36, p. 33-34.

Lydon, J.W. and Galley, A., 1986, Chemical and mineralogical zonation of the Mathiati alteration pipe, Cyprus, and its genetic significance, *in* Gallagher, M.J., Ixer, R.A., Neary, C.R. and Prichard, H.M., eds., Metallogeny of Basic and Ultrabasic Rocks: Institition of Mining and Metallurgy, London, p. 49-68.

Macdonald, K.C., Becker, K., Spiess, F.N. and Ballard, R.D., 1980, Hydrothermal Heat Flux of the "Black Smoker" Vents on the East Pacific Rise: Earth and Planetary Science Letters, v. 48, p. 1-7.

MacGeehan, P.J. and MacLean, W.H., 1980, An Archean sub-seafloor geothermal system, "calc-alkali" trends, and massive sulphide genesis: Nature, v. 286, p. 767-771.

McNitt, J.R., 1970, The geological environment of hydrothermal fields as a guide to exploration: Geothermics, Special Issue 2, v. 1, p. 24-31.

Miller, A.R., Densmore, C.D., Degens, E.T., Hathaway, J.C., Manheim, F.T., McFarlin, P.F., Pocklington, H. and Jokela, A., 1966, Hot brines and recent iron deposits in deeps of the Red Sea: Geochimica et Cosmochimica Acta, v. 30, p. 341-359.

Mortensen, J.K., 1987, Preliminary U-Pb zircon ages for volcanic and plutonic rocks of the Noranda-Lac Abitibi area, Abitibi Subprovince, Quebec: Geological Survey of Canada, Paper 87-1A, p. 581-590.

Mottl, M.J., 1983a, Hydrothermal processes at seafloor spreading centres: application of basalt-seawater experimental results, *in* Rona, P.A., Boström, K., Laubier, L. and Smith, K.L., Jr., eds., Hydrothermal Processes at Seafloor Spreading Centres: Plenum Press, New York, NATO Conference Series IV, v. 12, p. 199-224.

Mottl, M.J., 1983b, Metabasalts, axial hot springs, and the structure of hydrothermal systems at mid-ocean ridges: Geological Society of America, Bulletin, v. 94, p. 161-180.

Mottl, M.J., 1986, Chemical processes in submarine hydrothermal plumes near 21°N on the East Pacific Rise, (abstract): EOS, v. 67, p. 1027.

Mottl, M.J. and Holland, H.D., 1978, Chemical exchange during hydrothermal alteration of basalt by seawater: I. Experimental results for major and minor components of seawater and basalt: Geochimica et Cosmochimica Acta, v. 42, p. 1103-1116.

Nilsson, C.A., 1968, Wallrock alteration at the Boliden deposit, Sweden: Economic Geology, v. 63, p. 472-494.

Oftedahl, C., 1958, On exhalative-sedimentary ores: Geologiska Föreningens I Stockholm Förhandlingar, v. 8, p. 1-19.

Ohmoto, H., 1978, Submarine calderas: a key to the formation of volcanogenic massive sulfide deposits?: Mining Geology, v. 28, p. 219-232.

Ohmoto, H., Mizukami, M, Drummond, S.E., Eldridge, C.S., Pisutha-Arnond, V. and Lenagh, T.C., 1983, Chemical processes of Kuroko formation, *in* Ohmoto, H. and Skinner, B.J., eds., Kuroko and Related Volcanogenic Massive Sulphide Deposits: Economic Geology, Monograph 5, p. 570-604.

Ohmoto, H. and Rye, R.O., 1974, Hydrogen and oxygen isotopic compositions of fluid inclusions in the Kuroko deposits, Japan: Economic Geology, v. 69, p. 947-953.

Ohmoto, H. and Takahashi, T., 1983, Geologic setting of the Kuroko deposits, Japan - Part III. Submarine calderas and Kuroko genesis, *in* Ohmoto, H. and Skinner, B.J., eds., Kuroko and Related Volcanogenic Massive Sulphide Deposits: Economic Geology, Monograph 5, p. 39-54.

Oudin, E., 1981, Etudes mineralogique et géochemique des dépôts sulfurés sous-marins actuels de la ridge est-pacifique (21°N): Documents du BRGM 25, 241 p.

Oudin, E., 1983, Hydrothermal sulfide deposits of the East Pacific Rise (21°N) Part I: Descriptive mineralogy: Marine Mining, v. 4, p. 39-72.

Oudin, E. and Constantinou, C., 1984, Black smoker chimney fragments in Cyprus sulphide deposits: Nature, v. 308, p. 349-353.

Palmason, G., 1967, On heat flow in Iceland in relation to the mid-Atlantic ridge: Societas Scientiarum Icelandica, v. 38, p. 111-127.

Perfit, M.R. and Fornari, D.J., 1983, Geochemical studies of abyssal lavas recovered by DSRV "Alvin" from eastern Galapagos Rift, Inca Transform, and Ecuador Rift: 2. Phase chemistry and crystallization history: Journal of Geophysical Research, v. 88, p. 10530-10550.

Perfit, M.R., Fornari, D.J., Malahoff, A. and Embley, R.W., 1983, Geochemical studies of abyssal lavas recovered by DSRV "Alvin" from eastern Galapagos Rift, Inca Transform, and Ecuador Rift: 3. Trace element abundances and petrogenesis: Journal of Geophysical Research, v. 88, p. 10551-10572.

Pisutha-Arnond, V. and Ohmoto, H., 1983, Thermal history, chemical and isotopic compositions of the ore-forming fluids responsible for the Kuroko massive sulfide deposits in the Hokuroku District of Japan, *in* Ohmoto, H. and Skinner, B.J., eds., Kuroko and Related Volcanogenic Massive Sulphide Deposits: Economic Geology, Monograph 5, p. 523-558.

Plimer, I.R., 1978, Proximal and distal stratabound ore deposits: Mineralium Deposita, v. 13, p. 345-353.

Pottorf, R.J. and Barnes, H.J., 1983, Mineralogy, geochemistry, and ore genesis of hydrothermal sediments from the Atlantis II Deep, Red Sea, *in* Ohmoto, H. and Skinner, B.J., eds., Kuroko and Related Volcanogenic Massive Sulphide Deposits: Economic Geology, Monograph 5, p. 198-223.

Reed, M.H., 1983, Seawater-basalt reaction and the origin of greenstones and related ore deposits: Economic Geology, v. 78, p. 446-485.

Richards, H.G. and Boyle, J.F., 1986, Origin, alteration and mineralization of inter-lava metalliferous sediments of the Troodos ophiolite, Cyprus, *in* Gallagher, M.J., Ixer, R.A., Neary, C.R. and Prichard, H.M., eds., Metallogeny of Basic and Ultrabasic Rocks: Institition of Mining and Metallurgy, London, p. 21-31.

Richards, H.G., Cann, J.R. and Jensenius, J., in prep., Mineralogical and metasomatic zonation of the alteration pipes of Cyprus sulfide deposits, (submitted to Economic Geology).

Riverin, G. and Hodgson, C.J., 1980, Wall-rock alteration at the Millenbach Cu-Zn mine, Noranda, Quebec: Economic Geology, v. 75, p. 424-444.

Roberts, R.G., 1975, The geological setting of the Mattagami Lake Mine, Quebec: A volcanogenic massive sulfide deposit: Economic Geology, v. 70, p. 115-129.

Roberts, R.G. and Reardon, E.J., 1978, Alteration and ore-forming processes at Mattagami Lake Mine, Quebec: Canadian Journal of Earth Sciences, v. 15, p. 1-21.

Rona, P.A., 1984, Hydrothermal mineralization at seafloor spreading centres, Earth-Science Reviews, v. 20, p. 1-104.

Rona, P.A., Klinkhammer, G., Nelsen, T.A., Trefry, J.H. and Elderfield, H., 1986, Black smokers, masssive sulphides and vent biota at the Mid-Atlantic Ridge: Science, v. 321, p. 33-37.

Sangster, D.F., 1972, Precambrian volcanogenic massive sulphide deposits in Canada: A review: Geological Survey of Canada, Paper 72-22, 44 p.

Sangster, D.F., 1976, Sulphur and lead isotopes in strata-bound deposits, *in* Wolf, K.H., ed., Handbook of Strata-bound and Stratiform Ore Deposits: Elsevier, Amsterdam, v. 2, p. 219-266.

Sangster, D.F., 1980, Quantitative characteristics of volcanogenic massive sulphide deposits: 1. Metal content and size distribution of massive sulphide deposits in volcanic centres: Canadian Institute of Mining and Metallurgy, Bulletin, v. 73, p. 74-81.

Sangster, D.F. and Scott, S.D., 1976, Precambrian strata-bound, massive Cu-Zn-Pb sulfide ores of North America, *in* Wolf, K.H., ed., Handbook of Strata-bound and Stratiform Ore Deposits: Elsevier, Amsterdam, v. 6, p. 129-222.

Sato, T., 1972, Behaviours of ore-forming solutions in seawater: Mining Geology, v. 22, p. 31-42.

Sato, T., 1973, A chloride complex model for Kuroko mineralization: Geochemical Journal, v. 7, p. 245-270.

Sawkins, F.J., 1984, Ore genesis by episodic dewatering of sedimentary basins: Application to giant Proterozoic lead-zinc deposits: Geology, v. 5, p. 451-454.

Sawkins, F.J. and Kowalik, J., 1981, The source of ore metals at Buchans: magmatic versus leaching models, *in* Swanson, E.A., Strong, D.F. and Thurlow, J.G., eds., The Buchans Orebodies: Fifty Years of Geology and Mining: Geological Association of Canada, Special Paper 22, p. 255-267.

Schermerhorn, L.J.G., 1970, The deposition of volcanics and pyrite in the Iberian pyrite belt: Mineralium Deposita, v. 5, p. 273-279.

Schoell, M., 1976, Heating and convection within the Atlantis II Deep geothermal system of the Red Sea: Proceedings: Second United Nations Symposium on the Development and Use of Geothermal Resources, San Francisco, v. 1, p. 583-590.

Scott, S.D., 1978, Structural control of the Kuroko deposits of the Hokuroku district, Japan: Mining Geology, v. 28, p. 301-311.

Seyfried, W.E., Jr., Berndt, M.E. and Janecky, D.R., 1986, Chloride depletions and enrichments in seafloor hydrothermal fluids: constraints from experimental basalt alteration studies: Geochimica et Cosmochimica Acta, v. 50, p. 469-475.

Seyfried, W.E., Jr. and Bischoff, J.L., 1977, Hydrothermal transport of heavy metals by seawater: The role of seawater/basalt ratio: Earth and Planetary Science Letters, v. 34, p. 71-77.

Seyfried, W.E., Jr. and Bischoff, J.L., 1979, Low temperature basalt alteration by seawater: an experimental study at 70°C and 150°C: Geochimica et Cosmochimica Acta, v. 43, p. 1937-1947.

Seyfried, W.E., Jr. and Bischoff, J.L., 1981, Experimental seawater basalt interaction at 30°C, 500 bars, chemical exchange, secondary mineral formation and implications for the transport of heavy metals: Geochimica et Cosmochimica Acta, v. 45, p. 135-147.

Seyfried, W.E., Jr. and Janecky, D.R., 1985, Heavy metal and sulfur transport during subcritical and supercritical hydrothermal alteration of basalt: influence of fluid pressure and basalt composition and crystallinity: Geochimica et Cosmochimica Acta, v. 49, p. 2545-2560.

Shanks, W.C., III and Bischoff, J.L., 1977, Ore transport and deposition in the Red Sea geothermal system: a geochemical model: Geochimica et Cosmochimica Acta, v. 41, p. 1507-1519.

Shanks, W.C., III and Seyfried, W.E., Jr., 1987, Stable isotope studies of vent fluids and chimney minerals, Southern Juan de Fuca Ridge: sodium metasomatism and seawater sulfate reduction: Journal of Geophysical Research, v. 92, p. 11387-11399.

Shikazono, N, Holland, H.D. and Quirk, R.F., 1983, Anhydrite in Kuroko deposits: Mode of occurrence and depositional mechanisms, *in* Ohmoto, H. and Skinner, B.J., eds., Kuroko and Related Volcanogenic Massive Sulphide Deposits: Economic Geology, Monograph 5, p. 329-344.

Shirozo, H., 1974, Clay minerals in altered wall rocks of the Kuroko-type deposits: Society of Mining Geologists of Japan, Special Issue 6, p. 303-311.

Sibson, R.H., Moore, J.McM. and Rankin, A.H., 1975, Seismic pumping - a hydrothermal fluid transport mechanism: Geological Society of London, Journal, v. 131, p. 653-659.

Simmons, B.D. and Geological Staff, 1973, Geology of the Millenbach massive sulphide deposit, Noranda, Quebec: Canadian Institute of Mining and Metallurgy, Bulletin, v. 66, p, 67-78.

Solomon, M., 1976, "Volcanic" massive sulphide deposits and their host rocks - a review and explanation, *in* Wolf, K.H., ed., Handbook of Strata-bound and Stratiform Ore Deposits: Elsevier, Amsterdam, v. 2, p. 21-54.

Solomon, M. and Walshe, J.L., 1979a, The formation of massive sulfide deposits on the seafloor: Economic Geology, v. 74, p. 797-813.

Solomon, M. and Walshe, J.L., 1979b, The behaviour of massive-sulphide ore solutions entering seawater and the development of zoned deposits: Bulletin Mineralogique, v. 102, p. 463-470.

Solomon, M., Walshe, J.L. and Garcia Palomero, F., 1980, Formation of massive sulphide deposits at Rio Tinto, Spain: Institution of Mining and Metallurgy, Transactions, v. 89, p. B16-B24.

Speiss, F.N., Macdonald, K.C., Atwater, T., Ballard, R., Carranza, A., Cordoba, D., Cox, C., Diaz Garcia, V.M., Francheteau, J., Guerrero, J., Hawkins, J., Haymon, R., Hessler, R., Juteau, T., Kastner, M., Larson, R., Luyendyke, B., Macdougall, J.D., Miller, S., Normark, W., Orcutt, J. and Rangin, C., 1980, East Pacific Rise; hot springs and geophysical experiments, Science, v. 207, p. 1421-1433.

Spence, C.D., 1975, Volcanogenic features of the Vauze sulphide deposit, Noranda, Quebec: Economic Geology, v. 70, p. 102-114.

Spooner, E.T.C., 1977, Hydrodynamic model for the origin of the ophiolite cupriferous pyrite ore deposits of Cyprus, *in* Volcanic Processes in Ore Genesis: Geological Society of London, Special Publication 7, p. 58-71.

Spooner, E.T.C and Bray, C.J., 1977, Hydrothermal fluids of seawater salinity in ophiolitic sulphide ore deposits in Cyprus: Nature, v. 266, p. 808-812.

Spooner, E.T.C. and Fyfe, W.C., 1973, Subsea floor metamorphism, heat and mass transfer: Contributions to Mineralogy and Petrology, v. 42, p. 287-304.

Stakes, D. and Vanko, D.A., 1986, Multistage hydrothermal alteration of gabbroic rocks from the failed Mathematician Ridge: Earth and Planetary Science Letters, v. 79, p. 75-92.

Strauss, G.K., Roger, G., Lecolle, M. and Lopera, E., 1981, Geochemical and geologic study of the volcano-sedimentary sulfide orebody of La Zarza, Huelva Province, Spain: Economic Geology, v. 76, p. 1975-2000.

Strens, M.R. and Cann, J.R., 1986, A fracture-loop thermal balance model of black smoker circulation: Tectonophysics, v. 122, p. 307-324.

Styrt, M.M., Brackmann, A.J., Holland, H.D., Clark, P.C., Pisutha-Arnond, Eldridge, C.S., and Ohmoto, H., 1981, The mineralogy and isotopic composition of sulfur in hydrothermal sulfide/sulfate deposits on the East Pacific Rise, 21°N latitude: Earth and Planetary Science Letters, v. 53, p. 382-390.

Thompson, G., 1983, Basalt-seawater interaction, *in* Rona, P.A., Boström, K., Laubier, L. and Smith, K.L., Jr., eds., Hydrothermal Processes at Seafloor Spreading Centres: Plenum Press, New York, NATO Conference Series IV, v. 12, p. 225-278.

Thorpe, R.I., Franklin, J.M. and Sangster, D.F., 1981, Evolution of lead in massive sulphide ores of Bathurst district, New Brunswick, Canada: Institution of Mining and Metallurgy, Transactions, v. 90, p. B55-B56.

Tivey, M.K., 1986, Mineral precipitation in "black smoker" chimney walls: a 1-dimensional model of heat and mass transport, (abstract): EOS, v. 67, p. 1283.

Tivey M.K. and Delaney, J.R., 1986, Growth of large sulfide structures on the Endeavour Segment of the Juan de Fuca Ridge: Earth and Planetary Science Letters, v. 77, p. 303-317.

Tsuneishi, Y., and Nakamura, K., 1970, Faulting associated with the Matsushiro swarm earthquakes: Bulletin of the Earthquake Research Institute, v. 48, p. 29-51.

Turner, J.S. and Campbell, I.H., 1987, Temperature, density and buoyancy fluxes in "black smoker" plumes, and the criterion for buoyancy reversal: Earth and Planetary Science Letters, v. 86, p. 85-92.

Turner, J.S. and Gustafson, L.B., 1978, The flow of hot saline solutions from vents in the sea floor - some implications for exhalative massive sulfide and other ore deposits: Economic Geology, v. 73, p. 1082-1100.

Urabe, T. and Sato, T., 1978, Kuroko deposits of the Losaka mine, Northeast Honshu, Japan - Products of submarine hot springs on Miocene sea floor: Economic Geology, v. 73, p. 161-179.

Urabe, T., Scott, S.D. and Hattori, K., 1983, A comparison of footwall alteration and geothermal systems beneath some Japanese and Canadian volcanogenic massive sulfide deposits, *in* Ohmoto, H. and Skinner, B.J., eds., Kuroko and Related Volcanogenic Massive Sulphide Deposits: Economic Geology, Monograph 5, p. 345-364.

van Staal, C.R. and Williams, P.F., 1984, Structure, origin, and concentration of the Brunswick 12 and 6 orebodies: Economic Geology, v. 79, p. 1669-1692.

Von Damm, K.L. and Bischoff, J.L., 1987, Chemistry of hydrothermal solutions from the southern Juan de Fuca ridge: Journal of Geophysical Research, v. 92, p. 11334-11346.

Von Damm, K.L., Edmond, J.M., Grant, B., Measures, C.I., 1985, Chemistry of submarine hydrothermal solutions at 21°N, East Pacific Rise: Geochimica et Cosmochimica Acta, v. 49, p. 2197-2220.

Walford, P.C. and Franklin, J.M., 1982, The Anderson Lake Mine, Snow Lake, Manitoba, *in* Hutchinson, R.W., Spence, C.D. and Franklin, J.M., Precambrian Sulphide Deposits: Geological Association of Canada, Special Paper 25, p. 481-523.

Walker, R.N., Logan, R.G. and Binnekamp, J.G., 1977, Recent geological advances concerning the H.Y.C. and associated deposits, McArthur River, N.T.: Journal of the Geological Society of Australia, v. 24, p. 365-380.

Watanabe, M. and Sakai, H., 1983, Stable isotope geochemistry of sulfates from the Neogene ore deposits in the Green Tuff region, Japan, *in* Ohmoto, H. and Skinner, B.J., eds., Kuroko and Related Volcanogenic Massive Sulphide Deposits: Economic Geology, Monograph 5, p. 282-291.

Wilson, H.D.B., 1953, Geology and geochemistry of base metal deposits: Economic Geology, v. 48, p. 370-407.

Wilson, H.D.B. and Anderson, D.T., 1959, The composition of Canadian sulphide ore deposits: Canadian Institute of Mining and Metallurgy, Transactions, v. 62, p. 327-339.

Wolery, T.J. and Sleep, N.H., 1976, Hydrothermal circulation and geochemical flux at mid-ocean ridges: Jounal of Geology, v. 84, p. 249-275.

Zierenberg, R.A., Koski, R.A., Shanks, W.C., III and Rosenbauer, R.J., 1986, Form and composition of sediment-hosted sulfide-sulfate deposits, Escanaba Trough, Southern Gorda Ridge, (abstract): EOS, v. 67, p. 1282.

Zierenberg, R.A., Shanks, W.C., III and Bischoff, J.L., 1984, Massive sulphide deposits at 21°N, East Pacific Rise: chemical composition, stable isotopes, and phase equilibria: Geological Society of America, Bulletin, v. 95, p. 922-929.

Accepted, as revised, 6 January 1988
Originally published in
Geoscience Canada v. 15 Number 1
(March 1988)

INDEX